TECHNIQUES IN PROTEIN CHEMISTRY III

TECHNIQUES IN PROTEIN CHEMISTRY III

Edited by

Ruth Hogue Angeletti
Albert Einstein College of Medicine
Bronx, New York

ACADEMIC PRESS, INC.
Harcourt Brace Jovanovich, Publishers
San Diego New York Boston
London Sydney Tokyo Toronto

Academic Press Rapid Manuscript Reproduction

This book is printed on acid-free paper. ♾

Academic Press, Inc.
San Diego, California 92101

United Kingdom Edition published by
Academic Press Limited
24–28 Oval Road, London NW1 7DX

International Standard Book Number: 0-12-058755-6 (Hardcover)
International Standard Book Number: 0-12-058756-4 (Paperback)

PRINTED IN THE UNITED STATES OF AMERICA

92 93 94 95 96 97 EB 9 8 7 6 5 4 3 2 1

Contents

Section V
Amino Acid Analysis

Section VI
Structural Analysis of Glycoproteins

Foreword

My first graduate student learned quickly that if you want to draw a huge crowd around your poster at a meeting, don't show your latest results from an investigation of a biological problem, but present a hot new technique applicable to many problems. Advances in understanding biological systems draw the specialists, admittedly the most important people in your field; but making it possible for anyone to do something more quickly, at higher sensitivity and accuracy, and with less input of supplies and labor draws everyone.

I suppose that this is the wrong lesson for a graduate student to learn; *solving the problem* is the objective after all. But we are always pressing the limits of our assays, and each advance in sensitivity and speed opens new fields of research which were previously inaccessible. Thus, scientists will always be looking for better ways of doing things, will always cite useful technique papers more than those advancing their field of research (Oliver Lowry, where would we be without you?), and will always be interested in books reporting the latest in methodology.

This is the fourth volume resulting from the Protein Society symposia. The number of copies sold and the desire of investigators to contribute to these volumes (our editors always comment on the willingness of the authors to submit their manuscripts *on time!*) are indeed impressive. The Protein Society has always devoted a lot of attention at its meetings to new methodologies, and the Fifth Symposium in Baltimore, June 22–26, 1991, from which this book was compiled, was no exception. There were more posters than ever at the symposium, and their quality was outstanding. We are grateful for that continuing interest and will commit ourselves to fostering it in any way possible.

Special thanks are due Ruth Hogue Angeletti and her associates for editing this volume. Their diligence and the top quality work presented at the Fifth Symposium make this a most useful and informative book.

Mark Hermodson
President
The Protein Society

Preface

The pleasant surprise in editing this volume has been the enthusiastic cooperation of the members of the Protein Society. Without this support, it would have been impossible to produce a book whose ideas are still fresh and current at publication. This active participation indicates to me that the Techniques in Protein Chemistry volumes have truly become our forum for exchanging new methods and research approaches and reflect the spirited exchanges of the poster sessions.

My special appreciation is due those who made this book possible: the associate editors John Crabb, Steve Carr, Phil Andrews, and Ira Ropson; the special reviewers Al Smith and Ed Nieves; the staff of Academic Press; the presidents of the society Finn Wold and Mark Hermodson; my secretary Elyse Rizzo for her organization and attention to detail; Jim Sacchettini, who created the cover graphics; and the authors, who did the real work while tolerating my gentle, but persistent nagging. I hope each volume published becomes used and "dogeared."

Ruth Hogue Angeletti

Acknowledgments

The Protein Society acknowledges with thanks the following organizations who through their support of the Society's program goals contributed in a meaningful way to the fifth annual symposium and thus to this volume.

AAA Laboratory
Amgen, Inc.
Applied Biosystems, Inc.
AVIV Associates, Inc.
Beckman Instruments, Inc.
Boehringer Mannheim Corporation
Dionex Corporation
Hewlett-Packard
Immuno-Dynamics, Inc.
Merck, Sharp & Dohme Research Laboratories
Millipore Corporation
Molecular Dynamics, Inc.
Monsanto Co.
National Biomedical Research Foundation
PE/SCIEX
PerSeptive Biosystems
Pharmacia LKB Biotechnology
Pickering Laboratories, Inc.
Polygen Corporation
Proton Instruments, Inc.
Rainin Instrument Co., Inc.
Savant Instruments, Inc.
Vestec Corporation
Vydac, The Separations Group, Inc.

SECTION I

Protein Microsequence Methods

Protein and Peptide Recovery from PVDF Membranes

Susan Wong, Allan Padua and William J. Henzel

Department of Protein Chemistry, Genentech, Inc.
South San Francisco, CA 94080

I. Introduction

Electroblotting proteins from gels onto solid supports is a widely utilized technique for the microisolation of proteins. Several electroblotting membranes have been developed that allow direct sequence analysis (1-3). Polyvinylidene difluoride (PVDF) is the most frequently utilized support for sequence analysis of electroblotted proteins (4). A limitation of this technique is that N-terminal blocked proteins must be chemically cleaved or proteolytically digested to generate internal peptide fragments which can then be sequenced. *In situ* digestions on PVDF (5-8) and nitrocellulose membranes (9) have been described; however, these methods only result in partial digestion of the membrane-bound proteins. Methods that directly extract protein from PVDF have been developed (10,11), but are based on solutions containing detergents which can limit subsequent proteolytic digestions or interfere with HPLC separations. We have developed a simple extraction procedure using the solvent dimethyl sulfoxide (DMSO) that allows recovery of intact proteins from PVDF membranes. The DMSO can be easily removed, allowing the extract to be analyzed by protein sequencing, amino acid analysis, or by other biochemical analysis.

The extracted protein can be cleaved or digested and the resulting fragments can either be separated by HPLC or by tricine SDS-gels (12) followed by a second round of electroblotting. An additional advantage of this technique is that the protein may be directly digested in the extraction solvent after addition of an appropriate buffer.

II. Materials and Methods

Materials. Proteins were purchased from Sigma, with the exception of human growth hormone (hGH) which was obtained from Genentech, Inc. Lys-C was from Wako chemicals, and cyanogen bromide was from Pierce. Dimethyl sulfoxide (spectrometric grade) was from Aldrich. PVDF (Immobilon-P) membranes were obtained from Millipore and ProBlott membranes were from Applied Biosystems.

TECHNIQUES IN PROTEIN CHEMISTRY III

Electroblotting and elution. SDS-PAGE was done as described by Laemmli (13) using a BioRad minigel apparatus. Tricine minigels were obtained from Novex. Electroblotting was done in a BioRad transfer tank using 10 mM 3-[cyclohexylamino]-1-propanesulfonic acid, 20% methanol, pH 11.0 for 1 h at 250 mA (4). Coomassie stained bands from electroblotted proteins were extracted with 200 μl DMSO in a micro Eppendorf tube on a VWR vortexer-2 using a medium setting for 2 h. The DMSO was removed and another 200 μl of DMSO was added. The extraction was continued for an additional 2 h. The extracts were then combined and used for further experiments. Extractions were also performed using a single addition of 200 μl of DMSO, vortexed for 17h.

Fragmentation. Extracted proteins were dried in a Speed-Vac to remove DMSO and were cleaved with cyanogen bromide in 0.1 N HCl at 45°C for 3 h (14). Blots were prewetted with 5 μl of methanol prior to the addition of 0.1N HCl and cyanogen bromide.

Protease digestions were performed on electroblot extracts after partial removal of DMSO by Speed-Vac evaporation. Ammonium bicarbonate (0.1 M) was added to the residual DMSO (20-50 μl) resulting in a final solution of 200-500 μl of 10% DMSO. Digestion with Lys-C was carried out at 37°C for 17 h. with an enzyme to substrate ratio of 1:20 (w/w).

Electroblots were first treated with a modified methanol-chloroform precipitation (15) prior to DMSO elution in order to remove Coomassie blue and HPLC artifact peaks. Blots (wet) were placed in 100 μl of water in an Eppendorf tube and 400 μl of methanol was added. The solution was vortexed for 2 min. and 100 μl of chloroform was added. The solution was vortexed again for 2 min. and the liquid was removed. The membrane was then extracted with DMSO. This method effectively removed the Coomassie blue and minimized artifact peaks in the HPLC chromatograms.

Peptide separation. Peptides were purified by reversed-phase HPLC using a Synchropak 4000 Å C4 column (2 x 100 mm) on a Hewlett-Packard 1090 M liquid chromatograph equipped with diode array detector. Peaks were detected at 214 nm after elution with a linear gradient of 0-70% acetonitrile in 0.1% trifluoroacetic acid at a flow rate of 200 μl per minute.

Protein sequencing. Automated Edman degradation was performed with Applied Biosystems models 470A and 473A protein sequencers equipped with on-line PTH analyzers. Sequence interpretation was performed on a VAX 8750 as described (16).

Amino acid analysis. Extracted proteins were dried in a Speed-Vac to completely remove DMSO. Residual DMSO can result in the oxidation of some amino acids including methionine and cysteine. Blots were prewetted with 5 μl of methanol prior to hydrolysis. Proteins were hydrolyzed in the gas phase for 24 h with 300 μl 6N HCl containing 0.1% phenol at 110°C under vacuum using a Millipore Picotag workstation. Hydrolyzed blots were extracted twice with 100 μl of 6 N HCl. The extracts were dried and reconstituted with Beckman sample buffer and hydrolysates were analyzed on a Beckman Model 6300 amino acid analyzer.

III. Results and Discussion

In an effort to optimize recovery, a number of proteins were transferred to Immobilon-P or ProBlott membranes, stained with Coomassie and then extracted for various times with DMSO. The optimal extraction time was found to be 4 hours using 2 consecutive 2 hour extractions (of 200 μl DMSO each) or a single extraction of 200 μl for 17 h. Both extraction methods resulted in nearly identical protein recovery. Recovery values obtained from DMSO extractions of proteins on ProBlott and Immobilon-P extractions are shown in Table I. These values were obtained by using the 4 hour extraction procedure. The extraction efficiency from Immobilon-P was better than from Problott membranes. Other extraction solvents were examined including 50% mixtures of DMSO and methanol, and DMSO and water (data not shown). These solvents resulted in lower efficiency of extraction of protein from PVDF membranes compared to using DMSO alone.

Extraction efficiencies can be determined by analyzing an aliquot of the extract by amino acid analysis. Proteins which are found to be intractable to DMSO extractions can be cleaved with cyanogen bromide and the resulting peptides can then be efficiently extracted. Further purification of these peptides can be accomplished by SDS-PAGE or reversed-phase HPLC. Extraction of cyanogen bromide peptides from the membrane may be a preferred alternative to directly extracting intact proteins larger than 70 kDa. No indication of selective retention of hydrophobic peptides was seen when membranes were sequenced after DMSO extractions. The extraction efficiencies of cyanogen bromide cleaved proteins are shown in Table II. Peptide recovery from PVDF and ProBlott membranes were similar.

TABLE I. Recovery data of proteins eluted from PVDF electroblots. Proteins were separated on 12% SDS gel, electroblotted onto Immobilon-P or ProBlott and stained with Coomassie brilliant blue. Bands were excised and extracted with DMSO, transferred to hydrolysis tubes, and analyzed by amino acid analysis. Extracted blots were hydrolyzed to determine residual protein. Protein on the blot was calculated by adding the amount remaining on the blot after extraction to the extracted amount. All numbers represent the average of triplicate experiments.

Protein	Membrane	Amount on gel (pmol)	Protein on blot (pmol)	Protein eluted (pmol)	Percent recovery (± SD)
Carbonic anhydrase	Immobilon-P	100	45	31	69 ± 1
	ProBlott	100	44	31	70 ± 7
hGH	Immobilon-P	225	80	68	85 ± 10
	ProBlott	225	110	84	76 ± 6
Lysozyme	Immobilon-P	210	56	42	75 ± 7
	ProBlott	210	89	39	44 ± 9
BSA	Immobilon-P	108	50	18	36 ± 1
	ProBlott	108	38	12	32 ± 1

TABLE II. Recovery of peptides eluted from PVDF electroblots following CNBr cleavage. Proteins bound to PVDF membranes were cleaved with CNBr in 0.1 N HCl at 45°C, dried, then extracted with DMSO. The DMSO extracts were transferred to hydrolysis tubes and analyzed by amino acid analysis. All numbers represent the average of duplicate experiments.

Protein	Membrane	Protein on blot (pmol)	Protein eluted (pmol)	Percent recovery (± SD)
Carbonic anhydrase	Immobilon-P	61	55	90 ± 2
	ProBlott	50	45	90 ± 5
BSA	Immobilon-P	40	22	55 ± 1
	ProBlott	18	10	56 ± 4

The application of peptide mapping on proteins eluted from electroblots is shown in Figure 1. Human growth hormone was eluted from a ProBlott membrane. An aliquot was removed (25%) and analyzed by amino acid analysis, which indicated a total of 72 pmol was originally present in the extract. The remaining 54 pmol was cleaved with cyanogen bromide, dried in a Speed-Vac, electrophoresed on a 16% tricine SDS gel and electroblotted on a ProBlott membrane. The excised bands were sequenced by automated Edman degradation and revealed initial yields ranging from 2-16 pmol (Fig. 1).

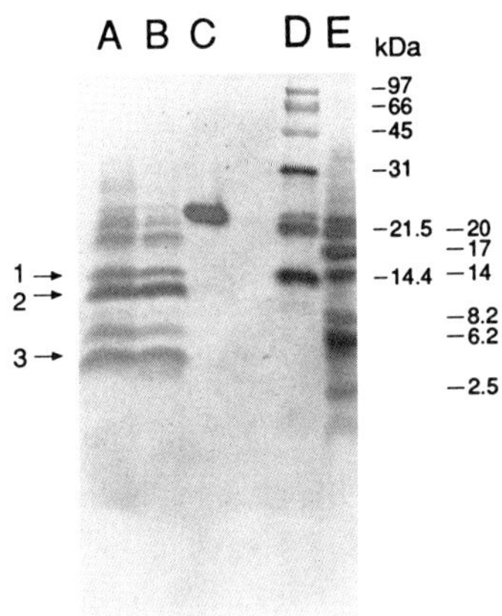

Band	Initial yield (pmol)	5 10 15 20
1	2.5	FPTIPLSRLFDNAMLRAHRL
2	2.0	LRAHRLHQLAXDTYQE
3	16.0	GRLEDGSPRTGQIFKQTYSKFDT

Figure 1. Coomassie blue stained ProBlott membrane of CNBr cleaved hGH after DMSO elution from a ProBlott membrane. (Lane A) Recombinant hGH was electrophoresed on a 12% SDS gel, electroblotted onto ProBlott, stained with Coomassie blue, the band excised and extracted with DMSO. The extracted protein (54 pmol) was cleaved with cyanogen bromide, and the peptides separated on a 16% tricine SDS gel. Bands 1-3 were sequenced (A); (B) control, hGH (100 pmol) + CNBr; (C) control, hGH (100 pmol); (D) LMW marker proteins (BioRad); (E) LMW marker (Diversified Biotech).

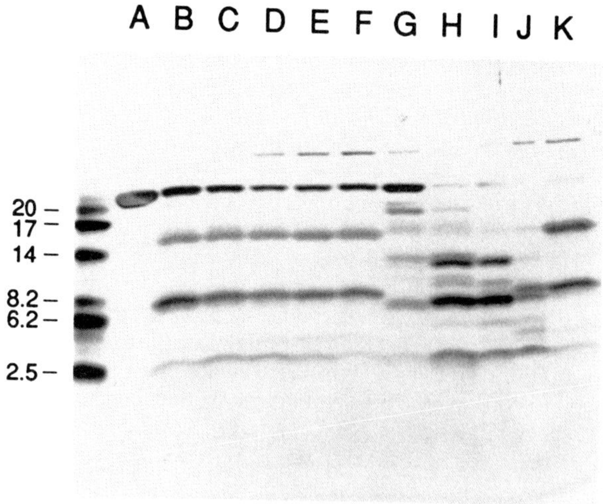

Figure 2. SDS-PAGE tricine gel (16%) of a Lys-C digest of hGH. Lane A is intact hGH, Lane B is a Lys-C digest of hGH, Lanes C-F are Lys-C digests of hGH with increasing amounts of DMSO added from 10 to 40% in 10% increments, Lane G is a Lys-C digest of hGH with 0.1% SDS. Lane H-K are the same conditions as in C-F except 0.1% SDS was included in the digestion buffer. All digestion were performed in 0.1M NH_4HCO_3 containing 0.01M DTT.

Proteins extracted from PVDF are difficult to resolubilize in aqueous buffers if DMSO is completely removed. To circumvent this problem proteins were digested with Lys-C in the presence of DMSO. Figure 2 shows the results of digesting hGH in the presence of increasing amounts of DMSO. Addition of DMSO to the digestion had little effect on proteolysis as shown in lanes C-F.

The addition of detergents such as SDS can be used to solubilize hydrophobic proteins and denatured proteins. The effects of increasing amounts of DMSO on the digestion of hGH by Lys-C in 0.1% SDS, 0.1 M ammonium bicarbonate is shown in lanes H-K in Fig. 2. DMSO significantly enhanced the activity of Lys-C in the presence of SDS (lanes H-K) resulting in greater proteolysis than seen with SDS alone (lane G) or in the absence of DMSO and SDS (lane B). Higher amounts of DMSO (30-40%) (lanes J-K) result in a similar cleavage pattern as observed when hGH is digested in the absence of SDS (lanes B-F). A similar effect with dimethylformamide was recently reported (17).

TABLE III. Sequence analysis obtained from an electroblot of a Lys-C digest of hGH eluted from a ProBlott membrane.

M.W.	Initial yield (pmol)	
		5 10 15 20 25
4 K	49	FPTIPLSRLFDNAMLRAHRLHQLAFDTYQE
5 K	5	XNLELLRISLLLIQSWLEPVQ

To demonstrate the utility of this procedure, extracted hGH was digested with Lys-C in 0.1 M ammonium bicarbonate in the presence of DMSO and SDS. The Lys-C digest was separated on a 16% tricine gel and electroblotted onto a ProBlott membrane. The sequence obtained for two of the resultant peptide bands is shown in Table III.

Protein digests often generate peptides smaller than 4 kDa and contain peptides having similar molecular weights. These peptides may be better resolved with higher recoveries when separated by reversed phase HPLC. This approach is demonstrated by a Lys-C digestion of DMSO-extracted hGH shown in Figure 3. The control Lys-C digest of hGH (Fig. 3B) appears similar to the digest of extracted hGH (Fig. 3A) except the earlier eluting peaks are less predominate. This is suggestive of more extensive cleavage for the digest of extracted hGH, which may be a result of protein denaturation resulting from the SDS gel electrophoresis. Despite staining the blot initially with Coomassie, no major artifacts are present. The Coomassie artifact peaks are minimized by

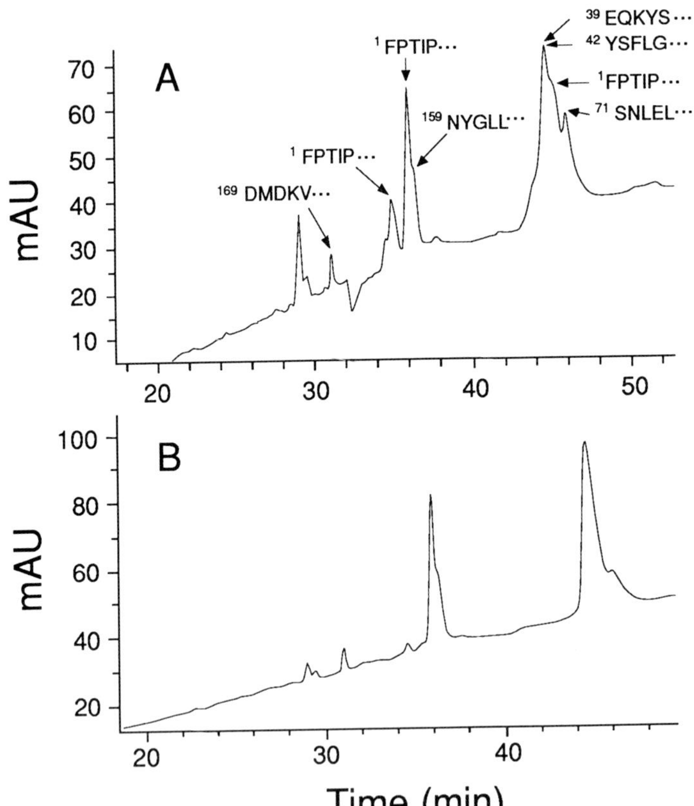

Figure 3. HPLC separation of Lys-C digest of hGH. (A) hGH eluted from a ProBlott membrane digested with Lys-C in 0.1 M NH_4HCO_3 containing 0.01M DTT, 0.1% SDS and 10% DMSO. Peptides were separated on a C4 column using a linear gradient of 0-70% acetonitrile in 0.1% TFA. (B) Control hGH digested with Lys-C using the same condition as in (A).

utilizing a chloroform-methanol precipitation (15) directly on electroblots before extraction. Amino acid analysis of the solvents used for Coomassie extraction on eletroblot membranes containing hGH, lysozyme and BSA resulted in less than 1% protein loss from the membranes. This method can also be used on the extracted protein prior to digestion. A disadvantage of performing precipitation after extraction is that significant losses may occur for proteins smaller than 20 kDa.

IV. Conclusions

This report describes a simple method for recovering intact proteins from PVDF membranes. Peptides and proteins can be recovered from both Immobilon-P and ProBlott membranes by DMSO extraction. Proteins obtained by DMSO extraction of PVDF membranes can be digested by proteases in solution, allowing more extensive cleavage than obtainable by *in situ* digestion. Extension of this work may allow additional methods of analysis of intact protein recovered from electroblots.

Acknowledgements

We thank Chris Grimley for providing sequencing analysis.

References

1. Vandekerchove, J., Bauw, G., Puype, M., Van Damme, J. and Van Montagu, M. (1985) *Eur. J. Biochem.* **152**, 9.
2. Abersold, R.H., Teplow, D.B., Hood, L.E. and Kent, S.B.H. (1986) *J. Biol. Chem.* **261**, 4229.
3. Eckerskorn, C., Mews, W., Goretzki, H., and Lottspeich, F. (1988) *Eur. J. Biochem.* **176**, 509.
4. Matsudaira, P. (1987) *J. Biol. Chem.* **262**, 10035.
5. Bauw, G.W., Van Damme, S., Puype, M., Vandekerchove, B.G., Ratz, G.P., Lauridsen, J.B., and Celis, J.E. (1989) *Proc. Natl. Acad. Sci. USA* **86**, 7701.
6. Simpson, R.J., Ward, L.O., Reid, G., Battenham, M.P., and Moritz, R.L. (1989) *J. Chromatogr.* **476**, 345.
7. Yuen, S.W., Chui, A.H., Wilson, K.J., and Yuan, P.M. (1989) *Biotechniques* **7**, 74.
8. Eckerson, C., and Lottspeich, F. (1989) *Chromatographia* **28**, 92.
9. Abersold, R.H., Leavitt, J., Saavedra, R.A., Hood, L.E., and Kent, S.B.H. (1987) *Proc. Natl. Acad. Sci. USA* **84**, 6970.
10. Vanfleteren, J.R.(1989) *Anal. Biochem.* **178**, 385.
11. Szewcyk, B., and Summers, D.F. (1988) *Anal. Biochem.* **168**, 48.
12. Schagger, H., and VonJagow, G. (1987) *Anal. Biochem.* **166**, 368.
13. Laemmli, U.K. (1970) *Anal. Biochem.* **155**, 23-27.
14. Zalut, C., Henzel, W.J. and Harris, W. (1980) *J. Biochem. Biophys. Methods* **3**, 11.
15. Wessel, D. and Flugge, U.I. (1984) *Anal. Biochem.* **138**, 141.
16. Henzel, W.J., Rodriguez, H., and Watanabe, C. (1987) *J. Chromatogr.* **404**, 41-52.
17. Houen, G., and Sando, T. (1991) *Anal. Biochem.* **193**, 186.

Automated C-Terminal Sequencing of Peptides

Jerome M. Bailey, Narmada R. Shenoy, and John E. Shively

Division of Immunology, Beckman Research Institute of the City of Hope, Duarte, CA 91010

I. Introduction

The development of a chemical method for the sequential degradation of a protein or peptide from the carboxy-terminus is a goal of our laboratory. Such a method, in addition to complementing existing N-terminal methods of degradation, would be invaluable for the sequence analysis of proteins with naturally occurring N-terminal blocking groups and for the detection of post-translational processing at the carboxy-terminus of expressed gene products. Although several methods for a sequential C-terminal degradation have been proposed (1,2), the thiocyanate method based on the procedure originally published by Schlack and Kumpf (3) has been the most widely studied. Recent work in our laboratory, introducing new reagents for the derivatization of the C-terminal amino acid (4) and for the specific cleavage of the derivatized amino acid (5,6), has suggested that an automated chemical method for the sequential degradation of polypeptides from the carboxy-terminus, analogous to the Edman method for amino terminal degradation, may be feasible.

The development of a C-terminal method involves the combination of several diverse areas of research, all of which must be considered concurrently. These include the design of an instrument capable of performing the sequencing chemistry, the choice of a solid support stable to the conditions used for sequencing, optimization of methodology to covalently couple polypeptide samples to the solid support in high yield, and the

TECHNIQUES IN PROTEIN CHEMISTRY III

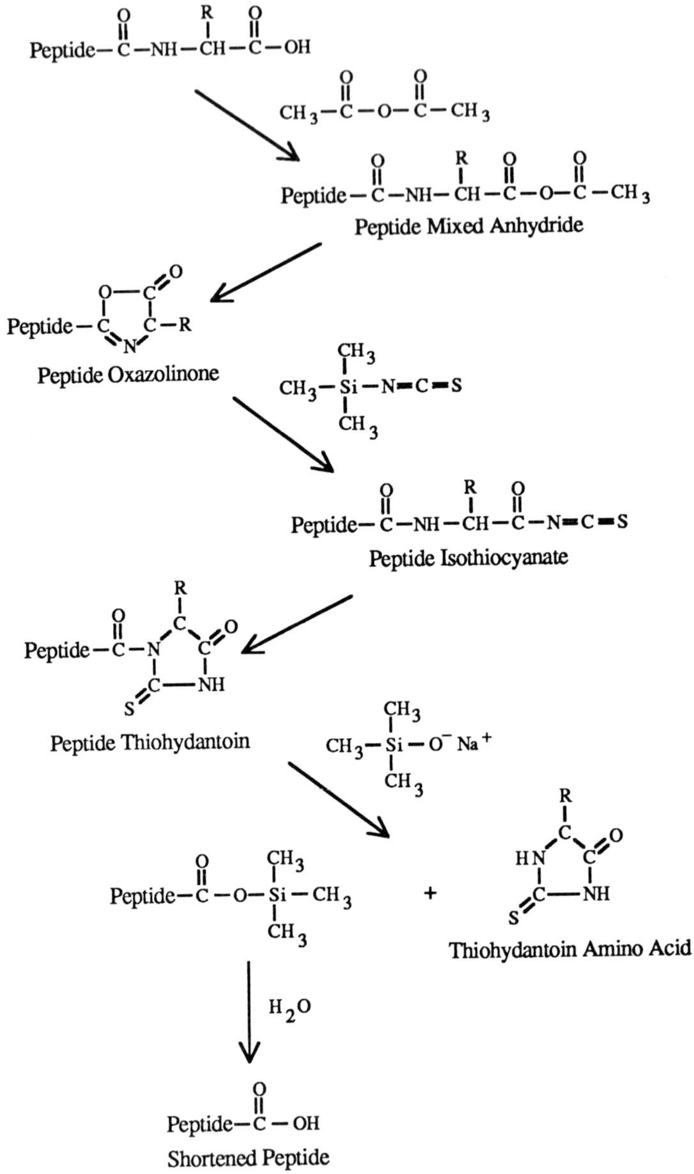

Fig. 1. Chemical Scheme for C-Terminal Sequencing.

development of chemistry to specifically derivatize and hydrolyze the C-terminal amino acid. As shown in figure 1, the chemistry currently used in our laboratory for C-terminal sequencing can be considered to consist of three steps: 1) activation of the C-terminal carboxylic group with acetic anhydride to form an oxazolinone, 2) derivatization of the C-terminal amino acid to a thiohydantoin with trimethylsilylisothiocyanate (TMS-ITC), and 3) specific cleavage of the derivatized amino acid with sodium trimethylsilanolate.

II. EXPERIMENTAL

The instrument used for automated C-terminal sequencing was constructed by the Biomedical Instrumentation Services of the City of Hope. This instrument is a modified N-terminal sequencer based on the design described by Hawke et al. (7) and includes the addition of a continous flow reactor (CFR) and an on-line HPLC for detection of the released thiohydantoin amino acids (8). The thiohydantoin amino acids were synthesized as described (5).

Reverse phase separation of the thiohydantoin amino acids was performed as described (6,9). Peptide samples to be sequenced were covalently coupled to a carboxylic acid modified polyethylene film (PE-COOH). This support is commercially available from the Pall Corporation (Long Island, NY). Strips of the polyethylene film (1 x 12.5 mm) were activated by an excess of 1,3-dicyclohexylcarbodiimide in anhydrous DMF (2g/2 ml) for 2 hr. Each strip was then washed with anhydrous DMF and dried in a vacuum centrifuge. The activated PE-COOH strips were then coupled to the peptides overnight (20 hr) at 22°C in a continuous flow reactor (CFR) (8). The microbore tubing on one end of the CFR was sealed by heating and then pinched closed with pliers. Peptide solutions were prepared in water containing 1% hexafluoroacetone. Aliquots of these solutions (2 µl to 5 µl) were transferred to the CFR. Ten microliters of a 1:1 solution of 25 mM HEPES buffer (pH 6.7) and DMF was added to each reaction mixture and the reaction volume was made up to 15 µl with 1% hexafluoroacetone trihydrate in water. The activated membrane was then inserted into each of these peptide solutions within a CFR as described above and the coupling reaction was carried out overnight (20 hr) at 22°C . At the end of the reaction, each membrane was washed with 2 ml of 25 mM Hepes buffer:DMF (1:1), water (2 ml), and methanol (2 ml) and dried in a vacuum centrifuge.

III. RESULTS AND DISCUSSION

The Solid Support. The need to apply the chemistry for C-terminal sequencing to the solid phase was recognized early on by a number of groups (4,10-14). Solution phase methods required laborious gel filtrations for purification of the peptidylthiohydantoin and evaporations to remove reagents (15). By covalently coupling a polypeptide to a solid support and thereby providing the ability to simply rinse away excess solvents without loss of the sample, the time required for a single cycle of degradation could be reduced to hours rather than the several days required by solution methods. Many of the initial studies involving the application of the thiocyanate chemistry for C-terminal sequencing to the solid phase involved the use of glass beads for the covalent immobilization of peptide samples (4,10-12,14). More recent work has involved the use of DITC-activated amino PVDF (13,16), carboxylic acid modified PVDF (6), and a disuccinimidoyl carbonate polyamide resin (17). The use of the glass and PVDF supports were found by our laboratory to be unsatisfactory for C-terminal sequencing (18). Only one cycle of degradation was obtained when the glass supports were used. Amino acid analysis of the glass support after 3 cycles of automated sequencing showed very little peptide remaining on the support. The loss of the peptide sample is most likely due to hydrolysis of the siloxane bonds formed on derivatization of the support during the basic conditions used for sequencing. This type of linkage is known to be base labile (19,20). Although the derivatized PVDF supports gave better sequencing yields than glass, these supports rapidly turned progressively blacker and more brittle with each cycle of sequencing. This is most likely due to dehydrofluorination of the support during the basic conditions of the cleavage reaction (21,22). This observation is consistent with the results of Shinohara (21) who showed that derivatized PVDF was much more susceptible to degradation (dehydrofluorination) by a basic hydroxide solution than was underivatized PVDF. A more suitable support was found to be a carboxylic acid modified polyethylene film. This novel solid support for C-terminal sequencing offers a number of advantages over the existing supports. These include 1) stability of the support to the conditions employed for C-terminal sequencing, 2) the ability, due to the hydrophilic nature of the surface groups, to use both aqueous and organic solvents for performing chemistry on covalently coupled polypeptide samples, 3) a high capacity (3.2 nmoles/mm^2) to covalently couple polypeptides, 4) the support is easily cut into 1 x 12.5 mm pieces for use in our continuous flow reactor (CFR) which is also used for automated N-terminal sequencing (8).

Covalent Immobilization of Peptides to PE-COOH. A number of reagents were tested for their ability to activate the surface carboxylic acid groups of PE-COOH. The highest yields of covalently attached peptide were obtained when DCC was used as an activating reagent. The optimal activation time was found to be 1 to 2 hours with an overnight coupling reaction with the peptide. Experiments to address the generality of this peptide coupling procedure are shown in Table I. Sixteen different peptides at two concentration levels (1 and 10 nmol) were applied to strips of PE-COOH (1 x 12.5 mm) activated with an excess of DCC in anhydrous DMF and coupled to the peptides overnight (20 hr) in a CFR as described above. A variety of peptides with differing chemical and physical properties were studied, and all of these coupled efficiently to the membrane surface (average yield was approx. 50%). A peptide with proline at the N-terminus (PFAL), which has a secondary amino group, was also found to couple to the membrane surface in relatively good yield. Experiments to determine the capacity of a 1 x 12.5 mm strip of PE-COOH (the size typically used for sequencing) for the covalent coupling of leucine enkephalin,YGGFL, revealed that the capacity of these small strips for short peptides is approximately 40 nmol (3.2 nmol/mm^2).

Automated Sequencing of Peptides Covalently Immobilized on PE-COOH. The composition of the reagents and solvents used and a description of the program for automated C-terminal sequencing is shown in Tables II and III, respectively. The total time required for each cycle was 120 min. The automated C-terminal sequencing of small peptides (3-10 residues, 3-20 nmol) routinely provided complete sequence determination. Typically the yield of sequencing decreased as the amino acid being sequenced became physically closer to the surface of the solid support. Amino acid composition of the support after several cycles of sequencing revealed that the C-terminal amino acids were completely removed while the N-terminal amino acid was

Table I Effect of Peptide Nature on the Coupling Yields of Carboxyl Activated PE-COOH

Peptides[1]	10 nmol applied		1 nmol applied	
	amt. coupled (nmol)	% yield	amt. coupled (nmol)	% yield
YGGFMRGL	1.79	18%	0.45	45%
KVILF	8.37	84%	0.74	74%
YGGFMRF	4.72	47%	0.76	76%
YGGFM	4.97	50%	0.48	48%
SIGSLAG	4.50	45%	0.62	62%
ADSGEGDFLAEGGVR	0.75	8%	0.06	6%
SYSMEHRWG	6.09	61%	0.66	66%
AFP	1.50	15%	0.14	14%
TVL	3.69	37%	0.29	29%
FAN	1.90	19%	0.23	23%
RGYALG	5.10	51%	0.48	48%
YGGFLRRIRP	4.95	50%	0.57	57%
YGGFMRRVGRPE	5.77	58%	0.62	62%
YKW	2.23	22%	0.56	56%
PFAL	3.68	37%	0.49	49%
YGGFL	2.38	24%	0.42	42%

only partially removed (50%). Since this is the amino acid from which the peptide was covalently attached to the solid support, it may be either difficult to derivatize this amino acid or subsequently hydrolyze it due to its proximity to the membrane surface. This surface effect was minimized by the introduction of spacers to the membrane surface prior to the attachment of peptide samples. This will permit the peptide sample being sequenced to be further from the surface of the solid support. Sequencing results, shown in figure 2, using 11-aminoundecanoic acid as a linker arm support this conclusion and resulted in significantly improved yields of sequencing with the pentapeptide, YGGFL. Amino acid analysis revealed only 12% of the tryosine remaining on the support after 6 cycles of sequencing.

The generality of our automated C-terminal sequencing methodology was examined by sequencing model peptides containing all twenty of the common amino acids. All of the amino acids were found to sequence in high yield (90% or greater) except for asparagine and aspartate, which could be only partially removed, and proline which was found to not be capable of derivatization. In spite of these current limitations, the methodology should be a valuable new tool for the C-terminal sequence analysis of peptides. Work is continuing to extend this technique to the sequence determination of covalently coupled proteins and toward improvement of the sequencing yields obtained when Asn, Asp, and Pro are encountered.

Table II Composition of Reagents and Solvents used in Automated C-Terminal Sequencing

R1	Acetic anhydride
R2	30% TMS-ITC in acetonitrile (v/v)
R3	0.10 M Sodium trimethylsilanolate in 50% methanol, 50% t-butyl alcohol
R4	Methanol
S1	5% Triethylamine in water
S2	1.0% Trifluoroacetic acid in water
S3	50% Methanol, 50% Water
S4	Acetonitrile

Table III C-Terminal Sequencer Program Summary

Continuous Flow Reactor (CFR) (65°C)	Conversion Flask (CF) (55°C)	Duration (sec)
S1 reaction		120
S1 reaction		120
	R4 rinse	120
Dry	Dry	100
R1 reaction		180
R2 reaction		450
S4 rinse		60
R1 reaction		180
R2 reaction		450
S4 rinse		60
R1 reaction		180
R2 reaction		450
S4 rinse		60
R1 reaction		180
R2 reaction		450
S4 rinse		120
S2 rinse		60
S4 rinse		120
S3 rinse		120
S4 rinse		120
R3 reaction		1200
R3 to CF		45
R3 reaction	Dry	240
R3 reaction	Dry	240
S2 delivery		12
S2 reaction		120
S2 to CF		45
S2 delivery		11
S2 reaction		120
S2 to CF		45
inject		7
pause		60

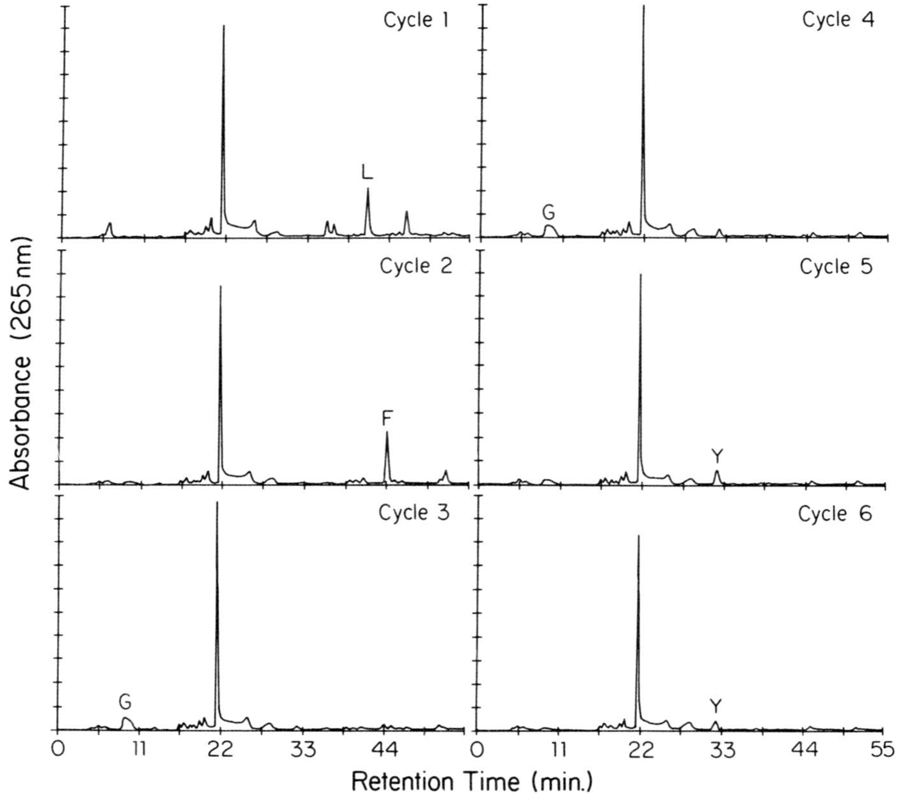

Fig. 2. Automated C-Terminal Sequencing of Leucine Enkephalin (3.7 nmol) Covalently Coupled to PE-COOH.

Acknowledgments

We thank the Tosoh Corporation for funding this research and Dr. Harvey Brandwein of the Pall Corporation for providing us with the carboxyl modified polyethylene films used in this study.

REFERENCES

1. Ward, C.W. (1986) Practical Protein Chemistry -- A Handbook (Darbre, A., Ed.) pp.491- 525, John Wiley and Sons, Ltd.
2. Rangarajan, M. (1988) Protein/Peptide sequence Analysis: Current Methodologies (Brown, A.S., Ed.) pp. 135-144, CRC Press, Bora Raton, FA.
3. Schlack, P., and Kumpf, W. (1926) Z. Physiol. Chem. **154,** 125-170.
4. Hawke, D.H., Lahm, H.-W., Shively, J.E., and Todd, C.W. (1987) Anal. Biochem. **166,** 298-307.
5. Bailey, J.M., and Shively, J.E. (1990) Biochemistry. **29,** 3145-3156.
6. Bailey, J.M., and Shively, J.E. (1991) in Techniques in Protein Chemistry: II (Villafranca, J.J., Ed.) pp. 115-129, Academic Press, Inc.
7. Hawke, D.H., and Harris, D.C. and Shively, J.E. (1985) Anal. Biochem. **147,** 315-330.
8. Shively, J.E., Miller, P., and Ronk, M. (1987) Anal. Biochem. **163,** 517-529.
9. Bailey, J.M., Shenoy, N.R., Ronk, M., and Shively, J.E. (1992) Submitted.
10. Williams, M. J., and Kassell, B. (1975) FEBS Lett. **54,** 353-357.
11. Rangarajan, M., and Dardre, A. (1976) Biochem. J. **157,** 307-316.
12. Meuth, J.L., Harris, D.E., Dwulet, F.E., Crowl-Powers, M.L., and Gurd, F.R.N. (1982) Biochem. **21,** 3750-3757.
13. Miller, C.G., Kong, C.-T., and Shively, J.E. (1989) Techniques in Protein Chemistry (Hugli, T. E., Ed.) pp. 67-78, Academic Press, Inc.
14. Inglis, A.S., Wilshire, J.F.K., Casagranda, F., and Laslett, R.L. (1989) Methods in Protein Sequence Analysis (Wittmann-Liebold, B., Ed.) pp.137-144, Springer-Verlag.
15. Stark, G.R. (1968) Biochemistry **7,** 1796-1807.

16. Inglis, A.S., Moritz, R.L., Begg, G.S., Reid, G.E., Simpson, R.J., Graffunder, H., Matschull, L., and Wittman-Liebold, B. (1991) Met. Protein Sequence Analysis (Jornvall/Hoog/Gustavsson, Eds.) pp. 23-34, Birkhauser-Verlag, Basel.
17. Hawke, D., and Boyd, V.L. (1991) Met. Protein Sequence Analysis (Jornvall/Hoog/Gustavsson, Eds.) pp. 35-45, Birkhauser-Verlag, Basel.
18. Shenoy, N.R., Bailey, J.M., and Shively, J.E. (1992) Submitted.
19. Iler, R.K. (1979) The Chemistry of Silica, Chap. 6, John Wiley and Sons, Inc., New York.
20. Taylor, R.F. (1991) Protein Immobilization: Fundamentals and Applications (Taylor, R.F., Ed.) pp. 139-160, Marcel Dekker, Inc., New York.
21. Shinohara, H. (1979) J. Polymer Science: Polymer Chem. Ed. **17**, 1543-1556.
22. Dias, A.J. and McCarthy, T.J. (1984) Macromolecules **17**, 2529-2531.

Elution and Internal Amino Acid Sequencing of PVDF-Blotted Proteins

Kathryn L. Stone[a], Dean E. McNulty, Mary L. LoPresti[a], J.M. Crawford, Raymond DeAngelis, and Kenneth R. Williams[a]

Yale University School of Medicine Protein and Nucleic Acid Chemistry Facility
New Haven CT 06510
and
[a]HHMI Biopolymer Facility
New Haven, Ct 06510

I. Introduction

SDS polyacrylamide gel electrophoresis (PAGE) followed by blotting onto polyvinylidene difluoride (PVDF) membranes has rapidly become one of the primary methods of choice for final or, in many instances, the total purification of proteins for microchemical sequence analysis. Because of the relatively high specificity with which PVDF membranes bind proteins, compared to most low molecular weight compounds such as salts, amino acids, and detergents, the resulting protein is isolated free of those contaminants that most commonly interfere with subsequent protein chemical analysis. Although PVDF-blotted proteins have been reported to sequence nearly as well as proteins that are applied directly from solution onto automated sequencers [1], many eukaryotic proteins have blocked NH_2-termini that prevent direct sequencing. In fact, chemical analysis of soluble proteins [2] and protein database compilations [3] suggest that as many as 80-90% of eukaryotic proteins have blocked NH_2-termini. In order to circumvent the problem of in vivo NH_2-terminal blocking requires that the electroblotted protein be cleaved into fragments, which then must be individually purified prior to sequencing. This has been accomplished by taking at least two different approaches. Either the remaining protein binding sites on the PVDF membrane have been blocked prior to in situ enzymatic digestion, which is then followed by extraction and subsequent HPLC purification of the resulting peptides [4], or the intact protein has been extracted from the PVDF membrane with a mixture of detergents [5] which have then been removed by "inverse gradient" reverse phase HPLC prior to enzymatic cleavage [6]. One drawback with both these approaches is a relatively low overall recovery. That is, the overall average recovery (calculated from the ratio of the yield of phenylthiohydantoin amino acid derivative as compared to the amount of protein initially applied to the SDS

TECHNIQUES IN PROTEIN CHEMISTRY III

polyacrylamide gel) of three peptides isolated by the detergent extraction approach was only 2.3%, even when as much as 1.1 nmol of protein (β-lactoglobulin) was applied to the SDS PAGE gel [6]. Similarly, the overall average recovery of 10 peptides isolated from 400 pmol of bovine serum albumin by the *in situ* approach was 6.1% [7]. Reasoning that "inverse gradient" reverse phase HPLC and the relative inaccessibility of PVDF-bound proteins to proteases (leading to the generation of over-lapping peptides in relatively low yield [7]), might significantly contribute to the low recoveries associated with these two procedures, we developed an alternative strategy that avoids these steps.

The method that will be described is an extension of the procedure of Yuen et al [7] and follows from their finding that there is an indirect relationship between the ease of eluting proteins (with organic solvents) from PVDF membranes and their molecular weights. Hence, by first carrying out an in situ cyanogen bromide cleavage, we have found that proteins can be eluted from PVDF membranes in high yield and in a form that is amenable to further enzymatic cleavage. Although the cyanogen bromide peptides could be directly fractionated and then individually sequenced [7], we have found it best to carry out a subsequent tryptic digest and then, following reverse phase HPLC, sequence one or more of the resulting cyanogen bromide/tryptic peptides. We have now used this approach on 26 different "unknown" proteins that have recently been submitted to the Yale University School of Medicine Protein and Nucleic Acid Chemistry Facility (in amounts that have ranged down to approximately 50 picomoles of PVDF-blotted protein) and have found it to have a 96% or greater success rate. In the course of these studies we have quantitatively examined each step in this procedure with the aim of identifying areas where additional research is needed and of helping to establish realistic expectations as regards high sensitivity sequencing of tryptic peptides obtained from sub-nanomole amounts of PVDF-blotted proteins.

II. Methods

A. *SDS PAGE Electrophoresis and Electroblotting*

SDS PAGE gels (12.5% acrylamide) were prepared according to Laemmli [8] and were cast and run the same day without pre-electrophoresis. Electrophoresis was carried out at 50 volts (constant voltage) for 16-18 hours. After electrophoresis, the gels were immersed in a Tris/glycine [9] electroblotting buffer for a few seconds prior to sandwiching next to 2 sheets of Immobilon-P membranes (that had been previously wet with methanol, then water and finally, the electroblotting buffer) and several sheets of Whatman 3MM chromatography paper. Electroblotting was done in a Bio-Rad Trans-Blot apparatus at 30 volts (constant voltage) for 2.5 hours. After electroblotting, the Immobilon-P membranes were washed with water for 5 minutes and then stained with 0.05% Coomassie Blue, 50% methanol for 3-5 minutes. Destaining was then carried out in 50% methanol for 3-5 minutes.

B. *Cyanogen Bromide Elution of PVDF-blotted Proteins*

The stained, blotted protein bands were cut into 1 x 2 mm pieces and submerged in 500 μl of ice-cold 95% acetone. After gentle agitation at 4°C for one hour, the acetone supernatant was removed and the PVDF-bound protein was washed with another 500 μl of 95% cold acetone. The membrane was then air dried prior to submerging in 50 μl of 70% formic acid. Cyanogen bromide was added according to the Yuen et al [7] procedure whereby 10 μl of 70 mg/ml CnBr is added per 10 μg of protein. After incubating in the dark at room temperature for 24 hours, the supernatant was removed to another 1.5 ml Eppendorf tube and the PVDF and supernatant were both dried. 100 μl of 40% acetonitrile was then added to the tube containing the PVDF pieces. After incubating at 37°C for 3 hours, the supernatant was removed and pooled in the tube containing the dried CnBr supernatant. Finally, the PVDF pieces were extracted at 50°C with 100 μl of 0.05% TFA in 40% acetonitrile. This supernatant was combined with the two prior extracts and dried in vacuo. All samples were then re-dried from 60 μl water.

C. *Reduction, Alkylation and Trypsin Digestion of Eluted CnBr Fragments*

The dried CnBr fragments were dissolved in 25 μl 8 M urea, 0.4 M NH_4HCO_3 prior to hydrolysis and ion exchange amino acid analysis of a suitable aliquot and then, following the addition of 5 μl 45 mM dithiothreitol, were reduced at 50°C for 15 min. After cooling to room temperature, the fragments were alkylated by adding 5 μl 100 mM iodoacetamide. Following a 15 min incubation at room temperature, the sample was diluted with water to a final urea concentration of 2 M, and trypsin was then added at a 1:25 (weight:weight) ratio of enzyme to protein. Proteolysis was allowed to continue for 24 hrs at 37°C prior to stopping the reaction by freezing or by direct injection onto the HPLC column. Additional details on this procedure can be found in Stone et al, [10-12].

D. *Peptide HPLC Isolation*

Peptides were separated on Vydac C-18 reverse phase HPLC columns (5 micron particle size, 300 angstrom pore size, 25 cm length) that were equilibrated in 0.05% TFA, 1.6% acetonitrile prior to eluting with increasing concentrations of buffer B (approximately 0.05% TFA, 80% acetonitrile) as follows: 0-63 min (2-37%B), 63-95 min (37-75%B), 95-105 min (75-98%B). In general, peptides obtained from amounts of protein that were above 250 pmol were fractionated at 0.5 ml/min on a Water's Associates HPLC System equipped with a 4.6 mm inner diameter column [13]. As previously recommended [11], peptides obtained from less than 250 pmol protein were separated at 0.15 ml/min on 2.1 mm inner diameter columns that were eluted with a Hewlett Packard Model 1090 HPLC System [12]. In either instance, all peptides were automatically collected (by peak) with the use of an ISCO peak detector [13]. The peptides were collected in 1.5 ml "capless" eppendorf tubes that had been positioned on the tops of 13 x 100 mm test tubes. Providing that these tubes are tightly capped within a few hours of their collection (to prevent acetonitrile evaporation) they may, in general, be stored for at least a year at 5°C. Peptides selected for re-purification were mixed with an equal volume of 8 M urea and then injected, in volumes that frequently exceeded 500 μl, onto the same HPLC system that was now equipped, as described previously [13], with a 1 mm inner diameter, Aquapore C-8 column . Whenever possible, peptides destined for sequence analysis were directly spotted onto Polybrene coated GF/C filters, which, in this case, were supplied by the manufacturer (Applied Biosystems) of the Model 470 and 477 protein/peptide sequencers that were used. Volumes larger than about 30 μl were multiply spotted onto the GF/C sequencing filter or, in those cases where the peptide volume exceeded about 300 μl, were dried in vacuo prior to loading in 30 μl 100 % TFA.

III. Results

A. Following In Situ Cleavage with Cyanogen Bromide, Proteins can be Eluted in High Yield from Immobilon-P PVDF Membranes

After testing several volatile solvents, including those recommended by Yuen et al, [7] we found (data not shown) that 40% acetonitrile was the most effective solvent tested at eluting PVDF bound proteins. On average, this solvent eluted 50.3% of 6 different blotted proteins (Table I). In agreement with Yuen et al, [7], there seemed to be an indirect correlation between the elution efficiency and the molecular weight of the protein. Thus, while β-lactoglobulin (18.4 kD) was quantitatively eluted, three proteins in the 42-75 kD range eluted with an average efficiency of less than 25%. Based on these data we reasoned that if an in situ cyanogen bromide digest was carried out prior to the elution then the elution efficiencies of larger proteins might be significantly improved. Indeed, this approach resulted in improving the elution efficiency for these same three, 42-75 kD proteins, from less than 25% to greater than 62% (Table I). The overall average elution efficiency for 11 different proteins following in situ cyanogen bromide cleavage was nearly 75%. Perhaps more importantly, for those 6 proteins that were subjected to both elution procedures (Table I), the inclusion of the in situ digest increased the lower limit for the average elution efficiency from only about 15% (for ovalbumin without prior cyanogen bromide cleavage) to 50% (for serum albumin with prior cyanogen bromide cleavage). The two fold decrease that in situ cyanogen bromide cleavage afforded in the variability of elution (as measured by the standard deviations shown in Table I), was an essential improvement in order to be

Table I. Efficiency of Elution of Proteins from PVDF Membranes[a]

Protein	MW	Acetonitrile Elution			CnBr Elution		
		#	Mean	S.D.	#	Mean	S.D.
β-Galactosidase	116000	-	-	-	2	35.6	6.58
Phosphorylase B	97058	-	-	-	6	79.7	15.5
Transferrin	75181	4	25.9	17.5	7	58.2	12.3
Serum Albumin	67500	4	30.7	7.13	10	50.4	12.0
Ovalbumin	42800	2	14.8	1.20	3	78.9	8.80
Carbonic Anhydrase	28980	2	106	6.20	13	76.8	10.1
Trypsinogen	23993	2	18.1	3.95	2	59.4	3.10
Trypsin Inhibitor	20100	-	-	-	13	94.7	11.0
β-Lactoglobulin	18363	2	106	7.75	2	91.8	8.25
Myoglobin	16950	-	-	-	2	106	11.4
α-Lactalbumin	14187	-	-	-	12	83.5	23.8
Total	-	20	50.3	39.8	72	74.1	20.0

[a]The average amount of blotted protein was 4.22 μg for the acetonitrile and 5.00 μg for the cyanogen bromide/acetonitrile experiments as determined by hydrolysis and amino acid analysis of duplicate PVDF-blotted lanes of protein. In general, no differences in elution efficiency were found over the range extending at least from 1.7 - 9.0 μg. Recoveries greater than 100% probably result from incomplete hydrolysis of PVDF-bound proteins.

able to routinely use this procedure in a core facility setting where a wide variety of "unknown" proteins would be encountered.

Because of the higher affinity of ProBlott compared to Immobilon-P PVDF membranes for lower molecular weight proteins (data not shown), we anticipated problems in eluting cyanogen bromide peptides from ProBlot membranes. Indeed, the average elution efficiency of four proteins that had been blotted onto ProBlott PVDF membranes and then cleaved with cyanogen bromide was only about 17% as compared to 76% for these same four proteins that had been blotted onto Immobilon-P PVDF membranes. Clearly, ProBlott PVDF membranes are not compatible with the cyanogen bromide elution procedure.

B. HPLC Separation and Amino Acid Sequencing of Tryptic Peptides Derived from Cyanogen Bromide Eluted Proteins

The upper panel in Fig. 1 shows an HPLC profile obtained from a tryptic digest of 100 pmol of ovalbumin (43 kD) that had been blotted onto PVDF, cleaved <u>in situ</u> with cyanogen bromide and then eluted with 40% acetonitrile

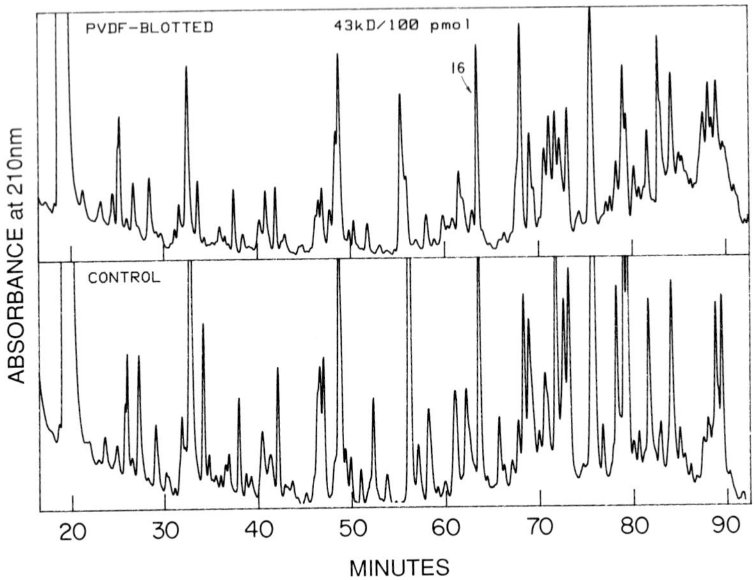

Figure 1. HPLC separation of tryptic peptides from ovalbumin that had either been blotted onto Immobilon PVDF and then cyanogen bromide eluted as described in Methods (top panel) or that had been cleaved in solution with cyanogen bromide and then trypsin (lower panel). In the case of the PVDF blotted sample, 100 pmol protein was digested while in the case of the lower chromatogram, 147 pmol protein was digested. These two chromatograms have been "normalized" by plotting the top panel with a full scale absorbance of 0.020 and the lower panel with a full scale absorbance of 0.029. Amino acid sequencing of those peaks that are numbered in this and the following two figures resulted in the indicated number of residues being positively assigned.

as described in Experimental Procedures. By comparing with the control digest (lower panel in Fig. 1), which was carried out on a sample of ovalbumin that had not been subjected to SDS-PAGE, it is apparent that a similar extent of cleavage is occurring in both instances and that there are few, if any, extraneous peaks present in the PVDF eluted sample. The acetone wash (see Methods) that is carried out just prior to the in situ cyanogen bromide digestion is thus effective at removing the excess Coomassie Blue that would have otherwise obscured the latter part of the chromatogram and that can also inhibit trypsin digestion (data not shown). The comparative HPLC profiles in Fig. 2 demonstrate the relatively low level of extraneous absorbance peaks that are present when a blank section of PVDF membrane is "eluted". The top profile in Fig. 2 shows a 206 pmol tryptic digest of a 52 kD cyanogen bromide-eluted protein, while the bottom profile was obtained by subjecting an equal size section of PVDF (that did not contain any blotted protein) to the same procedure. The peaks that are present in the "Blank" profile derive from the PVDF membrane itself as well as from the cyanogen bromide, reduction/alkylation reagents, and trypsin autolysis products. Since the acetone wash effectively eliminates absorbance

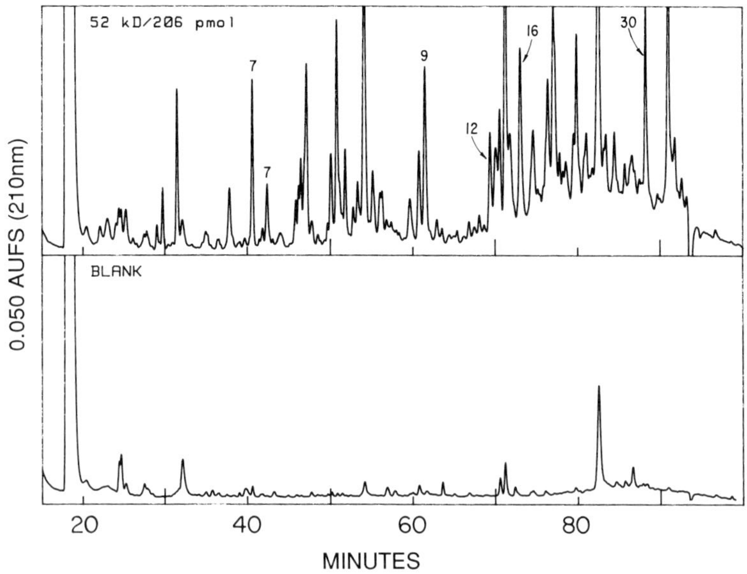

Figure 2. HPLC separation of tryptic peptides from 206 pmol of a 52 kD protein that had been cyanogen bromide eluted from Immobilon-P (top panel) and of trypsin autolysis products, reagent and other absorbance peaks obtained from an equal size piece of Immobilon-P that should not have contained blotted protein (lower panel).

peaks due to Coomassie Blue, the "blank" chromatogram in Fig. 2 readily identifies all other likely artifact peaks. Eight of the absorbance peaks shown in the upper panel of Fig. 2 were subjected to amino acid sequencing. While two of these peaks proved to contain approximately equimolar mixtures (and were thus of no use for designing oligonucleotide probes, but ultimately, will be able to be matched to the cDNA predicted sequence), the remaining six peaks resulted in 81 residues (as indicated by the numbers adjacent to these peaks) being positively assigned. The top two panels of Fig. 3 show that even when less than 50 pmol of protein are actually digested, it is still possible to automatically collect the resulting peptide peaks in individual eppendorf tubes and to then obtain as much internal sequence as is necessary to generate oligonucleotide probes. Since the top and bottom chromatogram in Fig. 3 were from proteins that had been stained with Ponceau S, this stain is also compatible with the cyanogen bromide elution procedure.

To ensure that the CnBr elution approach works as well with "unknown" as it does with "known" proteins, we have tested it on 26 different proteins that have been submitted recently for internal amino acid sequence analysis to the Yale University School of Medicine Protein and Nucleic Acid Chemistry Facility. As summarized in Table II, the overall success rate (as judged by the fraction of the proteins examined for which tryptic peptide sequences were obtained that were suitable for generating cDNA probes)

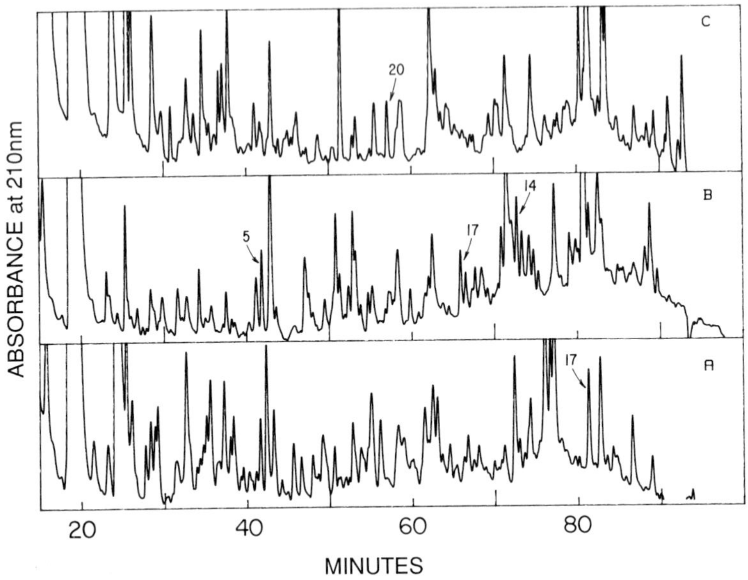

Figure 3. High sensitivity HPLC tryptic peptide mapping of three proteins that had been cyanogen bromide eluted from Immobilon-P. The size/amount of each protein that was digested with trypsin were as follows: panel A, 50 kD/100 pmol; panel B, 73 kD/45 pmol; and panel C, 95 kD/36 pmol. The following full scale absorbances were used: panel A, 0.014; panel B, 0.018, and panel C, 0.010. The two proteins in panels A and C had been stained with Ponceau S while the protein in panel B had been stained with Coomassie Blue.

was 96%. Four of the 26 proteins examined had been stained with Ponceau S while the remainder had been stained with Coomassie Blue. That at least all of the steps up to and including the cyanogen bromide elution can be carried out by staff having little or no training in protein chemistry is suggested by the fact that 8 of the 26 proteins summarized in Table II were investigator-eluted. Although the average amount of protein digested with trypsin was 307 pmol, the procedure appears to be applicable at least over the range extending from 36-2250 pmol of eluted protein. By always running a "blank", it is possible to avoid trying to sequence most artifact peaks and hence nearly 70% of the peptides chosen for analysis yielded unambiguous sequence. Of those absorbance peaks that failed to sequence, 17% were mixtures of peptides and the remaining 16% were either artifact peaks, peptides derived from the presumably blocked NH_2-termini of these proteins, or were lost prior to sequencing. When amounts of eluted protein above about 100 pmol are available, the data in Table II (on 7 re-purified peptides) indicate that individual peptides can then be re-purified in high yield or "screened for purity" (see Methods) prior to amino acid sequencing, thus minimizing the number of peptide mixtures subjected to sequencing. Based on the average abundance of arginine(5.4%), lysine(5.7%), and methionine(2.3%) in proteins that had been entered into the Protein Identification Resource by 7/1/90, the average length of a peptide following cyanogen bromide cleavage and trypsin digestion would be about 7-8

Table II. Summary of Results from Cyanogen Bromide/Acetonitrile Elution and Tryptic Digestion of 26 "Unknown" Proteins

Parameter	Range	Mean	Std. Dev.
Amount of Protein Digested (pmol)	36-2250	307	472
Mass of Protein (kilodaltons)	17-95	53.0	22.2
%Coupling Yield of 55 Peptides[a]			
7 Repurified Peptides	5.68-35.0	18.5	11.1
48 Non-repurified Peptides	0.78-76.3	20.9	18.9
Results of Sequencing 82 Peptides			
≥4 Residues Called (%)	-	67.1	-
Mixture (%)	-	17.1	-
No Sequence (%)	-	15.9	-
#Residues Called	4-30	11.7	5.72
Number of Peptides Sequenced/Protein	1-8	3.15	-
Overall Success Rate[b] (%)	-	96.2	-

[a]Does not include data from 14 peaks that amino acid sequencing demonstrated were mixtures of peptides. The % coupling yield was calculated based on the ratio of the first or second cycle phenylthiohydantoin derivative compared to the amount of protein that was digested, which was in turn based on hydrolysis and amino acid analysis of an aliquot of the cyanogen bromide-eluted protein.

[b]In terms of the fraction of the above 26 proteins for which tryptic peptide sequences were obtained that were suitable for generating cDNA probes.

residues. By selecting "later" eluting peptides we have achieved an average peptide length (in terms of positively called residues) of about 12 (Table II) which is usually adequate for generating oligonucleotide probes.

C. *Optimization of the Cyanogen Bromide Elution Approach for Obtaining Internal Amino Acid Sequences from Proteins*

Several parameters have been independently varied in an effort to improve the overall recoveries of CnBr-eluted peptides. One of the first areas of concern was loss of low molecular weight proteins during staining/destaining of the Immobilon-PVDF membranes. Even after removing the acetic acid from the staining/destaining solutions as recommended by Speicher [14], staining/destaining of 50-100 pmol amounts of several 14-20 kD proteins still resulted in 40-70% loss of protein from Immobilon-P membranes. In these cases, the use of Ponceau S (which does not require organic staining and destaining solvents) in place of Coomassie Blue may lead to decreased loss of protein. While oxidation of methionine residues during or after SDS PAGE is obviously a concern, at least based on studies on myoglobin, this phenomenon does not occur to any significant extent even in the absence of reducing agents in the gel and/or electrophoresis buffer. That is, when 15 μg myoglobin (888 pmol) was blotted onto PVDF, stained with Coomassie Blue, and stored at -20°C for 2 months prior to cyanogen bromide elution and amino acid sequencing, it was apparent that quantitative cleavage had still taken place. Hence, at cycle #2, the yield of leucine(2) was 187 pmol compared to 232 pmol for alanine(57), which resulted from cleavage following methionine(55). Although the exact elution times and temperatures cited in Methods have not yet been individually optimized, all of the major steps in the elution protocol appear to be essential based on tests that were carried out with 5 different proteins that ranged in size from 14-97 kD. Hence, while 70% formic acid itself (without prior cyanogen bromide cleavage) resulted in an average elution efficiency of 41%, the elution efficiency increased to 71% following cyanogen bromide cleavage and it further increased to 80% when the additional 40% acetonitrile elutions were carried out. Since excess Coomassie Blue can prevent trypsin digestion (data not shown), the PVDF-blotted protein is washed with acetone prior to in situ cyanogen bromide cleavage. While this step could lead to some loss of proteins that are both loosely bound to PVDF membranes and that are reasonably soluble in acetone, in practice, tests with 4-11 μg amounts of five PVDF-blotted proteins that range in size from 14-75 kD indicated that less than 13% of the protein was lost during acetone washing. Even this relatively modest loss could be avoided by using Ponceau S in place of Coomassie Blue. Finally, another important parameter is the ratio of protein to PVDF surface area. As might be predicted, this ratio should be as high as practical. When 15 μg of carbonic anhydrase was electrophoresed in one versus five lanes in an SDS-PAGE gel and then cyanogen bromide eluted and digested, the average peptide peak height for the one lane sample was almost twice that for the five lane sample. Every effort should therefore be made to minimize the PVDF surface area over which the protein will be distributed.

IV. Discussion

When coupled with previous micro-scale procedures that have been developed for isolating tryptic peptides from sub-nanomole quantities of proteins (Stone et al [10-13, 15], in situ cyanogen bromide cleavage followed by 40% acetonitrile elution offers a reliable approach for obtaining internal amino acid sequences from PVDF-blotted proteins. Although we have not specifically tried to answer it, the question invariably arises as to the minimum amount of protein that needs to be applied to an SDS-PAGE gel in order to succeed in obtaining reasonable internal amino acid sequences. Since in 9 of the examples shown in Table II, between 36 and 71 pmol of protein was subjected to trypsin digestion, it is possible to succeed with this amount of eluted protein. The relevant question, however, is not how much protein was eluted but rather, how much protein was actually loaded onto the gel. The average molecular weight of these 9 proteins was 63,000 (the range was 33,000-95,000) and the average amount of protein digested was 52 pmol. Assuming an average elution efficiency of about 74% (Table I) and an average blotting efficiency (onto the first membrane only) of about 40% (based on studies on four standard proteins in this molecular weight range (data not shown)), then it would appear that the average amount of these 9 proteins that had actually been subjected to SDS-PAGE was about 175 pmol (ie., 52/0.74/0.40). Although it is possible to succeed with less protein, for the present time at least, 175 pmol of protein applied to the SDS-PAGE gel or approximately 70 pmol of PVDF-blotted protein would seem to be a reasonable "target" in terms of the amount of protein that might be needed for internal amino acid sequencing.

Although extensive quantitative data (similar to that in Table II) on the overall yield of peptides obtained from the Aebersold et al [4] *in situ* digestion procedure has not yet been published, it appears from the limited data available that the CnBr elution approach described in the present study represents an alternative, rather than a clearly superior methodology. Hence, after correcting the average coupling yield of 20.9% (for 48, non-repurified tryptic peptides, Table II) for the 74% average PVDF elution efficiency (Table I) and the 34% average PVDF blotting efficiency (based on studies on 11 standard proteins - data not shown), we conclude that the overall average recovery of peptides from the CnBr elution procedure is only about 5%, which is nearly identical to the 6% value reported [7] for 10 tryptic peptides isolated from 400 pmol serum albumin by the Aebersold et al [4] procedure.

In terms of further improving the sensitivity of "internal" amino acid sequencing, it appears that further research might be warranted in terms of either improving the efficiency of PVDF blotting or of developing and improving procedures that avoid this step. The average PVDF elution efficiency of 74% (Table I) leaves relatively little room for improvement and the estimated recovery (assuming an average sequencing coupling yield of about 50%, [16]) of non-repurified tryptic peptides from the PVDF-eluted proteins that were sequenced in Table II, was about 42%. Since this latter value is close to the 52% value we obtained for a 100-fold larger scale (15 nmol) digest of a 36,000 dalton protein that had not been subjected to SDS PAGE (data not shown), it appears doubtful that significant further improvement can be obtained in this step. In terms of further improving the blotting efficiency of proteins, the simultaneous requirements of the cyanogen bromide elution procedure for tight binding of proteins to PVDF but relatively weak binding to cyanogen bromide peptides may stand in the way of further progress with regard to improving the sensitivity of this approach. For this reason, the alternative strategy of carrying out the digest in the gel (thus avoiding PVDF blotting) and then extracting the resulting peptides [6,17] would, at least in theory, seem to represent an attractive alternative. Although the overall recovery of 3 peptides isolated from 1.1 nmol β-lactoglobulin by the Ward et al in gel digest approach was only 2.2% [6], much of this loss may have occurred during the "inverse gradient" HPLC step that was used to remove SDS. The Kawasaki et al, [17] in gel digest approach avoids this inverse gradient step and indeed, parallel "in gel" versus *in vitro* experiments suggest the overall peptide yield from the Kawasaki et al procedure may be considerably higher than the 5-6% level. Since coupling yields were not given in the Kawasaki et al, [17] study, additional experiments are needed to confirm this hypothesis. Nevertheless, further study of in gel digestion approaches would seem to represent the best strategy for significantly improving the sensitivity of internal amino acid sequencing from SDS-PAGE-separated proteins. In the meantime, the data in Table 2 clearly demonstrate that CnBr elution/tryptic digestion already offers a well-proven approach for obtaining tryptic peptide sequences (suitable for designing oligonucleotide probe sequences) from 70 pmol or greater amounts of PVDF-blotted proteins.

References

1. Matsudaira, P. (1987) *J. Biol. Chem.* **262**, 10035-10038.
2. Brown, J.L. and Roberts, W.K. (1976) *J. Biol. Chem.* **251**, 1009-
3. Driessen, H.P., de Jong, W.W., Tesser, G.I. and Bloemendal, H. (1985). *In* "Critical Reviews in Biochemistry" (Fasman, G.D.,ed) Vol. **18**, pp. 281-325, CRC Press, Boca Raton.
4. Aebersold, R.H., Leavitt, J., Saavedra, R.A., Hood, L.E. and Kent, S.B. (1987) *Proc. Natl. Acad. Sci. USA* **84**, 6970-6974.
5. Szewczyk, B. and Summers, D.F. (1988) *Anal. Biochem.* **168**, 48-53
6. Ward, L.D., Reid, G.E., Moritz, R.L. and Simpson, R.J. (1990). *In* "Current Research in Protein Chemistry: Techniques, Structure, and Function" (Villafranca, J.J.,ed) pp. 179-190, Academic Press, New York.
7. Yuen, S.W., Chiu, A.H., Wilson, K.J. and Yuan, P.M. (1989) *BioTech.* **7**, 74-82.
8. Laemmli, U.K. (1970) *Nature* **227**, 680.
9. Towbin, H., Staehelin, T. and Gordon, J. (1979) *Proc. Natl. Acad. Sci. USA* **76**, 4350-4354.
10. Stone, K.L., LoPresti, M.B., Crawford, J.M., DeAngelis, R. and Williams, K. (1989). *In* "A Practical Guide to Protein and Peptide Purification for Microsequencing" (Matsudaira, P., ed) pp. 33-47, Academic Press, New York.
11. Stone, K.L., LoPresti, M.L., Williams, N.D., Crawford, J.M., DeAngeles, R. and Williams, K.R. (1989). *In* "Techniques in Protein Chemistry" (Hugli, T.E.,ed) pp. 377-391, Academic Press, New York
12. Stone, K.L., Elliott, J.I., Peterson, G., McMurray, W. and Williams, K.R. (1990). *In* "Methods in Enzymology" (McCloskey, J.,ed) Vol. **193**, pp. 389-412, Academic Press, New York.
13. Stone, K.L. and Williams, K.R. (1988). *In* "Macromolecular Sequencing and Synthesis" (Schlesinger, D.H.,ed) pp. 7-24, Alan R. Liss,Inc., New York
14. Speicher, D.W. (1989). *In* "Techniques in Protein Chemistry" (Hugli, T.E.,ed) pp. 24-35, Academic Press, New York
15. Stone, K.L., LoPresti, M.B., Crawford, J.M., DeAngelis, R. and Williams, K.R. (in press). *In* "HPLC of Peptides and Proteins: Separation, Analysis and Conformation" (Hodges, R.S., ed) CRC Press, Boca Raton
16. Niece, R.L., Williams, K.R., Wadsworth, C.L., Elliott, J., Stone, K.L., McMurray, W.J., Fowler, A., Atherton, D., Kutny, R. and Smith, A. (1989). *In* "Techniques in Protein Chemistry" (Hugli, T.E., ed) pp. 89-101, Academic Press, New York.
17. Kawasaki, H., Emori, Y. and Suziki, K. (1990) *Anal. Biochem.* **191**, 332-336.

Evaluation of Protein Sequencing Core Facilities: Design, Characterization, and Results from a Test Sample (ABRF-91SEQ)

Dan L. Crimmins[1], Gregory A. Grant[2], Liane M. Mende-Mueller[3], Ronald L. Niece[4], Clive Slaughter[5], David W. Speicher[6], and K. Ümit Yüksel[7]

[1] Howard Hughes Medical Institute Core Protein/Peptide Facility, Washington University School of Medicine; St. Louis, MO 63110

[2] Department of Biochemistry and Molecular Biophysics, Washington University School of Medicine; St. Louis, MO 63110

[3] Protein/Nucleic Acid Shared Facility, Medical College of Wisconsin; Milwaukee, WI 53226

[4] Protein/DNA Sequence/Synthesis Facility, University of Wisconsin Biotechnology Center; Madison, WI 53705

[5] Howard Hughes Medical Institute, University of Texas Southwest Medical Center; Dallas, TX 75235

[6] Protein Microchemistry Facility, Wistar Institute; Philadelphia, PA 19014

[7] Biopolymer Analysis Laboratory, Texas College of Osteopathic Medicine; Fort Worth, TX 76107

I. Introduction

Protein N-terminal sequence analysis has continued to play a major role in the elucidation of complex biological research issues. This sophisticated procedure is currently used to analyze a wide spectrum of samples, including: quality control of recombinant proteins and synthetic peptides; purity, identification, and evolutionary assessment of isolated proteins; analysis of protein fragments for the purpose of producing oligonucleotide probes and generating antibodies via the corresponding synthetic peptide; and, in favorable circumstances, identification of modified residues. Protein chemistry laboratories, as a result, are faced with challenging sample analyses on an ever increasing basis. Clearly, not every protein sequencing facility will possess the same degree of expertise and instrumentation as others. This has tended to promote unrealistic and uninformed expectations from users of these facilities.

A continuing effort to inform investigators about the average capabilities of protein sequencing core laboratories has been mounted by the Association of

TECHNIQUES IN PROTEIN CHEMISTRY III

Biomolecular Resource Facilities (ABRF) sequencing committee for the past several years. Samples with both specific design goals and well-characterized components were distributed to ABRF-member core facilities world-wide. The analysis of these data for STD-1 [1], ABRF-89SEQ [2], and ABRF-90SEQ [3] have provided valuable information to users regarding the performance of the average facility. Furthermore, these studies allow the participating facilities to objectively evaluate their strengths and weaknesses. In continuation of this effort, a test sample designated ABRF-91SEQ was distributed to 198 ABRF-member protein sequencing core facilities and this report describes the results from the analysis of the first 90 sets of data returned by May 15, 1991.

II. Materials and Methods

A. *Design of the ABRF-91SEQ Test Sample*

The test sample, designated ABRF-91SEQ, was designed to possess similarities to the previous year's sample, ABRF-90SEQ, but with several significant differences. The intention being to determine whether or not those differences had a practical effect on the sequencability of the sample. In 1990, the 29 residue synthetic peptide of ABRF-90SEQ was covalently coupled to transferrin, mixed with glucose-6-phosphate isomerase, and blotted onto polyvinylidenedifluoride (PVDF) membrane [3]. Since glucose-6-phosphate isomerase is naturally blocked and transferrin becomes mostly blocked during the first step of the coupling procedure, this produced a sample that would mimic a large molecular weight protein in both solubility properties and background. The present sample, ABRF-91SEQ contained all of the same components except that the peptide was not covalently linked to the transferrin and the sample was supplied in a form to be dissolved for loading. The transferrin was left unblocked in this case but was present in the sample at a level comparable to the level of unblocked transferrin in ABRF-90SEQ.

The two proteins, human transferrin (75.2 kDa) and yeast glucose-6-phosphate isomerase (E.C. 5.3.1.9; 2x63 kDa; naturally blocked N-terminus) were added at 0.3 μg (4 pmol) and 4.93 μg (38 pmol) respectively, to 0.27 μg (80 pmol) of a 29-residue synthetic peptide. The synthetic peptide itself (Fig. 1) possessed several features in common with the synthetic peptide portion of ABRF-90SEQ [3]. These include (i) a 29-residue length with identical overall amino acid composition; (ii) sequence identity at positions Trp-6, Leu-10, Arg-11, Pro-14, Ser-25, Phe-27, and Lys-28; (iii) an Asn-8:Asp-9 pair to evaluate possible effects of deamidation on sequence assignment; and (iv) a Leu-3:Leu-7:Leu-10 triplet early in the sequence to assess repetitive yield. Several new features were also incorporated into the synthetic peptide portion of ABRF-91SEQ. Glycine was chosen as the N-terminal residue to test the operator's ability to correctly assign this frequently occurring contaminating residue, particularly since the second amino acid, Arg, of ABRF-91SEQ is typically a "low-yield" residue. An unmodified cysteine was placed early in the sequence, Cys-5, but before any authentic Ser residues, to ascertain if this difficult residue could be tentatively inferred from the potentially diagnostic phenylthiohydantoin (PTH) adduct peaks. Lastly, a second potential deamidation pair, Gln-15:Glu-16, was incorporated into the sequence.

```
        5         10        15        20        25
G-R-L-K-C-W-L-N-D-L-R-S-Y-P-Q-E-M-T-H-V-A-S-I-D-S-V-F-K-V
```

Figure 1. Amino Acid Sequence of ABRF-91SEQ.

B. Peptide synthesis and characterization

The 29-residue synthetic peptide was synthesized with t-Boc chemistry on an Applied Biosystems Model 430A Synthesizer by the Protein Chemistry Laboratory at the Washington University School of Medicine and HF-cleaved from the resin using established procedures as previously described [2,3]. The crude product was subjected to preparative reversed-phase high-performance liquid chromatography (HPLC) and the purified peptide distributed to several of the authors for N-terminal sequence and compositional analysis. The peptide was judged to be pure by sequence analysis at sample loads of 50-200 pmol with initial yield (%) of 54.7 ± 21.3 (average ± sample standard deviation, seven independent determinations) with a range of 23 to 85%. Two independent Fast Atom Bombardment Mass Spectrometry determinations were performed on the putatively pure peptide yielding a m/z value of 3393.0 and 3393.4 (calculated m/z = 3394.96) as the major component (~90%). A minor component at 1106.0 (~10%) was observed and is consistent with an N-terminal acetylated 10-mer comprising residues 20-29. Two peaks consistent with this ratio and which eluted/migrated in the expected order for these species were also observed when the sample was analyzed by both analytical reversed-phase and strong-cation exchange HPLC, and free-solution capillary electrophoresis. This N-terminally blocked contaminant should not interfere with the sequence analysis, and was therefore not removed from the intended sequencable target peptide.

C. Preparation and distribution of the test sample

The three components were mixed in the proportions described above, aliquotted into pre-washed [1,2] 1.5 ml polypropylene microfuge tubes, and then vacuum dried. These samples were mailed to 198 core facility members of the ABRF along with a brief questionnaire and a detailed set of instructions for sample solubilization, loading, and data reporting. Data were returned to an independent third party who removed any identifying marks. These anonymous data were then forwarded to the authors for analysis.

III. Results and Discussion

A. Survey Responses of the Participating Facilities

The number of facilities participating in this year's study was about 1.6-fold greater (90 vs. 56) than in last year's study, yet the response rate was similar, 45.5% (90/198) compared to 42.4% (56/132) [3]. A short survey was included in the report with 86 of the 90 respondents completing this part of the questionnaire. There was no significant change in the average number of sequencers/facility (1.56, range 1-6) compared to the previous year [3] although the average age of the sequencers used to analyze the ABRF-91SEQ sample did decline to 4.2 years (range 0.8-9). Most of the facilities (78/86) have Applied Biosystems (ABI) sequencers (470-17, 473-6, 475-14, 477-41) with the remaining eight responses distributed between Porton (2090-5), Milligen (6625-2), and Beckman (890-1) sequencers. Approximately two-thirds (67.9%) of the sequencing laboratories purchase all of their sequencer/analyzer supplies from the manufacturer of their instrument(s).

The overwhelming majority of facilities (84/86) use on-line HPLC PTH analysis, mainly on ABI 120 instruments (77/85). Most perform the chromatography on a C18 column from ABI/Brownlee (72/85) with a 2.1 mm internal diameter column being most popular. The amount of sample injected onto the HPLC fell into three ranked groups of 50 µl (36/81), 100 µl (25/81), and 80 µl (7/81) [overall average of 54.8 ± 19.0 percent injected]. Not surprisingly, since the number of participating facilities increased so did the total number of samples analyzed in these laboratories (total 22,264; average 174, range 4-800/instrument; average 282, range 4-3000/facility). Both of these averages are down about 17% from 1990 [3]. Outside service to other investigators accounts for the bulk of the workload (66%) at the protein sequencing core facilities. The average site performs 25.6 (range 1-250) standard protein/peptide analyses per year per facility to routinely validate and trouble-shoot instrument performance. This translates to an average of one standard analysis per eleven sequencing runs.

Sequencing yield in each cycle from the ABRF-91SEQ sample was quantified after correcting for the fraction of injected sample (71/81) in units of picomoles (80/81) and in conjunction with an average loading of 49.8 pmoles of PTH amino acid standard but without the use of an internal standard (66/80). The sample was apparently loaded and analyzed as requested using standard cycles (74/86), on a glass-fiber filter (74/86) that had been treated with polybrene and pre-cycled (63/86). Only 3.7% of the facilities monitor absorbance at 313 nm, 6.2% utilize serine cycles, 56.3% do not utilize proline cycles although 32.5% do so ≤ 5 percent of the time, and 29.4% incorporate OPA blocking in their sequencing strategy. Virtually all laboratories sequence samples that have been electroblotted from one-dimensional gels (94%, 78/83) and almost one-half (46.7%, 35/75) analyze samples electroblotted from two-dimensional gels. The widespread success of the sequencing, and the relative ease of preparing these immobilized samples [4] is a likely factor in the near doubling over 1990 [3] of the facilities that offer electroblotting as a service (49% vs. 26.4%). Interestingly, this increase has occurred despite little change in the number of facilities that performed electroblotting this year (63.9%) compared to those who responded to the ABRF-90SEQ questionnaire (56%) [3].

B. Sequence Results for ABRF-91SEQ

A total of 1906 sequencing cycles were reported for the analysis of ABRF-91SEQ (Fig. 2) with an average of 22 cycles/report (range 1-30). There were 1421 (74.6%) positive and 200 (10.5%) tentative assignments, with 285 (15%) unassigned cycles ("holes"). Accuracy was 83.0% for positive assignments and 55.0% for tentative assignments. Thus, 1291 (67.7%) of the 1906 cycles were correctly identified. These values were all within a few percent of those reported for ABRF-90SEQ [3]. During this study, 9 (10%) instances of instrument malfunction were reported as compared to 8 (14.5%) for ABRF-90SEQ [3]. These included single instances of R2 delivery malfunction, computer failure, vacuum failure, and injection failure that terminated the run. There were also two undefined malfunctions and two injector malfunctions that did not result in run termination, and one fraction collector failure, on an instrument that did not employ on-line PTH-amino acid injection, that terminated the run.

Leu-3 was the most correctly identified residue overall (76 correct and 2 wrong positive; 3 correct and 2 wrong tentative; 0 holes), followed closely by Lys-4. Within the first 22 cycles, which was the average number reported, Cys-5 was the most difficult to identify as expected (6 positive and 5 tentative correct out of a total of 83 attempts). Position 5 was most often misidentified as threonine (29 positive and 9 tentative assignments) which was probably due to threonine being at position 5 in transferrin, although transferrin was present at only 5% of the synthetic peptide present in the sample. No other apparent misassignments related to the transferrin sequence were made although several responses correctly identified this as a minor secondary sequence.

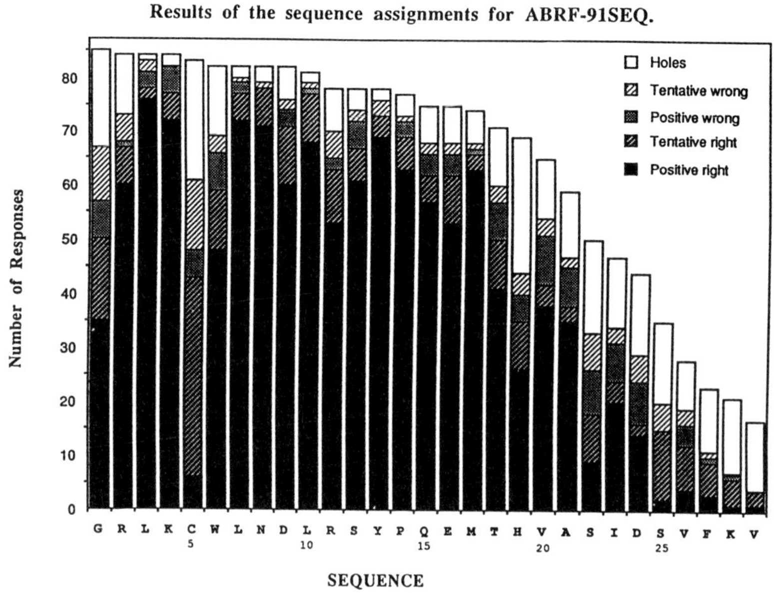

Figure 2. Results of Sequence Assignments of ABRF-91SEQ.
It was assumed that the data reported on the returned sequence data sheets were operator assigned sequence unless otherwise specified by the respondent. Scoring of residue assignments, exemplified by Cys-5, were as follows: C or C, (T) = positive correct; (C) or (C), (T) or (C,T) = tentative correct; T or T,C or T, (C) = positive wrong; and (T) or (T), (C) or (T,C) = tentative wrong. An "X", "?", "-", or greater than two residues being reported at a cycle was considered a "hole".

The single tryptophan residue, which was present at position 6 in both ABRF-90SEQ and ABRF-91SEQ, was identified with much greater success this year with 48 positive and 7 tentative correct (55/81, 67.9%) as opposed to 13 positive and 4 tentative correct last year (17/55, 30.9%) [3]. The reason for this greater than two-fold increase in Trp identification is not apparent, but may be attributable to adsorption to PVDF or an intrinsic recovery problem of Trp from PVDF.

Of the 3 additional residues that were present at identical positions within the first 22 residues in ABRF-90SEQ and ABRF-91SEQ (Leu-10, Arg-11, and Pro-14) only Arg-11 showed a significant difference in identification (70.5% positive and tentative correct out of 78 responses this year as compared to 35.4% out of 48 responses for ABRF-90SEQ [3]).

While the first residue is often difficult to identify because there is no previous cycle for comparison, the glycine at position 1 in ABRF-91SEQ was the least correctly identified N-terminus of all previous ABRF samples [1-3]. This may be due to the fact that glycine is often seen as a substantial contaminant at cycle 1 in experimental samples and may be dismissed as a valid sequence residue. Indeed, this was in fact the reason why glycine was placed at this position in ABRF-91SEQ and the results indicate that the coincidence of glycine contamination and its presence as the N-terminal residue present serious problems in assignment.

Table I. Sequencing results for ABRF-91SEQ (a)

Instrument		Repetitive	Number of	Accuracy of
Model (b)	# (c)	Yield (d)	Positive correct	Positive Calls (e)
470/120	17 [12]	87±4.9 (79-96)	15±6.5 (1-23)	83±31.0 (4-100)
475/120	14 [10]	87±5.5 (77-95)	11±8.0 (0-22)	65±39.4 (0-100)
477/120	41 [37]	81±10.0 (49-96)	14±7.0 (1-27)	77±28.1 (17-100)
473/120	6 [5]	75±11.0 (55-81)	17±6.2 (9-24)	87±11.3 (75-100)
2090	5 [4]	86±5.9 (77-90)	12±7.8 (0-21)	78±43.5 (0-100)
6625	2 [1]	87	10 (1-19)	65 (50-79)
890	1 [0]	-	6	35
All Instruments	86 [69]	83±9.0 (49-96)	14±6.7 (0-27)	83±21.8 (0-100)

(a) Values are expressed as mean ± standard deviation, with ranges in parentheses.
(b) 2090E is a Porton sequencer with a Hewlett Packard 1090 HPLC; 6625 is a Milligen 6600 sequencer with a Waters 625E HPLC; 890 is a Beckman 890M-2 seqencer with a Beckman 344 HPLC and all other instruments are from Applied Biosystems.
(c) Number of instruments for which data was reported. Numbers in brackets are those returns used for repetitive yield calculations.
(d) Expressed as percent. Based on background corrected yield of Leu-3 and 7.
(e) Expressed as percent. Based on correct calls for positively identified residues only. Cycles that were left blank or had tentative assignments were not scored.

Potential deamidation, as judged by identification of the Asn-Asp pair at positions 8 and 9, did not seem to be a problem. The Gln-Glu pair at positions 15 and 16 was consistent with this observation.

The average response reported identifications for 22 cycles including tentative assignments with an average of 14 positive correct assignments (see Table I). In comparison, eight responses (477-6, 473-1, and 470-1) identified at least 22 residues positively correct. Two of the longest and better responses reported 25 positive correct out of 28 calls, with 2 positive incorrect and 1 hole while another response reported 24 positive correct calls out of 25 attempts.

Analysis of the data for repetitive yield and accuracy were grouped by instrument type and are presented in Table I. Statistically there does not appear to be any significant difference in average repetitive yield, average number of positive correct, and average accuracy of positive calls between instruments.

Table II. Comparison of ABRF-91SEQ to Previous Test Samples (a)

Sample(b)	Description	Amount (pmol)	Repetitive Yield %	Number of Correct Res.	Accuracy % (c)
STD-1(1988)	single soluble peptide, 40 aa	100	88±3.9 (76-93)	24±7.1 (9-37)	95±7.0 (65-100)
ABRF-89SEQ	soluble peptides (5:1) 40 & 43 aa	240	89±2.4 (77-99)	30±6.5 (16-38)	95±6.6 (80-100)
ABRF-90SEQ	PVDF blot mimic 100 kDa protein, 29 aa	30	89±3.7 (82-95)	13±6.4 (1-27)	82±19.1 (36-100)
ABRF-91SEQ	soluble peptide, 29 aa,100 kDa protein	80	83±9.0 (49-96)	14±6.7 (0-27)	83±21.8 (0-100)

(a) Values are expressed as mean ± standard deviation, with ranges in parentheses.

(b) Data for STD-1, ABRF-89SEQ, and ABRF-90SEQ are from references [1], [2], and [3], respectively.

(c) Percentage of positive sequence assignments that were correct.

In previous ABRF sequencing studies, a comparison of the accuracy of automated, software assisted calls with manual, operator assignments indicated that the experienced sequencer operator is more accurate [1-3]. This year a similar analysis resulted in 54.0±29.0 percent accuracy for those respondents that reported software derived sequence assignments compared to a value of 83.0±21.8 percent accuracy (Table I) obtained from all responses to ABRF-91SEQ. These numbers and their differences are virtually identical to those reported for ABRF-90SEQ [3] (58.0 and 82.0 percent accuracy, respectively) and once again this demonstrates that more sophisticated versions of the software will likely be required to increase the accuracy of the automated calls. It should be pointed out however, that most versions of the current software, by default, make positive sequence assignments at most of the cycles. It is not surprising then, that these positive cycle assignments, on average, will statistically contain more positive wrong calls compared to an operator who can provide tentative cycle assignments.

Table II compares the results of ABRF-91SEQ to previous samples. Although the average repetitive yield for ABRF-91SEQ appears to be lower than in previous years, the range is also broader and there is some question as to whether this difference is significant. More importantly, perhaps, the table clearly indicates that the average number of positive correct and average percent accuracy are essentially identical for ABRF-91SEQ and ABRF-90SEQ. As discussed in the section on sample design, ABRF-91SEQ was designed as a soluble sample for comparison to ABRF-90SEQ to evaluate any differences in these parameters due to presenting the sample on PVDF.

The actual amount of sequencable peptide in both samples was approximately equivalent. Since it was not feasible to directly quantitate the amount of peptide present on the PVDF blot for ABRF-90SEQ, the 30 pmol so indicated in the table reflected the average initial yield produced by the samples. Initial experiments by the sequencing committee indicated that for an 80 pmol load of the peptide in ABRF-91SEQ, the average initial yield was 54.7±21.3% (range 23-85%) corresponding to an average initial yield of 44 pmol. This was verified by the data from the responses to ABRF-

91SEQ which indicated an average initial yield of 36±18.6 pmol (range 3-89 pmol). Thus, the average available sample for sequencing was similar for both ABRF-90SEQ and ABRF-91SEQ. Therefore, the values for these two samples in Table II are directly comparable and indicate that the number of identifiable residues and the accuracy of those identifications did not appear to be a function of immobilization onto PVDF or cross-linking to a carrier protein. In our view, this is not meant to indicate that there is no advantage to immobilizing samples on to PVDF, but rather, once equivalent amounts of sample are delivered to the sequencer, the analysis proceeds equally well in both cases. Indeed, many reports and individual experiences indicate that PVDF is an excellent medium for capturing and transferring small quantities of protein [4].

In general, considering the results from the four years that the ABRF has been distributing samples, it appears that the main factor in determining the number of residues that can be sequenced and the accuracy of that sequencing, is the total amount of sample delivered to the sequencer. Therefore, the results of ABRF-91SEQ are consistent with and further support the conclusions reached previously [5].

It must be pointed out that the data presented here represent the average for all facilities that reported data. This includes returns not only from experienced laboratories but also from laboratories that are just starting out and are still learning the techniques and art of sequencing. Moreover, these results are based on a one time analysis of a sample with little background information provided. Also, because it is an outside test sample, some operators might have been a little more daring in attempting to make positive calls than with actual experimental samples. As such, the results of this study should be viewed conservatively. The average results presented here, particularly the % correct positive calls, should not be regarded as representing an acceptable level of performance for a facility on a daily basis. On the contrary, facility personnel should strive to achieve the performance level of the best laboratories reported. There were 19 responses with 100 % accuracy, and 26 responses with 90-99.9 % accuracy. An experienced investigator will be very careful in assigning positive calls and will not hesitate to categorize a call as tentative if there is any doubt. Certainly, an accuracy of 90 % or better is desirable and realistic since 45 out of the 90 respondents achieved this level.

Acknowledgements

This work was partially supported by NSF grant DIR 9003100 to John Crabb (W. Alton Jones Cell Science Center) on behalf of the ABRF. We are grateful to the anonymous ABRF member facilities that participated in this study without whose interest and effort this project would not have been possible. The assistance of Melanie Budd (University of Wisconsin, Biotechnology Center) in coordinating the data return and providing the anonymity of the respondents is appreciated. Several members of the authors' laboratories contributed to various aspects of this project, including: Mark Frazier (Washington University School of Medicine, Protein Chemistry Laboratory) for synthesizing the peptide; Richard Thoma for amino acid analysis of the final test sample and David McCourt (Washington University School of Medicine, Howard Hughes Medical Institute) for initial sequence analysis; Brady Stoner and Rodney Prell (Medical College of Wisconsin) for preparation of the test sample; Jean Shelton and Peggy Vroman (Texas College of Osteopathic Medicine) for data entry; Carolyn Moomaw (University of Texas Southwest Medical Center, Howard Hughes Medical Institute) for sequence analysis of the test sample; and Lorraine Whiteley and Pat Parvin (Washington University School of Medicine, Howard Hughes Medical Institute) for expert secretarial assistance. Their efforts are gratefully acknowledged.

References

1. Niece, R.L., Williams, K.R., Wadsworth, C.L., Elliot, J., Stone, K.L., McMurray, W.J., Fowler, A., Atherton, D., Kutny, R., & Smith, A.J. (1989) *in* Techniques in Protein Chemistry, T.E. Hugli, ed., pp 89-101 Academic Press, San Diego.

2. Speicher, D.W., Grant, G.A., Niece, R.L., Blacher, R.W., Fowler, A.V., & Williams, K.R. (1990) *in* Current Research in Protein Chemistry, J.J. Villafranca, ed., pp 159-166 Academic Press, San Diego.

3. Yüksel, K.Ü., Grant, G.A., Mende-Mueller, L.M., Niece, R.L., Williams, K.R., & Speicher, D.W. (1991) *in* Techniques in Protein Chemistry II, J.J. Villafranca, ed., pp 151-162 Academic Press, San Diego.

4. LeGendre, N. (1990) Biotechniques **9** (6), 788-805.

5. Niece, R.L., Ericsson, L.H., Fowler, A.V., Smith, A.J., Speicher, D.W., Crabb, J.W., & Williams, K.R. (1991) *in* Methods of Protein Sequence Analysis, H. Jornvall, J.-O. Hoog, & A.-M. Gustavsson, eds., pp 133-141 Birkhauser Verlag, Basel.

OPTIMIZATION OF SOLID PHASE PEPTIDE SEQUENCING

Arie Admon and David S. King

Howard Hughes Medical Institute, Dept. of Molecular & Cell Biology, University of California, Berkeley, CA 94720

I. INTRODUCTION

As part of our attempt to sequence vanishingly small amounts of peptides, we tested solid phase (covalent) peptide sequencing technology as a way to improve the performance of our ABI 477A sequencer. Although the chemistry for solid phase sequencing was developed in the early seventies by Laursen (1971), only recently has it become commercially available for micro-sequencing of proteins and peptides by MilliGen/Biosearch. Milligen's kit includes either diisothiocyanate polyvinylidene difluoride (DITC-PVDF) membranes (Sequelon-DITC) for attachment via the primary amines of proteins, or arylamine-PVDF membranes (Sequelon-AA) for attachment via the carboxyl groups of peptides or proteins, after activation with a water-soluble carbodiimide (for recent review see Coull et al., 1991). Since our main interest is peptide sequencing, we tested only the arylamine-PVDF on the 477A sequencer. Covalent attachment was expected to improve sensitivity and shorten cycle times due to efficient removal of side products and complete extraction of amino acids. High repetitive yields can be obtained, since peptide wash-off is eliminated. Analysis of post-translational modification sites is possible due to flexibility of solvent use (Aebersold, 1990). Moreover, solid phase sequencing enables prior chemical and enzymatic modifications of the peptides while attached to the solid support (Andrews et al., 1991).

Here we describe optimized conditions for use of solid phase support on the 477A sequencer, and we will discuss the benefits of this combination.

II. MATERIALS AND METHODS

Sequencing was performed on an Applied Biosystems 477A sequencer with an on-line 120 PTH amino acid analyzer fitted with a 100 µl loop. We used modified ABI "Fast-1" cycles for reaction, conversion and analysis. All sequencing reagents and solvents were supplied by Applied Biosystems. S4 was diluted two fold with water to 10% acetonitrile in order to accomodate the 100 µl loop (Tempst and Riviere, 1989). Arylamine-PVDF covalent attachment

TECHNIQUES IN PROTEIN CHEMISTRY III

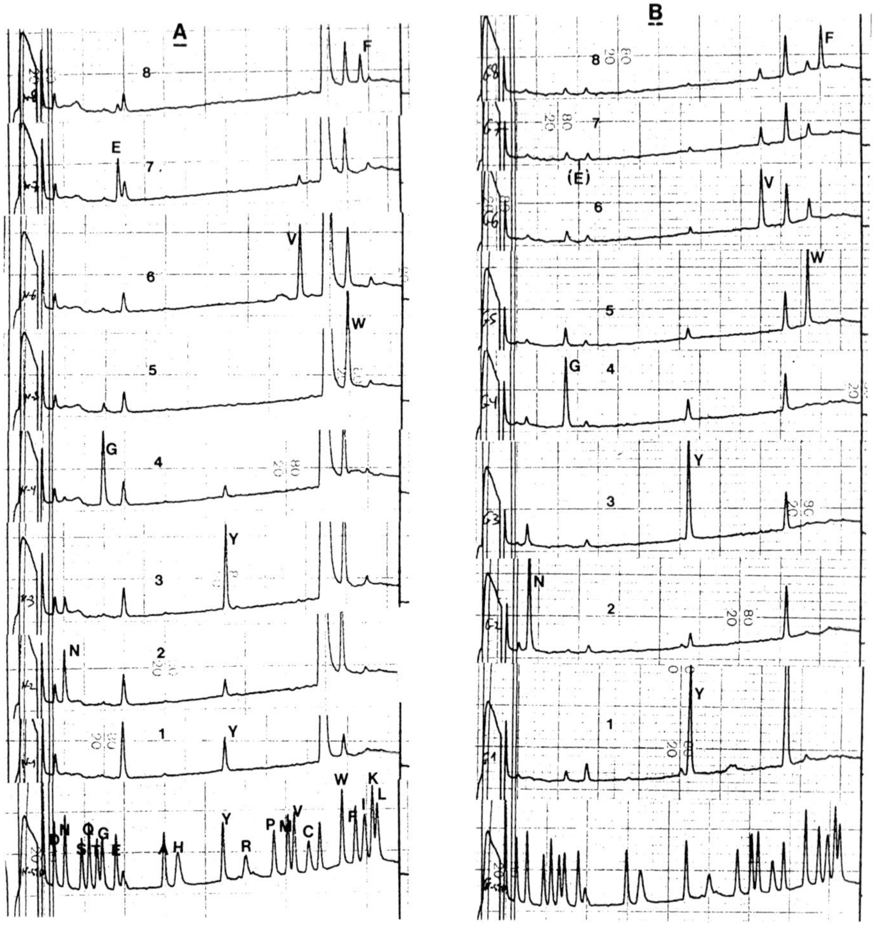

Fig 1. Chromatograms of 8 sequencing and analysis cycles of a test peptide YNYGWVEF. The peptide was dissolved in acetonitrile/TFA. Reaction, conversion and analysis cycles are "Fast-1". Sequencing was performed either on a precycled glass fiber filter coated with polybrene (A) or on arylamine-PVDF after covalent attachment (B). Standard PTH-aa are 15 pmole.

kits were obtained from MilliGen/Biosearch. Covalent attachment of peptides to arylamine-PVDF disks was performed according to Milligen's protocol. The peptides were spotted on the arylamine-PVDF disks and dried at 55°C for 15 min prior to covalent attachment of the peptides at room temperature with 2-5 µl of 10 µg/µl of 1-ethyl-3-(3-dimethylaminopropyl) carbodiimide (EDC) in 0.1M MES buffer pH 5.0.

Peptides were either synthetic or proteolytic products of proteins and all were purified on reversed-phase(RP) HPLC with a gradient of acetonitrile in 0.1% TFA.

III. RESULTS

A short synthetic peptide (YNYGWVEF) was used to compare sequencing with polybrene coated glass fiber filters (Fig 1A) and solid phase sequencing on arylamine-PVDF (Fig1B). In both cases a similar amount of the peptide and the same ABI Fast-1 reaction and conversion cycles were used. Polybrene modified glass fiber filter was used with the horizontal 8 mm cartridge, and arylamine-PVDF was used with the ABI vertical cross flow cartridge. The full size (8mm) arylamine-PVDF disks do not fit in the vertical flow cartridge, but as its peptide binding capacity is high, only part of the disk is required for sequencing amounts less than 100 picomoles. In those cases when relatively high concentrations of peptides were available, a quarter or a third of an arylamine-PVDF disk was used (2-3 µl spotted). For dilute samples, such as HPLC fractions, a slightly trimmed disk was used, enabling about 10 µl to be spotted at a time (Fig 2).

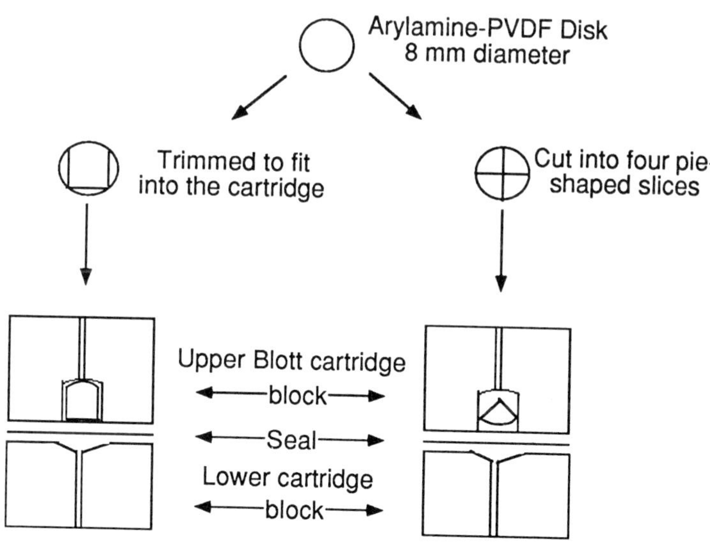

Fig 2. The arylamine-PVDF disks were cut to smaller pieces in order to fit in the vertical cross flow cartridge.

Table 1. The modifications to "Fast-1"cycle are illustrated. Cleavage in TFA is extended to 350 sec to reduce lag, similar to "Normal-2" from ABI (S. Yuen, personal communication) and wash times were shortened. Listed on the left are the reaction steps of "Seq100" and on the right the corresponding steps in "Fast-1".

- Applied Biosystems Reaction Cycle Listing -

Cycle Name : SEQ100 : FAST-1
Cycle Length : 45 steps : 34 steps
Runtime : 32 mins 24 secs : 36 mins 4 secs

Step	Function	Fxn#	Value	Fxn#	Value
1	Prep R2	4	6	4	6
2	Deliver R2	5	20	5	20
3	Prep R1	1	6	1	6
4	Deliver R1	2	2	2	2
5	Argon Dry	29	40	29	40
6	Deliver R2	5	250	5	250
7	Prep R1	1	6	1	6
8	Deliver R1	2	2	2	2
9	Argon Dry	29	40	29	40
10	Deliver R2	5	250	5	250
11	Prep R1	1	6	1	6
12	Deliver R1	2	2	2	2
13	Argon Dry	29	40	29	40
14	Deliver R2	5	250	5	250
15	Reaction Heater	32	48	32	48
16	Argon Dry	29	100	29	120
17	Deliver S1	14	10	14	60
18	Argon Dry	29	5		
19	Deliver S1	14	10		
20	Argon Dry	29	5		
21	Deliver S2	17	20	17	200
22	Argon Dry	29	5		
23	Deliver S2	17	20		
24	Argon Dry	29	100	29	120
25	Load R3	9	6	9	5
26	Argon Dry	29	4	29	4
27	Pause	33	350	33	300
28	Load S2	18	8	18	6
29	Block Flush	30	10	30	6
30	Argon Dry	29	60	29	120
31	Prep Transfer	22	30	22	30
32	Deliver S1	14	8	14	8
33	Pause	33	10		
34	Transfer w/ S3	23	15	23	52
35	Pause	33	15	33	20
36	Transfer w/ Argon	24	5	24	40
37	Transfer w/ S3	23	15		
38	Pause	33	15		
39	Transfer w/ Argon	24	40		
40	End Transfer	25	1	25	1
41	Reaction Heater	32	53	32	53
42	Deliver S3	20	15		
43	Argon Dry	29	5		
44	Deliver S3	20	15	20	30
45	Argon Dry	29	120	29	120

The initial yield was similar when arylamine-PVDF or polybrene glass fiber filters were used. When the solid phase approach was used, repetitive yields and recovery of the last amino acid (F) were higher and the background signals of the PITC side products were smaller or totally absent, enabling positive identification of tryptophan (cycle 5). However, with solid phase sequencing, Glu-PTH or Asp-PTH were not recovered efficiently, and sometimes not at all (cycle 7 in fig 1b and in fig 3). These results show the compatibility of the ABI 477A with solid phase sequencing without a need for modifying the instrument or changing the reagents.

Next, we attempted to shorten the cycle time and reduce reagent consumption by shortening wash and extraction times. Our goal was to achieve a reaction and conversion cycle as short as our fastest PTH-aa analysis time (table 1, 32 min including column equilibration). It became clear that total cycle times of 32 min can be easily obtained without sacrificing sensitivity or repetitive yields (Fig 3). The shortened cycle, here called "Seq100", also results in reduced lag and lower background signals.

We observed low recoveries of Asp-PTH and Glu-PTH when using arylamine-PVDF, and sometimes these amino acids were totally absent (compare Fig 1a and Fig 1b cycle 7). This occurred with all of the different extraction conditions we tested. These low yields could be due to covalent coupling of the arylamine filter to the carboxyl end of the peptide as well as to the carboxyl side chains. The signal in cycles following Asp or Glu returned to almost the same magnitude as before these amino acids. We attempted to modify the coupling conditions in order to limit coupling to the carboxyl end. A reduction in the concentration of EDC resulted in reduced coupling of the peptide to the solid support and did not seem to solve the problem (results not shown).

Another potential application of solid phase sequencing is its use for identification of the site and nature of various post-translational modifications. The problem with the use of polybrene coated glass fiber filters for this purpose is the low recovery of modified amino acids (such as phosphoamino acids). Using solid phase chemistry, MilliGen recommends transfer in neat TFA for their sequencer. We tried transfer with TFA on the 477A, which appears to function well. The background noise and lag were higher, but recoveries of amino acids were similar to transfer in *n*-butyl chloride (results not shown). We are now in the process of testing this technique for analysis of phosphorylated peptide.

When only very small amounts of peptides are available, collecting them directly off microbore RP-HPLC onto filters is very useful in order to increase yield and reduce contamination. The problem observed in this case is the limited capacity of the PVDF disk for liquid, and the need to first wet it with organic solvent when collecting peptides in aqueous buffer containing low concentration of organic solvent (Coull et al., 1991). For most peptides recovered off microbore RP-HPLC, the concentration of organic solvent is high enough to wet arylamine-PVDF. However, the limited liquid holding capacity of the disk (less than 10µl) is a problem for most reversed phase chromatography, as peptides normally elute in 30 to 60 µl fractions. The use of a 1 mm column run at 50 µl/min for peptide separation provided a solution to this problem, but only when extra column volumes were reduced to a minimum by use of a specialized microbore HPLC such as the Michrom HPLC, with peak volumes typically 6-15 µl (results not shown).

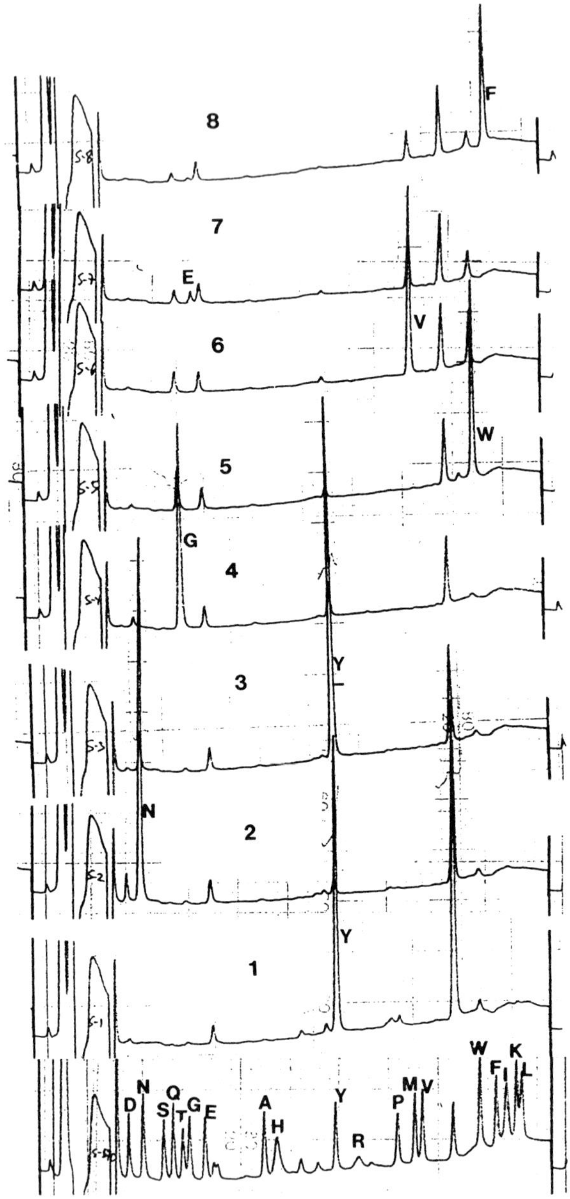

Fig 3. The same peptide as in fig. 1 was attached to arylamine-PVDF and sequenced using a shortened cycle "Seq100" (A) or using TFA transfer(B).

IV. DISCUSSION

In this report we demonstrate that solid phase sequencing can be done efficiently on the ABI pulsed liquid sequencer (477A) without any changes to the instrument. Sequencing covalently-bound peptides results in longer readable sequences due to the improved signal to noise and higher initial and repetitive yields. These improvements are most significant when attempting to sequence peptides at the 1 to 10 picomole range, or when complete sequence of the peptide is necessary. Identification of post translational modification sites (Aebersold, 1990) and post attachment chemical treatment of the linked peptides (Andrews et al., 1991) are other useful features of covalent sequencing. Shortening cycle time, reducing reagent consumption, and eliminating the need to precycle polybrene-coated filters has resulted in significant savings of time and expense. However, solid phase sequencing cannot be universally applied as many peptides do not have the available carboxyl groups to be covalently coupled to the solid support (such as peptides with a blocked carboxyl terminus). The reduced or absent signal of Asp and Glu is a major disadvantage we have found with the use of arylamide-PVDF for sequencing at the low picomole range. High carry over, which was thought to be a problem with solid phase sequencing, has been solved by the use of our optimised "Seq100" cycle. Similar to the report of Coull et al. (1991), we have not observed a significant difference of covalent coupling efficiencies between peptides with different amino acids at the carboxyl end (except Cys). Asp or Glu at the carboxyl end did not improve the coupling, nor did they recover well (results not shown).

Should one switch to solid phase sequencing? For sequencing more than 20 pmole, we believe the answer is yes, as the saving of time and expense are significant, recoveries of Asp and Glu are sufficient for positive identification, and sequencing through the carboxylic amino acid is very useful. For sequencing much smaller amounts, although positive identification of Trp is beneficial, the problem of low yield of Asp and Glu should be taken into consideration. If possible, one can first sequence the peptide with polybrene modified glass fiber filters, and then sequence again the same peptide with arylamine-PVDF as a way of obtaining longer sequence of the peptide, for Trp identification, or for post-translational modification analysis.

REFERENCES

Aebersold R. (1990) Nature **343**, 291-292.

Andrews D.W., M. Girardi, and J. Mark (1991) in Techniques in Protein Chemistry II (J.J. Villafranca, ed) pp. 75-78, Academic Press, San Diego.

Coull J.M., D.J. Pappin, J. Mark, R. Aebersold, and H. Koster. (1991) Analytical Biochemistry 194, 110-120.

Laursen R.A. (1971) Eur. J. Biochem. **20**, 89-102.

Tempst P. and L. Riviere (1989) Anal. Biochem. **183**, 290-300.

Evaluation of the Blott Cartridge for Enhanced Gas Phase Sequencing at Maximum Sensitivity

David F. Reim, Peter Hembach, and David W. Speicher
Wistar Institute, Philadelphia, PA

I. Introduction

The standard cartridge design used in most modern protein sequencers has been playing a dual role in recent years. It is used to sequence liquid samples, usually on TFA treated filters, and to sequence proteins electroblotted to PVDF membranes. The use of PVDF membranes has rapidly increased in popularity since its introduction as a sequencing support (1-3) to the point where a substantial number of protein sequencing facilities use this substrate for the majority of their facility's samples (4, 5). The major reason for this dramatic shift in sequence substrates over the past several years is that 1D and especially 2D polyacrylamide gel electrophoresis is the most powerful protein purification method for isolating picomole quantities of proteins with minimal loss.

The standard reaction cartridge design, conventional sequencer solvents, and sequencer programs that have evolved over the past 10 years for use with glass fiber substrates are not necessarily optimized for the strongly hydrophobic PVDF membranes. A major concern with the cartridge design when using PVDF membranes is the dead space above the membrane and the orientation of the membranes relative to the flow of solvents and reagents. Fluid deliveries to the protein bearing side of PVDF membranes is more problematic in that these membranes do not "wet" as efficiently as glass fiber filters. Another consideration is that the PVDF membrane is generally perpendicular to solvent flow in the standard cartridge and extractions may not be optimal. In addition, the hydrophobic character of PVDF and the large head volume frequently result in retention of gas bubbles above the membrane in the conventional cartridge leading to channeling of solvent around the membrane. Solvent channeling leads to variable extraction, especially at the ATZ derivative extraction step. This problem has previously been addressed by segmented or pulsed deliveries rather than one long or extended delivery for the S2 and S3 solvents (6). Other solutions to this problem of variable extraction include the continuous flow reactor incorporated into a compact sequencer design (7,8) and a cross-flow reactor (9).

Recently a new sequencer cartridge designed specifically for PVDF membranes has become commercially available (Blott cartridge, Applied Biosystems). This reaction cartridge orients the sample so that solvent flow is parallel to the protein loaded membrane surface. This

TECHNIQUES IN PROTEIN CHEMISTRY III

orientation together with the narrow width of the sample chamber slot, minimizes bubble trapping on top of the membrane and channeling of solvent around the sample is effectively eliminated or minimized. Since fluid movement relative to the membranes has been improved, wetting of the membrane should be improved, facilitating better extraction of the ATZ amino acids and byproducts.

We evaluated this new cartridge design using a program that had already been optimized for high sensitivity sequencing on PVDF membranes to determine whether the Blott cartridge would offer any further advantages.

II. Materials and Methods

A. *Sequencer Set Up and Gas Phase Delivery of TFA*

Experiments were performed using an updated ABI Model 475A protein sequencer connected to an ABI 120A PTH analyzer for on-line PTH amino acid analysis. The sequencer updates include a model 900A controller, bottle cap update, regulator update and liquid phase TFA delivery update. We converted back to gas phase TFA delivery after analyzing liquid phase delivery. Our earlier evaluations showed that gas phase was superior to liquid phase TFA delivery with higher repetitive yields in the gas phase mode due to reduced protein extraction from the membrane even when newer higher retention membranes were used. Both methods showed comparable lag (carryover). However, the small size of the TFA bottle used with the bottle cap update and on all newer sequencers required replumbing to deliver argon to the bottom of the TFA bottle rather than being swept over the TFA fluid surface. This technique delivers a more completely saturated gas stream to the reaction cartridge.

In these experiments the S3 extraction solvent which is typically butyl chloride has been replaced by a 1:1 mixture of heptane:ethyl acetate since this mixture wets PVDF membranes more efficiently than butyl chloride or ethyl acetate alone. Thc better solvent compatibility results in more complete extraction of the ATZ amino acids to the flask. All sequencer reagents and heptane were from Applied Biosystems. Acetonitrile was from J.T.Baker. Ethyl acetate and THF were from Burdick and Jackson.

A Supelco LC-18-DB 2.1mm X 25cm 3 micron column was used for identification of the PTH amino acids with the gradient shown in Table I. The HPLC mobile phase consisted of 100% acetonitrile as buffer B and buffer A was 4.25% THF/ 0.06% TMA/ 0.47% 3M sodium acetate buffer stock solution (21:4 of pH 3.8: pH 4.6). The gradient shown here includes a sharp slope increase in the latter part of the gradient to resolve DPU and tryptophan.

Table I. PTH Separation Gradient

Time	%B	μl/min	Time	%B	μl/min
0.0	12	110	22.5	42	220
0.1	12	220	33.0	45	220
2.0	16	220	33.2	90	400
14.5	25	220	37.3	90	400
20.1	35	220	37.5	90	5

B. *Sequencer Standards for Cartridge Evaluation*

Horse apomyoglobin was electrophoresed and blotted to BioRad or ABI PVDF membranes as previously described (10). We routinely use this protein rather than the more commonly used β-lactoglobulin as a sequencer standard because myoglobin contains three sensitive residues early in the sequence, i.e. serine in cycle 3 and tryptophan in cycles 7 and 14 that help analyze sequencer performance. Monitoring serine and tryptophan recoveries is especially critical when sequencing experimental samples at maximum sensitivity.

The apomyoglobin was prepared so that 300 pmoles could be loaded to each lane of a 13% gel giving a total of 15 lanes or samples for sequencing from a single aliquot. Using this approach, a single blot contained 15 highly consistent replicate standard samples for cartridge comparisons.

III. Results and Discussion

A. *Comparison of the Blott and Conventional Cartridges*

Table II summarizes the comparison of the Blott cartridge with the standard cartridge using the same sequencer program. The only preparation of the blotted protein prior to loading to the cartridge was a brief wash with Milli-Q water.

Table II. Sequence Performance of the Blott Cartridge[a]

		Standard	Blott
Initial Coupling(pmoles)		112±2	130±1
Repetitive Yield (%)		95.5±0.9	95.3±0.2
Lag (carryover)	A19	47%±10	29%±5
	A22	52%±11	31%±4
Tryptophan Yield (pmoles)			
	Cycle 7	16±14	64±1
	Cycle 14	8±6	48±15

[a] Values shown are the average of two replicate sequence runs with the standard cartridge and three sequence runs with the Blott cartridge ± standard deviation.

This sequence comparison showed that the Blott cartridge was superior to the standard cartridge with regards to lag (carryover) and tryptophan yields. Recovery of all amino acid residues with the exception of the serine in cycle 3 were slightly improved using the Blott cartridge (data not shown), while the serine yields and overall repetitive yields were comparable with both cartridges. The major advantages of the Blott cartridge are a significant decrease in lag or carryover and a marked improvement in tryptophan yield.

Figure 1 illustrates a further benefit of the Blott cartridge - reduction of typical byproducts resulting from the Edman degradation. In the example shown, DPTU has been decreased by almost 50% indicating more efficient extraction of byproducts from the reaction cartridge when the S2 (ethyl acetate) extraction times are held constant. Of course improved extraction means that extraction times can be shortened to maintain an acceptable background of reagent related peaks. Alternatively if the byproduct peak reduction is more desirable, longer solvent extraction times generally do not adversely affect the protein being sequenced due to its very strong affinity to PVDF membranes.

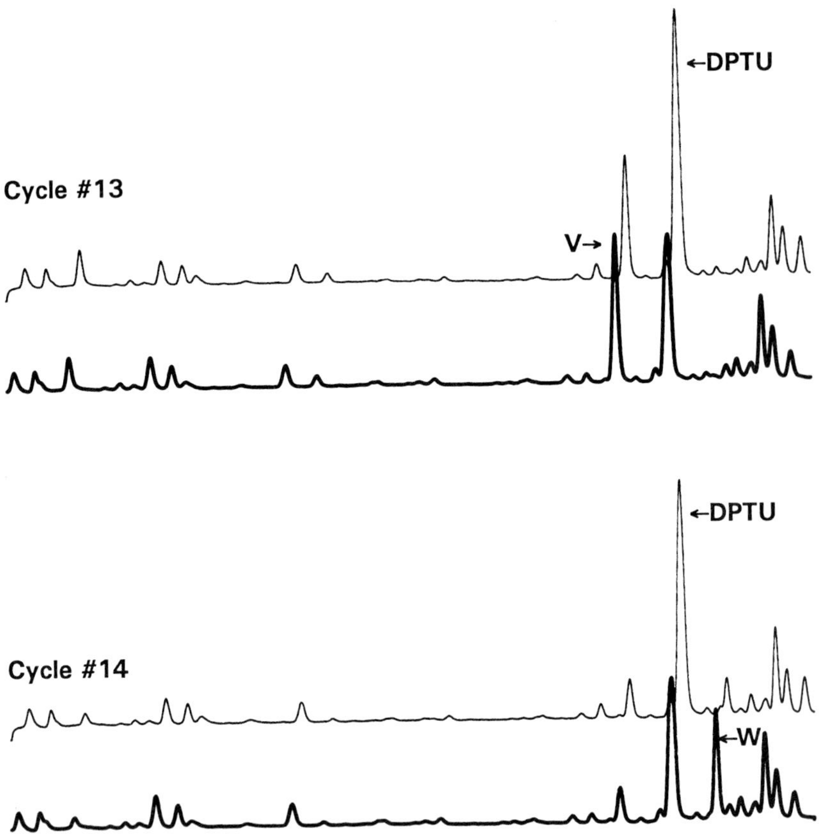

Figure 1. Decrease in DPTU and improved tryptophan recovery with the Blott cartridge. Cycles 13 and 14 from myoglobin standards blotted to a BioRad PVDF membrane. Duplicate lanes (300 pmoles/lane applied to the gel) from a single blot were compared. Blott cartridge - thick line, standard cartridge - thin line.

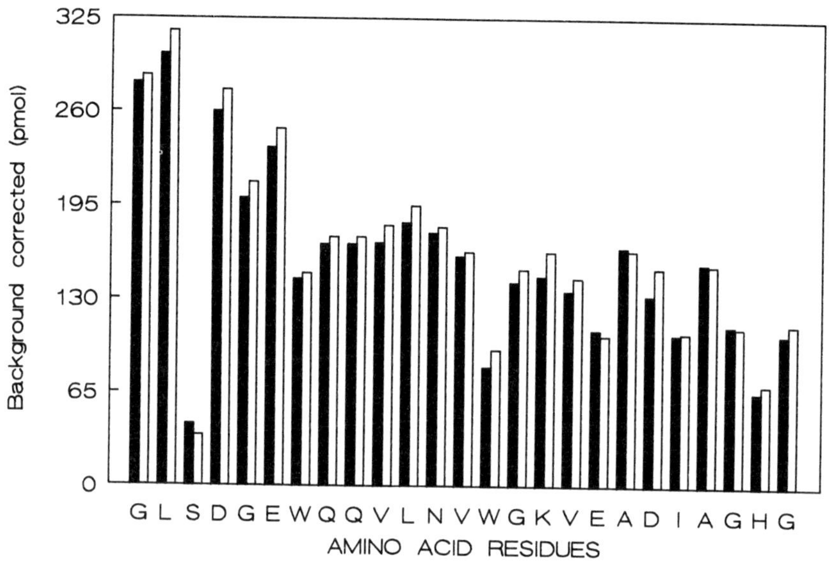

Figure 2. Effect of protein orientation in the Blott cartridge on sequence performance. Three replicate BioRad PVDF bands containing myoglobin standard (300 pmoles/lane applied to the gel) from a single blot were loaded in the Blott cartridge in each experiment. Solid bars - protein facing solvent flow; two bands were folded in half and inserted in cartridge with protein-containing face out(toward chamber walls), the third band was placed on top of the Zitex seal facing solvent flow. Open bars - bands were loaded in opposite orientation (folded with protein side of membrane inside - away from chamber walls).

B. Orientation of "Protein Side" of PVDF Membranes

As discussed earlier, proteins blotted to PVDF membranes produce in effect "one sided" membranes, ie. the majority of the protein is absorbed to the surface contacting the gel. The effects of orientation of this protein loaded face in the Blott cartridge were evaluated. In this experiment the excised protein bands were folded in half with either the protein side facing in or out (more directly exposed to the solvent stream). A large amount of PVDF membrane was loaded in this experiment to present the most difficult possible extraction requirements. Three pieces of PVDF membrane approximately 5 x 10 mm were loaded in each experiment with the first two bands inserted into the sample slot and a third band on top of the Zitex cartridge seal. In the protein "out" orientation the third band was placed protein side up or toward the solvent inlet and for the protein "in" experiment the third band was placed protein side down or toward the Zitex seal. In either case, with the protein side folded toward ("in") or away ("out") from the solvent/reagent flow, sequence performance was not affected nor does loading the Blott cartridge with a large amount of PVDF adversely affect performance (Figure 2). Repetitive yields for these two sequences (95.7% and 95.6%) and lag (carryover) for the alanines at positions 19 (29% and 30%) and 22 (32% and 31%) were nearly identical to those reported in Table II for the Blott cartridge runs with single band loads.

IV. Summary of Blott Cartridge Runs

Our evaluation of the Blott cartridge including experimental samples is summarized in Table III. As noted, 76 sequences were completed with the majority of the extended sequences for experimental proteins at the 5 pmole level or less. The use of the Blott cartridge and the StdPVDF program (Table IV) allowed us to definitively assign an average of 17 residues per sequence even with an initial yield of less than 5 pmoles. One experimental protein with an initial yield of approximately 5 pmoles was assigned through 30 residues. High sensitivity performance is particularly important as investigators study critically important proteins which are only available in low picomole quantities or less. High sensitivity or the ability to correctly assign low signals in a sequence run therefore becomes of increased importance to the sequencer operator.

Table III. Summary of Blott Cartridge Sequences

Classification		Total Sequences	Extended Sequences[d]	Residues Assigned Average	Range
Experimental[a];	1-5 pmoles	19	11	17	13-30
	5-10 pmoles	6	5	20	18-22
	10-50 pmoles	5	4	21	15-26
Blocked[b]		11			
No Sequence[c]		6			
Standards		29			
Total		76			

a. Experimental sequences where detectable signals were obtained are grouped by initial yields.
b. Protein was blocked - sufficient protein was on blot and no sequence was obtained.
c. Sequence could not be assigned due to either multiple signals or insufficient protein on blot.
d. Number of sequences where a maximum amount of sequence information was desired and the sequence was run as far as possible.

V. Reaction Program Modifications

Table IV illustrates our StdPVDF program. This program includes changes previously recommended by us (6) which include a short S2 delivery to clear argon from the E block (step 17) and pulsed or multiple deliveries of S2 and S3 and subsequently adopted by ABI (11). The other major modification in this program is the addition of a pause after the R1 delivery (steps 5, 9, 13) followed by a short argon dry. The pause provides time for the R1 to thoroughly wet the PVDF membrane. Immediately following the R1 delivery with an argon dry tends to blow the reagent (R1) out of the chamber rather than evaporating the heptane. An important consideration is the dramatically reduced solvent capacity of PVDF membranes compared with glass fiber filters.

Table IV. The StdPVDF Reaction Cartridge Program

STEP	FUNCTION	TIME (sec.)	STEP	FUNCTION	TIME (sec.)
1	Prep R2	6	23	Deliver S2	30
2	Deliver R2	60	24	Argon Dry	180
3	Prep R1	6	25	Deliver R3	800
4	Deliver R1	3	26	Argon Dry	40
5	Pause	20	27	Prep Transfer	30
6	Argon Dry	10	28	Transfer w/S3	11
7	Deliver R2	400	29	Pause	27
8	Deliver R1	3	30	Transfer w/S3	10
9	Pause	20	31	Pause	10
10	Argon Dry	10	32	Transfer w/argon	2
11	Deliver R2	400	33	Transfer w/S3	10
12	Deliver R1	3	34	Pause	10
13	Pause	20	35	Transfer w/Argon	2
14	Argon Dry	10	36	Transfer w/S3	10
15	Deliver R2	400	37	Pause	10
16	Argon Dry	60	38	Transfer w/Argon	2
17	Deliver S2	3	39	Transfer w/S3	10
18	Deliver S1	15	40	Pause	10
19	Deliver S2	30	41	Transfer w/Argon	30
20	Argon Dry	2	42	End Transfer	1
21	Deliver S2	30	43	Deliver S3	30
22	Argon Dry	2	44	Argon Dry	120
				TOTAL TIME	**48' 18"**

VI. Conclusions

The Blott cartridge is superior to the standard cartridge design when sequencing proteins absorbed to PVDF membranes as indicated by decreased lag or carryover and by improved tryptophan yields. Better solvent extraction reduces byproducts and/or solvent washes can be reduced. Improved solvent extraction also means that the protein side of electroblotted PVDF samples can be oriented in any direction and bands can be stacked if necessary to further increase sample loads.

Since more PVDF bands can be loaded, the amount of protein available for sequencing can be increased when working with very large proteins or diffuse bands. No program changes are required although wash times can be reduced and additional modifications can be made to further shorten sequence cycle times while maintaining high performance.

Acknowledgements

This work was supported by NIH grants AR39158 and HL38794. The authors gratefully acknowledge the editorial assistance of Lauren Kim.

References

1. Matsudaira, P.(1987). J. Biol. Chem. **261**, 10035-10038.

2. Bauw, G., De Loose,M., Inze, D., Van Montag, M., and Vandekerckhove, J. (1987). Proc. Natl. Acad. Sci. USA **84**, 4806-4810.

3. LeGendre, N., and Matsudaira, P. (1988). Biotechniques **6**, 154-159.

4. Yüksel, K.U., Grant, G.A., Mende-Mueller, L., Niece, R.L., Williams, K.R., and Speicher, D.W. (1991). Techniques in Protein Chemistry II, 151-162.

5. Crimmins, D.L., Grant, G.A., Mende-Mueller, L., Niece, R.L., Slaughter, C., Speicher, D.W., and Yüksel, K.U. (this volume).

6. Speicher, D.W. (1989). Techniques in Protein Chemistry, 24-35.

7. Calaycay, J., Rusnak, M., and Shively, J. (1991). Techniques in Protein Chemistry II, 131-137.

8. Xu, Q-Y., and Shively, J.E. (1989). Anal. Biochem. **163**, 517-529.

9. Reinke, H., Fischer, S.T., Reimann, F., and Tschesche, H. (1991). Methods in Protein Sequence Analysis, 55-66.

10. Mozdzanowski, J., and Speicher, D.W. (1990). Current Research in Protein Chemistry: Techniques, Structure and Function, 87-93.

11. Applied Biosystems User Bulletin Number 42, Rev. April 1991.

Reusing PVDF Electroblotted Protein Samples After N-Terminal Sequencing To Obtain Unique Internal Amino Acid Sequence

Cynthia L. Wadsworth[1,4], Mark W. Knuth[2], Laura W. Burrus[3], Bradley B. Olwin[3] and Ronald L. Niece[1]

[1]University of Wisconsin Biotechnology Center, University of Wisconsin-Madison, Madison, WI 53705

[2]McArdle Laboratory for Cancer Research, [3] Department of Biochemistry, University of Wisconsin-Madison, Madison, WI 53706

I. Introduction

Here we report a simple, high sensitivity method to obtain N-terminal and internal protein sequence from the same sample. The method is based on combining cyanogen bromide (CNBr) cleavage (1) and ortho-phthalaldehyde (OPA) blocking (2) essentially as proposed (3) except applying it to PVDF electroblotted samples which have already been sequenced, thus reusing the sample. The CNBr/OPA method to reuse PVDF samples involves three steps: 1) determination of the N-terminal sequence (if not blocked), 2) location of proline residues within CNBr fragments and 3) isolation of unique sequence by delivering OPA when a proline becomes N-terminal during sequencing. We have routinely employed the described CNBr/OPA method for samples submitted to the University of Wisconsin Biotechnology Center (UWBC) for protein sequencing.

II. Materials and Methods

Samples. All samples, proteins 1-6 and ^{125}I labeled basic fibroblast growth factor (FGF), were electroblotted to polyvinylidene difluoride (PVDF, Immobilon-P, Millipore). Electroblots from 1-D gels were prepared by the user essentially as described (4) or by Kendrick Laboratories (Madison, Wisconsin). Electroblots from 2-D gels (protein 2 only) were prepared by Protein Databases

---

[4] To whom correspondence should be addressed.

TECHNIQUES IN PROTEIN CHEMISTRY III

Inc. (PDI). Recombinant human basic FGF was purified from yeast and radioiodinated as described (5). Unlabeled basic FGF was mixed with radiolabeled protein to yield a final specific activity of 1000 cpm/pmol, with transfers to PVDF yielding 15,000 or 30,000 counts (two runs). All PVDF samples prior to and after sequencing were stored at -20°C.

Sample Handling and Sequencer Set-up. Prior to sequencing, the PVDF strips were wet with MeOH and rinsed alternately and repeatedly in MeOH and water. Multiple cuts, 1 mm apart, were made in the PVDF strip. The PVDF was always sequenced with a preconditioned Polybrene-coated glass fiber filter (PB-GF/C). The PVDF was placed between the upper cartridge block and the PB-GF/C. The side of the PVDF with the largest amount of protein (the side originally facing the gel) was placed toward the PB-GF/C. CNBr-treated PVDF was washed repeatedly in MeOH and then water. It was assembled in the reaction cartridge and washed again with a 120 second delivery of ethylacetate followed by a 120 second Ar dry.

Chemicals. CNBr (Pierce) and formic acid (HFo, J. T. Baker) were both A.C.S. grade. All sequencing and phenylthiohydantoin (PTH) analyzer chemicals and β-lactoglobulin A (BLG) were supplied by Applied Biosystems (ABI), except for the following: dithiothreitol (DTT) used in the heptane (S1), ethylacetate (S2) and butylchloride (S3) was ULTROL Grade (Calbiochem). The PTH solvent (S4) was 10% acetonitrile prepared from B&J HPLC grade acetonitrile. All water used was either Milli-Q or B&J HPLC grade. OPA (Fluoropa) and PTH-norleucine were grades appropriate for sequencing (Pierce).

Sequencing Cycles. Protein sequencing was performed on an Applied Biosystems (ABI) Pulsed-Liquid phase sequencer (model 477A) using the standard cartridge. The reaction cycles used for runs with PVDF were modified from the usual cycles used for glass fiber filters (6). The 100% trifluoroacetic acid (TFA) pulse and associated steps which comprise the first section of the usual BEGIN Cycle were replaced with a 300 second S2 delivery followed by a 120 second Ar dry. The reaction cycle for PVDF in the 477A was essentially that developed for the gas phase instrument (model 470A) (7) except that the midpoint step of the cycle was S1 (typical for the 477A) and not followed by a pause step. The S1 midpoint step (step 26 in the manufacturer's 477A NORMAL-1 cycle) was optimized prior to each run. Optimization adjusted the delivery of S1 to just beyond the PB-GF/C, compensating for the extra material (the PVDF) in the cartridge. The PVDF reaction cycles for proline residues used an extended cleavage time of 600 seconds. These cycles are referred to as BGNPVDF, PVDF and PROPVDF, respectively. The OPA cycle was modified to include washes appropriate for PVDF (7).

HPLC Separation of PTH-Amino Acids. The 477A sequencer was equipped with an ABI 120A analyzer modified with a 100 μl loop (8). PTH-norleucine was added as an internal standard to the S4 to give a peak size similar to leucine in the 8 picomole PTH-standard. Acetate buffers were stored at -20°C in small aliquots of the 3M stock solutions. Solvent A was prepared and diluted as recommended by the manufacturer except that it was prepared with a lower pH made from a solution of 5% pH 4.6 to 95% pH 3.8 acetate buffers.

CNBr Treatment. CNBr treatment was performed directly on PVDF placed on top of a PB-GF/C which was fitted into the upper cartridge block. All user samples were treated with CNBr after sequencing employing the PB-GF/C from that run. From 30 to 42 μl of 25% CNBr in 70% HFo was applied until the filter and PVDF were saturated. To prevent drying of the filters, the reaction cartridge was wrapped in parafilm, avoiding contact between the parafilm and the filters. The reaction proceeded for 24 hours in the dark at room temperature and was stopped by slicing a hole in the parafilm and vacuum drying the cartridge for 10 minutes. The PB-GF/C used in the CNBr treatment was not used again for sequencing.

OPA Treatment. OPA was prepared as recommended (6) on the day of the experiment and loaded at X1, the extra bottle position, prior to assembly of the sample in the cartridge. The BGNPVDF, PVDF, PROPVDF and OPA cycles were used in the order described (6) to automatically deliver OPA to the preselected position in the sequence.

Protein Tracking with ^{125}I. ^{125}I was counted in either a Packard Model C5301 or a LKB 1272 Clinigamma counter. The distribution of protein after CNBr/HFo treatment was determined by separately counting the treated PVDF and PB-GF/C.

Sequence Data and Interpretation. Sequence was read from the strip chart chromatograms produced from a Kipp and Zonen model BD41 duel pen chart recorder, 0.005 AUFS 10mV full scale, and from difference chromatograms generated using the 610A Data Analysis Software Package (ABI). Picomole tabulated data were obtained with the 900A Data Analysis Software (ABI) using the option "Baseline Correct". Protein sequence data were independently interpreted by two readers (CLW and RLN).

III. Results and Discussion

CNBr/OPA Method. The CNBr/OPA method has three steps and two general applications differing in how proline is located in step two. Step one, the determination of N-terminal sequence, generally uses 100% of the sample. After sequencing, 100% of the sample is treated with CNBr. Step two, the location of proline residues, has two applications. Application 1 is used to demonstrate that one protein is related to another characterized protein. The locations of prolines are predicted within the expected CNBr fragments using a sequence presumed to be the same or closely related to the sample. Application 2 is used to characterize an unknown protein. The locations and picomole recoveries of prolines are determined within generated CNBr fragments by rerunning a fraction (usually 25%) of the treated sample on the sequencer (called the screening run) for 15 cycles (plus the BGNPVDF cycle). Step three, the OPA-isolation run, determines unique internal sequence starting with a proline from one of the fragments. If there is an adequate amount of sample, several unique internal sequences can be isolated from different fragments in separate OPA-isolation runs. Generally, using the recoveries obtained in the screening run, the sample should be divided so that at least 10 picomoles of proline are expected to be isolated with OPA.

Obtaining internal sequence using the CNBr/OPA method is not significantly limited by the occurrence of proline following within 15 residues of methionine. The SWISS-PROT Protein Sequence Data Bank (9) was searched using the program FindPatterns (10) set to locate a proline up to the fifteenth residue past methionine, not counting methionine. A total of 15,952 out of 20,024 (80%) proteins in the database met the requirement.

In addition to the normal difficulties in determining PTH-S, PTH-T, PTH-H, PTH-R and PTH-W from very low quantities of sequencing protein, PTH-K and PTH-F are more difficult to observe. Presumably, unless special precautions are taken (11), PTH-K is oxidized during CNBr cleavage. Weak PTH-F peaks are difficult to see because PTH-F appears as a shoulder on an OPA-related artifact which remains equivalent to 2 picomoles throughout the run (Figure 1).

At the sensitivity limits of the instrument, a small increase in the picomoles of initial sequencing signal can have a large effect in the number of residues determined. Initial yield is lowered by protein removal from PVDF. Tracking with ^{125}I showed protein is removed from PVDF during CNBr treatment. However, sequencing prior to CNBr treatment significantly increases the amount of protein retained on the PVDF from 19% ± 3% (4 trials) without prior sequencing, to 73% ± 7% (4 trials) with prior sequencing. Therefore, even a protein known to be blocked should be coupled with PITC. We typically employ a BGNPVDF and two PVDF cycles. We propose that the reaction of the ε-amino group of the lysines and PITC improves the hydrophobic interaction between the protein and PVDF limiting its removal by CNBr/HFo.

Sample Results. Six samples submitted to the UWBC for protein sequencing are considered. The results of the three-step CNBr/OPA method are summarized (Table I). The data for the three steps of the CNBr/OPA method were obtained with one electroblotting preparation except for protein 6 which required two preparations and for protein 2 which required identical spots from two 2-D gels. Proteins 1-5 proved to be blocked, as shown by signals from multiple fragments after CNBr treatment. Protein 6 yielded N-terminal and internal protein sequence.

***Application 1* - predicting the location of prolines.** The sensitivity of the CNBr/OPA method is illustrated.

Protein 1 - a heterodimeric protein. This protein, Proximal Sequence Element-binding Protein 1 (PSE1), is a transcription activating protein (12). The sequencing goal was protein comparison to Ku (13) and to TREF (14, 15), both shown by immunological evidence to be related to PSE1 (12). The 63kDa subunit, PSE1-β, revealed no sequence as found for TREF (12). After the cDNA sequence of Ku was reported (13), the CNBr/OPA method was chosen to determine internal sequence so that the original PSE1-β sample could be reused. One year had elapsed since the N-terminal sequencing attempt. Possible outcomes for PSE1-β were that: 1) the sequence of PSE1-β did not match the sequence translated from the cDNA of Ku, and OPA either quenched all the signals or isolated a sequence other than the expected one or 2) the sequence matched as predicted.

Sequence deduced from the cDNA of Ku predicted that 6 CNBr fragments would have proline within the first 10 residues. The fragment with a proline at position 4 was selected for OPA treatment because: 1) the hydrophobic residues following proline would yield strong PTH signals, 2) the fragment would be long making it less prone to washout during sequencing and 3) the sequence would be unique since no other methionine was followed four residues later by a proline. A methionine followed immediately by a proline (sequence M-P) was not chosen for OPA-isolation because a result of no sequence would fail to distinguish between 1) not enough protein present on the PVDF to yield signal or 2) the sequence of the PSE1-β differed from the sequence deduced from the Ku cDNA and OPA quenched the signals. In the OPA-isolation run, recoveries for residues 1-3 showed that CNBr produced 2-3 picomoles of multiple fragments on the PVDF while OPA isolated 1.3 picomoles of proline at residue 4. Sequence was determined from signals as small as a few hundred femtomoles over background. The confident and tentative calls made matched the protein sequence deduced from the cDNA of Ku. According to the cDNA, the unidentified residues are lysine at positions 9 and 10, histidine at positions 11 and 12, arginine at position 15 and serine at position 17. The moderate rise in background which occurred in cycles immediately following the OPA block is shown by difference chromatograms which remove constant signals (Figure 1).

***Application 2* - determining the location of prolines.** The range of possible outcomes, advantages and disadvantages of the CNBr/OPA method are illustrated.

Protein 2 - a 55kDa newt protein. This protein from *Notophthalmus viridescens* is associated with changes which occur during limb regeneration (16). It was identified from a 2-D gel pattern using the PD QUEST (PDI) software (17). A PIR database search showed that the sequence is similar to type I B2 keratin from frog (*Xenopus laevis*), differing only at positions 2 and 5. Using the comparison, position 12 is probably leucine and position 17 is probably alanine.

Protein 3 - a 78kDa peptidase from human brain. After the sequence was determined, the cDNA sequence of the pig peptidase was published (18). The proline isolated with OPA proved to be 7 residues past the N-terminal methionine, differing at positions 7 and 15.

Protein 4 - a 60kDa protein. The 60kDa band from a crude column fraction showed immunological activity on Western blots. The sequence isolated from the 60kDa band proved to be identical to residues 338-344, including the methionine assumed from the CNBr cleavage, of the blocked protein catalase. Apparently, the protein with immunological activity was below the level that could be detected by sequencing and not separated from catalase which had been used as a marker. The length of sequence isolated at P1 was limited by preview. Preview at P1 can be avoided if a BGNPVDF cycle after the OPA cycle is not included (data not shown). Examination of the CNBr fragments predicted from the catalase sequence shows that 3 fragments have a proline located 9 residues past methionine. The quantities of proline in the screening run (Table I) are consistent with this prediction. Since yields decrease as sequencing continues, the yield for proline 9 is unusually high, equaling that of residue 1. OPA delivered to this position would not isolate unique sequence. In general, prolines with unusually high yields should not be used for isolation.

Protein 5 - a 60kDa retroviral receptor protein. This protein yielded no sequence after the OPA-isolated proline. Since 6 picomoles is an adequate signal, this result suggests that proline preceded methionine (sequence P-M) or was the C-terminal residue the protein.

Protein 6 - a 150kDa membrane protein. This protein, Cysteine Rich FGF Receptor (CFR), binds basic FGF (19). The sequencing goal was protein characterization and design of oligonucleotide probes to obtain the cDNA. N-terminal sequencing yielded residues 1-16, and with less confidence, residues 17-21 (20). The N-terminal sequence did not provide suitable information to permit use of PCR to amplify with oligonucleotide probes designed to each end, so internal sequence was pursued.

The CNBr/OPA method was chosen to avoid additional protein preparations which were lengthy and could not be scaled up. The screening run revealed two prolines with enough sample remaining to isolate at least 10 picomoles from each proline. Multiple PCR approaches would be possible utilizing the two internal sequences, N-terminal sequence and poly-A tail of the cDNA library. The earlier proline, P2, was chosen for OPA isolation for two reasons. 1) The recovery of P2 was not unusually high compared to P6 (Table I) suggesting that a single sequence would be isolated. 2) The two prolines could belong to the same fragment, and the most efficient way to determine this is to isolate the earlier proline first. OPA isolated a unique sequence containing only P2.

A clone was eventually identified using monoclonal antibodies. The N-terminal and internal sequences were used to verify that the correct clone had been obtained. The confident calls and 3 out of 5 of the tentative calls of the N-terminal sequence matched sequence deduced from the cDNA. The internal sequence differed at positions 7 (serine predicted) and 12 (histidine predicted). These differences have not been resolved. According to the cDNA, the unidentified residues are cysteine at position 8, which could not be detected in protein sequencing because the protein had not been reduced and alkylated, and arginine at position 11.

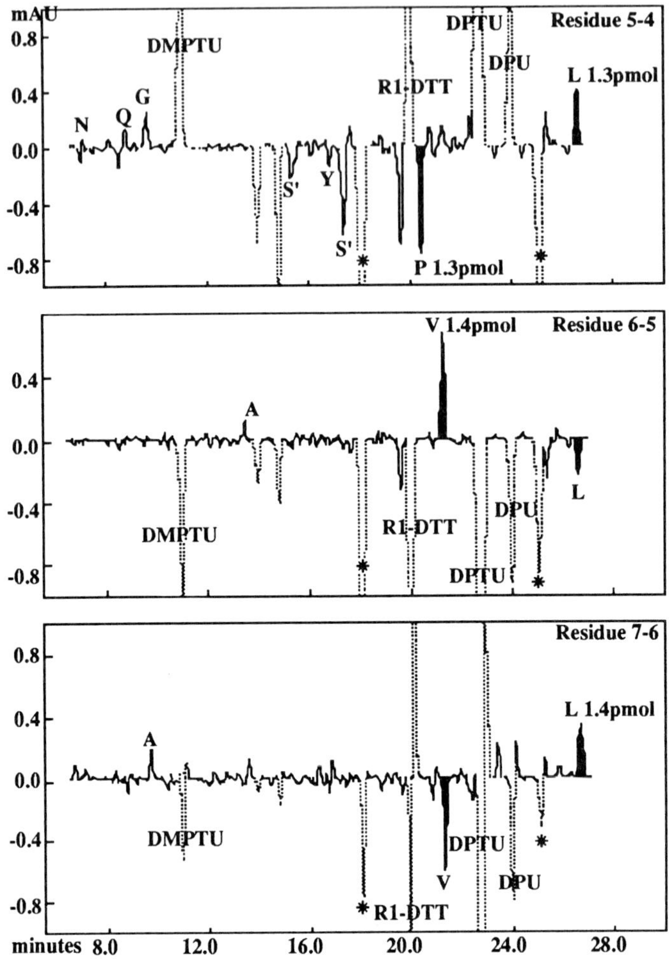

Figure 1. Difference chromatograms showing the sequence **P-L-V-L**. Signals are darkened. Artifacts are shown: sequencing (dashed lines), OPA (dashed lines and starred).

Table I. Results, proteins 1-6, for the three-step CNBr/OPA method[a]

Runs:	N-terminal	Screening		OPA-Isolation	
		Proline Position	Proline Yield[b]	Proline Yield[b]	Internal Sequence
Protein[c]					5 10 15 20
1	blocked	predicted[d]		1.3	P-L-V-L-L-X-X-X-X-Y-L-X-P-X-l-f-V-Y-P
					5 10 15 20
2		P4	2	10	P-A-V-D-L-G-X-I-t,l-S-D-M-R-a>t-Q-Y-X-T
					10 15
3	blocked	P6	6.5	11[e]	P-Q-V-Y-G-D-E-T-A-V-Q-T-Y
		P7	4.6		
		P9	4.3		
					5
4	blocked	P1	18	15	(M)-P-P-G-I-E-P[f]
	blocked	P6	12		
		P9	18		
5	blocked	P6	1.9	6	P.
		P9	1.5		
					5 10
6	16/16[g]	P2	25	8	P-G-A-V-F-R-X-L-Y-X-Q-A
		P6	21		

[a] All electroblots were from 1-D gels except protein 2 which used identical spots from two 2-D gels. Upper case letters indicate a confident identification. Lower case letters indicate a tentative identification. X indicates a position in which no identification could be made. Sequences begin with proline; numbering begins with the first residue of the fragment.

[b] Yields are picomoles for the fraction of the sample run.

[c] Protein 1 is PSE1-β. Protein 6 is CFR. All other protein samples are unnamed.

[d] No screening run was performed since the position of proline was predicted from another sequence, application 1.

[e] The results for the OPA isolation run, yields and sequence, are listed on the same line as the proline chosen for isolation when more than one proline was available.

[f] Methionine is assumed from the CNBr cleavage.

[g] Confident calls over the total length of readable sequence. The initial yield was 15 picomoles.

IV. Conclusions

CNBr in combination with OPA is a high sensitivity method to obtain N-terminal and limited internal sequence data. Applications I and II of the CNBr/OPA method have proven to be an excellent way to compare proteins or establish homology. In all the cases discussed in this paper, the protein could only be purified by gel electrophoresis, and in most cases, it was difficult to obtain a quantity sufficient for routine N-terminal sequencing. Methods which purify fragments for internal sequencing would have eliminated the N-terminal sequence determination necessary to verify a full-length clone and reduced the quantity available for sequencing even more. While some fragmentation/purification methods were not impossible on the quantities of protein available, they were not judged to be the best first approach. Reusing PVDF electroblotted protein samples allowed us to efficiently obtain N-terminal (when not blocked) and internal protein sequence data, consuming only protein that was dedicated to N-terminal sequence determination.

Acknowledgements

This work was supported by the University of Wisconsin Biotechnology Center, NIH grants 1 S10 RR02971, S10 RR03427 and DMB-8514305. We extend gratitude to all reviewers and to Xiaodun He who operates the protein sequencers at the UWBC.

References

1. Simpson, R. J., and Nice, E. (1984) Biochemistry International **8**, 787-791.
2. Brauer, A. W., Oman, C. L., and Margolies, M. N. (1984) Anal. Biochem. **137**, 134-142.
3. Yuen, S. W., Chui A. H., Wilson, K. J., and Yuan, P. M. (1988) Applied Biosystems User Bulletin No. 36.
4. Matsudaira, P. (1987) J. Biol. Chem. **262**, 10035-10038.
5. Olwin, B. B., and Hauschka, S. B. (1988) J. Cell Biol. **110**, 503-509.
6. Protein/Peptide Sequencer Users Manual Model 477A (1989) Applied Biosystems Inc.
7. Speicher, D. (1989) *in* Techniques in Protein Chemistry (Hugli, T. E., Ed.), pp. 24-35, Academic Press, San Diego.
8. Tempst, P., and Riviere, L. (1989) Anal. Biochem. **183**, 290-300.
9. Bairoch, A., and Boeckmann, B. (1991) Nucleic Acids Res. **19** supplement, 2247-2259.
10. Devereux, J., Haeberli, P., and Smithies, O. (11084) Nucleic Acids Res. **12**, 387-395.
11. Griffin, P. R., Kumar, S., Shabanowitz, J., Charbonneau, H., Namkung, P. C., Walsh, K. A., Hunt, D. F., and Petra, P. H. (1989) J. Biol. Chem. **264**, 19066-19075.
12. Knuth, M. W., Gunderson, S. I., Thompson, N. E., Strasheim, L. A., and Burgess, R. R. (1990) J. Biol. Chem. **265**, 17911-17920.
13. Mimori, T., Akizuki, M., Yamagata, H., Inada, S., Yoshida, S., and Homma, M. (1981) J. Clin. Invest. **68**, 611-620.
14. Roberts, M. R., Miskimins, W. K., and Ruddle, F. H. (1989) Cell Regul. **1**, 151-164.
15. Miskimins, W. K., McClelland, A., Roberts, M., P., and Ruddle, F. H. (1986) J. Cell Biol. **103**, 1781-1788.
16. Tsonis, P. A., submitted.
17. Garrels, J. I. (1989) J. Biol. Chem. **264**, 5269-5282.
18. Renner, D., Hemmings, B. A., Hofsteenge, J., and Stuart, R. S. (1991) Biochemistry **30**, 2195-2203.
19. Burrus, L. W., and Olwin, B. B. (1989) J. Biol. Chem. **264**, 18647-18653.
20. Burrus, L. W., in preparation.

CHROMATOGRAPHIC CARBON AS AN INERT SAMPLE ADSORBENT FOR PROTEIN SEQUENCING

Christopher Southan

Department of Chemistry Support,
SmithKline Beecham Pharmaceuticals,
Research and Development,
Welwyn, Herts, AL6 9AR, U.K.

I. INTRODUCTION

Samples for sequence analysis are often supplied at low concentrations in solutions that may have deleterious effects on Edman chemistry and sequencer performance. To concentrate and de-salt soluble proteins and peptides reverse-phase HPLC has become the method of first choice. Despite the general success of this strategy there remains a proportion of samples which may give variable recoveries or even be completely intractable to standard RP-HPLC protocols. This may arise with membrane proteins, aggregated synthetic peptides or samples with reduced solubility after denaturation, derivatisation or exposure to strong solvents. The difficulties of preparing this type of sample for protein sequencing are compounded if low microgram amounts have to be manipulated and solubilising agents such as chaotropic salts are present.

Current approaches to this problem include precipitation techniques, adsorbing to PVDF membranes, covalent immobilisation or specialised HPLC protocols. Such methods have been shown to work on a limited range of samples but each have associated difficulties. In this work a strategy was developed to non-covalently immobilise protein onto an HPLC column material that is stable to the conditions of Edman chemistry. The properties of a recently developed carbon reverse-phase packing seemed to be particularly suitable for this application. This material is a hydrophobic adsorbent with exceptional chemical and mechanical stability (1). After being used to adsorb protein from solution this packing material could be washed and transfered directly to the sequencer thus by-passing the problematic elution recovery step.

TECHNIQUES IN PROTEIN CHEMISTRY III

II. EXPERIMENTAL

A. Adsorbent

The Hypercarb HPLC packing (Shandon) consists of nominally 10µm spherical particles of porous graphitised carbon. Small amounts of loose material were kindly supplied be Dr Bulvinder Kaur (Shandon U.K)

B. Micro-adsorption column

This was constructed as shown in figure 1. A stainless steel 1/16" 2µm microbore frit (Upchurch) was fitted into a plastic micro pipette tip (Gilson or Alpha Laboratories) to form a column bed support. The horizontal orientation of the frit was ensured by using a flat-ended Reodyne syringe needle to seat the frit firmly into the tip. The top of the tip was cut back to accommodate a syringe leur fitting. An extended microtip of the type used for loading gels (Alpha Laboratories) was used for the addition or removal of slurry or solutions into the column. Once the column was full, a bubble-free connection could be made with a 1ml syringe for extensive washing. Loose Hypercarb packing, 0.5mg, was suspended in 50 µl of 100% acetonitrile in an eppendorf by aspirating with a pipette tip. This slurry was transfered to the column fitted with the frit. Using 5% acetonitrile, this was washed on to the frit to form a small column bed of approximately 2mm depth as illustrated in the upper part of figure 1 (N.B. some types of plastic syringe are not compatible with acetonitrile).

C. Protein Immobilisation

The sample loading protocol is outlined in the lower part of figure 1. By using air pressure from an empty syringe fluids could be passed slowly through the column until flush with the surface. The protein solution was added with a micropippete and slowly passed through the carbon packing. Washing was done with 5% acetonitrile, utilising the syringe as described above during packing. Trials were carried out with 5µl of lactoglobulin (Sigma) nominally at 25 pmoles/µl.

D. Peptide Immobilisation

A particularly hydrophobic 22 residue synthetic peptide, was found to be insoluble in aqueous buffers after freeze-drying. Some of this material was suspended as a slurry in 8M GuHCl at 10mg/ml. After loading 50µl of the slurry the Hypercarb bed was then washed with 1ml of 5% acetonitrile before re-suspension and transfer to the sequencer.

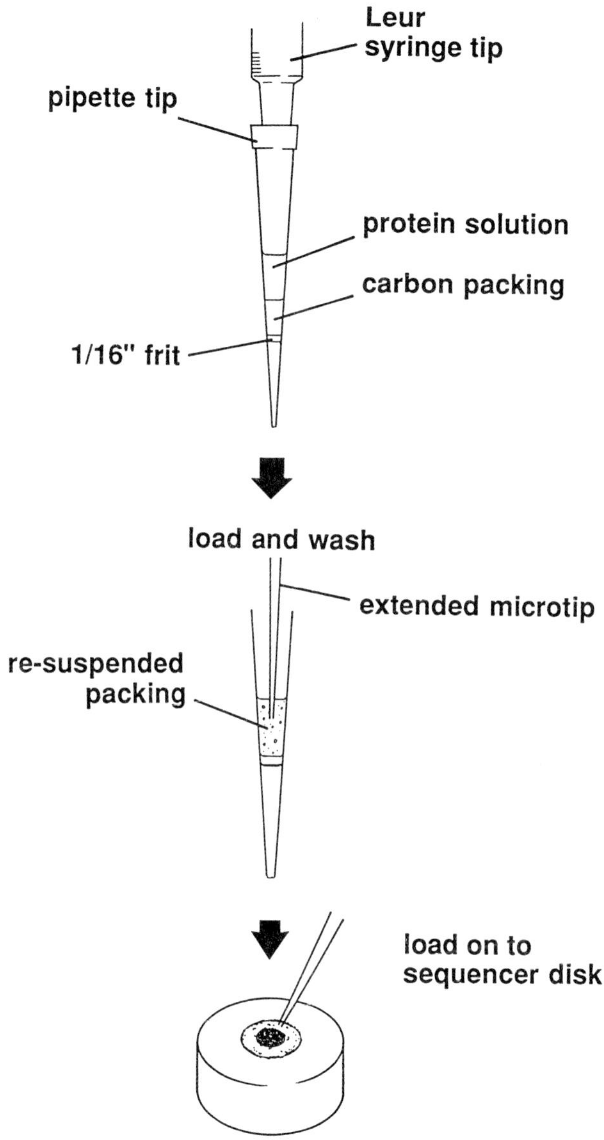

Figure 1. Outline of the micro-adsorption column design and the protocol for adsorption and transfer.

E. Sequencing

After adsorbtion and washing the carbon packing was then re-suspended in 25 ul of 100% acetonitrile and spotted onto the centre of a polybrened sample disk as outlined in figure 1. After drying the disk was inverted before re-assembling the cartridge of a 477A sequencer (Applied Biosystems). This ensured that the carbon particles lay on the upper surface of the sample disk with respect to the solvent flow. Control aliquot of lactoglobulin were run under identical conditions without any carbon beads using a BEGIN followed by 20 NORMAL-1 standard sequencing cycles. The plots and yield information were obtained using the latest version of the IBM-based sequence analysis software installed on the 477 instrument (N.B. the useful comparative plots obtained with the "compare plot" option from CHROMATOGRAM DISPLAY should be performed post-run as this may cause a touchpad lock-up if sequencing is in progress).

III RESULTS

A. Properties of the Carbon Packing

The particles formed a fine suspension in 100% acetonitrile so this solvent was utilised for packing. Washing was carried out in 5% acetonitrile to maintain particle wetting but also allow protein adsorbtion. Re-suspension could be carried out in 5% acetonitrile but this was distinctly flocculent so 100% acetonitrile was used to transfer the carbon packing to the sequencer. The particles also had the useful property of adhering to the standard glass fibre sample disk. This prevented particle loss during drying under nitrogen and inverting of the disk as described. Some preliminary experiments to test the effect on sequencer performance were performed by by pre-loading lactoglobulin solution in the normal way and subsequently adding small amounts of the carbon beads onto the sample disk. No obvious retardation of solvent flows could be observed and and only slight wash through of carbon particles through the sample disk onto the filter disk had occurred after 20 cycles.

B. Protein Immobilisation

Using the protocol described above lactoglobulin in solution was bound to the beads, washed, and transfered to the sequencer. The results in figures 2 and 3 compare the sequence information obtained from normal solution loading with adsorbtion of the same aliquot of protein solution onto hypercarb, respectively. The first three steps of lactoglobulin solution loaded (fig.2 top) Leu-Ile-Val- gave an initial yield of 81%. The histograms (fig.2 bottom) show repetitive yields for Ile (2,12) of 94% and Gln (5,13) of 94%. The corresponding first three steps from the hypercarb adsorbed sample (fig 3, top) gave an initial yield of 70% i.e. 11% lower. The histograms (fig.3 bottom) show repetitive yields for Ile (2,12) of 87% and Gln (5,13) of 88%.

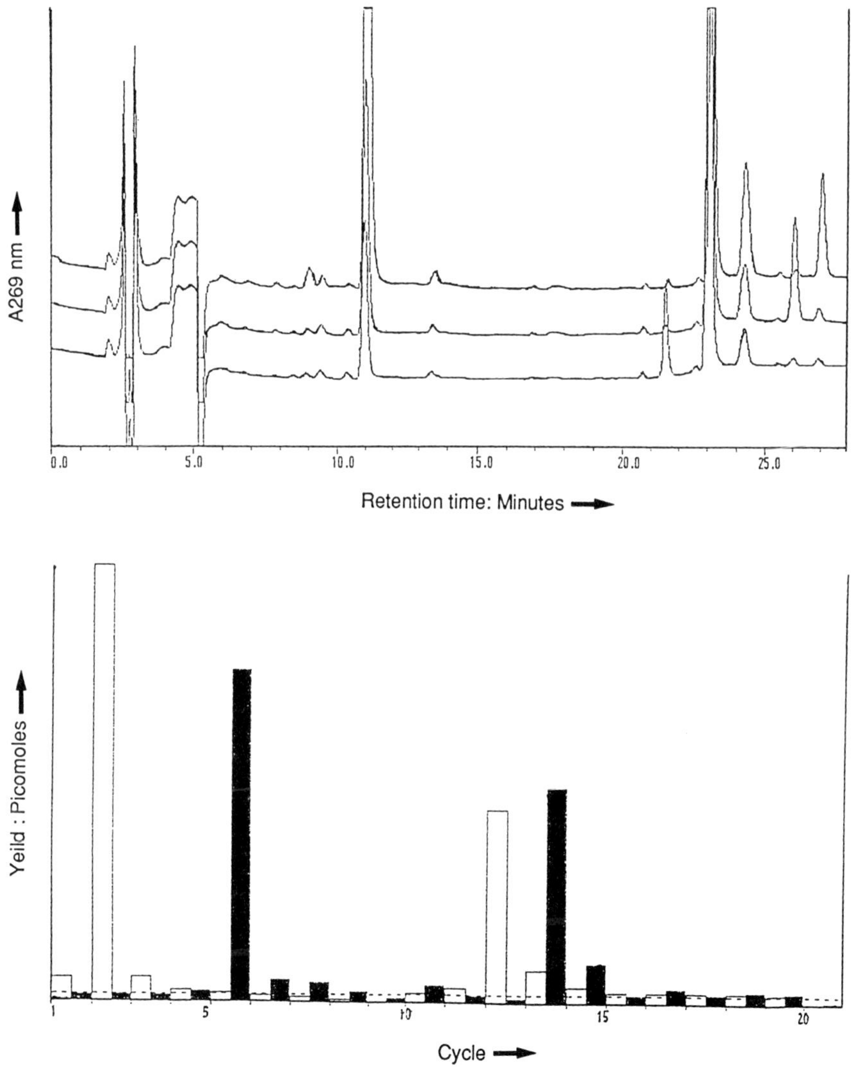

Figure 2. Sequencing results for 125 pmole of lactoglobulin control (normal solution loading).

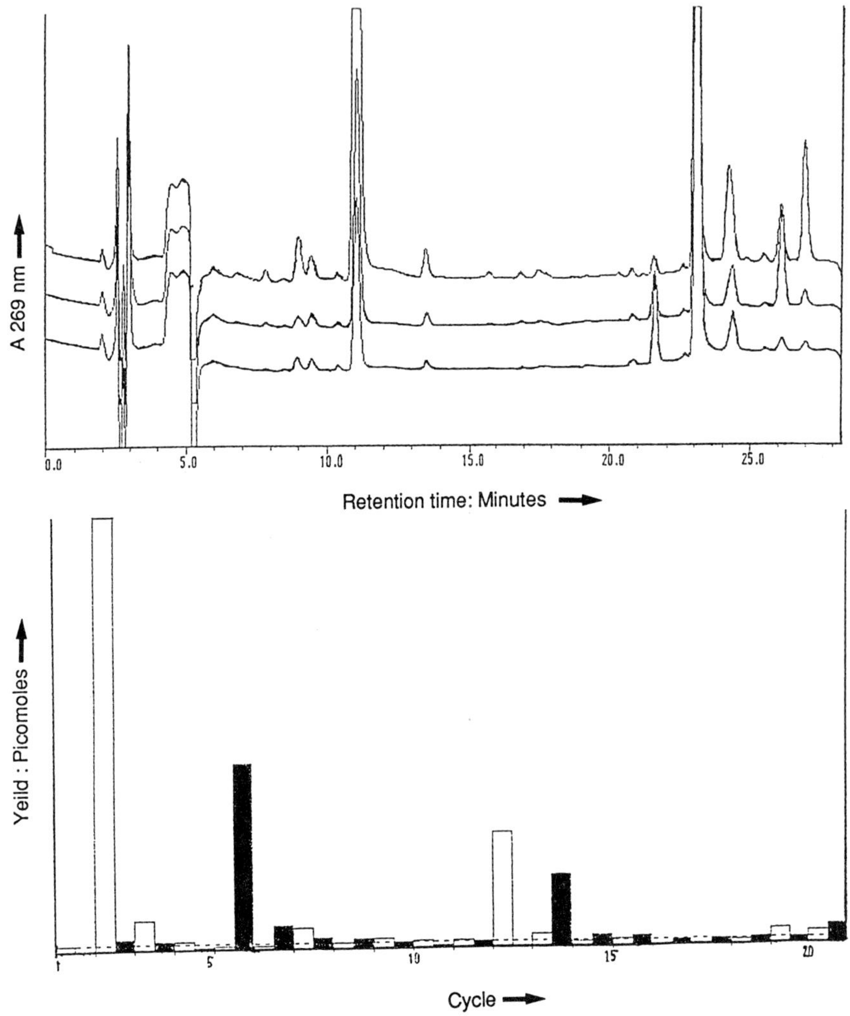

Figure 3. Sequencing results for 125 pmole of lactoglobulin immobilised on Hypercarb.

C. Peptide Immobilisation

Figure 4 shows the first six cycles, beginning with the PTH standard, from the analysis of the synthetic hydrophobic peptide. The N-terminal sequence was expected to be Asp-Phe-Ala-Asn-Arg-. However the detection of Phe at step 1 indicated the presence of a proportion of des-Asp peptide. This result was confirmed by FAB-mass spectrometry analysis. The total N-terminal yield indicated that 1.4 nmoles of peptide had been immobilised on 0.5mg of Hypercarb. This may represent the maximum binding capacity for this particular peptide on the micro column as the 50µl of slurry should have contained 500 nmoles. Step 5 (lowermost chromatogram in fig.4) shows recovery of 200 pmoles of PTH-Arg indicating that ATZ-Arg had been extracted from the sample disk in the presence of the carbon packing.

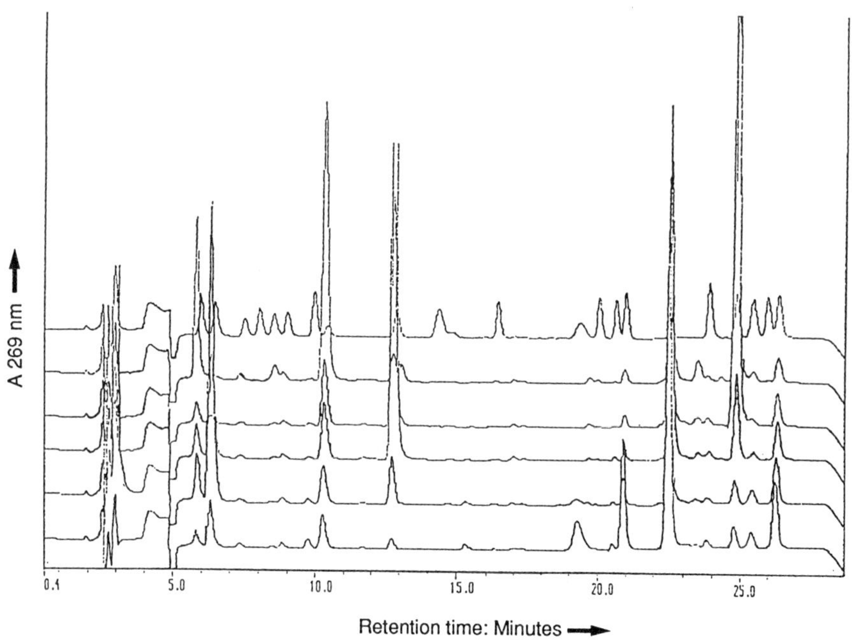

Figure 4. Sequencing results for a synthetic peptide immobilised on Hypercarb.

IV. DISCUSSION

The immobilisation column designed for this work could be applied to any other type of small scale solid phase adsorbtion-elution protocol using a range of chromatographic materials. The results reported demonstrate that both proteins and peptides can be non-covalently immobilised on chromatographic material that is compatible with Edman chemistry and subsequently be transfered to a sequencer sample disk. Although the performance with chromatographic carbon was inferior when compared to standard solution loading, extended N-terminal sequence information was obtained from the adsorbed samples.

The synthetic peptide was a realistic test in that the material was initially in the unsequencable form of a partially soluble slurry which gave very poor RP-HPLC recoveries. The sequence analysis obtained from the carbon-bound material was informative in revealing incomplete addition of the N-terminal Asp residue during synthesis.

The properties of the Hypercarb need to be explored further in order to establish the potential applications of this technique. It may be possible to improve the sequencing performance by cycle modifications and/or alternative sample cartridge configurations. However this approach certainly presents a feasible option for preparing samples from solutions that may be intractable to other clean-up techniques. An advantage associated with this method is that any material that fails to bind is not lost, as alternative recovery methods can still be applied to the eluate.

Footnote. During the limited number of experimental runs for this work there was no evidence of carbon particles washing through and blocking the sequencer transfer lines. However, pending further evaluation of this type of material the possibility of such problems should be considered and due caution exercised.

V. REFERENCES

1. Knox, J., and Kaur, B. (1986) J.Chrome., 352, 3-35.

N^{α}-ACYLAMINOACYL-PEPTIDE HYDROLASE: SPECIFICITY AND USE TO UNBLOCK N-ACETYLATED PROTEINS

RADHA G. KRISHNA
Department of Biochemistry & Molecular Biology
University of Texas Medical School
Houston, TX 77225

I. INTRODUCTION

Acylaminoacyl-peptide hydrolases have been isolated from a number of sources and been characterized with respect to substrate specificity and general molecular properties (1-5). Because of differences in assay and purification methods, it is difficult to establish the degree of identity amoung the different forms of the enzymes, but they all show very similar general properties in catalyzing the removal of Ac-Ala, Ac-Met, Ac-Thr, Ac-Ser and Ac-Gly from short N-acetylated peptides. Several attempts have been made to use these enzymes to unblock acetylated proteins for sequencing, but although the method has been successful with peptides up to 22 residues long (2-7), intact proteins are completely resistant to hydrolase action. We have isolated a hydrolase from rabbit muscle and established its specificity in considerable detail (1,6); like its liver and erythrocyte counterparts, the enzyme is inactive against intact protein, and to use it to unblock acetylated proteins for sequencing, it is thus necessary to fragment the protein before the unblocking can be carried out (7). In this paper we summarize briefly the key aspects of the enzyme specificity, and describe a method for fragmenting acetylated proteins, blocking the newly generated N-termini by succinylation before unblocking with hydrolase and seqencing the N-terminal peptide. The N-terminal acetylamino acid can finally be identified by treatment of the hydrolase-digest with acylase I and amino acid analysis.

II. EXPERIMENTAL

(A) Specific methods. The multistep protocol for the preparation of substrates from acetylated proteins and the details of the unblocking with N-acylaminoacyl-peptide hydrolase and acylase I are given in Figure 1. All the steps were carried out in a single test tube to ensure optimal yield. In a typical experiment, an aliquot of 5 nmol of protein was subjected to fragmentation, succinylation (8, 9), ether extraction (2 ml aliquots), hydrolase treatment and

TECHNIQUES IN PROTEIN CHEMISTRY III

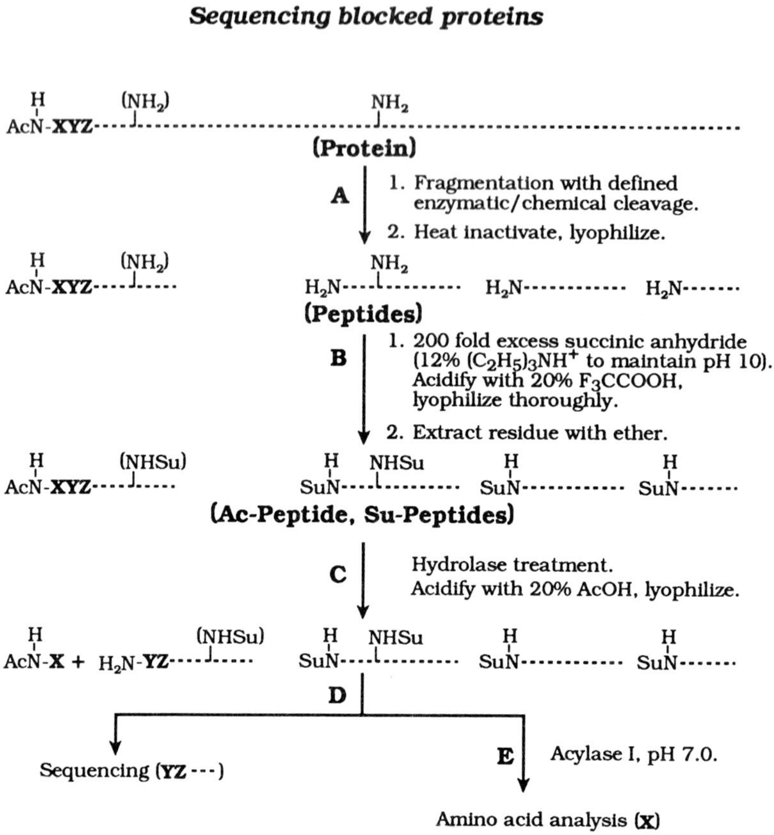

Fig. 1. The strategy for establishing N-terminal sequences in blocked proteins.
A1. Aliquots of 5 nmol of each protein were digested at 37° C and an enzyme:substrate ratio of 1: 20 by weight with trypsin/chymotrypsin in 50 mM NH_4HCO_3 buffer, pH 7.8-8.6 for 10 h, or with V8 protease in 50 mM phosphate buffer, pH 7.8 for 16 h. CNBr fragmentation was according to the reported procedures (17). 2. The proteolytic digests were heated at 100° C for 5 min and when NH_4HCO_3 was used, acidified with 20% CH_3COOH before lyophilizing.
B1. Solid succinic anhydride (200-fold molar excess to the calculated amount of α- and ε- amino groups in a given protein-protease/CNBr digest) was added slowly with vortex-mixing over a span of 1 h and left overnight at room temparature. 2. Multiple ether extractions (6-10 x 2 ml) were required depending on the amount of succinic anhydride used. (The extract was routinely dried and screened for peptides by amino acid analysis and reverse phase HPLC).
C. The blocked peptide mixtures (5 nmol from each protein) were incubated with 5 μg of hydrolase in 50 μl of 25-50 mM NH_4HCO_3 buffer (pH 7.0-7.4) containing 1 mM EDTA and 2 mM $MgCl_2$ for 12 h at 37° C. When needed, the pH of the reaction mixture was adjusted to 7.0 with 1-3 μl of 0.5 M NaOH before the enzyme was added.
D. The dried products were dissolved in 50 μl of 70% HCOOH and 1/10 (0.5 nmol) was used for sequencing. The remaining 9/10 was lyophilized and used for the digestion with acylase.
E. The remainder from D (above) was digested with 5 units of acylase I in 50 μl of 0.1 M (Na)phosphate buffer, pH 7.0 for 6 h at 37° C, acidified and applied directly to the amino acid analyzer.

acylase treatment in a 0.9 x 7.5 cm glass test tube. If larger quantities of known proteins (20 nmol) were processed to evaluate variations in the methodology, the fragmentation, succinylation and ether extraction (4 ml aliquots) were carried out in 1.1 x 10 cm test tubes; the residual peptide mixtures were dissolved in H_2O by adjusting the pH to 7.0 with 0.2 *M* NaOH and aliquots of 5 nmol were transfered to 1.5 ml Eppendorf tubes for the hydrolase and acylase treatments. The commercial samples of α-crystallin (10), enolase (11), myelin basic protein (12), superoxide dismutase (13), cytochrome c (14), ovalbumin (15) and parvalbumin (16) were dialyzed against 5 *mM* NH_4HCO_3 to eliminate salt contaminations, acidified with 5% CH_3COOH and lyophilized. Based on the known sequences, each protein was selectively processed with appropriate high specificity protease(s) or CNBr as indicated in Figure 3 to yield blocked N-terminal peptides of 6-20 residues in size which could possibly serve as good substrates for hydrolase.

(B) Materials and general methods. N-Acylaminoacyl-peptide hydrolase, the enzyme that was used in the present study for unblocking the acetylated N-terminal peptides was purified from rabbit muscle according to our previously reported procedures (1). The specific activity of the enzyme was determined using the substrate [^{14}C]Ac-MLG in 50 *mM* phosphate buffer, pH 6.9, containing 1 *mM* EDTA and 2 *mM* $MgCl_2$. Under standard assay conditions (200 nmol of substrate in 200 μl of buffer, 30 min at 37° C), 0.5 μg of hydrolase would release 45% of Ac-M from Ac-MLG, as ascertained by the quantitative HPLC assay (1). The specificity determinants for the enzyme were studied under these general conditions using a variety of synthetic acetylated peptides as substrates. For the unblocking experiments 5 μg of the 7000-7500 fold-purified hydrolase was incubated with the succinylated peptide mixtures at 0.1 *mM* substrate concentration. Hydrolase-treated, succinylated peptide mixtures of different proteins were sequenced on an Applied Biosystems 477A Gas Phase Protein Sequencer and the acylase treated hydrolase-digests were analyzed on the LKB Alpha Plus Amino Acid Analyzer. The reagents used for sequencing and amino acid analysis were ultra-pure. Standard proteins used as substrates for hydrolase (parvalbumin, enolase, α-crystallin, superoxide dismutase, cytochrome c, chicken egg albumin, and myelin basic protein), CNBr and TEA were from Sigma. Trypsin, chymotrypsin, V8 protease and acylase I were obtained either from Sigma or from Worthington Biochemical Corporation. Ethylether, TFA and acetonitrile of HPLC grade were purchased from Fisher and acetic and succinic anhydrides from Aldrich.

III. RESULTS AND DISCUSSION

Enzyme specificity. Some key aspects of the specificity of the rabbit muscle N^{α}-acylaminoacyl-peptide hydrolase are summarized in Figure 2. The most important conclusion from this data (1,6) is that in addition to the high degree of specificity for the first, acetylated, residue, the enzyme is strongly affected by the presence of certain other residues in the vicinity of the N-terminus. Positive charges from His as well as Arg and Lys (6) are potent inhibitors of hydrolase activity in position 2, 3 and 4 of the peptide substrates, and also show some inhibition as far out as in position 11. Pro is a strong inhibitor in position 2 and also in position 3 if a fourth amino acid is present in the peptide (Krishna, R. G., unpublished data). With the present data it should be possible to model an enzyme binding site which requires the first 4 residues of the substrate in an extended polypeptide chain and involves a topology which excludes positive charges, but not bulky side chains. This preliminary and descriptive picture appears to fit well with the relative abundance of certain amino acid residues in the N-terminal segments of acetylated and nonacetylated proteins.

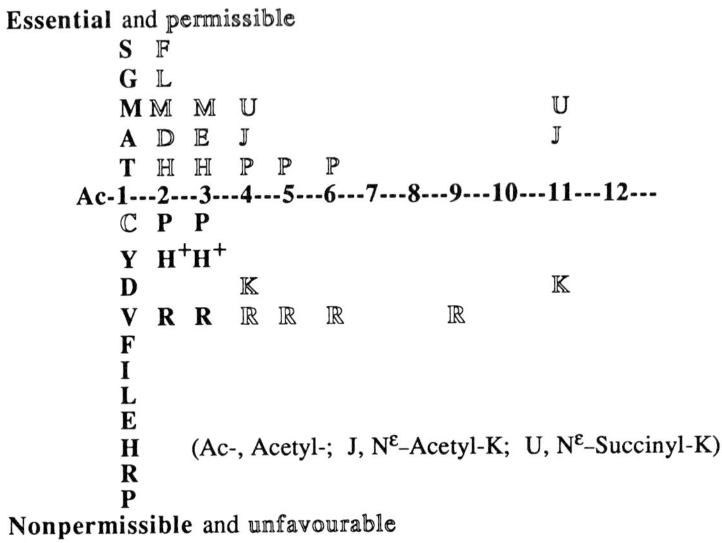

Fig. 2. Schematic representation of key specificity determinants of rabbit muscle N-acylaminoacyl-peptide hydrolase.

Sequencing the N-termini of Known Acetylated Proteins. The sequencing data for 7 known acetylated proteins after fragmentation, succinylation and unblocking with the hydrolase are reported in Figure 3. Several different fragmentation methods were used and the proteins contained either N-terminal Ac-Ala and Ac-Met, which are good substrates for the hydrolase, or Ac-Gly which is only an intermediate quality substrate, but it was still possible to obtain some sequence information in all cases. In the case of the tryptic peptides of parvalbumin, a sequence of 11 amino acid was established unequivocally, while in the case of myelin basic protein, only two residues could be identified. Based on our experience it appears that the established sequences often are shorter than those predicted from the known structures. The reasons for these reduced yields of information are not always obvious, but in most cases they can be explained as the products of several factors inherent in the methodology used. First of all, the shorter peptides are the best substrates for the hydrolase, and if a mixture of short and long peptides were produced, for example through incomplete protease cleavage, the short ones would predominate after hydrolase treatment. Secondly, as pointed out above, there is the strong inhibition of hydrolase activity caused by the presence of certain amino acid residues (i.e., Arg, His, Lys, Pro) in the proximity of the blocked N-terminus of the peptide substrate. It seems quite likely that the very limited sequence information obtained from the cyanogen bromide peptide from myelin basic protein reflects slow action of hydrolase on this large peptide containing Arg in positions 5 and 9. Both of the above factors reflect hydrolase activity, and can obviously be affected by increasing the amount of enzyme used in the unblocking step. To permit comparison of the different substrates used in this work, we arbitrarily fixed the amount of hydrolase at 5 μg per incubation; increasing this amount or using multiple additions of enzyme over a longer period of time will certainly improve the unblocking of the more resistant long peptides.

A third effect is based on interference with the automatic sequencing. As expected, the accumulation of succinate or phosphate in the final sample will

A. *(1) Parvalbumin (rabbit muscle).*

t

Ac-AMTELLNAEDIKKAIGAF·····

1.7

(2) Enolase (rabbit muscle).

V8

Ac-AMQKIFARE·····

2.0

(3) Myelin basic protein (bovine).

c

Ac-AAQKRPSQRSKYLASASTM·····

4.0

(4) Superoxide dismutase (bovine).

V8

Ac-ATKAVCVLKGDGPVQGTIHF·····

1.4

(5) α-crystallin A2 chain (bovine).

t

Ac-MDIAIQHPWFK·····

1.4

(6) Cytochrome c (horse).

c

Ac-GDVEKGKKIF·····

3.9

(7) Ovalbumin (chicken egg albumin grade VI).

n

Ac-GSIGAASMEFCFDVFK·····

3.2

B. *N^{α}-**Acylaminoacyl-peptide hydrolase** (rabbit muscle).*

c

Ac-X E R Q V L·····

(42) (11) (29) (41) (4)

M = (2000)

Fig. 3. The identified sequences of blocked N-termini of (A) different known proteins and (B) rabbit muscle N-acylaminoacyl-peptide hydrolase.
A. Each protein was subjected to several fragmentation reagents selected to give susceptible hydrolase substrates of optimal length. The underlined part of each known sequence shows the maximum sequence information obtained from a single peptide mixture for each protein; the arrow indicates the method of cleavage (t, trypsin; c, chymotrypsin; V8, V8-protease; n, CNBr) and the presumed cleavage site for that set of data. Some of the shorter sequences determined from other fragmentation methods ie., the chymotryptic peptides of parvalbumin, enolase, superoxide dismutase, α-crystallin and ovalbumin, the CNBr-generated peptides of myelin basic protein and the tryptic peptides of ovalbumin are indicated with plain letters. The number in parenthesis under the first residue shows the amount (in nmol) of the amino acid recovered from the acylase-treated hydrolase digests (theoretically 4.5 nmol).
B. For the unknown protein, N-acylaminoacyl-peptide hydrolase, only chymotrypsin was used for the fragmentation and the sequencing results are reported with the yields (in pmol) of PTH-amino acids obtained at each cycle. Methionine was identified as the blocked N-terminal residue from amino acid analysis after acylase-digestion; the recovery (in pmol) is also included.
Lysine was identified as succinyl-lysine in all cases.

interfere with the pH adjustment in the early sequencing cycles, and give rapidly declining yields of PTH-amino acids which together with high carry-overs severely limit the yield of useful sequence information. Generating volatile and/or ether-soluble salts and acids by using TEA to maintain pH during succinylation, acidifying with the strong acid, TFA, prior to the exhaustive ether extraction of succinic acid, and using bicarbonate buffer in the fragmentation procedures, the undesireable buffering salts can readily be eliminated. However, it also appears that for some high molecular weight proteins with high Lys content, the buffer capacity of the mixture of succinylated peptides itself may interfere with optimal automatic sequencing. To illustrate these effects, we have compared the sequencing of peptides in the presence of salts, and of a given peptide mixture with different levels (5x) of total peptide applied to the sequencer. The yield of useful sequence information was reduced from 11 cycles to 6-8 cycles for parvalbumin in the presence of salt, and the yield of useful information for α–crystallin/trypsin peptides and ovalbumin/CNBr peptides was reduced from 9 and 6 at a 0.5 nmol total load on the sequencer to 6 and 3, respectively at a 2.5 nmol load.

Identification of the N-Acetylamino Acid. In order to obtain as much information as possible from the N-terminal peptide, it is desirable to identify the blocked N-terminal amino acid released by hydrolase. This is readily achieved by digesting the hydrolase-treated peptide mixture with the enzyme acylase to release the free amino acid from acetylamino acid and thus permitting the identification of the N-terminal residue by direct amino acid analysis of the acylase digest. As expected, the yields of free amino acid varied considerably (from 1.4 to 4 nmol, corresponding to 30-90% of the 4.5 nmol starting material), but, most importantly, in all cases only a single amino acid peak was observed along with small amounts of ammonia. The main limitation of this approach is determined by the sensitivity of the amino acid analyzer technology; with the standard ninhydrin technology used here, 0.5 -1.0 nmol of amino acid represents a lower limit for standard detection.

Sequencing the N-terminus of Acylaminoacyl-Peptide Hydrolase. In order to apply the present method to an unknown protein, we subjected the rabbit muscle hydrolase itself to N-terminal sequencing. Preliminary studies demonstrated that the purified enzyme was blocked, and 5 nmol of enzyme was consequently subjected to chymotryptic digestion, succinylation, hydrolase treatment and sequencing according to the methods established with the known standards. The sequence obtained for 0.5 nmol of peptide is shown in Figure 3. When the remainder of the sample (4.5 nmol) was treated with acylase and subjected to amino acid analysis, 2.0 nmol of Met was obtained as the only free amino acid in the sample. The deduced sequences for the hydrolase from rat liver (18) and pig liver (19) are Ac-MERQVL--. In case of the rat liver enzyme, the sequence of several cyanogen bromide peptides were also determined, including the sequence starting with residue 2. Questions regarding the similarity of properties of the different tissue hydrolases have been raised on several occasions (1-3,18,19); it now appears that they may indeed be quite similar entities, in spite of the apparent differences in substrate specificity, molecular weight, metal ion requirements etc., reported from different laboratories for the different enzymes.

General discussion. The major challenge with the present method when applied to unknown proteins will be to find the optimal condition for producing the longest possible peptide for maximum sequence yield, and at the same time also have a good substrate for hydrolase. Both the initial method of fragmentation and the extent of the reaction will be mostly a matter of fortuitous choices, and may have to be followed up with alternative reactions based on the initial results. Regarding the possible application of this unblocking protocol to microsequencing, there will be some practical limitations to consider even for laboratories that are equipped for microanalysis and microsequencing. As already stated, 5 nmol of starting material was used for the experiments reported here primarily to satisfy the need for nanomolar quantities of amino acids for the amino acid analysis with ninhydrin detection; only 1/10 of the total sample was used for the sequencing, and it is clear that the method should be applicable to much lower levels of starting material. However, because of the strong buffering capacity of the incorporated succinyl groups in the petide mixture, it is necessary to determine, and as needed adjust, the pH of the hydrolase reaction mixture, and that usually requires a volume of at least 10-20 μl. Since it is impractical to consider substrate concentrations below 0.02 *mM* (1/50 K_m), especially for the relatively poor substrates that may be expected in this work, it will probably not be possible to go below about 200 pmol of starting protein.

The methodology used in this work differs only in detail from that reported recently by Tsunazawa *et al.* (7). They blocked the peptides produced in the fragmentation step with phenylisothiocyanate, and oxidized the resulting phenylthiocarbamoyl groups with performic acid to yield the phenylcarbamoyl derivatives of all the newly generated free N-terminals. The main reasons for using succinic anhydride in this work are the fact that it offered a simple and inexpensive single-step reaction, and that ε-succinyl-Lys was found to eliminate the inhibitory effect of Lys on hydrolase activity (Figure 2). Under the conditions described, the blocking of peptides was complete (>99%) and the stability and solubility of the derivatives were such that no peptides were lost during the ether extraction (the ether phase was completely devoid of amino acids) and no unblocked peptides were availabe for sequencing prior to the addition of the hydrolase.

IV. SUMMARY AND CONCLUSIONS

Acylaminoacyl-peptide hydrolase from rabbit muscle can be used to unblock the N-terminus of N-acetylated proteins for sequencing, after converting hydrolase-resistant intact proteins to hydrolase-susceptible peptides and blocking all the new N-termini by succinylation. After fragmentation of the blocked proteins with proteases or CNBr and succinylation, the large excess of succinate can be removed by ether extraction, and the total peptide mixture subjected to hydrolase digestion in the same reaction tube. Sequencing an aliquot of the digest yields the original N-terminal sequence starting with residue 2 as the only unblocked sequence. Treating the remainder of the hydrolase digest with acylase will cause the hydrolysis of the acetylamino acid, and permit the identification of the N-terminal amino acid by amino acid analysis. The method was applied to 7 known acetylated proteins, and permitted 5-12 amino acid sequences to be determined for 6 of these proteins; for one, myelin basic protein, only the first 3 amino acids could be identified. The hydrolase itself is a "blocked" protein; when it was fragmented with chymotrypsin, succinylated, and digested with hydrolase and acylase, the N-terminal sequence Ac-Met-Glu-Arg-Gln-Val-Leu- was determined.

ACKNOWLEDGEMENTS

I express my deep sense of gratitude to Prof. Finn Wold for his encouragement and helpful discussions in this study and his interest and help in the preparation of the manuscript. This work was accomplished in his laboratory with the support by his grants from USPH Service (GM-31305) and the Robert A. Welch Foundation (AU-916). My sincere thanks to C. C. Q. Chin and Li Zhao for their help in sequencing and amino acid analysis.

REFERENCES

1. Radhakrishna, G., and Wold, F. (1989) J. Biol. Chem. **264**, 11076-11081.
2. Kobayashi, K., and Smith, J. A. (1987) J. Biol. Chem. **262**, 11435-11445.
3. Jones, W. M. , and Manning, J. M. (1985) Biochem. Biophys. Res. Commun. **126**, 933-940.
4. Gade, W., and Brown, J. L. (1978) J. Biol. Chem. **253**, 5012-5018.
5. Tsunasawa, S., Narita, K., and Ogata, K. (1975) J. Biochem., **77**, 89-102.
6. Radhakrishna, G., and Wold, F. (1990) J. Prot. Chem. **9**, 309-310.
7. Tsunasawa, S., Takakura, H., and Sakiyama, F. (1990) J. Prot. Chem. **9**, 265-266.
8. Chu, F. S., Crary, E., and Bergdoll, M. S. (1969) Biochem. **8**, 2890-2896.
9. Riordon, J. F., and Vallee, B. L. (1964) Biochem. **3**, 1768-1774.
10. Van Der Ouderaa, F. J., Jong, W. W., and Bloemendal, H. (1973) Eur. J. Biochem. **39**, 207-222.
11. Chin, C. C. Q. (1990) J. Prot. Chem. **9**, 427-432.
12. Brostoff, S. W., Reuter, W., Hichens, M., and Eylar, E. H. (1974) J. Biol. Chem.. **249**, 559-567.
13. Steinman, H. M., Naik, V. R., Albernethy, J. L., and Hill, R. L. (1974) J. Biol. Chem. **249**, 7326-7338.
14. Margoliash, E., Smith, E. L., Kreil. G., and Tuppy, H. (1961) Nature, **192**, 1125-1127.
15. Nisbet, A. D., Saundry, R. H., Moir, A. J. G., Fothergill, L. A., and Fothergill, J. E. (1981) Eur. J. Biochem. **115**, 335-345.
16. Capony, J.-P., Pina, C., and Pechere, J.-F. (1976) Eur. J. Biochem. **70**, 123-135.
17. Gross, E. (1967) Methods Enzymol. **XI**, 238-255.
18. Kobayashi, K., Lin, L.-W., Yeadon, J. E., Klickstein, L. B., and Smith, J. A. (1989) J. Biol. Chem. **264**, 8892-8899.
19. Mitta, M., Asada, K., Uchimura, Y., Kimizuka, F., Kato, I., Sakiyama, F., and Tsunasawa, S. (1989) J. Biochem. **106**, 548-551.

SECTION II

Protein Microsequencing Workshop

Strategies for the Isolation of Peptides from Low-Abundance Proteins for Internal Sequence Analysis[3]

Ruedi Aebersold[1], Scott D. Patterson[2], Daniel Hess
The Biomedical Research Centre and Department of Biochemistry, University of British Columbia, Vancouver, B.C., Canada

I. Introduction

Sequence analysis is one of the most direct and revealing methods for the characterization of proteins at the molecular level. Empirical estimates from many laboratories involved in sequencing proteins and results from specific investigations (1) suggest that 50% or more of the naturally occurring proteins have modified (blocked) alpha amino groups and are therefore not directly accessible to amino terminal sequence analysis by the Edman degradation. As the nature of the amino terminal modification is generally not known and difficult to determine and reverse, the most general

[1] Address correspondence to this author at the Biomedical Research Centre,2222 Health Sciences Mall, University of British Columbia, Vancouver, B.C., Canada, V6T 1Z3

[2]Cold Spring Harbor Laboratory, Cold Spring Harbor, New York, USA.

[3]Work was supported in part by the Genetics Network of Centres of Excellence (Canada) and by the Biomedical Research Centre. R.A. was the recipient of an MRC of Canada Scholarship. D.H. was the recipient of a fellowship from the Swiss National Science Foundation

Abbreviations: **BNPS-skatole:** 2-(2'-nitrophenylsulfenyl)-3-methyl-3'-bromoindolenine; **BSA:** bovine serum albumin; **CAPS:** 3-[cyclohexylamino]-1-propanesulfonic acid; **CD:** Cationic Durapore; **CNBr:** cyanogen bromide; **GuHCl:** Guanidine hydrochloride; **IPG-IEF:** immobilized pH gradient isoelectric focusing; **NC:** nitrocellulose; **PVDF:** polyvinylidene fluoride; **PVP:** polyvinylpyrrolidone; **RP-HPLC:** reverse phase-high performance liquid chromatography; **SDS-PAGE:** sodium dodecyl sulfate polyacrylamide gel electrophoresis; **2DE:** two-dimensional gel electrophoresis; **TFA:** trifluoroacetic acid.

TECHNIQUES IN PROTEIN CHEMISTRY III

strategy for obtaining sequence information of blocked proteins is N-terminal sequence analysis of separated cleavage fragments. The generation and isolation of cleavage fragments from small amounts of proteins is technically challenging, mainly because unambiguous internal sequence information can only be generated from proteins of extraordinary purity and because many proteins are difficult to maintain in a solution compatible with proteolytic or chemical fragmentation and subsequent peptide separation. Recently, we and others have introduced methods for the generation and isolation of cleavage fragments from small amounts of proteins purified by various modes of polyacrylamide gel electrophoresis, including sodium dodecyl sulfate polyacrylamide gel electrophoresis (SDS-PAGE) , isoelectric focusing using immobilized pH gradients (IPG-IEF) and two-dimensional gel electrophoresis (2DE), thus overcoming the problem of purifying small amounts of proteins to homogeneity and eliminating to a large extent the idiotypic behavior of different polypeptides (Reviewed in 2). Most of these methods involve the electrophoretic transfer (electroblotting) of gel-separated proteins onto a suitable membrane, fragmentation of the protein on the membrane and release of the cleavage fragments for subsequent separation.

While these techniques have been very successful for generating limited amounts of sequence information, peptides derived from hydrophobic regions of the protein or very long peptides are generally only recovered in low yields due to the complexity of the protein/membrane interaction. This restricts the generality of the procedure, particularly in situations in which a specific peptide (e.g., carrying a regulatory phosphorylation site) is targeted for sequence analysis or in cases where long peptides generated by chemical fragmentation are to be analyzed.

We recently introduced a novel electroblotting support which is characterized by the defined nature of its protein/membrane interaction (3,4). Consequently, cleavage fragments as well as intact proteins can be recovered at high yields under mild conditions for further analysis.

In this manuscript we discuss optimal strategies for obtaining sequenceable fragments from gel separated proteins for a number of applications commonly faced by protein biochemistry laboratories.

II. Experimental Section

Materials. For gel electrophoresis, standard commercial sources of chemicals were used. Nitrocellulose (NC) (0.45μm pore size) was obtained from Schleicher & Schuell (Keene, NH) and Immobilon-P from Millipore (Bedford, MA). An experimental cationic polyvinylidene fluoride-based (PVDF) membrane (CD) was a generous gift of Dr. Jack Johansen (Millipore, Bedford, MA). Polyvinylpyrrolidone (average Mr=40,000; PVP), the protein stains

Amido Black 10B and Coomassie Blue R and octylglucopyranoside were from Sigma (St. Louis, MO). L-1-Tosylamido-2-phenylethyl chloromethyl ketone-treated trypsin was from Boehringer Mannheim (Indianapolis, IN) and bovine chymotrypsin was from Calbiochem. Fluorescamine was from Roche (Nutley, NJ). Cyanogen bromide (CNBr) was purchased from ICN (OH) and 2-(2'-nitrophenylsulfenyl)-3-methyl-3'-bromoindolenine (BNPS-skatole) was from Fluka (Switzerland). All chemicals and solvents used were of the highest available grade or purity.

Electrophoresis and transfer conditions. Electrophoresis and blotting was performed in a Mini-Protean II apparatus (Bio-Rad, Richmond, CA). Unless otherwise stated, proteins were separated in 12.5% T 2.67% C SDS-PAGE gels (1.0 mm) using the Laemmli (5) buffer system. Following electrophoresis, separated proteins/peptides were transferred to the respective membrane in either 10mM 3-[cyclohexylamino]-1-propanesulfonic acid (CAPS), pH 11, containing 10% (v/v)methanol (6) for the Immobilon-P membrane, or tris (25mM) glycine (192mM) pH 8.3 in 20% methanol for the CD membrane and nitrocellulose (7) at 70V for 70 min, without preincubation of the gel.

Enzymatic cleavage of proteins electroblotted onto NC. Essentially the procedure described previously (8) was used. Bands of Amido Black stained and destained NC pieces containing the protein were pooled in a single Eppendorf tube and incubated for 30 min at 37°C in 1.2 ml of 0.5% PVP dissolved in 100 mM acetic acid in order to prevent adsorption of the protease to the NC during digestion. Excess PVP was removed by extensive washing with water. NC strips were cut in small pieces of approximately 1 mm x 1 mm and put back into the same tube. The protein on the NC pieces was digested with trypsin in 100 mM $NaHCO_3$:acetonitrile, 95:5 (vol/vol) pH 8.2, at 37°C overnight, at an enzyme:substrate ratio of 1:20 (wt/wt). After digestion, the whole reaction mixture was frozen at -20°C or, after acidification, immediately loaded onto the HPLC column.

Enzymatic cleavage of proteins electroblotted onto CD membranes. Essentially the procedure described by Patterson (4) was used. Proteins electroblotted onto CD membrane were either detected by autoradiography or by limited fluorescamine staining following the procedure by Vandekerckhove (9). Elution of proteins from the CD membrane for proteolysis in a solution compatible with RP-HPLC, was effected using 4M guanidine hydrochloride (GuHCl) with 0.1% octylglucopyranoside. The protein containing CD membranes, cut into small pieces were sonicated (Branson, Model 1200) for 3 hours in 20µl of the GuHCl/detergent solution. The supernatent was removed and added to 20µl of 0.2M $NaHCO_3$. Two micrograms of bovine chymotrypsin were added to the solution and incubation was continued overnight at 37°C. The resulting peptide mixture was stored at -20°C until RP-HPLC analysis.

Chemical fragmentation of proteins electroblotted onto CD membrane. Proteins were cleaved on the CD membrane using the chemical reagents CNBr, specific for methionyl bonds and BNPS-skatole, specific for tryptophanyl bonds. Protein containing bands were excised and cut into small pieces. Twenty microlitres of reagent solution was added to the membrane pieces. Methionyl bond cleavage was effected using 20µg to10mg CNBr/20µl of formic acid (70%) for either overnight or 3 hours at room temperature in the dark. Tryptophanyl bond cleavage was achieved with BNPS-skatole at 20-40µg/20µl acetic acid (75%) for 1 hour at 47°C in the dark (10). After cleavage, the supernatant was removed and remaining peptide was eluted with 20µl of formic acid (70%) for one hour at room temperature. An equal volume of distilled water was added to the pooled supernatants and the samples taken to near dryness in a Speed-vac and stored at -70°C for further analysis.

Peptide separation. Peptides derived by chemical cleavage were either separated by tris-tricine SDS-PAGE (16.5% T, 2.67% C) (11) or by reverse phase-high performance liquid chromatography (RP-HPLC) using an Ultrafast Microprotein Analyzer (Michrom BioResources Inc., CA) equipped with a PLRP-S column with the dimensions 1x50mm. Electrophoretically separated peptides were blotted onto Immobilon-P and visualized by Coomassie staining (6).

Proteolytically derived peptides were separated by narrow-bore RP-HPLC using either a Waters peptide analyzer or a Waters MS-600. Following the addition of 10µl of 10% aqueous trifluoroacetic acid (TFA), the peptide solution was applied to the C_{18} column (2.1x150mm, 300Å), (Vydac, Hesperia, CA) at 50°C and developed over 100 min. with a 0-70% acetonitrile gradient. Eluting peptides were monitored by UV absorption at 215 nm and fractions collected manually for further analysis. The chromatography solvents for all RP-HPLC separations were: solvent A: 0.1% TFA in H_2O; and solvent B: 0.08-0.095% TFA in H_2O:acetonitrile (30:70, v/v). The UV absorbances of the solvents were matched at 215 nm by titrating the TFA concentration in solvent B.

III. Results and Discussion

A. *Proteolytic Cleavage on Nitrocellulose*

Proteolytic cleavage of proteins after electroblotting *in situ* on nitrocellulose introduced the concept of rapidly and efficiently generating cleavage fragments for internal sequence analysis of small amounts of gel-separated proteins. The scope of the method was subsequently extended by using different electroblotting matrices and chemical cleavage procedures (10,12, reviewed in 2) and shown to be compatible with a wide variety of proteolytic enzymes (13).

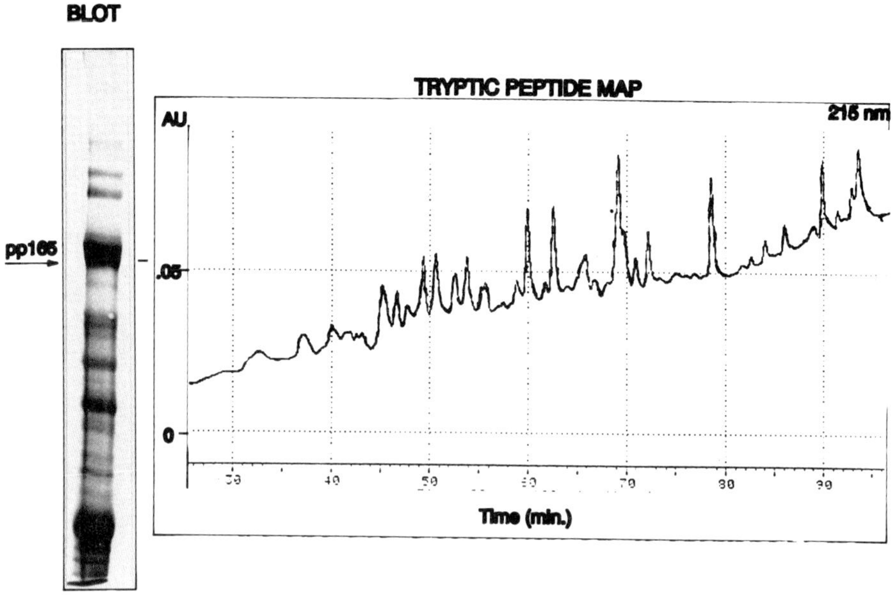

Figure 1. *Tryptic cleavage of pp165 on NC.* The extract from 100 10-cm tissue plates of cultured murine adipocytes was chromatographed on an affinity column with phosphotyrosine specificity (1G2 agarose). Proteins eluted with phenyl phosphate were concentrated by trichloroacetic acid precipitation, separated by SDS-PAGE and transferred to nitrocellulose and stained by Amido Black. The band corresponding to pp165 was cut out and the absorbed protein was cleaved with trypsin. Resulting cleavage fragments were separated by narrow-bore RP-HPLC and isolated peptides were collected for sequence analysis.

Comparative examination by RP-HPLC of peptide patterns obtained after cleaving the same protein in solution or on the membrane revealed differences in the respective patterns. This was most likely due to selective retention of a subset of peptides on the membrane. The nature of the interaction of polypeptides with nitrocellulose (and also with Immobilon-P) is complex and poorly understood (14). There is however a trend of increasing retention with increasing length and hydrophobicity of the polypeptide, so that longer and more hydrophobic cleavage fragments are selectively retained, whereas polar peptides are almost quantitatively released into the supernatant.

While selective retention of a subset of cleavage fragments is undesirable in cases where a specific region of the polypeptide chain is targeted for sequence analysis (e.g., in the analysis of post translationally modified amino acid residues or in the determination of active sites of an enzyme), a peptide map of significantly reduced complexity is of considerable advantage in cases where very large proteins are fragmented. A peptide map of reduced complexity decreases the chance of co-elution of unrelated peptides and therefore the possibility of obtaining mixed signals in peptide sequence analysis. This is illustrated by tryptic peptide mapping and sequence analysis of pp165 (Fig. 1). Pp 165 is a major substrate of the insulin receptor protein tyrosine kinase (15). A sample of pp165, partially purified by affinity chromatography using a column with phosphotyrosine specificity was separated by SDS-PAGE and electroblotted onto NC. Pp165 tryptic peptides derived as described in the experimental section were isolated and sequenced. Six peptide sequences ranging in length from 11 to 23 amino acid residues, covering a total of 102 amino acids could be established.

Obviously, *in situ* cleavage of proteins electroblotted onto nitrocellulose is equally well suited for generating peptide fragments from smaller proteins, as long as no specific region of the protein is targeted for sequence analysis.

B. Chemical and Proteolytic Cleavage on CD Membrane

The CD membrane is an experimental, PVDF-based electroblotting support, which we have recently introduced for applications involving the recovery of proteins and peptides at high yields and under mild conditions (3,4). The membrane has a cationic surface and appears to quench non-specific protein/membrane interactions, so that the nature of protein binding is defined and reversible.

We have used this membrane for generating samples for internal sequence determination. Because the protein/membrane interaction is essentially based on ion-exchange principles, peptides covering the entire primary sequence of a protein are recovered in high yields, irrespective of their length and hydrophobicity, as long as sufficiently strong peptide solvents are used. This is illustrated in Fig 2. Bovine Serum Albumin (BSA) (10μg) was electrophoresed and blotted onto CD and cleaved with the reagents CNBr or BNPS-skatole. Resulting cleavage fragments were recovered, separated by gel electrophoresis and detected by Coomassie blue staining (Fig. 2). Essentially quantitative cleavage was observed in three hours using 10mg CNBr/20μl formic acid (Fig. 2C). BNPS-skatole cleavage even at 40μg/20μl acetic acid was not complete but at least as efficient as that reported by Crimmins (Fig. 2E) (10). In both experiments the control samples (BSA recovered from CD membrane and incubated in the respective solvents devoid of the specific cleavage reagent) showed negligible cleavage (Fig. 2, Lanes B and D). Using radiolabeled samples, the elution efficiency of CNBr and BNPS-skatole induced cleavage fragments has been determined to be between 65 and 84% for a range of proteins (4).

Proteins electroblotted onto CD membrane are also compatible with direct enzymatic fragmentation on the support. Furthermore, peptide maps of the same protein cleaved on the membrane and in solution are very similar, indicating that peptides spanning the entire sequence of the protein are recovered. This is illustrated in Fig.3. Bovine alpha lactalbumin was cleaved by

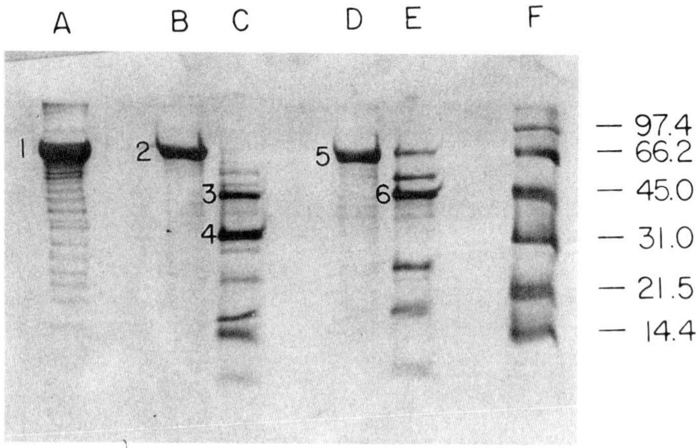

Figure 2. *Chemical cleavage on the CD membrane.* Chemical cleavage of CD blotted/eluted BSA as described in the text. The lanes are as follows: (A) 10μg BSA (control), (B-E) CD eluted BSA; (B) formic acid control, (C) CNBr cleaved, (D) acetic acid control, (E) BNPS-skatole cleaved, (F) mass standards (M_r x10^{-3}, indicated to the right). The numbered bands were subjected to automated N-terminal sequence analysis.

chymotrypsin in solution (Fig. 3A) or after electroblotting onto CD membrane on the support (Fig. 3B).

Cleavage of proteins on CD membranes is therefore ideally suited for the preparation of peptides in cases where a specific fragment is to be isolated for sequence analysis. Isolation of a specific peptide cleavage fragment is of particular importance for the determination of the site and nature of post translationally modified amino acid residues. Furthermore, the method is well suited to the isolation of relatively large peptide fragments, such as the ones generated by specific chemical cleavage at the rare amino acids methionine and tryptophan. The availability of long peptides has long been recognized as advantageous in comparative peptide mapping experiments (16) and has more recently gained renewed interest for molecular cloning of genes using amplification methods based on the polymerase chain reaction (17).

IV. Conclusions

The successful solution of different problems confronting the protein biochemist requires an arsenal of different methods, each one optimized for a particular application. It is predominantly the choice of the right method, in addition to the correct execution, which in most cases determines success or failure of an experiment.

In this overview we have summarized some of the methods available for the determination of internal sequence information of

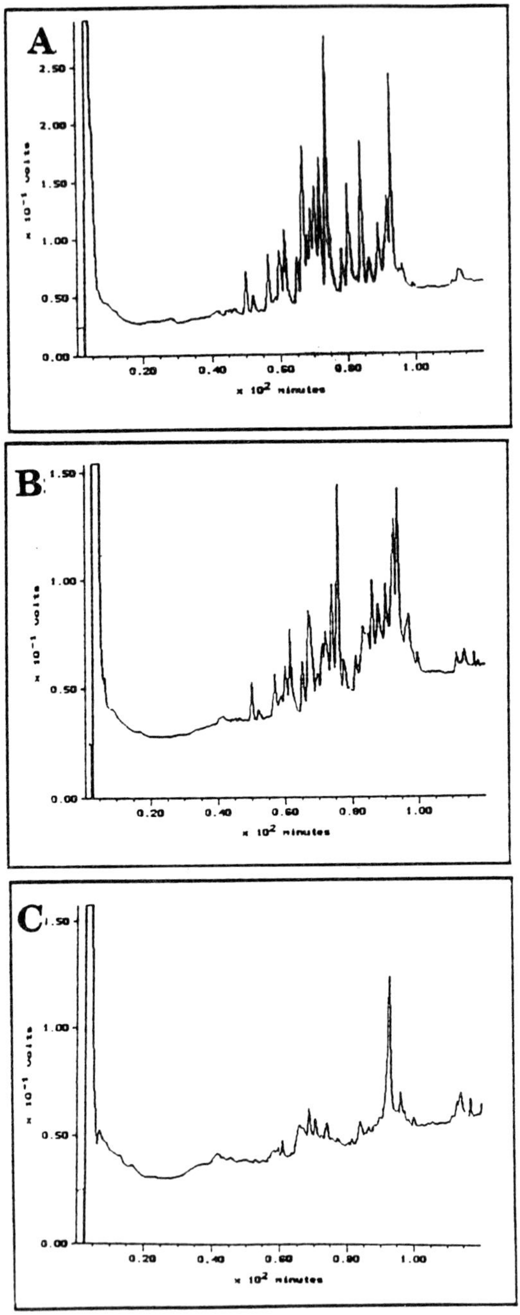

Figure 3. *Enzymatic cleavage on the CD membrane.* Ten micrograms of bovine alpha lactalbumin was either digested by chymotrypsin in solution (A) or after gel electrophoresis and electroblotting onto the CD membrane on the support (B). Generated peptide fragments were separated by RP-HPLC. (C) represents an enzyme blank reaction without substrate, but in the presence of CD membrane.

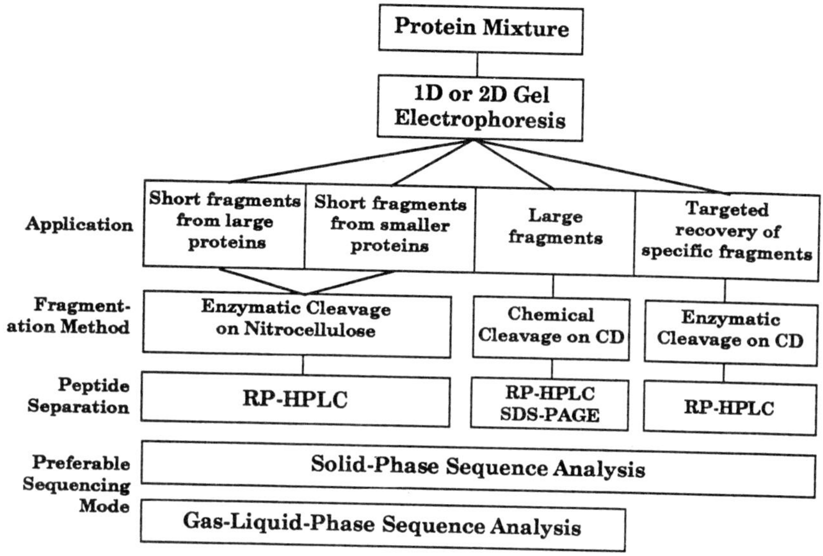

Figure 4. *Schematic summary of methods for generation of cleavage fragments from gel separated proteins for sequence analysis.*

small amounts of proteins separated by polyacrylamide gel electrophoresis (Fig. 4). The methods have in common that they are based on the isolation of the separated proteins by electroblotting. During the whole isolation and fragmentation process, proteins don't have to be maintained or brought back into solution. Therefore, the idiotypic behavior of proteins is to a large extent eliminated and the methods are applicable to a wide variety of proteins.

If no particular region of the protein primary sequence is targeted for sequence analysis (e.g., if the sequence is used for molecular cloning of a gene), enzymatic fragmentation on NC is a suitable method. This is particularly true for large proteins, since a subset of the generated fragments generally are not released from the membrane. In cases which require sequence analysis of one particular peptide fragment or if a large peptide has to be recovered, fragmentation of the protein on CD membrane seems to be the method of choice.

Peptide fragments released from both types of membrane are compatible with RP-HPLC separation and direct sequence analysis in the gas-liquid phase and solid-phase sequencing modes. Recently, protocols based on high-sensitivity solid-phase sequence analysis for the determination of the sites of protein serine/threonine as well as protein tyrosine phosphorylation at the low picomole level have been introduced (18,19). Therefore, compatibility with solid-phase sequencing protocols is of particular importance for the analysis of peptides containing post-translationally modified amino acids.

Acknowledgements

We thank Dr. Jack Johansen (Millipore, Bedford, MA) for providing us with the experimental CD membrane. We also thank Dr. Gustav Lienhard, (Dartmouth Medical School, Hanover, NH) for providing us with the pp165 sample.

References:

1. Brown, J.L. and Roberts, W.K. (1976) J. Biol. Chem. **251,** 1009-1014
2. Aebersold, R. (1991). Advances in Electrophoresis (Chrambach, A., Dunn, M.J. and Radola, B.J. Eds), VCH Verlag, Weinheim, Germany, Vol. 4, pp. 81-168.
3. Patterson, S.D. and Aebersold R. (1991). in: 2D-PAGE '91. Dunn, M.J. (ed), in press.
4. Patterson, S.D., Hess, D., Yungwirth, T. and Aebersold, R. submitted.
5. Laemmli, U.K. (1970). Nature **227,** 680-685.
6. Matsudaira, P. (1987). J. Biol. Chem. **262,** 10035-10038.
7. Towbin, H., Staehelin, T. and Gordon, J. (1979). Proc. Natl. Acad. Sci. USA **76,** 4350-4354.
8. Aebersold, R., Leavitt, Saavedra, R.A. and Hood, L.E. (1987). Proc. Natl. Acad. Sci. USA **84,** 6970-6974.
9. Vandekerkhove, J., Bauw, G., Puype, M., Van Damme, J. and Van Montagu, M. (1985). Eur. J. Biochem. **152,** 9-19.
10. Crimmins, D.L., McCourt, D.W., Thoma, R.S., Scott, M.G., Macke, K. and Schwartz, B.D. (1990). Anal. Biochem. **187,** 27-38.
11. Schagger, H. and von Jagow, G. (1987). Anal. Biochem. **166,** 368-379.
12. Scott, M.G., Crimmins, D.L., McCourt, D., Tarrand, T.T., Eyerman, M.C. and Nahm, M.H. (1988). Biochem. Biophys. Res. Comm. **155,** 1353-1359
13. Tempst, P., Link, A.J., Riviere, L.R., Fleming, M. and Elicone, C. (1990). Electrophoresis, **11**,537-553.
14. Gershoni, J.M. and Palade, G.E. (1983). Review: Protein blotting: Principles and applications. Anal. Biochem. **131,** 1-15.
15. Keller, S.R., Kitagawa, K., Aebersold, R., Lienhard, G.E. and Garner, C.W. (1991). J.Biol. Chem., in press.
16. Cleveland, D.W., Fischer, S.G., Kirschner, M.W. and Laemmli, U.K. (1977). J. Biol. Chem. **252,** 1102-1106..
17. Gautam, N., Baetscher, M., Aebersold, R. and Simon, M. (1989). Science. **224,** 971-974
18. Wettenhall, R.E.H., Aebersold, R.and Hood, L.E. (1991). Methods Enzymol. **201,** in press
19. Aebersold, R., Watts, J.D., Morrsion, H.D. and Bures, E.J., submitted.

High-Sensitivity Peptide Mapping Utilizing Reversed-Phase Microbore and Microcolumn Liquid Chromatography

Robert L. Moritz, Larry D. Ward
and Richard J. Simpson

Joint Protein Structure Laboratory
Ludwig Institute for Cancer Research (Melbourne Branch)
and The Walter and Eliza Hall Institute of Medical Research
P.O. Royal Melbourne Hospital, Victoria 3050, Australia

I. Introduction

Peptide mapping by RP-HPLC is an extremely powerful technique that is now widely applied in the structural analysis of proteins. When used in conjunction with Edman degradation procedures, peptide mapping provides an avenue for obtaining internal amino acid sequence information - the mainstay of many gene cloning strategies. Other important applications for peptide mapping include the detection of post-translational modifications, epitope mapping, disulfide bond assignment, the identification and localization of genetic variants and the analysis of purity and quality control of genetically engineered protein products.

Crucial to obtaining high-sensitivity peptide maps is the ability to isolate and manipulate (e.g., concentrate, buffer-exchange, reduce and alkylate, fragment) microgram quantities of proteins. For this reason, it is important to interface peptide mapping strategies with high-resolution protein and peptide separation techniques such as RP-HPLC, affinity chromatography and two-dimensional gel electrophoresis (2-DE).

Recent reports from our laboratory describe the use of short microbore column (5-10cm x 1-2.1mm i.d.) liquid chromatography for the peptide mapping of proteins isolated by RP-HPLC [1,2]. We now extend these studies to include the peptide mapping of proteins isolated by (i) affinity chromatography, and (ii) 2-DE. The potential of microcolumn (≤0.5mm i.d.) liquid chromatography in peptide mapping and microsequencing strategies is also assessed.

TECHNIQUES IN PROTEIN CHEMISTRY III

II. Experimental

1. **Peptide Mapping of Proteins Purified by Affinity Chromatography.** Rabbit liver UDP-N-acetylglucosamine: α-3-mannoside β-1,2-N-acetylglucosaminyltransferase I (GnT1), purified in the laboratory of H. Schachter [3], was eluted from the final affinity column in 1.5ml of 20% glycerol, 1% Triton X-100 and 0.3M NaCl. For structural analysis, GnT1 (~15μg) was recovered from this solution by 'inverse-gradient' RP-HPLC [4,5]. Briefly, the GnT1 solution (100μl) was diluted to 1.2ml with n-propanol in a sample-loading syringe and loaded at 1ml/min onto a VeloSep C8 cartridge (3-μm d_p, 30 x 2.1mm i.d.; Applied Biosystems) that had been equilibrated with 100% n-propanol. GnT1, which bound tightly under these conditions, was eluted at 0.1ml/min by a linear gradient (5%/min) of decreasing n-propanol concentration (containing 0.4% TFA). GnT1 fractions were adjusted to 0.02% Tween-20, concentrated to ~100μl and then diluted to 1.5ml with 5% formic acid. GnT1 was initially digested with pepsin for 1h at 37°C. Undigested core GnT1 (~10μg) was S-carboxymethylated (SCM-) [3] and digested further with trypsin, in 1ml 1% NH_4HCO_3/1mM $CaCl_2$/0.02% Tween 20, for 16h at 37°C. Undigested core SCM-GnT1 (~8μg) was further digested with thermolysin in 1ml 1% NH_4HCO_3/0.02% Tween 20, for 2h at 50°C. Enzyme/substrate ratios were 1:20. Peptides were fractionated by microbore column RP-HPLC [6,7].

2. **Peptide Mapping of Proteins Resolved by 2-DE.** Total cellular protein was extracted from a murine plasmacytoma cell line (TEPC 2027) according to the method of Garrels [8]. Sample preparation, in 9.0M urea, 2% NP-40, 2% ampholytes (LKB) and 50mM DTT, 2-DE and SDS-PAGE were performed as previously described [9,10].

After 2-DE, acrylamide gels were stained with 0.1% Coomassie brilliant blue R-250 (CBR250) in 50% methanol/10% acetic acid for 1h and then destained in 7% acetic acid/12% methanol. Protein spots (x8) from identical gels were excised, placed in the sample well (7mm wide, 23mm high) of a second SDS-acrylamide gel (1.5mm thick, 11cm long) and re-electrophoresed in a Molecular Dynamics (Palo Alto, CA) electrophoresis system using the Laemmli discontinuous buffer system [11]. Proteins were visualized, without staining, by high-resolution dynamic imaging (HRDI), excised from the gel and then passively eluted by soaking with either 20mM Tris-HCl buffer, pH 7.4 or 1% NH_4HCO_3 containing 0.02% Tween-20 (3 x 1ml) at 25°C for ~20h [10,12]. For peptide mapping, proteins were concentrated and buffer-exchanged either by microbore column RP-HPLC or 'inverse-gradient' RP-HPLC.

3. Reversed-Phase HPLC.

Microbore column chromatography. RP-HPLC was carried out on a Hewlett-Packard liquid chromatograph (1090A/1040A) as described elsewhere [6]. **Microcolumn chromatography.** The overall schematic of the gradient microcolumn HPLC system is shown in Fig. 1. The following columns were used for peptide separation, - (i) Brownlee RP-300 (300Å 7-µm d_p, 30 x 2.1mm i.d. cartridge; Applied Biosystems), - (ii) ODS-Hypersil (120Å 5-µm d_p, 100 x 2.1mm i.d. column; Hewlett-Packard, Waldbronn, FRG), and - (iii) ODS-Hypersil (120Å 3-µm d_p 50 x 0.3mm i.d., LC Packings, Amsterdam, The Netherlands). For PTH-amino acid analysis, the following columns were used (i) ABI-PTH-C18 (100Å 5-µm d_p, 200 x 2.1mm i.d. cartridge, Applied Biosystems) and (ii) ABI-PTH-C18 (100Å, 5-µm d_p, packed into a 150 x 0.3mm i.d. column, LC Packings, Amsterdam, The Netherlands).

4. Amino Acid Sequence Analysis. Automated Edman degradation was performed on an Applied Biosystems model 470A sequencer fitted with a model 120A PTH-amino acid analyzer modified as described elsewhere [13].

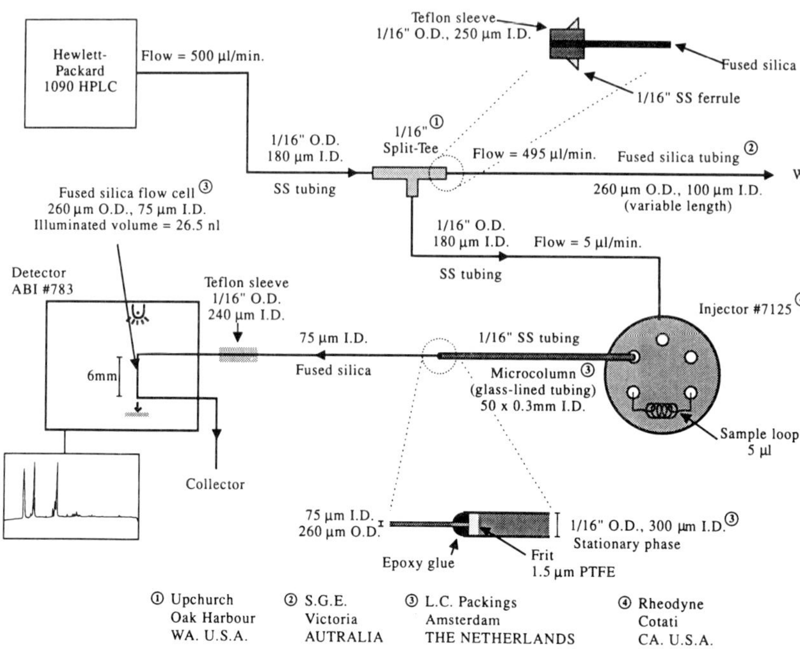

Figure 1. Schematic diagram of gradient microcolumn RP-HPLC system.

III. Results and Discussion

Peptide Mapping of Proteins Purified by Affinity Chromatography.

GnT1 is a medial Golgi-localized glycosyltransferase that catalyses an essential first step in conversion of high mannose N-glycans to hybrid and complex N-glycans. Since GnT1 is a membrane-associated enzyme, detergents were employed throughout the purification procedure, including the final affinity chromatography steps. This posed a strategic problem, since, for structural analysis GnT1 had to be recovered from a 1.5ml solution containing Triton X-100, NaCl, and glycerol. Initial attempts to desalt GnT1 using traditional RP-HPLC were unsuccessful due to the high-absorbing characteristics of Triton X-100. However, by using RP-HPLC in the 'inverse-gradient' mode, a procedure that we previously developed for recovering proteins from SDS-PAGE electroeluates, GnT1 (~15μg) was isolated free from detergent and salts (Fig. 2, panel A). Recovered GnT1 was digested with pepsin and the digest fractionated by RP-HPLC (Fig. 2, panel B). Undigested 'core' GnT1 remaining after pepsin digestion was reduced with dithiothreitol and alkylated with iodoacetic acid to give 'core' SCM-GnT1, which was desalted by RP-HPLC on the same column (Fig. 2, panel C).

Since treatment of 'core' SCM-GnT1 with trypsin did not yield a limit digest (Fig. 2, panel D) the 'trypsin-core' material was treated with thermolysin. Generated peptides (Fig. 2, panel E) were purified to homogeneity using additional chromatographic steps with different mobile phases (such as 1% aq. NaCl (Fig. 2, panel F) or 50mM sodium phosphate, pH 6.5 to achieve changes in chromatographic selectivity. Using this approach, 10 peptides were subjected to sequence analysis resulting in 103 amino acid assignments (~30% of the molecule). This data led to the successful cloning of GnT1 [3].

Peptide Mapping of Gel-resolved Proteins.

A number of strategies have been documented for obtaining peptide maps and internal sequence data from gel-resolved proteins. These include (i) recovery of proteins electroblotted onto polyvinylidene difluoride (PVDF) membrane using detergents followed by enzymic digestion and separation of peptides by RP-HPLC, (ii) in situ protease digestion of proteins electroblotted onto PVDF, nitrocellulose or derivatized glass followed by extraction of the peptides and subsequent separation by RP-HPLC, and (iii) in situ protease digestion or chemical

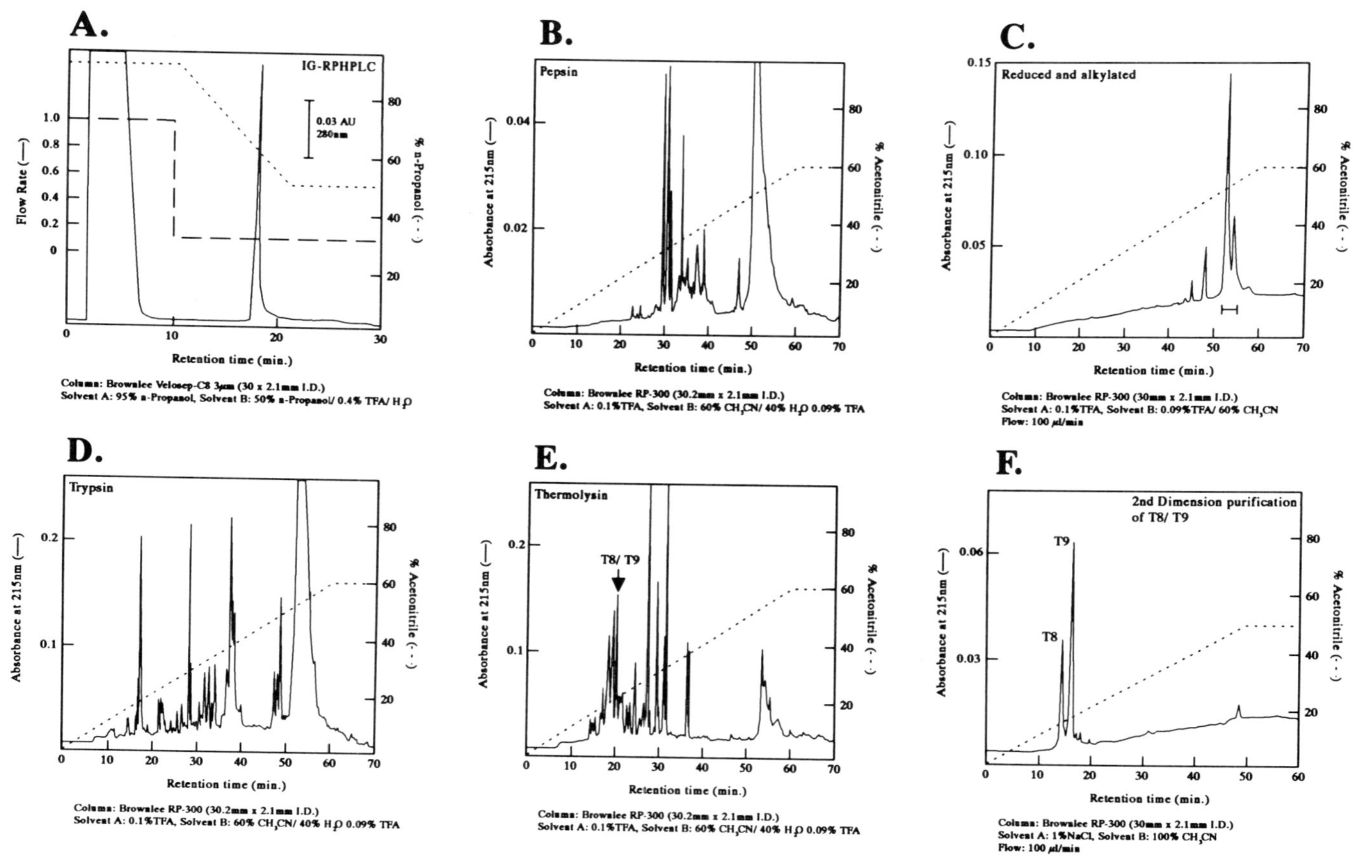

Figure 2. Strategy for high-sensitivity peptide mapping of rabbit liver GnT1.

fragmentation of proteins in the gel matrix followed by separation of generated peptides by RP-HPLC or a second SDS-acrylamide gel [1,10,14,15 and references therein].

Notwithstanding the success of peptide mapping strategies that include an electroblotting step, low sequencing yields of isolated peptides are frequently encountered. This is largely attributed to low yields from electrotransfer. Although many of the parameters that contribute to low transfer yields (e.g., nature of the immobilizing membrane, concentration of SDS and methanol used in electrotransfer buffers) have been addressed in the literature, it would appear that no one set of experimental electrotransfer conditions is generally applicable for all gel-resolved proteins and that conditions need to be tailored for an individual protein.

An alternative approach for peptide mapping gel-resolved proteins, that does not rely on an electroblotting step, is outlined in Fig. 3. The example given is for an N-terminally blocked protein (TEPC #8), a protein of M_r ~70K, isolated from TEPC 2027 cells by 2-DE. TEPC #8 (8 spots from identical gels, 150µg total protein/gel), visualized by CBR250, was excised, inserted into the sample well (7mm wide x 23mm high) of a 1D-acrylamide slab gel (1.5mm thick, and 11cm long) and the gel electrophoresed at 100V constant voltage to concentrate the protein and remove CBR250, the latter interferes with high-sensitivity RP-HPLC peptide mapping. TEPC #8 was detected by HRDI during electrophoresis, and excised from the gel as described elsewhere [12]. Excised gel pieces were placed in 1ml of elution buffer (20mM Tris-HCl, pH 7.4, 0.02% Tween 20) in a 1.5ml Eppendorf tube and TEPC #8 was passively eluted by gentle shaking at 25°C for 16 h.

TEPC #8 was recovered from the elute by RP-HPLC (Fig. 3). This HPLC step fulfills two objectives - firstly, it allows concentration of the protein and buffer-exchange into an appropriate buffer for subsequent protease digestion and, secondly, it removes acrylamide-related contaminants that may interfere with the interpretation of the peptide map. If the protein cannot be recovered by RP-HPLC, as indicated by a pilot chromatographic run using a 1mm i.d. column, then 'inverse-gradient' RP-HPLC can be employed. A tryptic digest of TEPC #8 is shown in Fig. 3. To obtain homogenous peptides for sequence analysis, it is essential that peptide-containing fractions be subjected to further chromatographic steps. For proteins of high M_r, this is particularly critical. Representative sequence data for a number of peptides (T1-T3), together with yields (5-26 pmol) obtained in the first

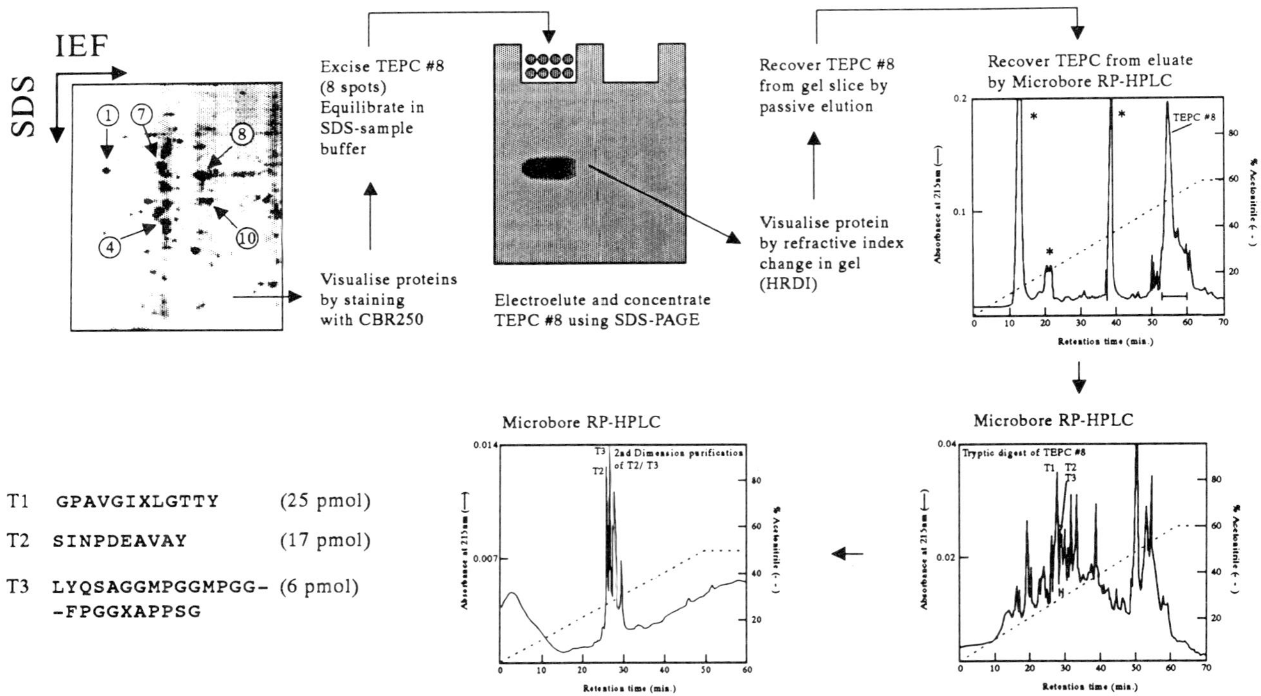

Figure 3. Strategy for high-sensitivity peptide mapping of a 2-D gel resolved protein (TEPC #8)

cycle of the Edman degradation, are given. These yields, obtained from 8 protein spots, compare favorably with typical N-terminal sequence yields (5-20 pmol) obtained for 2-DE resolved proteins (4 protein spots) recovered from gels by electroblotting [9,10,14]. A comparison of this sequence data with sequences tested in the available protein databases revealed that TEPC #8 was the murine form of heat shock protein 70 (hsp 70).

Application of Microcolumn RP-HPLC

Microbore column (1-2.1mm i.d.) RP-HPLC [16,17] is now an extensively used technique for the high-sensitivity peptide mapping of rare proteins [1,2,6,7]. In the past few years, there has been an exciting trend towards further column miniaturization. Advances in this area include (i) the development of packed microcolumns (0.2-0.5mm i.d.) [18], (ii) the construction of long pathlength (6-20mm), microvolume (25-100nL) axial-beam detectors [19,20], and (iii) the ability to produce precision gradient flows [21]. Compared to microbore column (2.1mm i.d.) HPLC, these advances offer several distinct advantages (i) improved mass sensitivity and (ii) recovery of sample in reduced volumes.

In the examples given, the mass sensitivity of the 0.3mm i.d. column is ~20 times better than that of the 2.1mm i.d. column for both synthetic peptides (Fig. 4) and PTH-amino acids (Fig. 5). The separation efficiencies for both 0.3mm and 2.1mm i.d. columns are comparable as evidenced by the peak profiles; indeed, for the PTH-amino acid analysis, the separation is marginally better for the 0.3mm i.d. (see Fig. 4 and 5).

These studies suggest that microcolumn RP-HPLC offers a valid method for separating trace amounts of peptides and - given the low flow rates (≤5μl/min) and sample volumes ~1μl - for direct interfacing to mass spectrometers.

The enhanced mass sensitivities of PTH-amino acids achieved with microcolumn RP-HPLC (Fig. 1) provide far-reaching consequences for microsequencing. Obviously, sample volume limitations exist. This is due to the low concentrations of organic solvent required to solubilize the PTH-amino acid derivatives. However, for protein and peptides, sample volumes are not critical since they can be diluted below their critical organic solvent concentration required for retention. In the present study samples of ≤5μl were employed. Miniaturized instrumental design for microsequence analysis appears to be essential

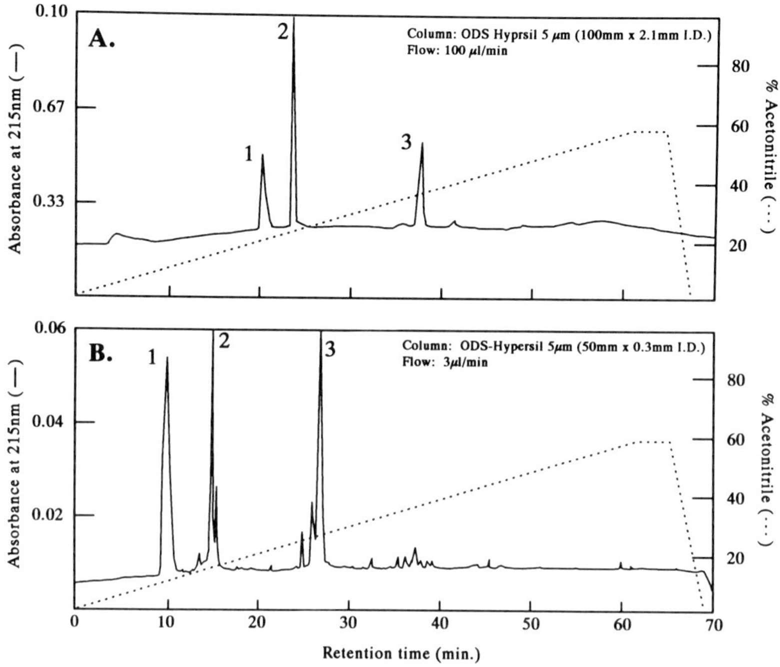

Fig. 4. Comparison of microbore (1.0μg) (A) and microcolumn (0.1μg) (B) RP-HPLC separation of synthetic peptides.

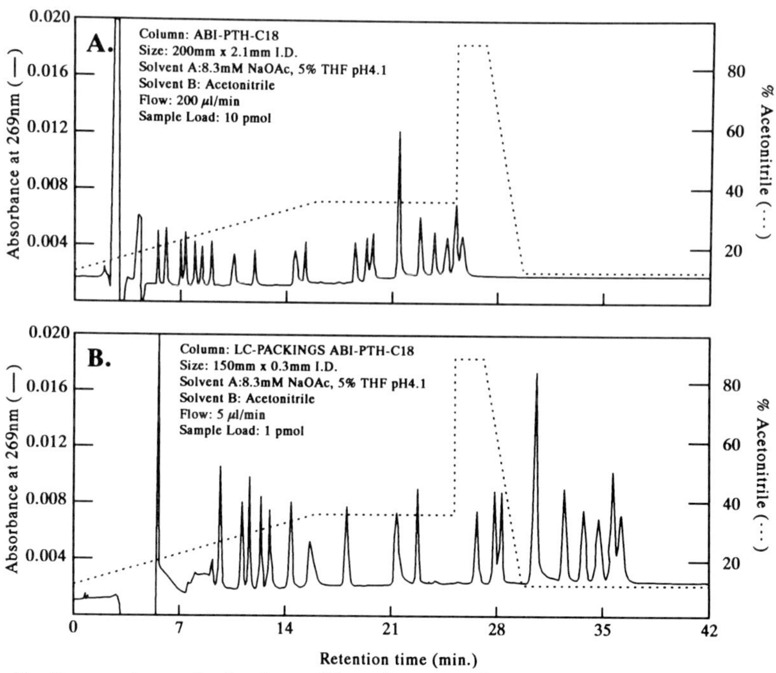

Fig. 5. Comparison of microbore (A) and microcolumn (B) RP-HPLC separation of PTH-amino acids.

to take advantage of the ultra-high-sensitivity offered by microcolumn HPLC (and capillary electrophoresis) PTH-amino acid analysis.

IV References

1. Simpson, R.J., Moritz, R.L., Begg, G.S., Rubira, M.R. and Nice, E.C. (1989) Anal. Biochem. 177, 221-226.
2. Simpson, R.J., Begg, G.S., Reid, G.E., Rubira, M.R., Ward, L.D. and Moritz, R.L. In: Wittmann-Liebold, B. (Ed.), Methods in Protein Sequence Analysis, Springer-Verlag, Berlin, (1989) pp. 256-263.
3. Sarkar, M., Hull, E., Simpson, R.J., Moritz, R.L., Dunn, R. and Schachter, H. (1991) Proc. Natl. Acad. Sci. USA 88, 234-238.
4. Simpson, R.J., Moritz, R.L., Nice, E.C. and Grego, B. (1987) Eur. J. Biochem. 165, 21-29.
5. Simpson, R.J. and Moritz, R.L. (1989) J. Chromatogr. 474, 418-423.
6. Simpson, R.J., Moritz, R.L., Rubira, M.R. and Van Snick, J. (1988) Eur. J. Biochem. 165, 21-29.
7. Simpson, R.J., Moritz, R.L., Gorman, J.J. and Van Snick, J. (1989) Eur. J. Biochem. 183, 715-722.
8. Garrels, J.I. (1979) J. Biol. Chem. 254, 7961-7977.
9. Ward, L.D., Hong, J., Whitehead, R.H. and Simpson, R.J. (1990) Electrophoresis 11, 883-391.
10. Ward, L.D., Hong, J. and Simpson, R.J. In: Villafranca, J.J. (Ed.), Techniques in Protein Chemistry II, Academic Press (1991), pp. 21-30.
11. Laemmli, U.K. (1970) Nature 227, 680-685.
12. Ward, L.D. and Simpson, R.J. (1991) Peptide Res. 4, 187-193.
13. Begg, G.S. and Simpson, R.J. In: Hugli, T.E. (Ed.), Techniques in Protein Chemistry, Academic Press, (1989) pp. 79-88.
14. Ward, L.D., Reid, G.E., Moritz, R.L. and Simpson, R.J. In: Villafranca, J.J. (Ed.), Current Research in Protein Chemistry: Techniques, Structure and Function. Academic Press (1990), pp. 179-190.
15. Matsudaira, P.T. (Ed.), A Practical Guide to Proteins and Peptide Purification for Microsequencing, Academic Press, (1989).
16. Scott, R.P.W. and Kucera, P. (1976) J. Chromatogr. 125, 251-264.
17. Ishii, D., Asai, K., Hibi, K., Jonokuchi, T. and Nagaya, M. (1977) J. Chromatogr. 144, 157-168.
18. Yang, F.J. (1982) J. Chromatogr. 236, 265-277.
19. Chervet, J.P., Ursem, M., Salzmann, J.P. and Vannoort, R.W. (1989) J. High Resolution Chromat. 12, 278-281.
20. Xi, X. and Yeung, E.S. (1990) Anal. Chem. 62, 1580-1585.
21. Coutant, J.E., Chen, T-M. and Ackermann, B. (1990) J. Chromatogr. 529, 265-275.

MICROSCALE PROTEIN ISOLATION AND MICROSEQUENCING WORKSHOP: AN OVERVIEW

John E. Shively

Beckman Research Institute of the City of Hope, Duarte, CA 91010

I. INTRODUCTION

This is an overview of the Microscale Protein Isolation and Microsequencing Workshop held at the Fifth Symposium of the Protein Society in Baltimore on June 23-26, 1991. Previous workshops had focussed on the analysis of small peptides by microsequence analysis or mass spectrometry, and had included the distribution of an unknown peptide whose structure was analyzed by many of the workshop participants. This year's workshop was aimed at the problems involved in coordinating protein isolation and structural analysis at the micro-scale. Several experts presented major topics including protein isolation by micro-LC and two dimensional gel electrophoresis. Peptide mapping of soluble proteins or those blotted onto membranes was presented, including structural analysis by mass spectrometry. The mass spectrometric methods included electrospray mass spectrometry (LC/MS) and FAB/MS (fast atom bombardment mass spectrometry). Many pertinent questions were introduced by the audience, resulting in a lively discussion of the problems faced by the modern protein chemist.

II. ANALYSIS OF PURIFIED PROTEINS

The discussion began with criteria for identifying purified proteins. The most common criteria include a single peak on chromatography (reversed phase, ion exchange, or gel permeation) and a single band (unless multiple subunits are present) on SDS gel

TECHNIQUES IN PROTEIN CHEMISTRY III

electrophoresis. It was pointed out that molecular weight determination is a critical aspect of this analysis, providing information which will be used throughout the planning of structural analysis and for correlation of other analytical results. In this regard, mass spectrometry is playing an important new role, with UV matrix assisted, laser desorption, time of flight mass spectrometry (LD/TOF/MS) taking center stage. The reasons for this is that LD/TOF/MS can obtain accurate mass information (up to 0.01% accuracy) on a wide variety of samples in the low picomole to femtomole range, and in a wide variety of buffers including phosphate buffered saline, glycerol, detergents, and denaturation agents such as guanidine hydrochloride. Jack Shively gave several examples of samples (glycoproteins, membrane proteins, etc.) analyzed in this way, including samples spotted onto PVDF membranes. Brian Chait presented a more thorough introduction to the method during the regular presentations at the Symposium. It was concluded that this simple method will become a major tool for the protein chemist.

Another important mass spectrometric method for mass determination of proteins is via the electrospray ion source which allows proteins to be analyzed as a cluster of multiply charged peaks directly from liquid solutions. The electrospray ion source is most commonly used on quadrupole analyzers at this time, but many researchers are also implementing this source on magnetic and ion trap instruments. Examples of these instruments and samples analyzed by them are given elsewhere in this volume. The main advantage of the method is that highly accurate mass determinations are possible (usually within 0.01%). The main disadvantage is that salts severely interfere with the electrospray source. Thus, samples must be desalted prior to analysis. One way to accomplish this is to submit the sample to on-line reversed phase liquid chromatography interfaced to the electrospray source. Jack Shively gave several examples of this approach, and noted that many labs have little or no experience with micro-LC (flow rates in the range of 1-5 µl/min). He discussed simple methods for implementing the micro-LC method and packing capillary columns.

It was also stressed that the the amount of purified protein obtained must be quantified by amino acid analysis. Lamentably, this step is often ommitted in many labs. Samples which are subjected to further analysis are usually overestimated leading to poor planning in subsequent steps. In the worst scenario, a large contaminating protein may be mistaken for the biologically active protein of interest which is present in much smaller amounts. Although current methods of amino acid analysis require as much as 0.1-1.0 μg of protein, this analytical step is perhaps one of the most crucial to be performed. This discussion concluded with the declaration of need for more sensitive methods for mass quantitation. It was noted that mass spectrometric and SDS gel staining methods are notoriously poor at mass quantitation.

III. PEPTIDE MAPPING

In order to obtain peptide maps on protein samples, the protein may require prior desalting, or S-reduction and alkylation followed by desalting. These methods were discussed in detail by Richard Simpson. The key to the HPLC methods is to perform the analysis on a small amount of sample prior to subjecting a precious sample to the method. In order to increase the sensitivity of the method into the low micromole range, it is necessary to convert to micro-LC using capillary columns. Peptide maps may be analyzed on-line by electrospray mass spectrometry. This method will usually give molecular weights for each peptide, include mixtures of peptides. By applying MS/MS techniques many of the structures can be directly determined at the 10-20 picomole level. Unfortunately, not all structures can be obtained by this technique, necessitating conventional microsequence analysis or FAB/MS/MS analysis. In any case the Edman chemist is not out of a job yet. There is still a need for both approaches, mass spectrometry and Edman chemistry, to work together.

A second approach was described by Ruedi Aebersold who separates the proteins by two dimensional gel electrophoresis and then performs peptide mapping on the electroblotted samples. This approach has the advantage of including a denaturation step in the method, uses simple, inexpensive equipment, and works for most proteins. However, the overall yields are probably in the 10-30% range, necessitating further improvements in sensitivity (blotting, staining, Edman chemistry).

Steve Carr focussed his presentation on the analysis of highly glycosylated proteins which present many challenges, especially molecular microheterogeneity. He demonstrated that electrospray mass spectrometry is a particularly powerful approach for assessing the degree of microheterogeneity, and in many cases can be used to quantitate the amount of each species present. Although the fragmentation patterns and mass information do not constitute a proof of structure for glycosyl units, this information, together with our current understanding of the glycosyl units known to exist on glycoproteins, allows a confident prediction of structure. It can be concluded that the ES/MS/MS approach on the triple quadrupole mass spectrometers will be the method of choice for the analysis of glycopeptides and glycoproteins for the foreseeable future.

The overall conclusion is that it should be possible to perform peptide mapping on 5-20 pmole amounts of protein, and obtain significant structural information. In the case of mass spectrometric analysis, the analysis can be performed in several hours to several days, and include significant posttranslational information on the location of glycosylation and phosphorylation sites. In the case of Edman chemistry, the results may take several weeks to months, but these instruments are more widely available. In the future we look forward to more widely available (and perhaps more inexpensive) mass spectrometers. In optimal cases, especially for low molecular weight proteins (10-30,000), it is possible to determine entire structures on 100-200 picomoles of sample without resorting to molecular cloning studies. However, for the majority of proteins, combined protein and molecular cloning studies remain the method of choice.

SECTION III

Peptide/Protein Separation Techniques

Micropreparative Separation of Tryptic Digests by Capillary Electrophoresis and Characterization by Protein Sequencing

Alan Smith,
James W. Kenny

Beckman Center,
Stanford University Medical Center,
Stanford, CA.

J. I. Ohms

Beckman Instruments Inc.
Palo Alto, CA.

Introduction

Since its introduction in 1981 by Jorgenson and Lukacs, capillary electrophoresis (CE) has been used primarily as an analytical technique for a wide variety of molecules of biological interest (1). The very high theoretical plate count that can be achieved by CE makes it ideal for this application. The use of small diameter capillaries of 50-75 μm i.d. means that only nanoliter volumes of sample are required for analysis. However, these small volumes require relatively high (μg/μL) starting concentrations. This has limited the utility of CE in high sensitivity protein and peptide analysis where larger volumes of lower concentrations are more commonly available.

Relatively little work has been reported on the micro-preparative capabilities of capillary electrophoresis for proteins and peptides (2,3). The most frequent approach for isolating peptides is the pooling of peaks from a number of individual analytical separations (4). In this manner, sufficient quantities of peptides can be obtained for characterization by microsequence analysis. It would be of considerably greater utility to be able to collect a series of peptides from a single separation that could be characterized directly by protein sequencing. The utilization of larger bore capillaries is one method to achieve this goal.

TECHNIQUES IN PROTEIN CHEMISTRY III

We have investigated the use of 200 μm capillaries for the micropreparative separation of peptides. Our initial report demonstrated the viability of this approach (5). In this study, we demonstrate the feasibility of isolating peptides from a model peptide mixture and from a complex protein digest in single micropreparative separations. The peptides are recovered in sufficient purity and yield for characterization by the standard protein sequencing methodology.

Materials and Methods

The test mixture of three peptides contained ACTH fragment 4-10, angiotensin I and angiotensin II (Sigma) at a concentration of 0.6 mg/mL in 0.1% TFA:ethylene glycol (70:30). A tryptic digest of β-lactoglobulin (Sigma) was prepared and diluted to 6.0 mg/mL in 0.1% TFA:ethylene glycol (70:30). Aliquots of these mixtures were used for micropreparative injections into the separatory capillary.

The CE instrument used in this study was a P/ACE™ System 2000 (Beckman Inst.) programmed for automated fraction collection. The fused silica capillaries were obtained from Polymicro Technologies. The protein sequencer used in these studies was a Model 475 gas phase sequencer (Applied Biosystems).

The analytical and micropreparative CE separation conditions were as shown in Table I. The electrolyte used in all separations was 50 mM sodium phosphate, pH 2.70.

Table I. Separation Conditions

	Analytical	Micropreparative
Capillary diam. (μm)	75	200
Capillary volume (μL)	2.5	17.8
Capillary length (cm)	57	57
Flowrate (nL/sec)	5	250
Voltage (kV)	15	7.5
Current (μA)	45	175
Sample vol. injected (nL)	25	500

For micropreparative separations, fractions were collected by reduction of the field strength similar to the method of flow programming in HPLC chromatography. An initial electrokinetic separation was performed until the desired peaks were at the detector or just prior to exiting the capillary. The voltage was reduced by approximately half (3.5 kV) and fraction collection was started. The fraction collection vials contained 5 to 10 μL of electrolyte. The CE instrument was programmed to collect fractions for three minutes before moving to another vial. Sets of eight fractions could be programmed repeatedly with manual change between each of four sets for a total of 32 fractions.

Fractions that were collected from the micropreparative separation were screened directly by analytical CE or alternatively, dried and resuspended in 0.1% TFA:ethylene glycol (70:30) immediately prior to analysis. Those fractions containing peptides were then further characterized by protein sequence analysis. Recoveries were calculated as the percentage of the initial protein sequence yield relative to the quantity of peptides applied to the capillary.

Results

The test mixture of ACTH fragment 4-10, angiotensin I and angiotensin II was selected for the initial micropreparative separation since these peptides were well resolved with analytical CE. They were also well resolved on the 200 μm micropreparative capillary at an approximate load level of 100 pmol (Figure 1). The presence of 30% ethylene glycol in the sample solution was essential for optimal resolution (6). The protein sequence results of the micropreparative separation yielded the expected sequence. Each of the peptides eluted as single fractions and no cross contamination was observed (Figure 1C). The electrolyte did not interfere with sequence analysis. The highest contaminant (glycine) was <5 pmol. Recoveries for each peptide were in the 50% range (*i.e.* approximately 50 pmol of sequenced peptide).

These initial results encouraged us to attempt the micropreparative separation of a more complex mixture of peptides. We selected a tryptic digest of β-lactoglobulin. The analytical and micropreparative separations are shown in Figure 2. As in the separation of the test peptides, the resolution of the tryptic digest on the 200 μm capillary was qualitatively comparable to the analytical separation.

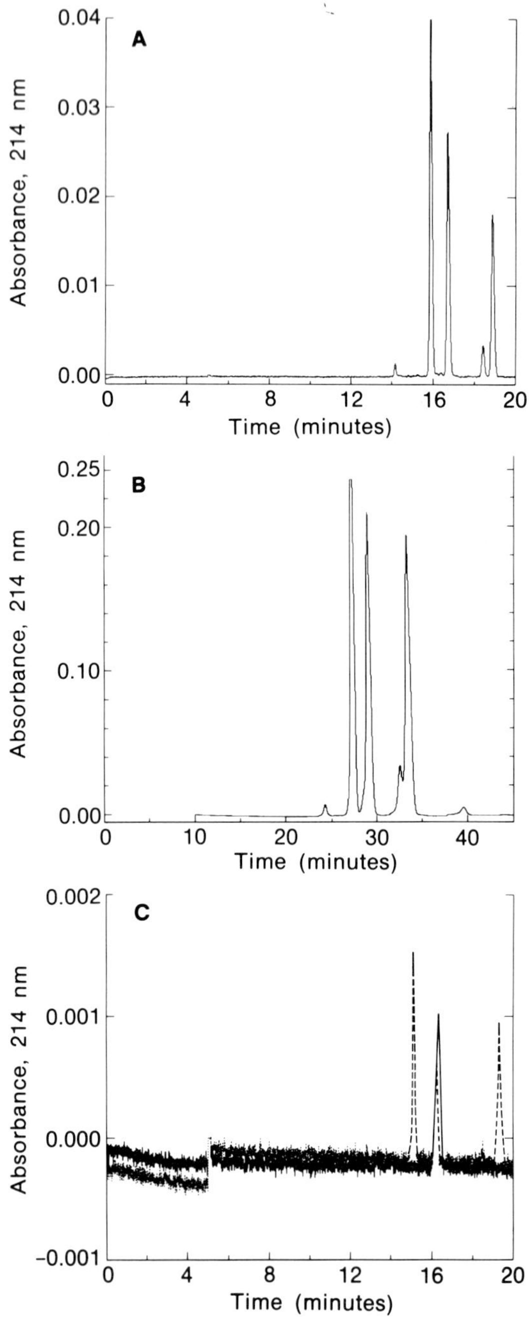

Figure 1. Capillary electrophoretic separation of test peptide mixture. **(A)** Analytical CE of test peptide mixture. **(B)** Micropreparative CE of test peptide mixture. **(C)** Analytical check of angiotensin I isolated by micropreparative CE.

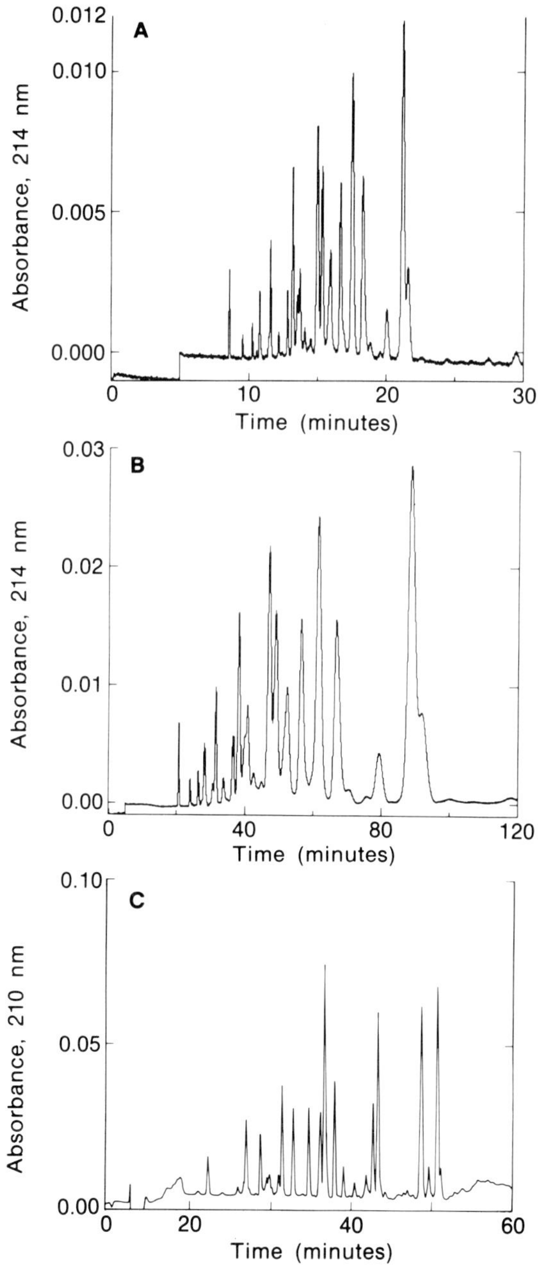

Figure 2. CE separation of a complex protein digest peptide mixture. **(A)** Analytical CE of the protein digest. **(B)** Micropreparative CE of the digest. **(C)** Narrowbore RP-HPLC analytical separation of the digest.

Initial collection over the first hour of the separation produced peptides of good yield and purity (Table II). However, these peptides all appeared to be short, positively charged species. In order to determine whether longer peptides could be recovered by this technique, we developed an alternative procedure. The separation was repeated, but fraction collection was not initiated until 50 minutes into the run and then continued for the next 72 minutes. Fractions were screened by analytical CE as before to identify fractions to be pooled for protein sequence analysis. The results shown in Table II clearly demonstrate the feasibility of this approach for obtaining peptides that elute later in the run. The levels of protein sequence recovered from these peptides represent the minimum recovery levels since only those fractions that appeared to contain single peaks were pooled prior to sequencing. Recoveries of peptides from approximately 100 pmol of digest were in the 10-75% range.

Table II. Recovery of Peptides from β-Lactoglobulin Tryptic Digest

Frac. #	Amino acid sequence	Yield pmol	Recovery %	Purity %
8	TK	70	70	>90
11	ALK	35	35	50
	TMK	35	35	50
13	LQK	50	50	80
15	FPK	40	40	>90
18	VYVEELKPTPEGDDLEILLQK	25	25	>90
21	TPEVDDEALEK	40	40	75
28	SLAMAASDISLLD(AQSAPLR)	35	35	65
	VAGT(W)Y	15	15	35

Conclusions

Analytical peptide maps of proteins have been well documented for both narrowbore RP-HPLC and free solution CE (Figure 2). The quantities required for these analyses are impressively small, *i.e.* in the low to sub-picomole range. However, the utilization of these techniques in a preparatory application is dependant on both the recovery of the peptides and on the sensitivity of the characterization method (protein sequencing). In our hands, adequate sequence information from narrowbore RP-HPLC separations requires the loading of at least 50 pmol of digested protein. Under these circumstances, we can expect to recover peptide sequences at the 5 to 10 pmol level. The micropreparative CE recovery of peptides from a test peptide mixture and from a tryptic digest of β-lactoglobulin are in the range of 10-75% at a 100-150 pmol load level.

Resolution on the micropreparative 200 μm capillary is adequate to obtain the same qualitative profile as that obtained with analytical capillaries. Using on-instrument collection, peptides are separated in comparable purity to those obtained from preparative HPLC separations. There is considerable spreading of peaks later in the CE micropreparative separation, which may require the pooling of more than one fraction. However, there appears to be minimal coelution of peptides in this region, and the sequence of the major peptide is easily identified.

Most peptides of interest for subsequent oligonucleotide production (larger fragments) occur in the later eluting region. For these peptides, we have modified our collection protocol to initiate fraction collection partially into the run. At present, using this approach, it is not possible to collect all of the peptides eluting from a complex digest during a single run.

The starting peptide concentrations we used in this preliminary study were in the μg/μL range. When more dilute samples were used, they were preconcentrated before running on CE. The samples were subjected to a mild drying and were resolublized on-instrument in a small (μL) volume of electrolyte by reversing the electrolyte flow of the P/ACE unit briefly immediately prior to injection.

The recovery levels of peptides from micropreparative CE at load levels of approximately 100-150 pmol are encouraging. Fractions obtained from micropreparative separations appear to contain relatively low levels of background contamination. This should permit sequence analysis at low picomole levels. Since the recovery of peptides in this study is well above this level, it should be possible to perform micropreparative CE separations at much lower levels than the current study.

References

1. Jorgenson, J. W., Lukacs, K. D. *Anal. Chem.* **53,** 1298-1302 (1981)
2. Guttman, A., Paulus, A., Cohen, A. S., Karger, B. L. *Electrophoresis* **'88,** Proc. Int. Electrophoresis Society, 6th, Copenhagen, 1988, pp.151-159. Edited by C. Schafer-Nielsen. Weinheim, VCH Verlagsellschaft, 1988.
3. Hecht, R. I., Coleman, J. F., Jr., Morris, J. C., Stover, F. S., Demarest, C. *Prep. Biochem.* **19,** 363-366 (1989)
4. Bergman, T., Agerberth, B., Jornvall, H. *FEBS Letters* **283,** 100-103 (1991)
5. Ohms, J. I., Kenny, J. W., Smith, A. *HPCE '91,* Poster Int. Symp. High Perform. Capillary Electrophor., 3rd, San Diego, 1991.
6. Gordon, M. J., Lee, K.-J., Arias, A. A., Zare, R. N. *Anal. Chem.* **63,** 69-73 (1991)

Estimation of Protein Isoelectric Points by Capillary Electrophoresis Using MICRO-COAT™

Michael F. Rohde
Kendall S. Stoney
Amgen, Inc., Thousand Oaks, CA

and John E. Wiktorowicz
Applied Biosystems Inc., San Jose, CA

Introduction

Information on the isoelectric points of proteins is important for a number of reasons. Among these are discrimination of potential isoforms of proteins of the same apparent molecular weight, as a guide to suitable buffers for handling and storage, as well as detection of deamidation or other unexpected changes in the charge state of a purified protein.

Traditional methods for obtaining the isoelectric point of proteins are well established, but are often labor intensive. The most commonly used method involves electrophoresis of the protein of interest in a polyacrylamide gel containing an assortment of smaller charged species (usually ampholytes) which are designed to produce a pH gradient; the use of urea as a denaturant is the common mode. Once the gradient is established and the proteins migrate to their position of no net charge, the gel is usually stained and destained to detect the proteins. To assign an isoelectric point, one must either measure the pH gradient in the gel at several points, or should include markers of known pI values in the same gel.

The ability of capillary electrophoresis to provide separations of charged molecules rapidly, with a minimum sample requirement, and in the absence of denaturant would make it a likely candidate for a simpler and quicker method. This potential is complicated by the

TECHNIQUES IN PROTEIN CHEMISTRY III

tendency of the capillaries above pH 4 to bind positively charged proteins via the unprotonated silanol groups of the fused silica wall. A solution to this complication has been offered via the method of reversing the charge on the capillary wall by applying a coating of a polycationic reagent MICRO-COAT™. When a capillary is treated in this manner, the walls become positively charged at neutral and low pH, allowing proteins below their pI to move through the capillary without being attracted to the wall via electrostatic interaction.

Materials and Methods

Capillary electrophoresis was performed on an Applied Biosystems Model 270A-HT instrument. Capillaries were conditioned for running with MICRO-COAT (Applied Biosystems, Foster City, CA) according to the procedures recommended by the manufacturer. A new capillary was conditioned with a 1 N sodium hydroxide wash for 20 min, followed by a 5 min wash with 0.1 N sodium hydroxide, rinsed with water for 5 min and running buffer for 5 min. The newly conditioned capillary was then coated with MICRO-COAT solution, and equilibrated with a final buffer wash. Run cycles were programmed to consist of a 1 min strip with 0.1 N sodium hydroxide, a 1 min replenishment with MICRO-COAT, a 3 min equilibration with running buffer, and vacuum injections of neutral marker and of protein. Solutions for capillary pretreatment, marker and samples were loaded into the 50 sample positions of the 270A-HT, leaving the 8 buffer positions for buffers of differing pH.

Proteins were obtained from Sigma Chemical and were either IEF standards or were of the highest purity available from Sigma; proteins were dissolved in water and clarified by centrifugation for 10 min in an Eppendorf Microfuge. Neutral marker was a 0.1% solution of Mesityl Oxide in water.

Buffers were made up from the corresponding acids or free amines and titrated with sodium hydroxide or hydrochloric acid. Citric acid and PIPES were used as buffering agents. To simplify data interpretation for a given set of analyses, the same buffering ions were used as a set, then the next pH range was done with a new set of buffer ions. Each was made isoconductive by dilution with water until all had the same conductivity; this allowed run times to be similar for the range of pH values in a set. Ethylene glycol (2% v/v) was added to the diluted buffers and the pH redetermined prior to running.

Data was collected and analyzed on a MicroVAX 3300 using Beckman PeakPro software. Migration times for the protein and marker peaks were combined to give a reduced mobility value; this was calculated as the difference between the inverse of the respective migration times. The isoelectric point is thus determined as the point of zero mobility, or net neutral charge.

Results and Discussion

Figure 1 is a combined plot of myoglobin run in PIPES buffers at five different pH values. The principles of the method are evident here; as the pH increases, the net positive charge on the protein decreases, and the peaks of the protein and the neutral marker move closer to each other. Between pH 7.41 and 7.65, the charge on the protein goes from net positive to net negative, and the myoglobin emerges before the neutral marker. Close examination of the electropherograms reveal a second minor species migrating slightly earlier than the main myoglobin peak. This peak crosses the neutral marker peak between pH 6.94 and 7.20.

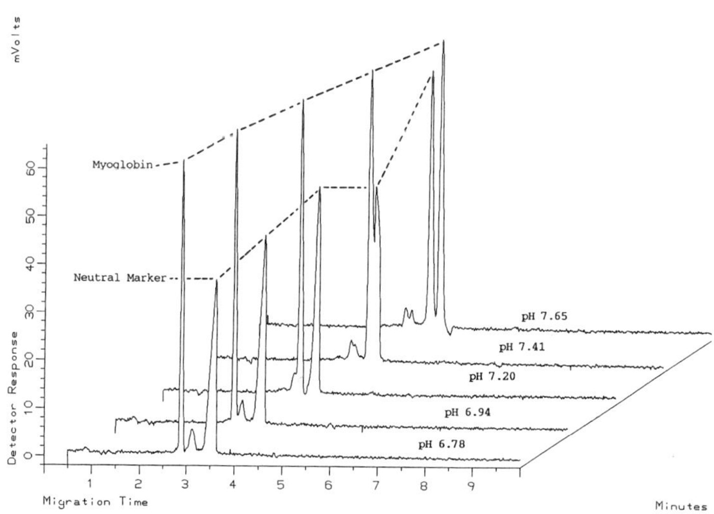

Figure 1 Capillary electrophoresis of Myoglobin and neutral marker between pH 6.78 and pH 7.65. Capillary was coated with MICRO-COAT and run at -30 KV.

Figure 2 graphs these data as mobilites versus pH. Here the isoelectric points are estimated as the pH where the mobilities change sign, from positive to negative. For the major myoglobin peak this is pH 7.5; for the minor species this is 7.1. These values are in close agreement with the pI values given by the supplier of 7.2 and 6.8.

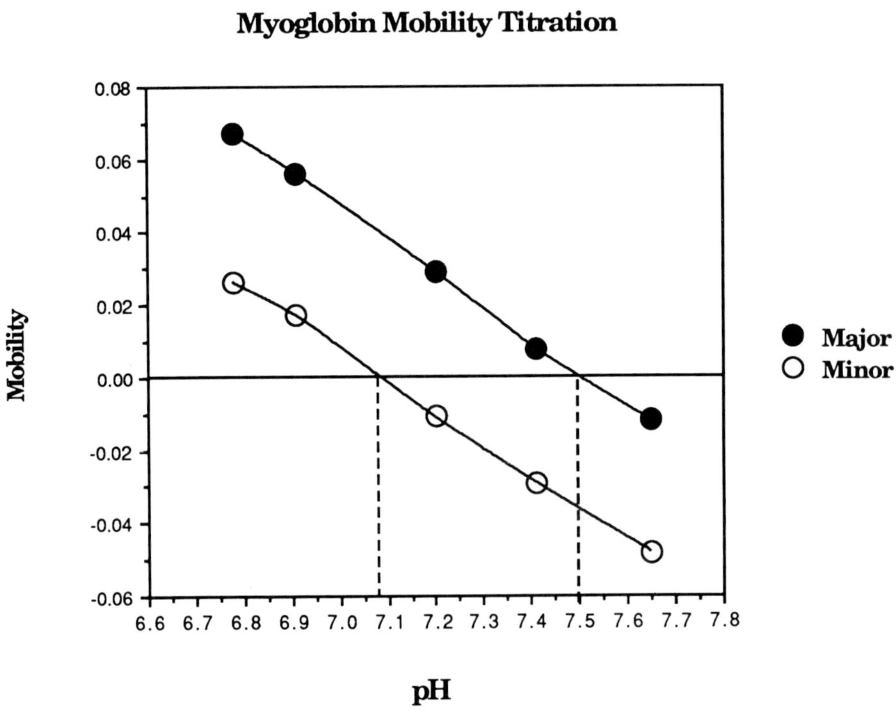

Figure 2 Plot of Myoglobin relative mobility as a function of pH. Data from Figure 1 was reduced from elution times to Mobility and plotted versus pH. The pI is determined as the point where the protein mobility goes to zero.

Four additional proteins were examined: Amyloglucosidase, Glucose Oxidase, Soybean Trypsin Inhibitor and beta-Lactoglobin A. Since these proteins had expected pI values between 3.0 and 5.5, citrate buffers were chosen for this range. Figure 3 contains the mobility plots for these four proteins, for the major species; in two cases, additional minor pI components were observed and are reported in Table I.

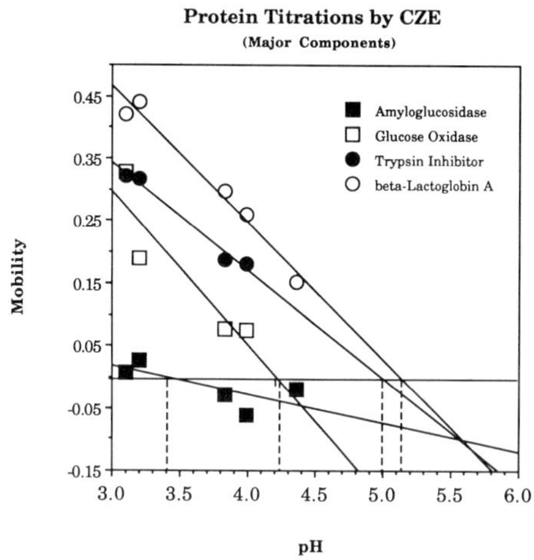

Figure 3 Isoelectric point determinations for four proteins. Mobility is plotted as a function of pH to obtain the pI where the mobility changes sign relative to a neutral marker.

Table I

Protein	Isoelectric Point CZE	Isoelectric Point Reported
Amyloglucosidase (*Aspergillus niger*)	3.4 & 3.6	3.6
Glucose Oxidase (*Aspergillus niger*)	4.2 & 5.4	4.2
Trypsin Inhibitor (Soybean)	5.0	4.6
beta-Lactoglobin A (bovine milk)	5.1	5.1
Carbonic Anhydrase I (bovine erythrocytes)	5.1	5.4
Carbonic Anhydrase II (bovine erythrocytes)	6.0	5.9
Carbonic Anhydrase I (human erythrocytes)	6.3	6.0
Carbonic Anhydrase I (human erythrocytes)	6.7	6.6
Myoglobin (horse heart)	7.1 & 7.5	6.8 & 7.2

Next, Figure 4 presents the mobility plots of four isoforms of Carbonic Anhydrase with differing isoelectric points. The buffers used in this set of experiments were PIPES buffers and all were above the isoelectric points of each of the isoforms. For the higher two pI isoforms, the buffers used were near the neutrality points and the extrapolation approaches linearity. The other two required longer extrapolations with a second order curve approximating the trend of the mobilities. While there is no physical basis for this type of curve fitting, there is good agreement with the predicted and observed pI values of all four isoforms.

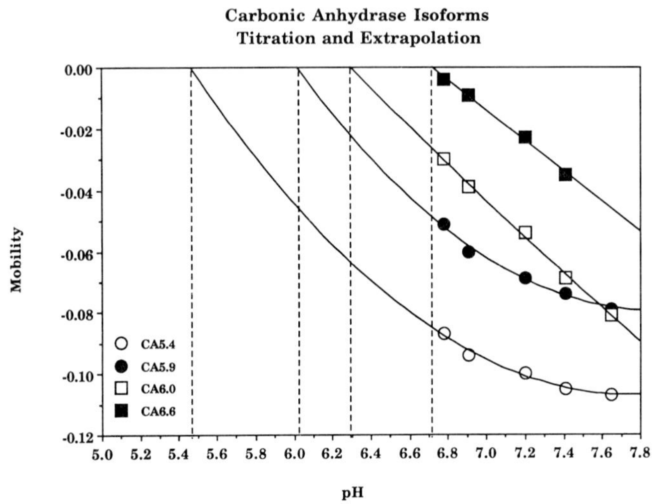

Figure 4 Extrapolation of Carbonic Anhydrase isoform mobilities to estimate their pI. The two upper curves were linear fits, while the two lower curves were second order polynomials.

The optimal use of the method would be the type of experiment where the pI of the protein is straddled by buffer solutions in a relatively narrow range of pH. This avoids long extrapolations which may give way to errors. Also it is important to realize that when the pH is far above the pI of the protein, the protein will carry a net negative charge which may be sufficient to cause it to bind to the coated capillary in a manner analogous to the interaction of positively charged proteins with an uncoated capillary. Lastly, it should be noted that some proteins may not benefit from the change in the surface charge on the capillary wall. Specifically, one can envision that proteins with uneven surface charge distributions may have large enough patches of titratable groups without opposing charges in that region on the surface; this would explain why a protein could have a net charge near neutrality and still not migrate near the neutral marker. However, if the method is used with these provisions in mind, quick estimates of protein isoelectric points are possible without the usual requirements of IEF gels, and has the potential to provide determinations of protein pI values in the absence of denaturant.

References

P. G. Righetti (1983) *Isoelectric Focusing: Theory, Methodology and Applications* (Eds. T.S. Work and R.H. Burdon), Elsevier, Amsterdam.

J. E. Wiktorowicz and J. C. Colburn (1990) Electrophoresis **11**, 769-773.

CAPILLARY ELECTROPHORESIS FOR PREPARATION OF PEPTIDES AND DIRECT DETERMINATION OF AMINO ACIDS

Tomas Bergman and Hans Jörnvall

Department of Chemistry I, Karolinska Institutet,
S-104 01 Stockholm, Sweden

INTRODUCTION

Capillary electrophoresis in different separation modes, here capillary zone electrophoresis, offers a powerful analytical approach for small peptides at the fmol level [1,2]. The short separation time (minutes) and low consumption of material can be conveniently used to test the purity of peptide fractions separated by HPLC [cf. 3]. Although the high resolution would make capillary zone electrophoresis extremely powerful for peptide purifications in structural analyses, the small amounts and volumes (nl) have been considered a difficulty for use in preparative work. We therefore studied the preparation of peptides by capillary zone electrophoresis for direct analysis of both amino acid sequences and compositions [4]. Highly reproducible separations are obtained, allowing repetitive runs for collection of sufficient material. Surprisingly, only four to five repeat runs can yield sufficient material for ordinary sequence analysis in gas-phase or solid-phase instruments. Utilizing sensitive phenylthiocarbamyl amino acid analysis [5], enough material for reliable total compositions is obtained in single runs [4]. In addition, we found capillary zone electrophoresis useful for direct determination of underivatized amino acids and of value for analysis of carboxypeptidase-liberated residues [4,6]. Detection at low levels, less than 200 fmol for aromatic amino acids, is possible and amidated amino acids can easily be distinguished from the unmodified forms [4].

These applications of capillary zone electrophoresis highlight the potential of the technique for direct peptide analysis. They indicate that capillary zone electrophoresis in the 1990s may become for sample preparation in protein structural analysis what HPLC has been during the 1980s.

TECHNIQUES IN PROTEIN CHEMISTRY III

MATERIALS AND METHODS

[^{14}C]carboxymethylated bovine insulin B chain was prepared as described [4]. Peptide digestions were performed with the Glu C endoprotease from *Staphylococcus aureus* [7] and with carboxypeptidase Y [6].

Capillary zone electrophoresis employed 50 mM phosphate buffer, pH 2.5. The system was a Beckman P/ACE 2000 equipped with an IBM PS/2 Windows-based control including the System Gold (Beckman) software. Capillaries were of fused silica with inner diameter 75 μm and total length 57 cm. The ultraviolet detector was positioned 7 cm from the outlet end. Samples were injected by pressure and electrophoresis was conducted at 20-25 kV. Temperature was thermostatically controlled at 25°C. In preparative runs, separated peptides were recovered by a programmed positioning at the capillary outlet of conically shaped microvials containing 5-10 μl phosphate buffer.

Amino acid analysis of peptides prepared by capillary zone electrophoresis was performed as described [5] and sequence analysis with an Applied Biosystems 470A instrument, utilizing PTH analysis by reverse phase HPLC [8].

RESULTS AND DISCUSSION

Sensitivity and resolution by capillary zone electrophoresis of a Glu C protease digest of [^{14}C]carboxymethylated insulin B chain is shown in Figure 1. Only 200 fmol peptide was injected and separation is complete within 15 min, confirming the usefulness of capillary zone electrophoresis for analytical purposes. However, the technique is useful also for direct analyses in preparative applications. Thus, an aliquot of the digest was concentrated 24-fold from initially 1.5 pmol/μl to 36 pmol/μl final solution. Since injections can be made from only a few μl of sample solution, concentration is easily accomplished through lyophilization and resolubilization in less volume of electrolyte buffer. Peptides 2 and 3 (Figure 1) were collected in four and five preparative runs, respectively (Table I). Amounts recovered were 19 pmol of peptide 2 and 24 pmol of peptide 3, showing that 5 pmol readily can be injected without overloading the system. The isolated peptides were directly applied to sequencer degradation and interpretation was possible for both peptides up to the C-terminus. The data from the sequence analysis reveal pure preparations and repetitive yields were > 90% (Table I). Consequently, provided an adequate sample concentration (about 40 pmol/μl) can be obtained, it is possible to perform sequence analysis of peptides prepared by a limited number of capillary zone electrophoresis runs. Similarly, reliable total compositions can be obtained utilizing phenylthiocarbamyl amino acid analysis with samples separated and prepared in single runs [4].

An interesting application tested is the separation and direct determination of underivatized amino acids. Ultraviolet detection at the

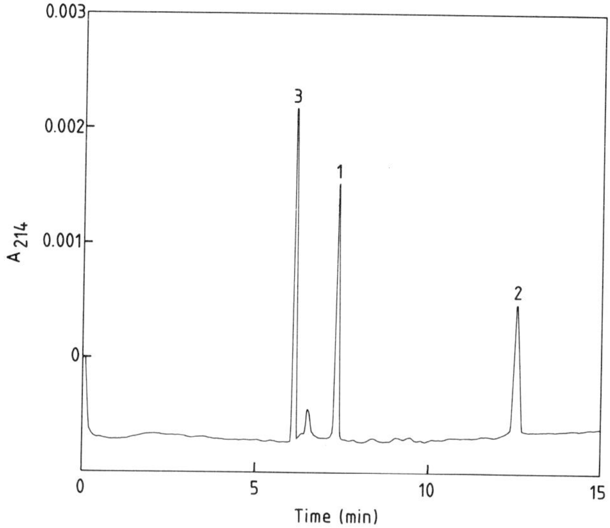

Figure 1. Separation by capillary zone electrophoresis of a Glu C endoprotease digest of [^{14}C]carboxymethylated insulin B chain. The peaks correspond to: 1, residues 1-13; 2, residues 14-21; 3, residues 22-30. Amount was 200 fmol injected by 30 sec pressurization. Conditions for electrophoresis: 25 kV and 25°C.

Table I. Results from N-terminal sequence analysis of peptides 2 and 3 (Figure 1) prepared by capillary zone electrophoresis. Before preparative runs, the digest (cf. legend Figure 1) was concentrated 24-fold by volume reduction. Conditions for capillary zone electrophoresis were as described (legend Figure 1). The total amount from each preparation was applied to a gas-phase sequencer.

	Peptide 2	Peptide 3
No of preparative runs	4	5
Amount obtained (pmol)	19	24
Initial coupling (pmol)	10	7
Repetitive yield (%)	91	92

pmol level or even in the fmol-range (aromatic amino acids) is possible [4]. Separation time is short and structurally similar amino acids or amino acids with modifications not always easily distinguished with conventional methods are well resolved [4]. The latter observation is shown in Figure 2 for Pro and $ProNH_2$ which are resolved and separated by more than 10 min. This direct analysis of amino acids and amides can be utilized for rapid interpretations of results from carboxypeptidase digestions for characterization of polypeptide C-termini. In Figure 3 this is demonstrated with a 39 amino acid porcine intestinal peptide [cf. 4,6]. Aliquots after different times of digestion were lyophilized and redissolved in electrolyte buffer followed by capillary zone electrophoresis. The released amino acids were identified by mixing of the aliquots with standards and re-separation. Phe, Arg, Pro and $ProNH_2$ could thus be detected (Figure 3).

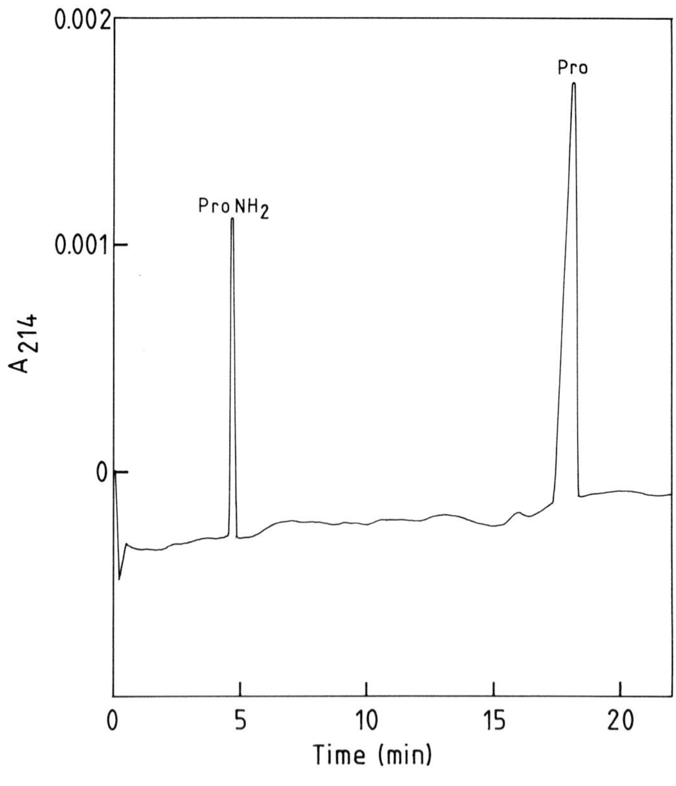

Figure 2. Capillary zone electrophoresis of proline and proline amide. Amounts: $ProNH_2$ 23 pmol; Pro, 190 pmol. Conditions: 20 sec pressure injection, 25 kV and 25°C.

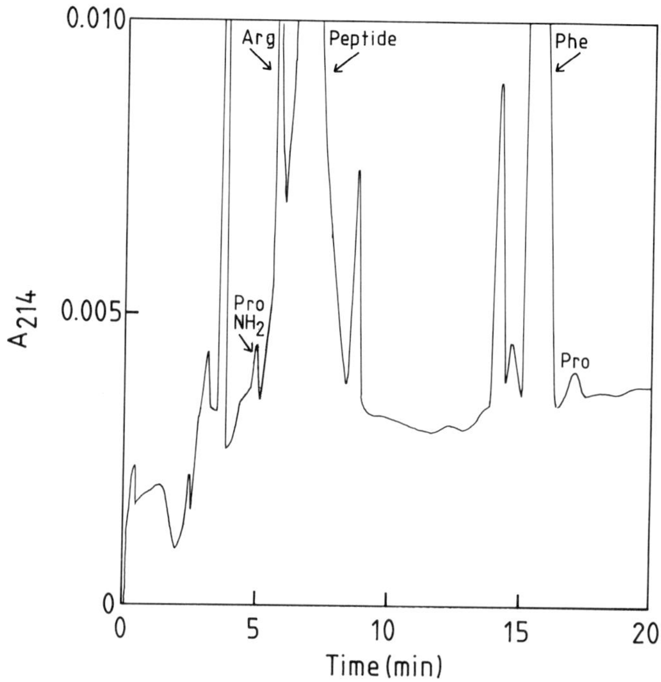

Figure 3. Analysis of an 1 h aliquot from carboxypeptidase Y digestion of a 39-residue peptide (50 pmol) by capillary zone electrophoresis. Conditions: 20 sec pressure injection, 25 kV and 25°C. For amino acid standard separations, cf. Figure 2 and [4].

CONCLUSION

Preparation of peptides by capillary zone electrophoresis for direct analysis of both amino acid sequences and compositions can readily be obtained through repetitive collections. Enough material for sequence analysis is obtained after just a few runs, while single runs suffice for total compositions. A further application of capillary zone electrophoresis is the separation and direct determination of underivatized amino acids. C-terminal residues from carboxypeptidase digestions can be analyzed in the pmol-range or for strongly absorbing aromatic amino acids, at the fmol level. Convenient distinction of amidated versus free residues is also obtained. All these applications demonstrate the usefulness of capillary zone electrophoresis for direct peptide analysis.

ACKNOWLEDGMENTS

This work was supported by grants from the Swedish Medical Research Council (project 03X-3532), the Swedish National Board for Technical Development (project 90-00439P), the Swedish Cancer Society (project 1806), the Knut and Alice Wallenberg Foundation, and the Bengt Lundqvist and Anders Otto Swärd Foundations.

REFERENCES

1. Gordon, M.J., Huang, X., Pentoney, Jr., S.L. and Zare, R.N. (1988) *Science* **242**, 224-228.
2. Karger, B.L., Cohen, A.S. and Guttman, A. (1989) *J. Chromatogr.* **492**, 585-614.
3. Large, T. (1991) In *Methods in Protein Sequence Analysis* (Jörnvall, H., Höög, J.-O. and Gustavsson, A.-M., eds.) Birkhäuser, Basel, pp. 79-90.
4. Bergman, T., Agerberth, B. and Jörnvall, H. (1991) *FEBS Lett.* **283**, 100-103.
5. Bergman, T., Carlquist, M. and Jörnvall, H. (1986) In *Advanced Methods in Protein Microsequence Analysis* (Wittmann-Liebold, B., Salnikow, J. and Erdmann, V.A., eds.) Springer, Berlin/Heidelberg, pp. 45-55.
6. Agerberth, B., Lee, J.-Y., Bergman, T., Carlquist, M., Boman, H.G., Mutt, V. and Jörnvall, H. (1991) *Eur. J. Biochem.*, submitted.
7. Chen, Z-w., Agerberth, B., Gell, K., Andersson, M., Mutt, V., Östenson, C.-G., Efendić, S., Barros-Söderling, J., Persson, B. and Jörnvall, H. (1988) *Eur. J. Biochem.* **174**, 239-245.
8. Kaiser, R., Holmquist, B., Hempel, J., Vallee, B.L. and Jörnvall, H. (1988) *Biochemistry* **27**, 1132-1140.

PERFUSION CHROMATOGRAPHY: A NOVEL TOOL FOR PROTEIN PURIFICATION AND ANALYSIS

Noubar B. Afeyan, Neal F. Gordon, Scott P. Fulton
and
Fred E. Regnier[+]

PerSeptive Biosystems, Cambridge, MA 02139
and
[+]Department of Biochemistry, Purdue University, West Lafayette, IN 47907

I. INTRODUCTION

Liquid chromatography (LC) is, without question, the most important and widely used technique for the recovery and isolation of proteins, peptides and other biomolecules. LC has proven to be not only highly selective, but also extremely flexible and very gentle. This unique combination allows good yield of both mass and biological activity together with extremely high purity.

However, LC as currently practiced has a number of critical limitations. Cycle times for chromatography are relatively long compared to other separations methods. This has tended to restrict the use of LC in time-critical applications such as on-line monitoring. In addition, the time required per run restricts the amount of work done in method development and optimization. Due to the complexity of chromatographic chemistry, many current applications, from routine analyses to full-scale production processes, are run under sub-optimal conditions. Finally, the restricted throughput of chromatography columns has to date limited the use of LC for selective adsorption of material from dilute feed solutions.

II. CHROMATOGRAPHIC MASS TRANSPORT

Most of the problems inherent in LC stem from restricted mass transport within packing materials.[6,7,8] In order to maximize the amount of available surface area (and therefore the binding capacity), highly porous particles with a high surface-to-volume ratio have generally been

TECHNIQUES IN PROTEIN CHEMISTRY III

employed. A primary challenge in chromatography is to find a means of rapidly and efficiently exposing the mobile phase and sample molecules to the entire available surface area within the column, including the area inside the pores.

In conventional "diffusion chromatography" media (Figure 1A), transport through the column to the outer surface of the packing particles is by rapid convective flow. Once the molecules reach the outer surface of the particles, however, mass transport to the inner surface area (where most of the binding capacity lies) is entirely by molecular diffusion, which is very slow relative to convection, especially for proteins and peptides.

Two serious problems can occur if the time required to diffuse into and out of the pores is much greater than the time required to flow past the particle. The first problem is a dramatic increase in bandspreading (loss in resolution), since the molecules still in the convective stream or binding near the outside of the particles will elute long before those that have diffused to the centers of the particles. The second problem is a loss in binding capacity, caused by the great range in access times to the active surface within the particles. It is possible for some of the sample to pass fully through the column well

Figure 1A: Schematic diagram of "diffusion-based" chromatography packing particle.

before all of the binding sites deep within the particles are occupied. Because of these problems, conventional chromatography has always involved a tradeoff between speed, capacity and resolution, primarily caused by the potential mismatch between mass transport rates inside (diffusion) and outside (convection) the particles.

III. PRINCIPLE OF PERFUSION CHROMATOGRAPHY

In Perfusion Chromatography™ flow-through particle chromatography, transport into the particles occurs through a combination of convection and diffusion in a way which eliminates the need for this compromise (Figure 1B). Transport to the surface of the particles is by convective flow through a packed bed, just as in a conventional column. The particles themselves, however, contain two distinct classes of pores—throughpores, large enough to allow some convective flow through the particles and smaller diffusive pores lining the throughpores, which provide high adsorptive surface area.

Above a relatively low critical flow rate, convection dominates over diffusion in the throughpores, allowing efficient access to the high surface area diffusive pores throughout the entire particle. The rapid "perfusive" transport within the throughpores and the ultra-short diffusion path lengths (<1 μm) of the diffusive pores combine to make both resolution and capacity essentially independent of flow rate. Detailed theory, experimental verification and comparisons of this technology with conventional chromatography have been published elsewhere.[1-3,5,13,14]

IV. CONVENTIONAL APPROACHES

Perfusion chromatography contrasts sharply with alternative ways used to solve the mass transport problems of chromatography. With conventional diffusion-limited media, the standard approach is to reduce the packing particle diameter, thereby reducing the diffusion path and thus the time required for molecules to diffuse in and out of the particles. This is the general principle of HPLC.[12,15] The constraint on this approach is caused by the fact that reducing the particle size dramatically increases the pressure

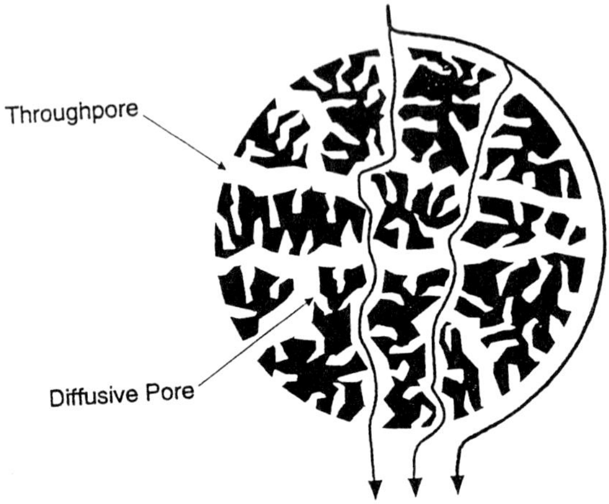

Figure 1B: Schematic diagram of Perfusion Chromatography packing particles showing throughpores for fast convective intraparticle mass transport and diffusive pores for high surface area and binding capacity.

drop needed to obtain a given flow rate. Current HPLC technology with 3-5 μm particles is at or near practical operating limits. These materials are quite useful for analysis but are not practical for large-scale, preparative applications.

Two more recent approaches to solving the problem of slow diffusive mass transport have been the development of non-porous media and so-called membrane chromatography. In non-porous media (Figure 1C) the packing contains no pores, and all the binding occurs on the outer surface, entirely accessible by rapid convective transport.[9,10,16] In order to partially reduce the loss in surface area, very small particles (1-3 μm) are used. This allows higher flow rates without significant band-spreading, but creates a serious pressure drop problem, as well as limited capacity. Often these columns are run at elevated temperature to reduce the mobile phase viscosity to allow reasonable flow. These factors have restricted the non-porous particle approach to analytical applications.

In membrane chromatography (Figure 1D), microporous or ultra-filtration membranes are surface-derivatized to give selective binding. As with non-porous particles, the surface area is entirely accessible via convective flow through the membrane, allowing for rapid binding and elution, but also suffering from low capacity per unit volume. The very short 'bed length' makes complete capture of binding molecules a problem.

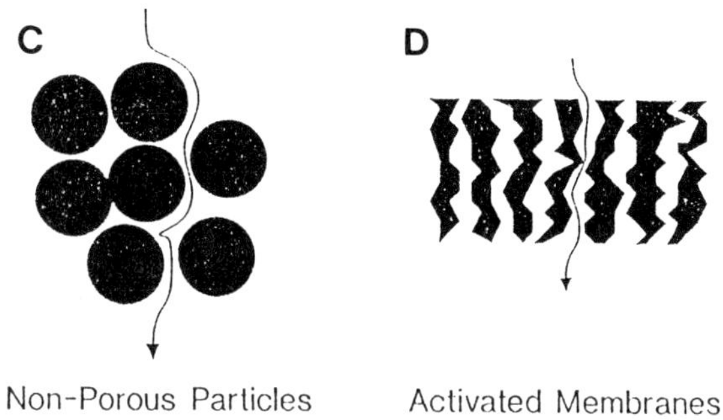

Figure 1C,1D: Schematic diagrams of "convection" chromatography media.

Membrane systems also usually have an extremely high ratio of extra-bed volume to bed volume, giving large dilution factors and high bandspreading, which severely limit application in high resolution separations. Flow rate declines due to membrane fouling and concentration polarization effects are also obstacles in this approach.

V. IMPLICATIONS OF FLOW-THROUGH PARTICLES ON CHROMATOGRAPHIC PERFORMANCE

In standard diffusion-based media, as flow rates increase, the bandspreading (or theoretical plate height) increases linearly (i.e. chromatographic efficiency decreases linearly). This classical behavior is due to the so-called "C-term" in the VanDeemter equation, which accounts for mass transport into and out of the particle. With Perfusion Chromatography media, this relationship between flow rate and bandspreading no longer applies. Once the rate of perfusive transport exceeds that of diffusion (typically below 300 cm/hr linear velocity), bandspreading becomes virtually independent of flow rate, up to at least 7000 cm/hr linear velocity[4-8] (figure 2). This contrasts sharply with the significant increase in bandspreading that occurs with conventional media. A comparison between a conventional 300Å Reversed-Phase separation of proteins with POROS perfusable supports is shown in figure 3. Clearly, the high speeds associated with the separation are not compatible with conventional supports.

An additional effect of convective Flow through the particle is that the plate height or bandspreading becomes dependent only upon the particle diameter, and not the particle diameter squared, as with conventional diffusive supports 5. This is shown in Figure 2, where 10 and 20 μm perfusive supports show the same reduced plate height (plate height divided by particle diameter), which remains constant with flow rate. This is especially important for large scale preparative applications, since it allows very high resolution to be achieved with relatively large particles that give low pressure drop at high flow rate.

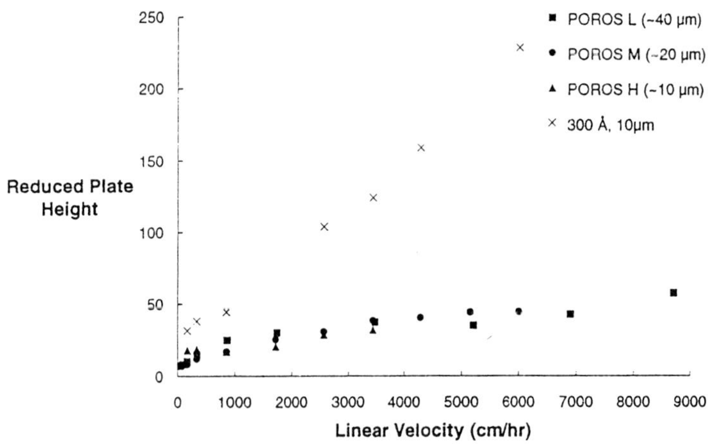

Figure 2: Reduced plate height vs. linear velocity for conventional (10 μm, 300 Å) and Perfusion Chromatography media of various particle sizes. All columns are polymeric reversed-phase and were tested with insulin under non-retained conditions.

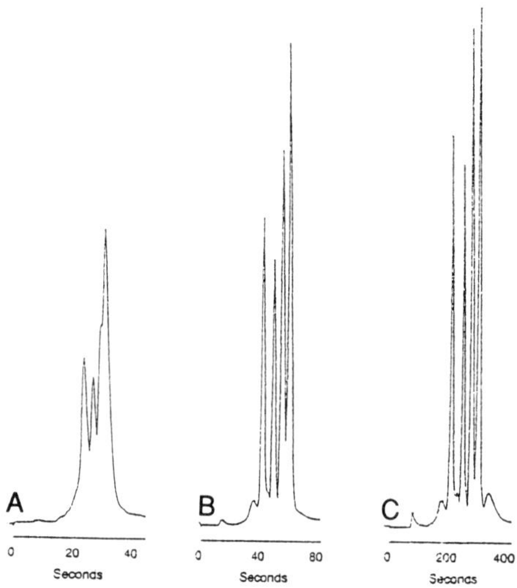

Figure 3: Comparative analytical separations with conventional perfusable supports. 3A shows the separation of a protein mixture on a 300Å polymeric reversed phase column at 1800 cm/hr. Figure 3B shows the same separation on a POROS column at the same speed while 3C shows the POROS run at 360 cm/hr (typical of conventional HPLC).

A:

Column: PLRP-S 300Å 4.6mmD/50mmL

B,C:

Column: POROS R 4.6mmD/100mmL
Sample: 1 Ribonuclease
2 Lysozyme
3 ß-Lactoglobulin A,B
4 Ovalbumin
Eluent: 0.1% TFA/water
0.1% TFA/acetonitrile
Gradient: 0-70% in 3 column volumes
Flowrate: A - 5ml/min
B - 5ml/min
C - 1ml/min

In preparative chromatography, throughput (amount of product produced per unit time) is as important as resolution. One important variable affecting throughput is the dynamic loading capacity, normally measured by applying a sample feed continuously to the column at a given flow rate until a significant concentration is detected in the eluent ("breakthrough"). The amount of feed bound divided by the column volume gives the unit dynamic loading capacity. This technique is called frontal chromatography.

Dynamic capacity, as measured by frontal chromatography, is expected to have a dependence on flow rate. It has been shown that with "diffusion-limited" media, increasing the loading flow rate causes substantial loss in dynamic capacity as feed molecules begin emerging prematurely from the column due to slow mass transport.[4,11] Perfusion media have virtually the same dynamic capacity as conventional supports at very low flow rates but show extremely sharp breakthrough curves with no change in dynamic capacity over a very broad flow rate range[1-3,5] (Figure 4).

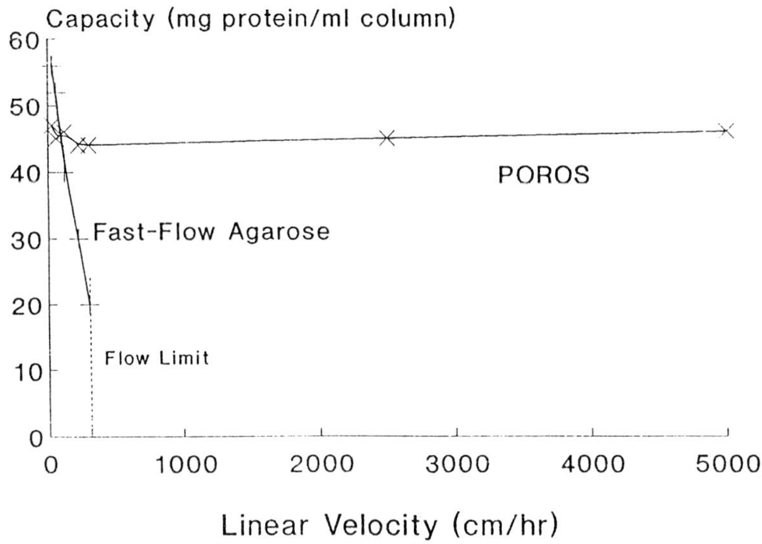

Figure 4: Dynamic binding capacity vs. linear velocity for conventional and Perfusion Chromatography media. The conventional support (fast-flow agarose, 50-100 μm) and perfusive support (20 μm) were both strong anion exchangers. Capacity was measured by frontal analysis using BSA as the test solute.

Table 1 summarizes the relative merits of each of the available approaches to chromatographic mass transport.

Table 1

	Speed	Capacity	Resolution	Scalability	Pressure
Low Pressure LC	-	+	-	+	Low
Conventional HPLC	+/-	+	+	+/-	High
Non-Porous HPLC	+	-	+	-	Very High
Membranes	+	-	-	+	Low
Perfusion	+	+	+	+	Low-High

VI. APPLICATIONS

A. High Speed Analysis and On-Line Monitoring

Virtually any analytical method benefits from higher speed, since answers can be obtained more quickly and the number of samples processed per unit time can be increased. Perfusion Chromatography enables the reduction of chromatographic analysis times from the current 3 60 minute range down to a few minutes or less without loss in resolution (Figure 5A).

In addition to high resolution gradient separations on reversed-phase or ion exchange columns, Perfusion Chromatography can also be used for rapid analysis in the affinity mode (Figure 5B). This allows rapid quantitation, for example, of antigens and antibodies. In addition to being far more rapid than conventional ELISA techniques, this technique is also extremely reproducible, in part because of the excellent repeatability and precision of HPLC pumps, injectors and detectors.

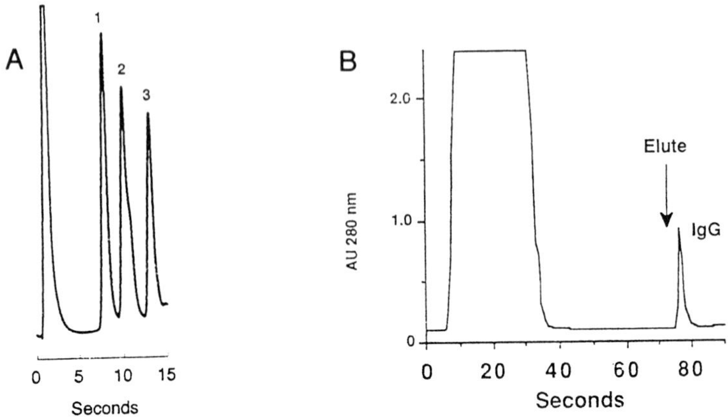

Figure 5: Analytical separations on Perfusion Chromatography columns. 5A shows an extremely rapid gradient reversed-phase separation. 5B shows a rapid affinity analysis of antibody concentration using a protein A column.

A:

Column:	POROS R/M 2.1 mmD/30 mmL
Sample:	10 μl 5 mg/ml each of 1. Ribonuclease 2. Cytochrome C 3. a-Lactoglobulin
Eluent:	A-0.1% TFA/water B-0.1% TFA/acetonitrile
Gradient:	20-60 %B in 0.2 minutes (10 column volumes)
Flow Rate:	5.0 ml/min (9000 cm/hr)
Detection:	OD 220 nm

B:

Column:	POROS A/M 2.1 mmD/30 mmL
Sample:	1 ml cell culture supernatant
Buffer:	10 mM phosphate pH 7.4, 0.15 M NaCI
Eluent:	0.3 N acetic acid (2% v/v), 0.3 M MgCl2
Flow Rate:	2.0 ml/min (3600 cm/hr)
Detection:	OD 220 nm

B. Rapid method Development

The capability to complete runs in a few minutes also significantly affects the development of chromatographic methods. Entire studies of variables such as mobile phase composition, sample loading, etc. can be completed in the time ordinarily taken for a single run. Figure 6A shows a rapid gradient run used to measure the binding strength of two test proteins. When many of these runs are done in sequence at different pH's, the binding strength can be plotted (Figure 6B) as a function of pH for the protein of interest (in This case a monoclonal IgM antibody) and for the major contaminants (in this case transferring and insulin). These "pH maps" are extremely useful in the development of a purification or analytical process. However, the time required to produce them by conventional methods is quite prohibitive.

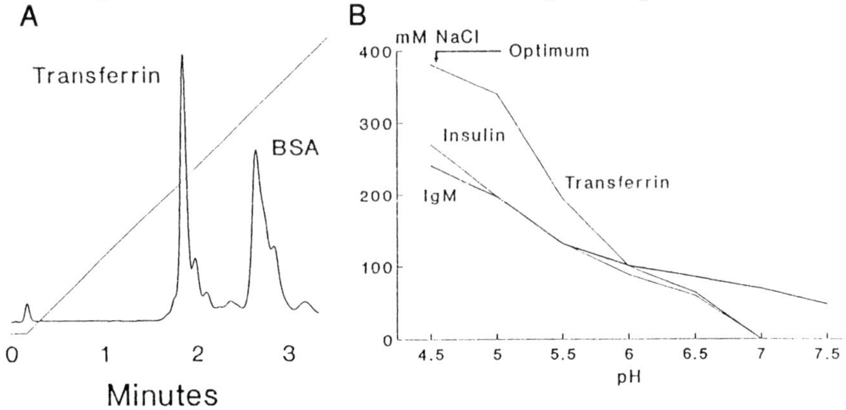

Figure 6: Development of a pH map for an IgM antibody purification on POROS S/M strong cation exchanger. 6A shows a typical analytical-scale gradient run at pH 4.5 run with insulin and transferrin, the two major contaminants. Similar runs were done at 0.5 pH intervals from pH 4.5 to 7.0, and with a purified antibody standard. Figure 6B shows the pH map constructed by plotting the NaCl concentration at elution as a function of pH for each of three tested proteins. The optimal pH was the one with the greatest difference in binding strength (NaCl concentration for elution) between the antibody and the contaminants.

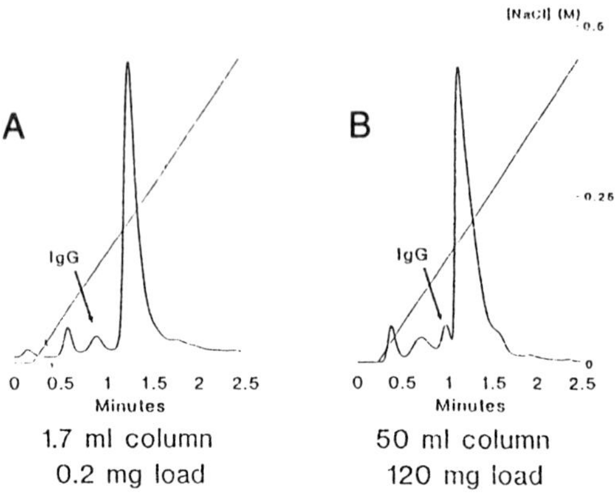

Figure 7: 600x scaleup from analytical to preparative of high speed, high resolution anion exchange purification of IgG from hybridoma cell culture supernatant on POROS Q/M columns. The sample load per ml column volume was increased 20-fold and the column volume increased 30-fold.

Columns: A- 4.6 mmD/100 mmL (1.7 ml) POROS Q/M
B-25.4 mmD/100 mmL (50.7 ml) POROS Q/M
Sample: 4 mg/ml total protein hybridoma cell culture supernatant
Loading A--50 μl (0.2 mg total protein)
B-30 ml (120 mg total protein)
Eluent: 20 mM Tris/HCl pH 8.0
Gradient: 0-0.5 M NaCl in 25 ml (15 column volumes)
Flow Rate: A--10 ml/min (3600 cm/hr)
B--340 ml/min (4000 cm/hr)
Detection: OD 280 nm

D. Dilute Feed Adsorption (Capture)

The limited throughput of conventional chromatography, especially with soft gels, has caused a need for upstream preconcentration steps (usually using ultrafiltration or precipitation) for large volume, dilute feed solutions such as cell culture supernatants. These steps both add time and cause product loss, while adding little to product purity. With Perfusion Chromatography columns, the combination of high flow rates with very high capture efficiency allows concentration and preliminary purification to be achieved in a single, rapid step.

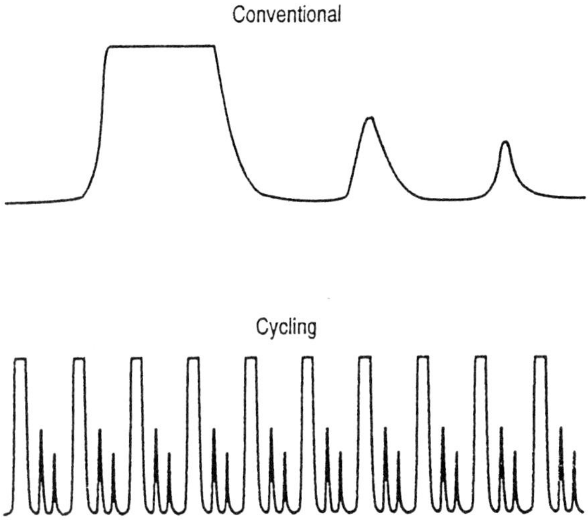

Figure 8: Comparison of a conventional chromatographic process with a rapid cycling process. In rapid cycling chromatography a smaller, very high speed column is used repeatedly in a given time period to produce the same amount of product as a larger, low speed column can produce in the same period. In this example, a rapid cycling column 10 times faster can be 1/10 the size and still have the same throughput.

Figure 9 shows an example with a test protein (BSA). In this case a 20 ml column was able to capture all the protein (0.3 g) from 3 liters of dilute feed prior to breakthrough. This was accomplished in less than 15 minutes, operating at less than 7 bar (100 psi) pressure. This same column could also perform the same application in less than 30 minutes using a peristaltic pump.

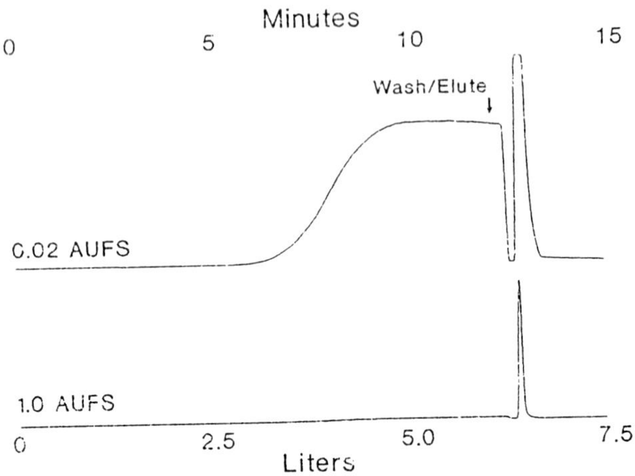

Figure 9: Frontal uptake curve of dilute protein solution on POROS anion exchange column. The tracing at 0.02 AUFS shows the frontal breakthrough of the feed stream while the tracing at 1.0 AUFS shows the concentration of the eluted peak. Concentration factor was approximately 50-fold.

Column: 50 mmD/10 mmL (20 ml) POROS Q/M
Sample Feed: 0.1 mg/ml bovine serum albumin in 20 mM Tris/HCl pH 8.0
Eluent: 0.5 M NaCl in buffer
Flow Rate: 500 ml/min (1500 cm/hr)
Pressure: 6.7 bar (100 psi)
Detection: OD 280 nm

VII. CONCLUSIONS

Perfusion chromatography represents a fundamentally new approach to solving the "mass transport problem" of liquid chromatography. Perfusion allows the use of 10- to 100-fold higher flow rates (and lower cycle times) without significant loss in either resolution or capacity compared to conventional soft gel and HPLC columns. This technique can be applied to all conventional adsorption modes of chromatography for proteins and peptides, including ion exchange, reversed phase, hydrophobic interaction and affinity. Perfusion chromatography has significant implications for analysis, on-line monitoring, method development, scaleup and processing of dilute feed streams.

REFERENCES

1. Afeyan, N. et al., Bio/Technology,8:203 (1990).
2. Afeyan, N. et al., J.Chromatogr.519:1(1990).
3. Afeyan, N., Fulton, S. and Regnier, F. J. Chromatogr. 544:267 (1991).
4. Chase, H., J. Chromatogr.297:179 (1984).
5. Fulton, S., Afeyan, N., Gordon, N. and Regnier, F. J., Chromatogr.547:452 (1991).
6. Giddings, J. C., Dynamics of Chromatography, Part I, Principles and Theory. Marcel Dekker, New York (1965).
7. Horvath, Cs. and H.-J. Lin, J. Chromatogr. 149:43 (1978).
8. Huber, H., Ber. Bunsenges. Phys. Chem. 77:179 (1973).
9. Kalghatgi, K. and Cs. Horvath, J. Chromatogr. 398:335 (1987).
10. Kalghatgi, K. and Cs. Horvath, J. Chromatogr. 443:343 (1988).
11. Kopaciewicz, W. et al., J. Chromatogr. 409:111 (1988).
12. Majors, R., Anal. Chem. 44:1722 (1972).
13. Regnier, F., Nature 350:634 (1991)
14. Spalding, B.J., Biotechnology 9: 229 (1991).
15. Unger, K. et al., J. Chromatogr. 99:435 (1974).
16. Unger, K. et al., J. Chromatogr. 359:61(1986).

Comparison of Coupling Procedures for Development of Affinity Membranes:Optimization of the CDI Method

Pamela W. Feldhoff
Department of Medicine, University of Louisville School of Medicine, Louisville, KY

I. INTRODUCTION

In developing affinity membranes, the goal is to confer biospecificity to membranes with various permeabilities. There are variables which must be considered in order to produce an optimal matrix for the isolation of biomolecules based on general or specific ligand:ligate interactions. These variables include choice of polymer and membrane configuration; the chemistry of membrane activation and ligand attachment; the biochemistry of the ligand:ligate interaction and the complexity of the feed solution.

The ideal membrane would be hydrophilic to eliminate non-specific adsorption and uncharged to prevent non-specific ionic interactions. The required pore size of the native membrane is dependent on the complexity of the starting material and the size of the ligand:ligate pair. A microporous membrane with large pores, high porosity and thick walls would be ideal in providing high surface area while still permitting permeation of very large proteins. An extended surface is needed to provide binding sites for ligands which can be accessed in flow conditions. Additionally, the membrane must have sufficient functionality for ligand binding and be stable to anhydrous solvents used for activation and covalent attachment of the ligand.

A reactive and efficient linking chemistry is essential in the development of affinity membrane technology. Two chemical processes are required to chemically immobilize a ligand onto an inert matrix. First, a reactive group of defined chemical functionality is introduced into the

TECHNIQUES IN PROTEIN CHEMISTRY III

inert matrix. The reactive group can then function either as a leaving group, or participate in the covalent attachment of the ligand to the matrix. The first step, matrix activation, is accomplished by introducing an electrophilic group into the matrix to make the membrane more reactive toward the nucleophiles of the ligand. I have utilized three different activation chemistries to introduce reactive groups into polyamide and cellulosic membranes: carbonyldiimidazole (CDI) (1), 2 fluoro-1-methyl pyridinium (FMP) (2) and cyanuric chloride ($CyCl_3$) (3). Each leads to the formation of a different type of covalent linkage between the ligand and the matrix.

The CDI reacts with the membrane to give an imidazolyl intermediate. The imidazole ring is then displaced by the amine function of a protein ligand (Fig 1). CDI activation yields a covalent N-alkyl carbamate (urethane) linkage between the amine of the protein ligand and a cellulosic polymer and an urea linkage between the ligand and the polyamide polymer. FMP, also, reacts with a cellulosic to form a polymer intermediate (Fig 2). The ring structure of the FMP acts as a leaving group and is displaced by the N nucleophile of the peptide or protein to yield a stable amide bond between the carbon of the matrix and the amine of the ligand. Thiols of the ligand will also bind to the FMP-activated matrix. In the case of $CyCl_3$ (Fig 3), a chlorine is displaced in the activation of the matrix and the $CyCl_2$ is attached to the matrix. A second chlorine then acts as a leaving group when the ligand is attached to the activated matrix. The third chlorine is displaced with ethanolamine to eliminate unwanted reactions, leaving the inert ring structure of the CyCl as part of the matrix-ligand linkage.

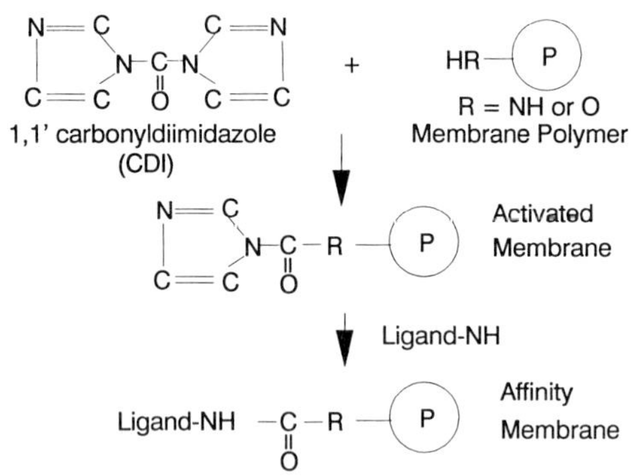

Figure1. Carbonyldiimidazole activation chemistry.

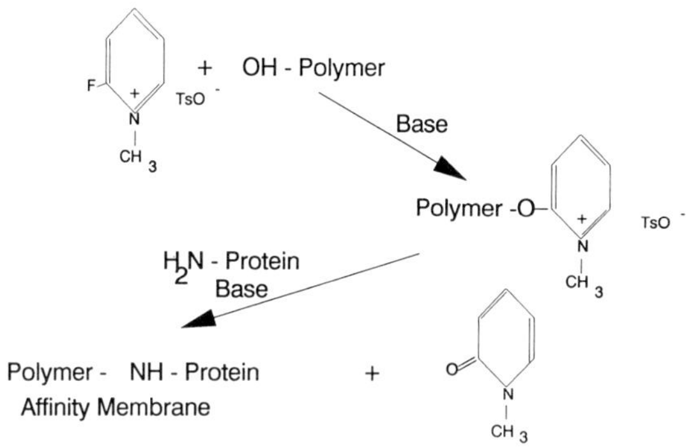

Figure 2. 2-fluoro-1-methyl pyridinium toluene-4-sulfonate activation chemistry.

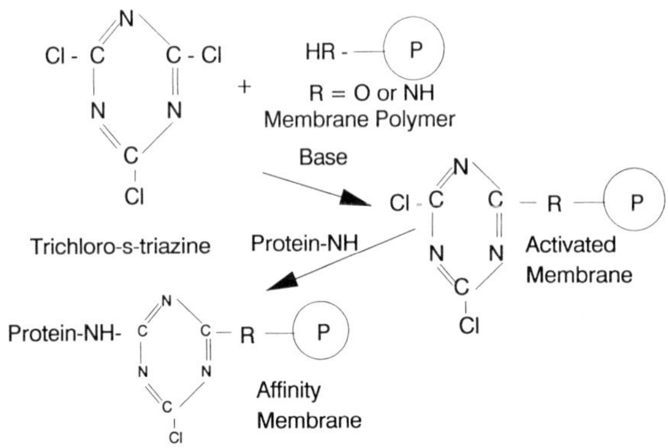

Figure 3. Cyanuric chloride activation chemistry.

II. METHODS

General Procedure for CDI Activation of Membranes for Ligand Coupling:

1. Sheet membrane reactions were carried out in pyrex dishes with shaking. Membranes in a hollow fiber configuration were placed in a flow-through tubular column (borosilicate, with polypropylene fittings and a porous polyethylene disc base support), with solutions circulated from the base of the column at 1.5 ml/min. All activation procedures were at room temperature (RT).

2. Membranes were rinsed with H_2O, stepwise with acetone/H_2O and finally dry acetone. How the membrane is dehydrated and later rehydrated (Step 4) is determined by the physical characteristics of the membrane and the integrity of the pore structure.
3. Activation: 0.1 M CDI in dry acetone, 30 min.
4. Rinse with acetone to remove CDI. Depending on the solubility and stability of the ligand, the attachment of the ligand can be carried out directly from organic solvent or from aqueous buffer. In the latter case, the membrane can be rehydrated stepwise with acetone/H_2O or in one step with H_2O, saline or 0.01 M bicarbonate buffer, pH 8.5. The imidazolyl carbamate intermediate is water-labile. Therefore, exposure to aqueous environment or buffer solutions without ligand should be limited.
5. Attachment of ligand or spacer: The membrane is reacted with 1% ligand in 0.1 M bicarbonate buffer, pH 8.5 overnight at RT or at 4°C depending on the stability of the ligand. The pH range for ligand attachment to CDI activated membrane is 8 to 10. A lower concentration of ligand is used for ligands >55 KDa, e.g., IgG, 0.5%; IgM, 0.1%.
6. Wash membrane with 0.1 M bicarbonate buffer, pH 8.5; H_2O; 0.1 M acetate buffer, pH 4.0; H_2O and phosphate-buffered saline/0.02% sodium azide (PBS). The washed membrane is stored in PBS or working buffer containing azide at 2-8 °C.

FMP Activation of Membranes for Ligand Coupling:

Steps 1 and 2 as for CDI activation

3. Activation: 0.3 g FMP in 15 ml dry acetone/0.15 ml triethylamine (TEA) is added slowly over 15 min to membrane in 100 ml dry acetone/1 ml triethylamine. React at RT for 60 min.
4. Rinse stepwise with acetone/H_2O/2 mM HCl solutions. Store in 2 mM HCl at 4°C.

Steps 5 and 6 as for CDI activation. The pH range for ligand attachment to FMP activated membrane is 6.5 to 9.5, optimum 8 to 8.5.

Cyanuric Chloride Activation of Membrane for Ligand Coupling:

1. Wash membrane with 1 N NaOH, 15 min, and drain.
2. Activation: Add washed membrane to $CyCl_3$ acetone solution (1g/25 ml) at 50°C, add 25 ml H_2O, 50°C. React 5-60 min, quench with 20% acetic acid.
3. Attachment of ligand and washes as for CDI activation (steps 5 and 6). The pH range for ligand attachment to $CyCl_3$ activated membrane is 9 to 11. Attachment can be carried out at a pH >8.6 for shorter time periods if the ligand is stable to conditions.

III. RESULTS AND DISCUSSION

The stability of these three types of linkages leads to the production of affinity membranes which retain capacity and effective long term operation. Additionally, these chemistries do not introduce charged groups, thus eliminating ion exchange effects which can cause non-specific binding. When charged membranes are used for affinity separations, such non-specific binding (ion exchange) reduces product purity.

The favored method of activation for formation of stable non-ionic bonds between the matrix and the ligand will depend on the reactivity, stability and local ionization atmosphere of the particular membrane/ligand. All three reagents are chemically reactive yielding efficient activation, form stable covalent linkage between the ligand and the matrix and do not introduce any charge or hydrophobic groups to the matrix. For these reasons the methods are superior to alternative coupling chemistries utilizing CNBr, tresyl chloride, chloroformate or hydroxysuccinimide.

A study of ligand attachment was conducted using different concentrations of CDI. Results of 3 such experiments utilizing 1,6-diaminohexane (DAH) as ligand show the reaction to be reproducible and the optimal concentration of CDI to be 0.1 M (Fig 4). A time course of CDI activation showed that 15 min is adequate and ligand addition was not improved by longer CDI activation times. This result was found to be independent of ligand molecular weight and pI.

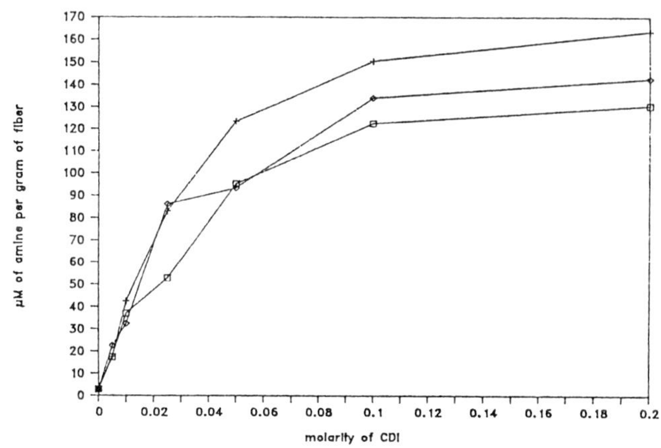

Figure 4. Carbonyldiimidazole activation utilizing increasing concentrations of CDI for 15 min. Extent of activation quantified by ninhydrin analysis of the covalently-attached 6 carbon spacer, 1,6-diaminohexane.

In the case of all three activation agents the matrix activation is carried out in a non-aqueous environment. To date, we have utilized acetone, dioxane, acetonitrile, cyclohexane, freon or DMSO as solvents. All lead to satisfactory activation. The concentration of activating agent, the solvent used for activating the fiber and the time and temperature of activation are variables to consider in order to achieve an activation level which will yield the best protein capacity and greatest purification factor. The integrity of the membrane must be considered in choosing the solvent and the conditions of activation so that good mechanical and chemical stability to flow and elution conditions are maintained. The freely permeable pore structure of the matrix should not change due to cross-linking of matrix surface or pore collapse. Non-specific binding, integrity of the fiber to the chosen chemistry and manual manipulation, and stability of both the activated intermediate and the ligand matrix linkage must be assessed to maximize yield, integrity and purity of the desired product.

The density of ligand attachment to the fiber will depend on the concentration of activated functional groups of the fiber, the size of the ligand, the concentration of ligand offered and the reaction efficiency of the matrix intermediate with the ligand. The groups of the polyamide matrix have been amplified by covalent attachment of a neutral hydrophilic carbohydrate layer, increasing the matrix affinity capacity. The chemistry of this functional-group amplification utilizing $CyCl_3$ in the initial activation is illustrated in Figure 5A. The polyglucose layer can then be activated and ligand attached (Fig 5B). Alternatively, the polyglucose can be activated and added to the polyamide matrix.

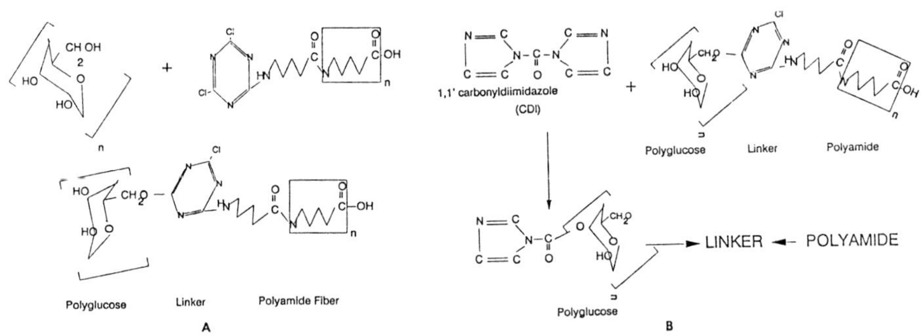

Figure 5. Amplification of the functional groups of the polyamide matrix. Polyglucose is covalently bound to the activated matrix. Activation with cyanuric chloride is illustrated (A). The amplified polyamide matrix is then activated for ligand attachment. Activation with carbonyldiimidazole is illustrated (B).

Ligand attachment to the activated matrix can be carried out in an aqueous environment. The three chemistries described allow ligand to be coupled under mild conditions which maintain the biological activity of the ligand yet afford high binding capacity. The coupling reaction does not denature, cross-link or hydrolyze the ligand. However, multi-point attachment of ligand to the activated matrix sites may occur leading to improper orientation or altered conformation of the ligand. This could decrease function. It is important to note that the most highly activated matrix may not provide the optimal protein capacity or lead to the most efficient isolation of ligate, i.e., purification factor. The degree of matrix activation that will give optimal ligand attachment and ligand:ligate interaction will depend on the (molecular weight and the number of groups available for attachment to the activated matrix) and the nature of the ligand:ligate interaction. The desired density and orientation of the ligand is that which will allow good ligate interaction with the ligand. Time, temperature and pH can be manipulated to optimize the ligand attachment. The described activation chemistries allow for fast coupling of the ligand to the activated membrane, minimizing protein:protein crosslinking or chemical interactions. A time course of DAH attachment utilizing two concentrations of CDI indicated that a long incubation, traditionally 16-24 hr, of activated fiber with the ligand is not necessary to displace the imidazole ring and bind the ligand. Results indicate that 15 min is adequate for DAH attachment (Fig 6). However, the displacement/ attachment reaction is dependent on the molecular weight and pI of the ligand.

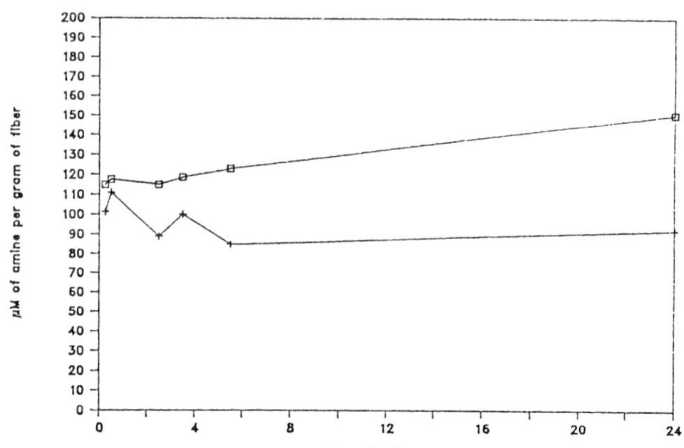

Figure 6. Time course of small ligand attachment to activated membrane. Activation with 0.05 M (+) or 0.1 M (□) CDI for 15 min, ligand is 1,6-diaminohexane (180 daltons), quantified as in Figure 4.

A time course of bovine serum albumin attachment indicated that 4 hr was necessary for maximal attachment (Fig 7).

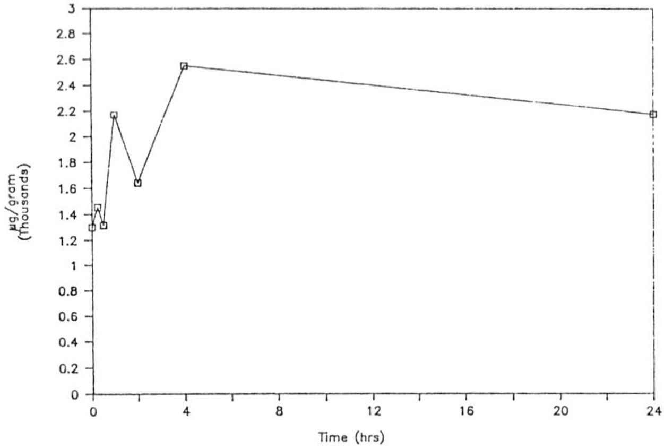

Figure 7. Time course of large ligand attachment to activated membrane. Activation with 0.1 M CDI for 15 min, ligand is bovine serum albumin (66,200 daltons), attachment quantified by BCA protein analysis.

The stability of the modified membrane with storage over a 6 month period is shown in Figure 8 for 3 independently processed samples. The results indicate good stability independent of ligand density.

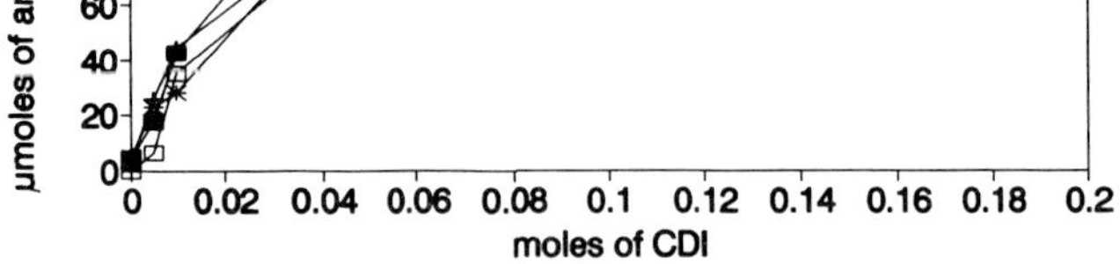

Figure 8. Stability of cellulose-1,6-diaminohexane membrane. Cellulose membranes were activated with a range of CDI concentrations for 15 min and the ligand attached to the activated membrane from 0.1 M bicarbonate buffer, pH 8.5. The membranes were then washed extensively and stored at RT in saline/0.02% sodium azide. Diaminohexane was quantified by ninhydrin immediately (■), and at 0.5 (+), 2 (✱) and 6 (□) months.

The ligand:ligate affinity interaction must be biospecific and reversible. In conferring specificity to microporous membranes the ligand:ligate interaction can be a general/class type (i.e., Protein A:IgG, lectin:sugar) or be more narrow and defined in nature (i.e., antigen: antibody, enzyme:inhibitor, protein:receptor). Once the ligand sites are saturated, the affinity membrane can be cycled by eluting the ligate, equilibrating with the starting buffer or ionic strength, and reloading the membrane. The elution method depends on the ligand:ligate interaction. Elution should be under conditions that do not denature the ligate. Since the ligand:ligate interaction is based on stereospecific recognition the ligate can be eluted by altering the conformation. Common methods utilized are to increase eluant salt concentration, to decrease pH, or to displace the ligate with an excess of a competitive analogue. If pH is used to elute the ligate the pI of the ligand and ligate should be avoided and the eluant should be immediately neutralized.

Depending on the ligand:ligate involved it may be desirable to extend the ligand from the fiber. The ligand can be attached to the matrix through a hydrocarbon chain of 2-10 carbons with a reactive terminal group. Examples of such spacers are 1,6-diaminohexane, 6-aminohexanoic acid and adipic acid dihydrazide. By limiting the length of the hydrocarbon chain the hydrophobic nature of the spacer will be masked by the covalently attached ligand. The spacer is covalently attached to the matrix by one of the chemistries described leaving, respectively, a free amino, carboxyl or hydrazine group. The ligand is then covalently linked to the spacer. If utilizing 1,6-diaminohexane or 6-aminohexanoic acid the ligand is linked to the spacer by a condensation reaction utilizing a water soluble carbodiimide, such as 1-ethyl-3-(3-dimethylaminopropyl) carbodiimide (EDAC). Adipic acid dihydrazide is often used as a spacer in the attachment of glycoproteins to an affinity matrix through the sugar moiety of the carbohydrate portion of the glycoprotein. The hydrazine reacts with the sugar aldehyde and is then reduced with sodium cyanoborohydride. This spacer has been used successfully with IgG, providing IgG attachment without interfering with antigen recognition. Unreacted spacer sites are blocked with ethanolamine if a carboxyl is used to couple ligand or acetic acid if an amine is used to couple ligand. The choice of spacer is determined by the biospecificity of the ligand:ligate interaction. In most interactions any of the spacers would allow proper orientation of the ligand for interaction with the ligate. However, in some cases, there is a strict requirement for availability of the ligand carboxyl or amino terminal for ligate binding and this would dictate the choice of spacer. If there is no orientation requirement, then 6-aminohexanoic acid should result in the most efficient coupling of amine containing ligands to the matrix. This spacer is covalently linked to the matrix and activated with EDAC on the matrix, excess EDAC washed out and the ligand attached. The ligand is not present in solution with the EDAC during the condensation reaction, the carboxyl groups of the ligand are not activated and there is no possibility of ligand crosslinking. This leads to better coupling efficiency of ligand to the fiber. The use of a spacer is critical in ligand:ligate interactions of low affinity or when a small molecular weight ligand, such as a hapten, interacts with a large ligate, such as an antibody. Such a spacer allows for better ligand:ligate interaction to increase ligate yield and shorten residence time. Additionally, the matrix-spacer intermediate yields alternate functional groups on the matrix allowing alternate attachment chemistries, so one is not limited to the chemistry of the base polymer.

Affinity membranes combine specificity of a biological molecule with transport properties of thin film membranes allowing high separation efficiency with high throughput. The convective high speed separations of membranes avoid the diffusive mass transfer limitations of gel chromatography, facilitating rapid large scale separations. The possibility of cross-flow operation, combined with the ability to handle dispersions, may allow one to condense multiple purification steps to a few simple steps which could be easily scaled-up for industrial processes.

REFERENCES

1. Bethell GS, Ayers JS, Hancock WS, Hearn MTW. (1979) A novel method of activation of cross-linked agaroses with 1,1'-carbonyldiimidazole which gives a matrix for affinity chromatography devoid of additional charged groups. J Biol Chem **254**: 2572-2574.

2. Ngo TT (1986) Facile activation of sepharose hydroxyl groups by 2-fluoro-1-methylpyridinium toluene-4-sulfonate: preparation of affinity and covalent chromatography matrices. Bio-Technology **4**: 134-137.

3. Kay G Crook EM (1967) Coupling of enzymes to cellulose using chloro-s-triazines. Nature **216**: 514-515.

SECTION IV

Peptide Synthesis

Randomization of Amino Acid Sequence: An Important Control In Synthetic Peptide Analogue Studies of Nucleic Acid Binding Domains

Steven G. Nadler, Janet L. Kapouch, James I. Elliott, and Kenneth R. Williams[a]

Dept. of Molecular Biophysics and Biochemistry, Yale University School of Medicine, New Haven, CT 06510
and
[a]Howard Hughes Medical Institute
New Haven, CT 06510

I. Introduction

Over the last several years, improvements in automated peptide synthesis have helped to account for the increasing use of synthetic peptides for mapping epitopes, "active site" regions, and binding domains in proteins (1-3). In the case of several nucleic acid binding proteins, synthetic peptide analogues have been shown to recognize specific sequences(4-6). Hence, a 9 residue synthetic peptide analogue of the HIV TAT protein was still able to specifically recognize TAR RNA (6). In most of these and other synthetic peptide analogue studies that we have found in the literature, the control experiment involved binding studies (or other assays) on a peptide of unrelated sequence and amino acid composition. The assumption has almost always been that the measured binding affinity was a result of the amino acid sequence, as opposed to the amino acid composition of the synthetic peptide analogue. By characterizing the nucleic acid binding properties of control peptides with the same composition as the actual peptide analogue, but with randomized sequences, we have found that in at least one system the amino acid composition is more important than the amino acid sequence in determining the nucleic acid binding properties of the synthetic peptide. Our data suggest that a "randomized" peptide, which has the same composition but a different sequence than the "correct" peptide, might serve as a better control than a completely unrelated peptide does for these studies. In this way, effects due to amino acid composition can begin to be differentiated from those resulting from amino acid sequence.

The system that we chose to study is the 90 amino acid, eukaryotic RNA binding motif that has now been found in one or more tandemly arranged copies in numerous proteins that bind hnRNA, mRNA, snRNA, and pre-ribosomal RNA (7-9). The synthetic peptides used in this study were based on the sequence of the fourth RNA binding domain in the yeast poly (A) binding protein (YPAB) (10). Using synthetic peptides corresponding to the two halves of this domain, we found that while the 45 residue, NH_2-terminal peptide analogue had no detectable affinity for nucleic acids, the 44 residue, COOH-terminal peptide bound well to RNA and retained many of the characteristic binding properties of the 577 residue YPAB protein, including a preference for binding RNA versus DNA. Interestingly, two peptides with randomized sequences, which had the same amino acid composition as the 44 residue, COOH-terminal peptide, had similar binding characteristics as the peptide with the "correct" sequence.

II. Materials and Methods

Wild type yeast poly (A) binding protein from *Saccharomyces cerevisiae* was a generous gift from Dr. Alan B. Sachs. Polynucleotides were obtained from Pharmacia. Poly r(ϵA) was synthesized as described (11). All peptides were synthesized on an Applied Biosystems Inc. 430A Solid Phase Peptide Synthesizer using standard t-BOC chemistry. Unless otherwise noted, all peptides contained a free NH_2-terminus and an amidated COOH-terminus.

All fluorescence titrations were carried out by adding protein directly to a stirred cuvette containing the fluorescent polynucleotide analogue, poly etheno(adenylic acid) (poly r(ϵA) or poly d(ϵA)). Fluorescence studies were performed on an SLM 8000C spectrofluorometer interfaced to an IBM AT computer. All titrations were carried out at 25°C, in a 10 mM Tris-HCl, 0.1 mM EDTA, pH 8.0 buffer with the indicated concentration of NaCl. Titrations were performed using an excitation wavelength of 315 nm and an emission wavelength of 400 nm. To determine binding constants, all data were first visually and then statistically curve fit using the equations developed by McGhee and von Hippel (12).

III. Results

A. *Interaction of Yeast Poly (A) Binding Peptides and Protein With Poly r(ϵA) and Poly d(ϵA)*

Figure 1 shows a partial list of sequences of proteins containing the 90 amino acid RNA binding domain described above. The peptides YPAB(313-357) and YPAB(359-402) correspond to the NH_2- and COOH-terminal halves of the fourth domain of yeast poly (A) binding protein (numbering based on YPAB residue numbers). Since these peptides had low intrinsic fluorescence, binding studies were carried out with the fluorescent analogue polyetheno(adenylate), poly r(ϵA). The fluorescence of this polynucleotide was found to be increased from 200-250% by binding

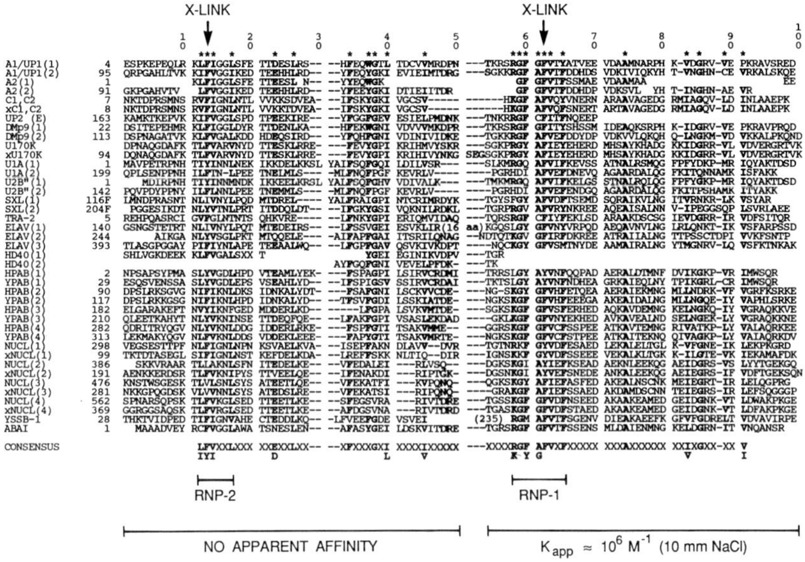

Figure 1. Alignment of eukaryotic RNA-binding protein sequences to show protein:protein and internal repeat sequence homologies. Highly conserved residues are indicated in bold face. Phenylalanine residues that have been shown by UV cross-linking studies to be at the interface of the A1 hnRNP:oligonucleotide complex are indicated by the two arrows at the top of the figure. The lines below the figure depict the two synthetic peptide analogues used in this study. The sequences of the latter were based on the fourth RNA binding domain in the yeast poly (A) binding protein. The protein sequences shown are all in the PIR Database and are the A1, A2, C1 and type E hnRNP; DMp9, Drosophila clone p9; U170, U1A, U2B" snRNP proteins; Drosophila Tra-2 and sex lethal (SXL); Drosophila neuron-specific elav protein; Xenopus HD40 hnRNP-like protein; yeast and human poly (A) binding proteins; nucleolins; yeast SSB-1; and the abscisic acid induced protein (ABAI). The lower case "x" preceding several of the above sequences indicates Xenopus.

of the YPAB(359-402) peptide. As shown in Table I, the YPAB(359-402) peptide binds well to poly r(ϵA) in the range extending from 10 to at least 100 mM NaCl. In 10 mM NaCl, this peptide has an affinity of about 10^6 M^{-1} for poly r(ϵA). In contrast, the 45 residue peptide corresponding to the NH_2-terminal half of this RNA binding domain, YPAB(313-357), had an affinity for poly r(ϵA) of less than $10^3 M^{-1}$ in 10 mM NaCl, with no detectable binding at higher salt concentrations. For comparison, binding of the intact YPAB protein was also measured. Because the affinity of YPAB protein for poly r(ϵA) was too high below about 0.1 M NaCl to be directly measured, YPAB binding constants in this salt range were extrapolated from a log K_{app} versus log [NaCl] plot obtained at higher salt concentrations. In 100 mM NaCl, we calculated a binding constant of 7.9 x 10^8 M^{-1} for YPAB protein binding to poly r(ϵA). Using the curve fitting procedure described in Materials and Methods, we also determined a cooperativity parameter which reflects the strength of protein:protein

Table I. Summary of Peptide Binding Data[a]

Peptide	Poly r(εA)			Poly d(εA)
	10 mM NaCl	50 mM NaCl	100 mM NaCl	10 mM NaCl
YPAB(359-402)	1.0×10^6	7.8×10^4	1.8×10^4	3.0×10^4
YPAB(313-357)	$<1.0 \times 10^3$	N.D.	N.D.	-
Random 1	1.6×10^6	1.1×10^5	6.5×10^4	9.1×10^4
Random 2	1.1×10^6	5.5×10^4	6.7×10^3	1.9×10^5
YPAB Protein[b]	6.3×10^{12}	1.3×10^{10}	7.9×10^8	8.0×10^6 (100 mM NaCl)

[a]All binding constants (K_{app}) were determined in duplicate and are M^{-1}. N.D. indicates that a titration was performed but binding was not detected.

[b]YPAB affinities for poly r(εA) were extrapolated from a plot of log K_{app} versus log [NaCl] because at salt concentrations below about 0.1 M, YPAB binding was too strong to be directly measured.

interactions occurring between adjacent YPAB molecules bound to the poly r(εA) lattice. The overall apparent affinity of YPAB for poly r(εA) that is given in Table I therefore represents the product of the cooperativity parameter and the intrinsic YPAB affinity for the nucleic acid lattice. While we determined that the YPAB protein binds cooperatively with a parameter of about 23, the synthetic peptide analogues listed in Table I bound non-cooperatively. Based on these data, the intrinsic affinity of YPAB protein for poly r(εA) in 100 mM NaCl is approximately 3.4×10^7 M^{-1} ($K_{intrinsic} = K_{apparent}$/cooperativity parameter). Therefore, in 100 mM NaCl the 44 residue YPAB(359-402) peptide retains approximately 57% (log 1.8×10^4/log 3.4×10^7) of the intrinsic binding free energy of the entire 577 amino acid protein. Using an analysis described by Record, et al (13) and Lohman, et al (14), we determined that in 100 mM NaCl, approximately 41% of the binding free energy of YPAB(359-402) for poly r(εA) is derived from electrostatic interactions. This is virtually identical to a value of 44% for the YPAB protein binding to nucleic acids.

Since the RNP type RNA binding proteins bind RNA in vivo, they would be expected to retain this specificity for RNA in vitro. Indeed, based on the relative affinities for poly r(εA) compared to poly d(εA) (see Table I), YPAB has nearly a 100-fold higher affinity for RNA than for DNA. Remarkably, this property is largely retained by YPAB(359-402) which, in 10 mM NaCl, has a 33-fold higher affinity for RNA than for DNA (Table I).

B. *Binding of Randomized and Protease-treated Peptides to Poly r(εA)*

In order to estimate the relative importance of primary sequence versus amino acid composition to the binding of YPAB(359-402) to nucleic acids,

```
YPAB(359-402)
  GKSKGFGFVT  FSTPEEATKA  ITEKNQQIVA  GKPLYVAIAQ  RKDV

Random 1
  FKVKTGATNK  QKTTIEKVFG  AIEADYIVLA  PSSPKAGGEQ  FRQV

Random 2
  PETNVSQAYE  KDKVTFQAIQ  RIKATKPVET  GAFGVSLKKI  GFGA
```

Figure 2. Amino acid sequences of synthetic peptide analogues of the poly (A) binding protein. The two "random" peptides have the same amino acid compositions but different amino acid sequences from the YPAB(359-402) peptide.

two peptides were synthesized which had the same composition, but different sequences from YPAB(359-402). These sequences (see Fig. 2) were chosen by first randomly generating sequences of a large number of 44 residue peptides which had the same composition as YPAB(359-402). The secondary structures of these peptides were then predicted using a Chou-Fasman analysis. The Random 1 peptide was chosen since it had approximately the same percentage of predicted secondary structure but only 23% sequence identity with YPAB(359-402). A second peptide, Random 2, had only minimal predicted secondary structure and no detectable features in common with YPAB(359-402). Table I shows that YPAB(359-402) as well as both its randomized controls all had indistinguishable affinities for poly r(ϵA) in 10 mM NaCl. Only in 100 mM NaCl were small differences in affinity seen, with Random 1 having a three-fold higher and Random 2 a three-fold lower affinity than YPAB(359-402) for poly r(ϵA). The most notable difference that accompanied randomization of the YPAB(359-402) sequence was the consequent 2-6 fold decrease in the specificity of their binding to RNA. That is, the RNA/DNA relative affinity ratio for Random 1 was only about 6 as compared to about 33 for YPAB(359-402).

To determine whether the length of the YPAB(359-402) peptide was a critical factor in accounting for its affinity for nucleic acids, this peptide was digested with chymotrypsin or pepsin prior to determining its affinity for poly r(ϵA). Chymotrypsin digestion of YPAB(359-402), which yields 5 major peptides, decreased the affinity for poly r(ϵA) by 25 fold. Pepsin, which cleaves less specifically than chymotrypsin, completely abolished binding to nucleic acids.

C. Nucleic Acid Base and Single-strand Specificity of YPAB Peptides and Protein

Since poly (A) binding protein binds to poly (A) tracts in vivo, it was of interest to determine if both the intact protein as well as its synthetic peptide analogues would bind preferentially to poly (A) in vitro. By carrying out poly r(ϵA) titrations in the presence of competing homopolynucleotides, it was possible to derive apparent affinities for poly (C), poly (U) and poly

(A). As shown in Table II, YPAB protein does indeed have a 6-30 fold preference for binding poly (A) as compared to the other homopolynucleotides tested. Although the degree of specificity for binding poly (A) is decreased upon going from YPAB to the synthetic peptides, the order of preference is still largely the same, with both YPAB(359-402) and its Random 2 control having highest affinity for poly (A) (Table II).

Yeast poly (A) binding protein, as well as the other RNP type proteins that share the 90 amino acid RNA binding domain shown in Figure 1 are thought to bind preferentially to single versus double stranded nucleic acids. Since proteins that bind preferentially to single-stranded nucleic acids lower the melting temperature of double-stranded polynucleotides such as poly (A-U), the latter provides a convenient assay for "helix-destabilizing activity". In 10 mM NaCl, the YPAB protein decreased the T_m of poly (A-U) by approximately 30°C (data not shown). The YPAB(359-402) peptide also had considerable helix-destabilizing activity as it decreased the T_m by 16°C. Using the method described by Crothers (15) to calculate apparent affinities from thermal melting data, we determined a K_{app} for single-stranded poly (A-U) of about 2.2 x 10^{12} M^{-1} for the YPAB protein and of about 3.6 x 10^5 M^{-1} for the YPAB(359-402) peptide in 10 mM NaCl. Hence, using a different technique, and still assuming a cooperativity parameter of 23 for YPAB, we again conclude that the YPAB(359-402) peptide retains approximately 50% of the <u>intrinsic</u> free energy of binding of the YPAB protein. Since both of the randomized peptides formed insoluble peptide:nucleic acid complexes under the conditions of the melting experiments, their helix-destabilizing activities could not be assayed.

IV. Conclusions

To our knowledge, no attempt has yet been made to compare the relative contribution of amino acid composition versus amino acid sequence towards the affinity of synthetic peptides for nucleic acids or, for that matter, towards other activities that have been found to be retained by synthetic peptide analogues of proteins. We hypothesized that in short peptides, where there is typically little secondary and tertiary structure, the relative lack of

Table II. YPAB Peptide and Protein Binding to Homopolynucleotides[a]

Peptide/Protein	[NaCl] (mM)	Poly r(C)	Poly r(U)	Poly r(A)
YPAB Protein	300	≤5.0 x 10^3	2.6 x 10^4	1.5 x 10^5
YPAB(359-402)	10	3.3 x 10^4	8.0 x 10^4	9.5 x 10^4
Random 2	10	5.1 x 10^3	2.0 x 10^4	1.9 x 10^4

[a]All of the binding constants are M^{-1} and were determined using a competition assay with poly r(εA). In each titration, binding to 1 μM poly r(εA) was competed with 100 μM of the various polynucleotides. Note that because of its higher affinity, the YPAB titrations were carried out in 300 mM as compared to 10 mM NaCl for the peptide titrations.

topological constraints might allow specific amino acid:nucleic acid contacts to occur with little positional preference. Hence, a lysine:phosphodiester backbone contact might be just as favored if the lysine were to occupy residue 4 in the peptide as if it were to be in position 15. In fact, the relative importance of amino acid sequence might be so reduced that the amino acid composition would become as important or even, a more important determinant of the peptide's "activities" than the actual sequence of the peptide. Such appears to be the case with the YPAB(359-402) synthetic peptide analogue.

Several of the nucleic acid binding properties of the YPAB(359-402) synthetic peptide suggested that it could serve as a reasonable model for the YPAB protein. Hence, like the native protein, the peptide analogue bound preferentially to

1. single as opposed to double-stranded nucleic acids
2. RNA versus DNA
3. the poly (A) homopolymer

Furthermore, despite the fact that the peptide represented less than 8% of the YPAB sequence, it nonetheless retained greater than 50% of the free energy of YPAB binding to poly r(ϵA). That binding of the YPAB(359-402) synthetic peptide was not simply the result of non-specific ionic interactions was suggested by its preferential affinity for single-stranded nucleic acids (since the charge density is higher on double-stranded nucleic acids, non-specific ionic interactions might be expected to be favored on double-stranded nucleic acids) and for RNA in general and for poly (A) in particular (both of which would be expected to have similar charge densities as single-stranded DNA and other single-stranded homopolynucleotides). In addition, salt sensitivity studies indicated that only 40% of the free energy of binding of the YPAB(359-402) peptide derived from ionic interactions.

Intriguingly, randomization of the sequence of the YPAB(359-402) peptide had relatively little effect on any of its characteristic nucleic acid binding properties. The inescapable conclusion is that these latter properties derive more from the amino acid composition of the YPAB(359-402) peptide than from its sequence. Some caution, therefore must be exercised in interpreting results of structure/function studies on the YPAB(359-402) peptide. While, for instance, the 33-fold specificity of the YPAB(359-402) peptide for RNA, as compared to DNA, might provide an attractive model for determining which particular kinds of amino acid residue:sugar interactions are required to confer binding specificity for RNA, the YPAB(359-402) randomization studies clearly indicate that it would be inappropriate to extrapolate conclusions regarding sequence specific structure/function relationships in the YPAB(359-402) synthetic peptide back to the native YPAB protein. Since the 90 amino acid, eukaryotic RNA binding domain has been shown to have a highly ordered structure in solution, consisting of two α-helices and a four stranded β-sheet (16), randomization of the YPAB sequence spanning residues 359-402 would almost certainly have a more pronounced effect on nucleic acid binding and on protein folding within the context of the native YPAB than it appears to have had in the YPAB(359-402) synthetic peptide analogue. The results presented in this study strongly suggest that randomization of amino acid sequence is an appropriate control that should be considered prior to interpreting synthetic peptide analogue studies in terms of the function of the primary sequence of the peptide.

References

1. Lenert, P., Kroon, D., Spiegelberg, H., Golub, E.S., and Zanetti, M. (1990) *Science* **248**, 1639-1643.
2. Hahn, K.W., Klis, W.A., and Stewart, J.M., (1990) *Science* **248**, 1544-1547.
3. Zlotnick, A., and Brenner, S.L. (1989) *J. Mol. Biol.* **209**, 447-457.
4. Bruist, M., Horvath, S., Hood, L., Steitz, T., and Simon, M. (1987) *Science* **235**, 777-780.
5. Archer, T., Hager, G., and Omichinski,J. (1990) *Proc. Natl. Acad. Sci.* **87**, 7560-7564.
6. Calnan, B.J., Biancalana, S., Hudson, D., and Frankel, A.D., (1991) *Genes and Dev.* **5**, 201.
7. Bandziulis, R.J., Swanson, M.S., and Dreyfuss, G. (1989) *Genes and Dev.* **3**, 431-437.
8. Merrill, B.M., and Williams, K.R., (1990). *In* "The Eukaryotic Nucleus, Molecular Biochemistry and Macromolecular Assemblies" (Strauss, P.R., and Wilson, S.H., eds.) 579-604.
9. Keenan, D.J., Query, C.C., and Keene, J.D. (1991) *TIBS* **16**, 214-220.
10. Sachs, A.B., Bond, M.W., and Kornberg, R.D., (1986) *Cell* **45**, 827-835.
11. Kohwi-Shigematsu, T., Enomoto, T., Yamada, M.A., Nakanishi, M., and Tsuboi, M., (1978) *Proc. Natl. Acad. Sci.* **75**, 4689-4693.
12. McGhee, J.D. and von Hippell, P.H., (1974) *J. Mol. Biol.* **86**, 469-489.
13. Record, M.T. Jr., Lohman, T.M. and deHaseth, P., (1976) *J. Mol. Biol.* **107**, 145-158.
14. Lohman, T.M., deHaseth, P.L., and Record, M.T. Jr. (1980) *Biochemistry* **19**, 3522-3530.
15. Crothers, D. (1971) *Biopolymers* **10**, 2147-2160.
16. Nagai, K., Oubridge, C., Jessen, T., Li, J., and Evans, P. (1990) *Nature* **348**, 515-520.

Comparative Analysis of Synthetic Peptides by Free-Solution Capillary Electrophoresis (FSCE) and Strong-Cation Exchange (SCX)-HPLC

Dan L. Crimmins
Howard Hughes Medical Institute Core Protein/
Peptide Facility, Washington University
School of Medicine, St. Louis, MO 63110

I. Introduction

Synthetic peptides have become essential reagents in diverse areas of biomedical research. Much of the current interest in synthetic peptides is due in part to the successful and almost routine implementation of automated solid-phase peptide synthesis techniques. The synthetic peptides are typically used as: antigens, enzyme substrates/ inhibitors, affinity reagents, protein folding mimics, and pharmaceutical agents. This widespread use of synthetic peptides has placed increased demands on the analytical techniques available to assess sample purity with the degree of acceptable purity being dictated by the intended use of the synthetic peptide. For example, peptides to be used as antigens in the production of polyclonal antibodies will require far less purity compared to those used for medicinal/drug purposes. It is unlikely however, that a single method alone can provide the data required to determine sample purity although in practice, mass spectrometry approaches this goal [1-4]. Sequence analysis, high-performance liquid chromatography (HPLC), and capillary electrophoresis (CE) are additional, potentially definitive, analytical procedures used to assess synthetic peptide sample purity.

HPLC, and recently CE, are separation methods available to most researchers to evaluate synthetic peptide purity. Reversed-phase (RP)-HPLC is the most popular chromatographic method for the analysis of synthetic peptide components. This is largely due to

TECHNIQUES IN PROTEIN CHEMISTRY III

the proven record of successful applications, the volatile nature of the mobile phases thus facilitating sample recovery, and wide selection of columns [5,6]. An alternative form of chromatography based on a different separation mechanism is strong-cation exchange (SCX)-HPLC. SCX-HPLC is generally performed near pH 3 and sample separation therefore, mainly results from differences in the number of positively charged residues. Experimentally, this hypothesis has been verified for proteolytic digests [7-10], synthetic peptide mixtures [11-14], N-terminally blocked peptides [13,15] and disulfide-linked peptides [8,16]. Similar applications of free-solution CE (FSCE) have also been reported [17-23]. Low sample consumption, sensitivity, speed of analysis, and automation all combine to make FSCE a valuable analytical separation technique [24]. In this report, synthetic peptides with known modifications at the N-terminus, C-terminus, methionine residues, and truncated species are subjected to SCX-HPLC at pH 3 and, FSCE at pH 2 and 3 using uncoated silica capillaries. At these pH's, the electroendosmotic flow should be minimal [17,18] and the peptides are expected to electrophoretically separate according to their net positive charges. A comparison of these two analytical separation techniques in terms of rapidity of analysis, sensitivity, robustness, and preparative applications is also presented. Theoretical, operational, and practical aspects of HPLC and CE have been compared in a recent review [25].

II. Materials and Methods

A. *Chemicals and Reagents*

The synthetic peptides used in this study (see Table I) were purchased from either Sigma Chemical Co. (St. Louis, MO) or Peninsula Laboratories (Belmont, CA) and used without further purification. Synthetic peptides were reconstituted with 1% (v/v) AcOH to ~1 mg/ml and stored at 4°C until use. Sample quality assurance procedures, the sources of HPLC columns and solvents, water purification, and additional reagents have been described [9,13,15,16]. Stock solutions of sodium phosphate buffer were prepared from 14.7 M H_3PO_4 ("HPLC-grade", Fisher Scientific, St. Louis, MO) by appropriate aqueous dilution, pH titration with 10 N NaOH (Fisher Scientific), then vacuum-filtered through 0.2 μm Nylon-66 membranes (Rainin Instrument Co., Woburn, MA).

B. HPLC and FSCE Analysis

The chromatographic hardware and data acquisition procedures have been described [9,13,15,16]. SCX-HPLC was performed with a sulfoethyl aspartamide column (200 x 4.6 mm, 5 μm, 300Å) and separation effected by NaCl gradient elution from 0 to 60% B in 60 min at 1 ml/min ([A] 5 mM sodium phosphate pH 3, 25% (v/v) CH_3CN; [B] 5 mM sodium phosphate, 1 M NaCl pH 3, 25% (v/v) CH_3CN) and monitored at 214 nm. FSCE was performed with a Beckman P/ACE System 2000 (Palo Alto, CA) using an uncoated silica capillary (75 μm x 57 cm; 50 cm from inlet to detector), applied voltage of +25 kV, temperature control at 20°C, a measured current of ~150 μA, and monitored at 200 nm. Samples were diluted 1/5 with H_2O, pressure-injected for 5 s and the capillary was washed between runs with electrophoresis buffer (100mM sodium phosphate pH 2 or 3). Peptide samples (Table I) were mixed in different proportions for each analytical technique to expedite peak assignment.

III. Results and Discussion

The sequences, position of residue modifications, and nominal net charge at pH 2-3 for the synthetic peptides used in this study are listed in Table I. The peptides were placed into five groups and for each, the modifications were chemically, relatively simple and at known sites, thus facilitating data interpretation. For example, the first four peptides are related to Leu-enkephalin and all variations occur at the C-terminus. Likewise, the next set of three peptides belong to the Met-enkephalin family and differences are again at the C-terminus with an additional modification of Met to Met[O]. The third group of peptides further studies the effect of a free-acid vs. amidated C-terminus for Luteinizing Hormone-Releasing Hormone. Modifications at the N-terminus are evident for the last four peptides which are divided into two groups for a Thymus Factor/Serum Thymic Factor pair and a Neurotensin Fragment (8-13) pair.

Figure 1 displays a composite chromatographic (top panel) and electrophoretic (middle and bottom panels) analysis for the first three groups of peptides listed in Table I. SCX-HPLC at pH 3 shows that in all cases, the C-terminally amidated version of a given peptide elutes later than the free-acid counterpart (top panel: A,B, and C). This result is in complete agreement with previous data [12,13] and

Table I
Synthetic Peptide Characterization

Sequence[a]	Modifications N-terminus	C-terminus	Other	~ Charge @ pH 2-3[b]
YGGFL(OH)	-	-	-	+1
YGGFL(NH_2)	-	amide	-	+1
YGGFLR(OH)	-	-	-	+2
YGGFLRR(OH)	-	-	-	+3
YGGFM(OH)	-	-	-	+1
YGGFM(NH_2)	-	amide	-	+1
YGGFM[O](OH)	-	-	met-sulfoxide	+1
pEHWSYGLRPG(OH)	pyroglutamyl	-	-	+2
pEHWSYGLRPG(NH_2)	pyroglutamyl	amide	-	+2
QAKSQGGSN(OH)	-	-	-	+2
pEAKSQGGSN(OH)	pyroglutamyl	-	-	+1
RRPYIL(OH)	-	-	-	+3
N-Ac-RRPYIL(OH)	N-α-acetyl	-	-	+2

[a] The standard single-letter code is used for the amino acids; M[O] = met-sulfoxide; pE = pyroglutamic acid; N-Ac = N-α-acetyl.

[b] Nominal net charge at pH 2-3 is computed from the total number of R, K, and H residues and the "free" amino terminus each assigned a value of +1.

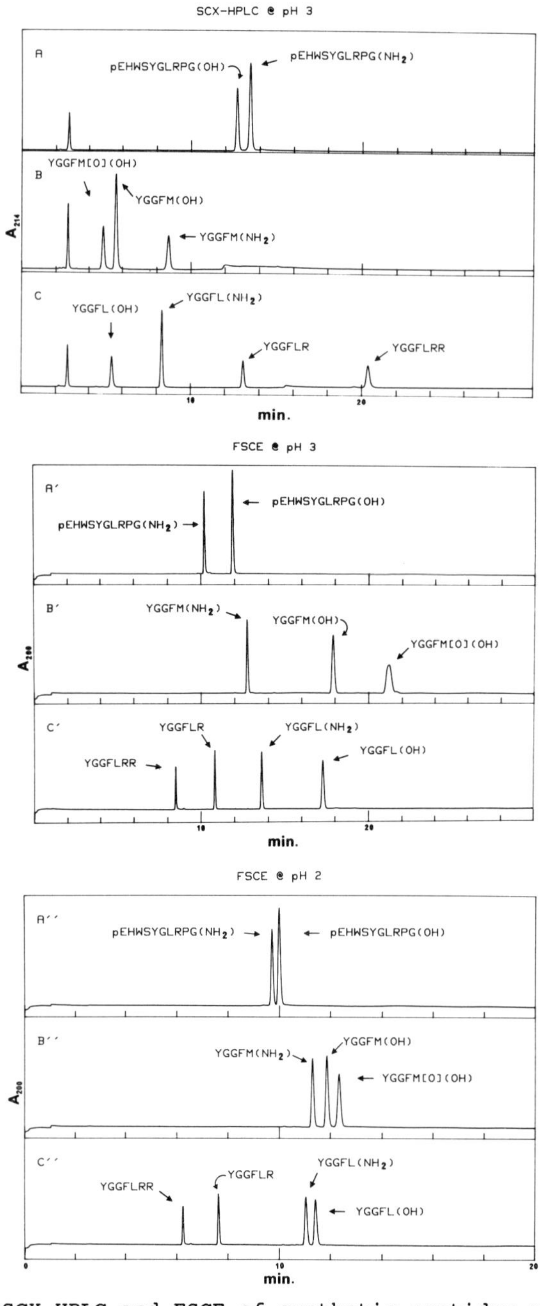

Figure 1. SCX-HPLC and FSCE of synthetic peptides with C-terminal and methionine modifications, and truncated species. SCX-HPLC at pH 3, unprimed panels, A-1250 mV full scale (mVFS), B and C-500mVFS; FSCE at pH 3, single-primed panels, all-325mVFS; and FSCE at pH 2, double-primed panels, all-325mVFS.

a likely explanation for this observation is that the amidated derivative is slightly more positive in character than the free-acid peptide and would therefore be expected to elute later. Addition of one and two Arg residues to Leu-enkephalin, as expected, results is a longer retention time for these two species (top panel: C). The more polar Met[0] derivative of Met-enkephalin elutes earlier than the unmodified parent (top panel: B) and this has been attributed to a mixed-mode (ionic and hydrophilic/hydrophobic) separation mechanism [12,13].

FSCE for the exact same set of peptides was performed at pH 3 (Fig. 1, middle panel) and pH 2 (Fig. 1, bottom panel). It is clear from the pH 3 electrophoresis experiments that the overall separation for a given peptide group is better for FSCE than SCX-HPLC. Furthermore, although the displayed time scale is similar for each, in reality the gradient re-cycle time to initial conditions makes sample-to-sample analysis in this case ~3 times longer for HPLC than CE. The absolute sensitivity observed here for CE is at least 100 times greater than HPLC since 1 to 10 ng was injected on the capillary and 1 to 10 μg injected on the column. It should be noted however, that a fairly concentrated stock peptide solution is required for FSCE analysis because of the small injection volumes used. FSCE at pH 2 (bottom panel) shows the migration trend, albeit with poorer separation, for a given set of peptides to be similar to the pH 3 runs, but that the analysis times are ~1.5-fold faster at the more acidic pH. Importantly, at either pH for FSCE, the migration order for a family of peptides is precisely the exact opposite to that observed for SCX-HPLC further indicating that both electroendosmosis and capillary wall adsorption effects, as stated [17,18], are minimal at these pH's. These results indicate that the order of cathodic migration is strongly correlated to the net positive charge of the peptides at pH 2-3 [17-19].

A similar type of analysis for N-terminally modified synthetic peptides is illustrated in Figure 2. In this case, the modified species possessed a blocked N-terminus either as a pyroglutamyl- or acetyl- function resulting in a decreased net nominal positive charge at pH 2-3 compared to their respective free N-terminal counterparts. SCX-HPLC at pH 3 (top panel) shows the expected elution order and excellent resolution for these peptide pairs [15]. Unlike the enhanced resolution at pH 3 for the SCX-HPLC/FSCE comparison shown in Figure 1, such improvement is not always the case for these species. Indeed, the separation of the Neurotensin Fragment pair (8-13) (middle panel: A') is actually less than

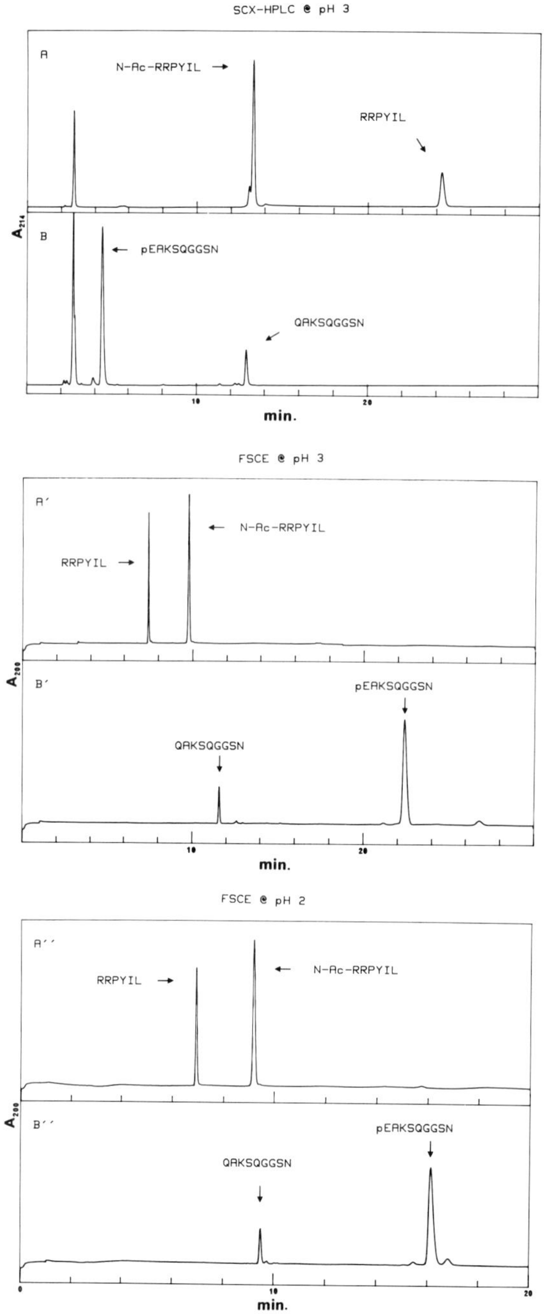

Figure 2. SCX-HPLC and FSCE of synthetic peptides with N-terminal modifications. SCX-HPLC at pH 3, unprimed panels, all-500mVFS; FSCE at pH 3, single-primed panels, all-300mVFS; and FSCE at pH 2, double-primed panels, all-300mVFS.

that demonstrated in the corresponding SCX analysis (top panel: A). Interestingly, and in contrast to the data from Figure 1, little change in the absolute difference in migration times are observed for FSCE at pH 3 (middle panel) and pH 2 (bottom panel) for either group of peptide pairs modified at the N-terminus. Collectively then, this suggests that small changes in this acidic pH range can be used to selectively exploit minor differences in C-terminal residues.

A particularly informative and relevant example comparing the analytical potential of SCX-HPLC at pH 3 and FSCE at pH 2 is illustrated in Figure 3. This analysis shows the chromatographic (A) and electrophoretic (A'') profiles for a 29-residue synthetic peptide designated ABRF-91SEQ (see D.L. Crimmins et al., this volume). The major peak at ~34', and which comprises ~90% of the integrated area, in the SCX chromatogram (A) is the desired peptide product. The peak at ~4' is an N-terminal acetylated failure sequence peptide (confirmed by mass spectrometry) accounting for about 9% of the area. A tentative assignment of the late-eluting species at ~58', assumed to be the disulfide-linked 29-residue dimer, was made based on a decreased area for this peak after sample treatment with dithiothreitol (data not shown) and from previous observations [16]. All of these three species then, elute in the expected order on SCX-HPLC at pH 3. Three peaks are also observed following FSCE at pH 2 (A'') with essentially identical area proportions. The 29-residue peptide has a calculated net nominal

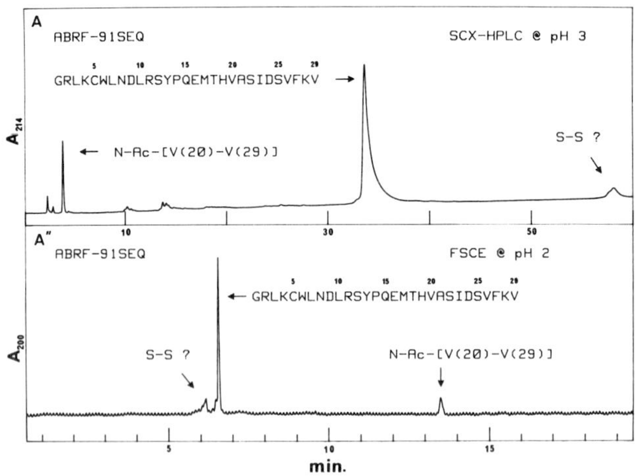

Figure 3. SCX-HPLC at pH 3 and FSCE at pH 2 of the synthetic peptide portion of ABRF-91SEQ. SCX-HPLC at pH 3, A-250 mVFS; and FSCE at pH2, A"-35mVFS. Sequence assignment of major peaks are shown and S-S? refers to tentative assignment of the disulfide linked dimer [16].

charge at pH 2 of +6 and the N-terminal acetylated failure sequence peptide a value of +1 (see footnote b, Table I) such that the 29-residue peptide will migrate much faster towards the cathode and therefore be detected earlier. This is precisely the experimental observation for FSCE at pH 2. The early-migrating species at ~6' is tentatively identified as the disulfide-linked 29-residue dimer, again, because the area of this peak decreased following sample incubation with dithiothreitol (data not shown).

It is clear from the data shown in Figure 3 that each analytical technique was able to successfully resolve and provide positive peak assignment for the major 29-residue synthetic peptide, the primary contaminant, and possibly, a disulfide-linked 29-residue dimer. Although in this example, the resolution afforded by SCX-HPLC is slightly greater than for FSCE both yield exceptional resolution. More importantly, the chromatographic procedure could easily be adapted to a preparative scale to provide large quantities of extremely pure 29-residue synthetic peptide. Obviously, this would be a formidable and time-consuming task for FSCE.

The commercial adaptation of early CE research [26-28] into cost-effective and versatile instruments is proving to be a valuable addition to the current list of bioanalytical procedures. CE applications are developing at a particularly opportune time as the need for rapid, sensitive, and accurate sample characterization in all areas of biomedical research is clearly increasing. The use of CE in a combined mode with liquid/gas chromatography and mass spectrometry along with even more sensitive detection systems is expected to streamline the entire sample characterization process.

In this report a comparison of the analytical capabilities of SCX-HPLC and FSCE at acidic pH for the characterization of synthetic peptides is presented. During the initial phase of this study, buffers covering the pH range of 2-10 were used for FSCE analysis of some of the synthetic peptides (data not shown). These pH experiments can be performed in a automated-overnight fashion [24]. This type of experiment would not be possible with silica-based HPLC columns, and even for polymeric resins the buffer (mobile phase) changing process would be extremely tedious. The majority of analyses however, were performed at low pH because of the numerous successful applications and predictable elution profiles for peptides subjected to SCX-HPLC [12,13] and since the potentially undesirable effects observed at higher pH's for FSCE should be minimized at pH 2-3 [17,18]. Furthermore, many peptides are expected to be positively charged at this pH so the procedures described herein should be of general utility. For the majority of the peptide groups analyzed, each analytical technique was equally proficient in peak assignment and resolution of within group peptide components (Figs. 1-3). Not surprisingly, CE exhibited far greater absolute sensitivity with the added bonus of rapid analysis time [24] compared to HPLC. Thus for many analytical applications, CE should be given primary consideration. On the other hand, it is virtually impossible and cumbersome at best, despite claims to the contrary, with the present configuration of CE instrumentation to perform CE in the "preparative" mode. Future technological innovations appear necessary before this becomes a reality.

Acknowledgements

The CE data presented in this report were generated during an instrument evaluation process and I thank the following vendors for participating in this study: Beckman (Palo Alto, CA), Applied Biosystems (Foster City, CA), Dionex (Sunnyvale, CA), Waters (Milford, MA), and Spectra-Physics (San Jose, CA). I am grateful to Gregory A. Grant (Washington University School of Medicine) for critical reading of the manuscript, the Association of Biomolecular Resource Facilities (ABRF) executive committee for permission to use the data presented in Fig. 3, and Lorraine Whiteley and Pat Parvin for expert secretarial assistance.

References

1. R.B. Merrifield, J. Singer, & B.T. Chait (1988) Anal. Biochem. **174**, 399-414.
2. S.W. Chowdhury & B.T. Chait (1989) Anal. Biochem. **180**, 387-395.
3. W.T. Moore & R.M. Caprioli (1991) *In* Techniques in Protein Chemistry II, J.J. Villafranca, ed., pp. 511-528 Academic Press, San Diego.
4. C.C. Chang, P.A. Babayan, L. Chen. & A.L. Hong (1991) *In* Techniques in Protein Chemistry II, J.J. Villafranca, ed., pp. 529-536 Academic Press, San Diego.
5. G. Szokan, A. Torok, & B. Penke (1987) J. Chromatogr. **387**, 267-280.
6. C.T. Mant & R.S. Hodges (1990) *In* HPLC of Biological Macromolecules: Methods and Applications, K.M. Gooding & F.E. Regnier, eds., pp 315-325 Marcel Dekker, Inc., New York.
7. H. Kawasaki, S. Imajoh, & K. Suzuki (1987) J. Biochem. **102**, 393-400.
8. P.C. Andrews (1988) Pept. Res. **1**, 93-99.
9. D.L. Crimmins, R.S. Thoma, D.W. McCourt, & B.D. Schwartz (1989) Anal. Biochem. **176**, 255-260.
10. T.D. Schlabach, J.C. Colburn, R.J. Mattaliano, & S. Yuen (1989) *In* Techniques in Protein Chemistry, T.E. Hugli, ed., pp. 497-505 Academic Press, San Diego.
11. C.T. Mant & R.S. Hodges (1985) J. Chromatogr. **327**, 147-155.

12. A.J. Alpert & P.C. Andrews (1988) J. Chromatogr. **443**, 85-96.
13. D.L. Crimmins, J. Gorka, R.S. Thoma, & B.D. Schwartz (1988) J. Chromatogr. **443**, 63-71.
14. T.W.L. Burke, C.T. Mant, J.A. Black, & R.S. Hodges (1989) J. Chromatogr. **476**, 377-389.
15. D.L. Crimmins, D.W. McCourt, & B.D. Schwartz (1988) Biochem. Biophys. Res. Commun. **156**, 910-916.
16. D.L. Crimmins (1989) Pept. Res. **2**, 395-401.
17. P.D. Grossman, K.J. Wilson, G. Petrie, & H.H. Lauer (1988) Anal. Biochem. **173**, 265-270.
18. R.M. McCormick (1988) Anal. Chem. **60**, 2322-2328.
19. P.D. Grossman, J.C. Colburn, & H.H. Lauer (1989) Anal. Biochem. **179**, 28-33.
20. P.D. Grossman, J.C. Colburn, H.H. Lauer, R.G. Nielsen, R.M. Riggin, G.S. Sittampalam, & E.G. Rickard (1989) Anal. Chem. **61**, 1186-1194.
21. T. Dulffer, R. Herb, H. Herrman, & U. Kobold (1990) Chromatographia **30**, 675-685.
22. G.L. Hortin, T. Griest, & B.M. Benutto (1990) Biochromatogr. **5**, 118-120.
23. T.W. Wheat, P.M. Young, & N.E. Astephen (1991) J. Liq. Chromatogr. **14**, 987-996.
24. G. McLaughln, R. Palmieri, & K. Anderson (1991) *In* Techniques in Protein Chemistry II, J.J. Villafranca, ed., pp. 3-20 Academic Press, San Diego.
25. H.J. Issag, G.M. Janini, I.Z. Atamna, & G.M. Muschik (1991) J. Liq. Chromatogr. **14**, 817-845.
26. F.E.P. Mikkers, F.M. Everaerts, & Verheggen, Th. P.E.M. (1979) J. Chromatogr. **169**, 11-20.
27. J.W. Jorgenson & K.D. Lukacs (1981) Anal. Chem. **53**, 1298-1302.
28. J.W. Jorgenson & K.D. Lukacs (1983) Science **222**, 266-272.

Immunoreactive Synthetic Peptide Epitope Mapping with Structural Validation Using Electrospray Mass Spectrometry

William T. Moore, Jerry S. Wolinsky*, Marc J.-F.Suter, Terry B. Farmer, and Richard M. Caprioli

The Analytical Chemistry Center, the Department of Biochemistry and Molecular Biology, and the *Department of Neurology
The University of Texas Medical School, Houston, Texas 77030

I. Introduction

A. Background

Rubella virus (RV), the cause of German measles, is a Togavirus that has been extensively characterized. The virion contains three structural proteins: two envelope glycoproteins (E1 & E2) and a 300 amino acid residue capsid protein (C) associated with the genomic RNA[1]. The nucleotide sequence of the entire genome has been determined and from this information the amino acid sequences of the structural proteins have been deduced[2]. RV infection in man has been causally implicated with the onset of immune mediated disease[3]. In order to understand and explore the presumed autoimmune basis for the disorders associated with prior RV infection, a recent and extensive study was undertaken by Wolinsky et al.(1991)[4] with the goal to map the epitopes of the structural proteins of RV recognized by in vivo immune responses. An extensive library of monoclonal antibodies (mAbs) was generated after infecting mice with RV. A subset of mAbs that recognized "conformation independent" epitopes were selected on the basis of being active in immunoblots of fractionated and denatured viral proteins derived from either RV infected eukaryotic cells or recombinant hybrid

TECHNIQUES IN PROTEIN CHEMISTRY III

proteins expressed by prokaryotic cells containing plasmid constructs coding for different overlapping portions of the viral proteins (see ref. 4).

In the present study we focus on the fine-structure mapping of an epitope on the capsid protein of RV that binds to one mAb designated C9. This epitope is of interest since mAb C9 has been shown to cross-react with a 52kD protein in rat pancreatic islet cells[5]. The 52kD protein also binds antibody from sera of diabetic mice that have spontaneously developed anti-islet cell antibodies[6]. Human sera from some diabetic patients (type II diabetes mellitus) also cross-react with the 52kD protein.

This interesting cross-reactivity suggests that the C9 mAb defined RV capsid protein epitope may be analogous to a site on the 52kD protein and may be the basis for some form of molecular mimicry[7] that may be responsible for one of the late consequences of congenital rubella.

B. *Objective & Strategy*

Immunoblot analysis of biosynthetically (recombinant) expressed proteins suggested that the binding domain for mAb C9 was present within a region bounded by RV capsid protein C64-C97, a stretch of 34 amino acid residues. The precise boundaries of this RV C segment was determined by limited sequencing of the plasmid-RV cDNA insertion sites and the intervening RV segment[4]. No protein chemistry was performed on the expressed recombinant proteins. For additional verification, a 34 residue peptide designated RV C64-97 was prepared by solid phase peptide synthesis techniques and was shown to be mAb C9 reactive by standard ELISA. Prior to the present study the structure of the peptide was partially validated with fast atom bombardment mass spectrometry and micro-Edman sequencing[4].

In this study we used a synthetic peptide analog approach in an attempt to further delimit the mAb C9 reactive epitope on the RV capsid protein existing somewhere from residue number 64 to 97. We conveniently synthesized several derivatives of the RV C64-97 region using micro-methodology suitable for subsequent mass spectrometric analysis and immunoanalysis by standard and competitive ELISA. The experimental approach that we took was a logical extension of work that we have previously[8] described involving the preparation of analytical amounts of peptide from 1-10 mg of peptide-resin that were automatically removed after every residue addition cycle by an instrumental resin sampling feature of the synthesizer. In the previous study we used the analyte peptides for mass spectrometric analysis so that the stepwise assembly of a peptide in an automated run could be carefully evaluated. The analyte levels of synthetic peptides were prepared by a micro-TFMSA cleavage and deblocking method. In the present study we used the peptides in ELISA immunoassays for evaluation of mAb immunoreactivity. In order to rule out "false" negative solid-phase ELISA results due to incorrect assembly of synthetic peptides or chemical modification, stringent assessment of the structural status of the synthetic peptides was performed by electrospray ionization mass spectrometry

(ESIMS)[9,10], a mass spectrometric method that permitted analysis of the full length product and all derivatives. ESIMS was the method of choice since FABMS analysis did not readily permit structural evaluation of the full length product of the synthetic RV C64-97 peptide.

II. Experimental

A. Peptide Synthesis

All peptides were synthesized with an Applied Biosystems Inc. (ABI) Model 430A automated peptide synthesizer using Merrifield solid phase t-BOC(benzyloxycarbonyl) methodology[11]. The starting material was a BOC-C-terminal amino acid-OCH_2-Pam-copoly(styrene-1%-divinylbenzene) resin obtained from ABI. Approximately 10 mg peptide-resin aliquots were automatically collected after each coupling. All BOC amino acids were obtained from Peninsula Laboratories Inc. The side chain protecting groups were Bzl (Ser & Thr), Cl-Z(Lys), CHO (Trp), OBzl (Asp & Glu) and both TOS and MTS for Arg. All of the peptide synthesis chemicals were from ABI except trifluoroacetic acid (Halocarbon Products Corp., Hakensack, NJ) DCM and DMF (Burdick and Jackson). Preparative amounts of peptide were deblocked and released from resin by HF cleavage (Immuno-Dynamics, Inc., La Jolla, CA). Since the peptides contained formyl(CHO) Trp, they were deformylated in 10% piperidine in DMF.

B. Micro-TFMSA Cleavage

As previously described[8] 1-10 mg amounts of peptide-resins were treated with trifluoromethanesulfonic acid(TFMSA) in TFA in the presence of scavengers for 15 min at RT. Released peptides were precipitated with ether and dissolved in 0.1% TFA and further diluted in either 50% methanol - 5% HOAc for ESIMS or appropriate buffers for ELISA immunoanalysis. Since the peptides contained formyl-Trp, deformylation was performed prior to TFMSA treatment. The deformylation conditions were a 15 min/RT treatment of the peptide-resin in 10% piperidine in DMF followed by several DCM washes and a Speed-vac removal of residual DCM.

C. Electrospray Ionization Mass Spectrometry

Electrospray ionization was performed on a Finnigan MAT TSQ 70 quadrupole mass spectrometer equipped with a Finnigan MAT ESI ion source. Samples were analyzed by direct infusion at a flow rate of 2.25 μl/min using a sample matrix consisting of 50% methanol - 5% acetic acid. The typical operating conditions were

Q3MS scanning from m/z 200-2000, 1s/scan
Acquisition time < 5 min @ 20 pmol/μL or less
Unit resolution
N_2 drying gas flow setting 10, 60°C
EMULT 1.5 kv, electrometer gain (ELGAIN) 10^{-8}A/V
ESI voltage = 5 kV

D. Immunoanalysis

1. **Monoclonal Antibody C9 (mAb C9):** mAb C9 was generated and characterized by procedures previously described[4]. Briefly 2-4 week old male BALB/c mice were immunized by infection with one or more intraperitoneal injections of live RV. An additional intrasplenic injection was performed 4 days prior to spleen harvest for the preparation of mononuclear cells for hybridoma partner fusion. Hybridoma selection was biased using SDS/DTT treated virions as the target antigen in standard ELISA. The hybridoma cell line was established by cloning in soft agar. Ascites was produced by injecting hybridoma cells into the peritoneal cavity of primed BALB/c mice. mAb C9 was found to be immunoblot reactive with whole C protein derived from RV and C protein derivatives expressed in E.coli transformed with plasmids containing information coding for different portions of the RV C protein. One plasmid pBstUI coded for RV C64-97, a 34 residue region of the RV capsid protein.

2. **Standard ELISA:** ELISA immunoanalyses were performed as previously described[12] and entail the following steps.

   a. 40-100 pmol of synthetic peptide derivative diluted into 0.05 M sodium bicarbonate buffer pH 9.5 with overnight adsorption in polystyrene microtiter plates at 4° C.
   b. Plate wells washed with 0.1% Tween 20 in PBS pH 7.0 3 times.
   c. Nonspecific protein binding sites present in the wells were blocked by treating with 0.5% BSA, 0.1% Tween, PBS.
   d. mAb C9 (1/1000 dil. ascites) was added to the peptide coated wells and incubated for 90 min at RT.
   e. Plate wells were washed again 3 times.
   f. HRP-conjugated secondary antibody was added and allowed to incubate for 90 min at RT.
   g. Wells were exhaustively washed (5X) and color development was measured at A_{492} after addition of HRP-substrate: 200 μL 2.3 mM ortho-phenylenediamine, 0.37 M citrate, 0.8 M phosphate, pH 5.0 and 0.3% H_2O_2. The enzymic conversion of substrate (color development) was allowed to proceed for 5 min at RT and stopped with the addition of 25 μL of 5 N sulfuric acid.

3. **Competitive ELISA:** A competitive ELISA assay was also performed to evaluate the synthetic peptide derivatives. This approach was included to avoid "false" negative results if some peptide derivatives were preferentially removed from the adsorptive surface of the microtiter plate well. In addition, there has been the suggestion that there might not be sufficient amounts of epitope especially if short peptides are adsorbed to polystyrene[13]. In the competitive ELISA all steps were identical as described above except that the target antigen adsorbed to the plates was exclusively the RV C64-97 34 aa residue synthetic peptide preparation, and the mAb C9 solution was equilibrated with a large excess of synthetic peptide derivative prior to addition to the microtiter plate. Also, while color development in the well in the standard ELISA indicated synthetic peptide immunoreactivity with bound mAb C9 and was taken as a positive result, color development in the well for the competitive ELISA indicated nonreactivity with soluble mAb C9 and was taken as a negative result. Attenuated color development or absence of color development was an indication of immunoreactivity with mAb C9 in the competitive ELISA analysis and taken as a positive result.

III. Results

A. ESIMS of HF/RV C64-97 Synthetic Peptide Derivatives

A summary of the synthetic peptide derivatives analyzed both by ESIMS for structural status and immunological reactivity to mAb C9 is shown in Figure 1. Part A of the Figure shows the desired sequences for four preparative HF-derived synthetic products representing different portions of the full length sequence in question. The upper sequence designated HF/C64-97 is the full length synthetic product that had been previously characterized and shown to be reactive with mAb C9 in standard ELISA[4]. The ESI mass spectrum of this product (data not shown) enabled a clear evaluation of this peptide. The expected molecular weight was 3883.2 and observed molecular species were 3884.1, 3865.6, 3998.9, 3981.1, 3788.2, 3767.9. This complex ESIMS spectrum indicated that this synthetic peptide preparation is a mixture of three peptide derivatives; 3884.1 (desired product), 3998.9 (Asn^{65} insertion derivative), and 3788.2 ($Pro^{78\text{-}81}$ deletion derivative); and three companion peaks all 18 mass units lower than the previous molecular species. The exact location or the chemical mechanism for the 18 mass unit loss has not yet been determined.

Preparation HF/C78-97 was also assessed by ESIMS. The major ion in the spectrum was that of the desired C-terminal portion of the peptide. A pro-deletion product was evident here also. It should be noted that no product showing a loss of 18 mass units was detectable. When these two products were subjected to standard ELISA analysis, the full length derivative

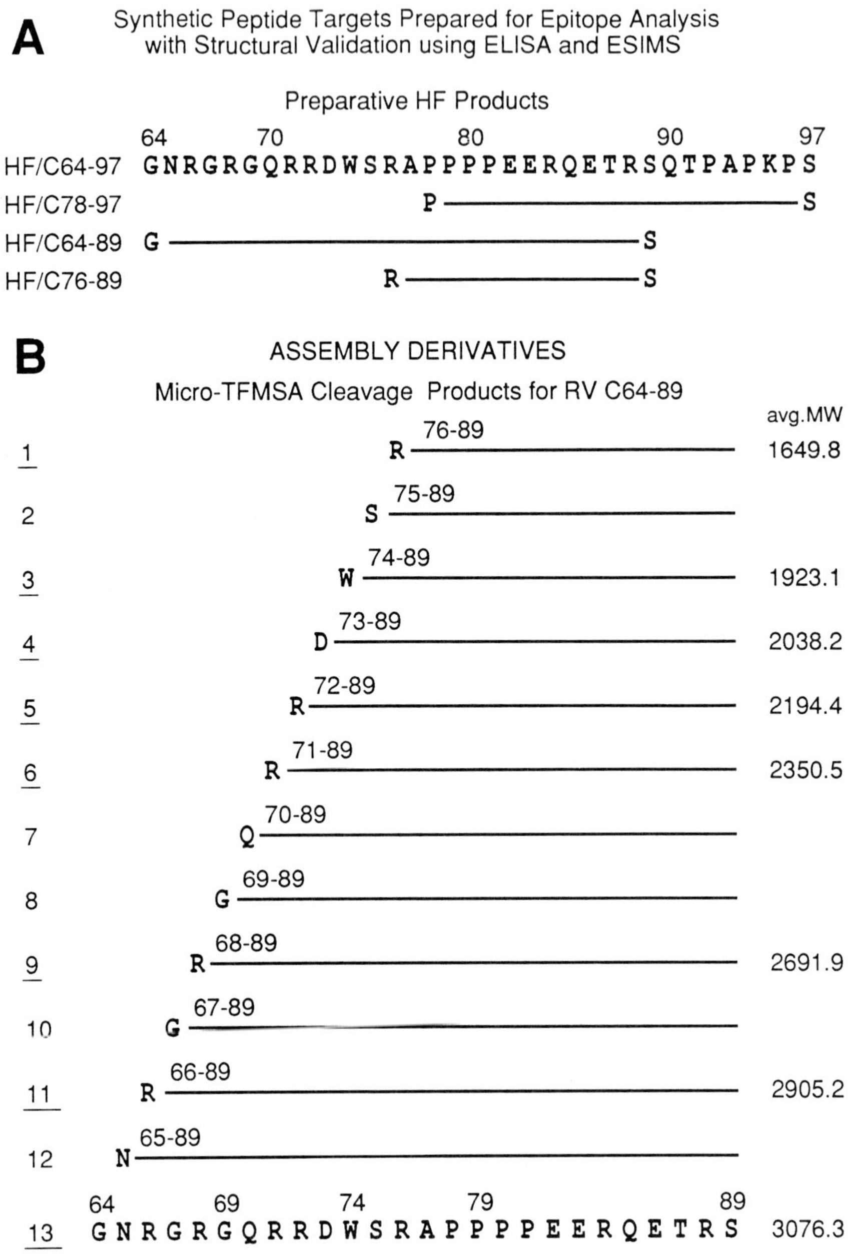

Figure 1. Rubella virus capsid protein synthetic peptide analogs from region C64-97. Panel A - HF-derived peptide analogs. Panel B - micro-TFMSA-derived peptide analogs (assembly derivatives).

was mAb C9 reactive and the C-terminal portion was nonreactive[4]. The expected molecular weight was 2229.4, and measured masses were 2228.0, 2383.0, and 2133.0. The ESIMS spectrum for this peptide indicated that the preparation is a mixture containing three major products. The desired product (2228.0 da.) is present as +4, +3, and +2 charge state ions, the +4 ion being the most abundant. An ion having a mass consistent with the presence of a tosyl group on either R^{70} or R^{72} is also indicated. It is of interest to note that the most abundant ion is the +3 charge state for this species indicating that the tosyl blocking group prevents one of these arginines from acting as a proton acceptor. A Pro deletion product was also observed in this precursor as expected since this was seen in the final preparation. The most abundant charge state for the desired product and the Pro-deletion peptide were both +4 suggesting that all expected basic residues are protonated.

ESIMS data for the next two shorter HF-derived derivatives are shown in Figure 2A and Figure 3A. These data are presented since the longer derivative was found to be mAb C9 reactive enabling the removal of some of the C-terminal region residues from consideration as part of the epitope involved. The synthesis of this product also yielded the stepwise assembly products listed in Part B of Figure 1.

The deconvoluted ESIMS spectrum for precursor HF/RV C76-89 presented in Figure 2A indicates the presence of two species, the desired product at 1650.0 and a Pro-deletion product. These two species are present as +3 and +2 charge state ions. The maximum charge state being +3 suggests that perhaps only one of the basic groups of the N-terminal Arg holds a proton. The ESIMS spectra for HF/RV C64-89 is shown in Figure 3A. The deconvoluted spectrum derived from the ESI mass spectrum indicates the presence of a mixture of at least 4 major components, the desired product at 3078.0 and product from loss of 18 mass units at 3059.0, and a Pro deletion product at 2980 and its product from loss of 18 mass units.

To summarize the results for the full length HF-derived products, RV C64-97 and the C-terminal truncated analog RV C64-89 were positive for interaction with mAb C9 as demonstrated by standard ELISA. The major component of the peptide mixtures in these preparations were shown by the ESIMS analysis to be the desired products. The fact that the precursor derivatives were negative in ELISA suggests that critical residues contributing to the epitope or antibody binding region of the peptide are localized in the N-terminal portion of the synthetic peptide derivatives. This set of derivatives narrows the region for antibody binding on RV C64-97 to the area underlined below, a region comprising a sequence of 26 N-terminal residues.

<u>GNRGRGQRRDWSRAPPPPEERQETRS</u>QTPAPKPS

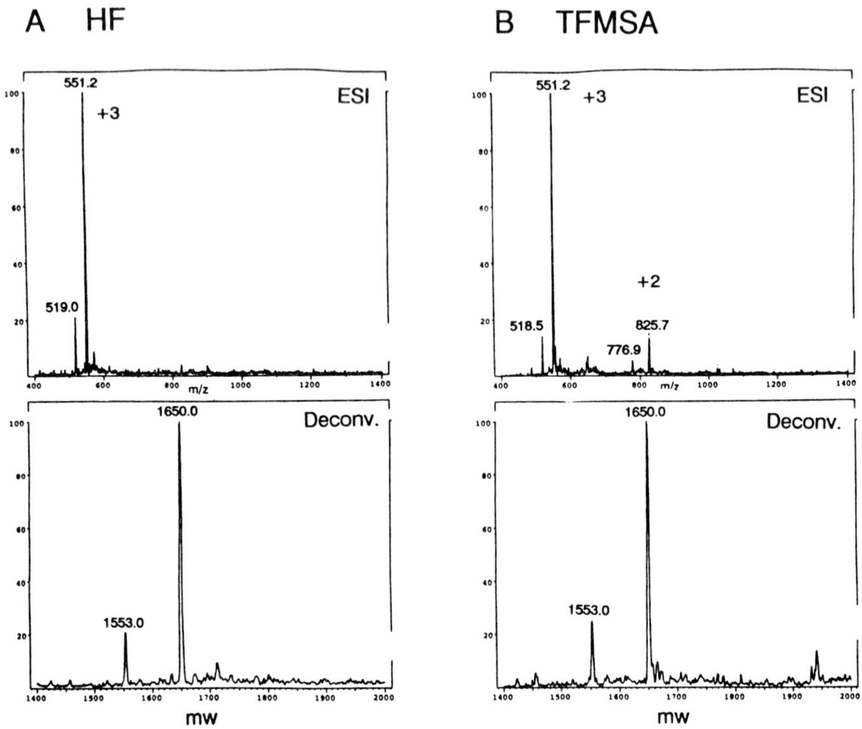

Figure 2. ESIMS analysis of non-immunoreactive precursor synthetic peptide RV C76-89 derived from HF (A) and TFMSA (B) treatments. The upper spectra are ESI spectra and the lower spectra are deconvoluted spectra. The RV C76-89 sequence is RAPPPPEERQETRS and the expected average MW is 1649.8. Charge states are indicated in the ESI spectra (+3 & +2).

B. Epitope Mapping Using the Micro-TFMSA Cleavage Method to Generate Synthetic Peptide Derivatives with ESIMS Structure Verification

Before the assembly derivatives were analyzed by either ELISA or ESIMS, the full-length product RV C64-89 that was obtained by the micro-TFMSA cleavage chemistry was evaluated. The ESIMS spectra for the micro-TFMSA precursor peptide and the full-length product are presented along with the HF-derived preparations in Figures 2 and 3. The spectra for the TFMSA-derived preparations are shown in Part B of both Figures. Note that the micro-TFMSA derived product spectrum is essentially identical to that of the HF-derived material indicating that cleavage and deblocking in TFMSA was as efficient as HF for this peptide. Note however that in Figure 3, that the spectra for the TFMSA-derived full-length RVC64-89 derivative (Panel B) are quite different from that of the HF-derived material (Panel A). The

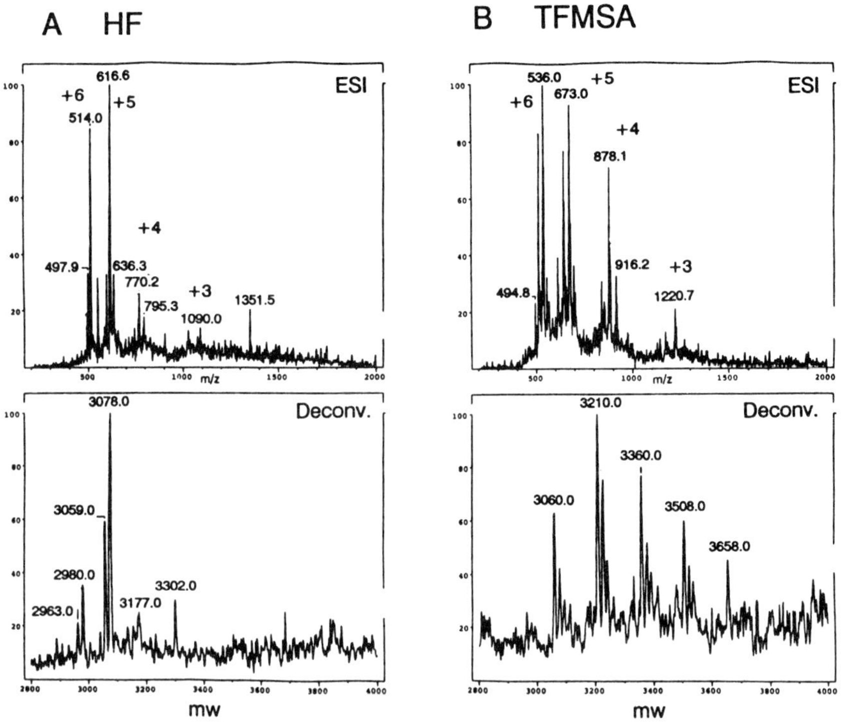

Figure 3. ESIMS analysis of mAb C9 reactive synthetic RV C64-89 preparations derived from HF (A) and TFMSA (B) treatment. The upper spectra are ESI spectra and the lower spectra are deconvoluted spectra. The RV C64-89 sequence is GNRGRGQRRDWSRAPPPPEERQETRS and the expected average MW is 3076.3. Charge states are indicated in the ESI spectra (+3, +4, +5 etc.)

ESIMS spectra for the micro-TFMSA derived material contain more lower charge state clusters and the deconvoluted spectra contain a series of signals starting at 3060.0 extending to 3658. The series are spaced equally at 150 masses and have been seen previously in FABMS analysis of similar samples[8]. The series arise from noncovalent adduction of TFMSA to the basic residues of the peptide (ion pairing that survives the ESI process). Note the inverse relationship between the abundance of the respective charge states and the extent of TFMSA adduction. It was found that the full length derivatives gave a positive result upon ELISA analysis with mAb C9. It should be noted as seen in Figure 3B, the predominant ion signal is that of the desired peptide mass minus 18 mass units. It appears that this product was preferentially produced unlike that seen for the HF-derived material. It appears that the loss of 18 mass units has little effect on the immunoreactivity of the peptide.

Another synthesis was performed from which a set of assembly derivatives was derived by a separate deformylation treatment and micro-TFMSA cleavage treatment. The products were found to be nonreactive with

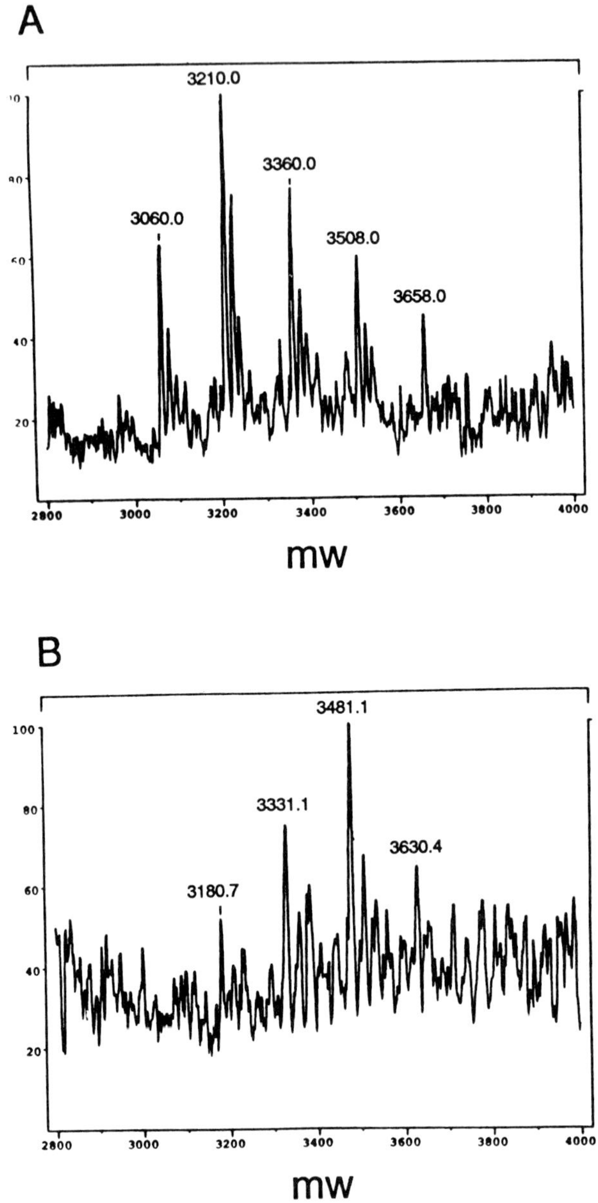

Figure 4. Comparison of the deconvoluted ESIMS spectrum of the mAb C9 reactive RV C64-89 synthetic peptide (A) with that for the nonreactive preparation that was derived from a separate micro-deblocking and TFMSA cleavage treatment (B).

mAb C9. The ESIMS spectra for the full-length derivative is compared with that for the mAb C9 reactive material in Figure 4. Panel A is the deconvoluted spectrum for the active TFMSA-derived material previously described. Panel B is the deconvoluted spectrum for the inactive product. In Figure 4B, note that there is a mass difference of about 121 amu over that of the major product in Figure 4A. But because the predominant product shown is the -18 mass product, the difference is really 106 amu over that of the desired product.

C. *Assessment of Assembly Derivatives by ELISA and ESIMS*

The synthesis that provided the mAb C9 reactive full-length micro-TFMSA derived product also provided assembly derivatives that could be assessed by both ELISA and ESIMS. All the assembly derivatives examined are schematically presented in Figure 1B and each is assigned a number. Each derivative was immunoanalyzed by standard ELISA and competitive ELISA and the results are shown in Figure 5. The results indicate that mAb C9 reactivity is detected minimally when derivative number 8 is encountered

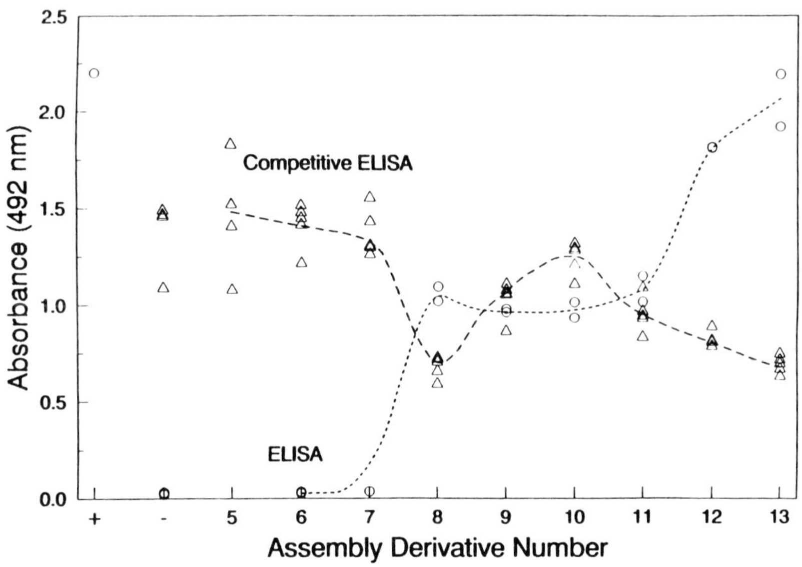

Figure 5. Immunoanalysis of the synthetic peptide RV C64-89 assembly derivatives. Open circles represent the standard ELISA assay result for assembly derivatives 5-13 (see Fig. 1). Open triangles represent competitive ELISA results. The + and - on the abscissa indicate the positive and negative controls for the standard ELISA. All data points from replicate determinations are presented.

and thereafter, the full length product being the most active. Similar results were obtained for both the standard ELISA and the competitive ELISA. Only those derivative numbers underlined in Figure 1B were assessed by ESIMS. A limited set of these data is presented in Figure 6. The deconvoluted spectra for assembly derivatives 3, 4, and 5 are shown because they are the most interesting members of the set. These three spectra indicate the locus of the loss of 18 mass units that was seen in the full length derivative spectrum shown in Figure 3B. As can be deduced from Figure 6 this loss occurs when Asp^{73} is present in the derivative. These three spectra and the spectrum in Fig. 4A for derivative number 13 indicate the successful synthetic history for this peptide beyond the Asp^{73} coupling step. Although the data is not shown, it was determined mass spectrometrically that the +106 change on the set of assembly derivatives that was nonreactive with mAb C9 also was connected to the presence of Asp^{73}. The characterization of the nature of this particular modification believed to take place around the locus of Asp^{73} awaits tandem MS/MS analysis. The location of this modification site is important since this series of assembly derivatives was inactive in ELISA suggesting that the site of modification is within the epitope structure. If this modification does indeed occur at the Asp or the adjacent Trp, we feel that this region should be considered part of the epitope and could constitute the C-terminal limit.

Conclusions

1. ESIMS analysis yields a clear and realistic picture of the outcome of a particular automated synthetic run and effectively assesses the structures of the analyte peptides derived from the micro-TFMSA method.

2. Either the micro-deformylation or micro-TFMSA treatment induced some changes in the peptide structure. In one micro-TFMSA generated series that contained mAb C9 reactive species, a loss of 18 amu was observed from the point where Asp^{73} was coupled. That assembly derivative and all descendants show this same 18 mass unit loss. It should be noted that there is also a small amount of this product generated by HF-derived product which had also undergone deformylation with piperidine treatment. Loss of water from Asp^{73} suggests a side reaction is taking place that has been described in peptides which contain aspartyl residues having the beta carboxyl group blocked as a benzyl ester. This reaction is a base catalyzed ring closure[14] resulting in the formation of a beta-aspartyl residue succinimide, a consequence of an intramolecular attack of a proton abstracted from the adjacent amide nitrogen. It is noted, however, that this reaction is considered to be slow if the residue on the alpha-carboxyl side of the Asp residue has a bulky R-group which is the case in the peptide analogs under study. These peptide-resins were treated

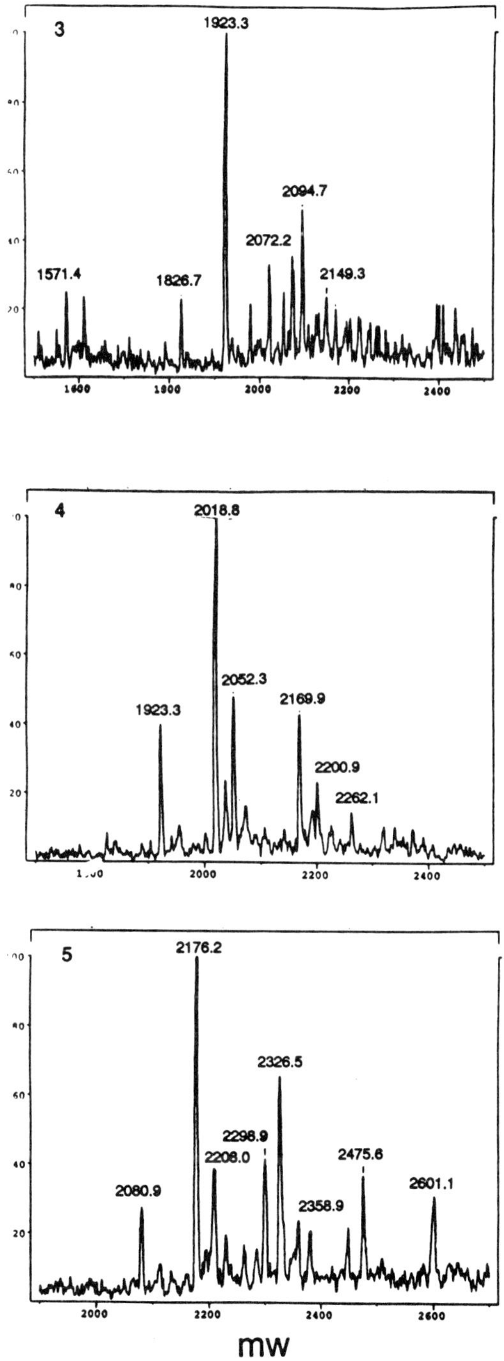

Figure 6. Deconvoluted ESIMS spectra for synthetic RV C64-89 assembly derivatives, TFMSA derived peptides 3, 4 and 5. See Fig. 1 for sequences and expected mw.

however with 10% dipiperidine in DMF to deformylate the adjacent Trp before micro-TFMSA cleavage. Verification that this chemical modification occurred at the expected site awaits confirmation by tandem MS/MS analysis. In another micro-deformylation/TFMSA generated series, a gain of 106 amu was noted occurring at the same site in the synthesis. Again the Asp^{73} assembly derivative and the descendants that followed inherited the +106 change.

3. Noncovalent TFMSA adducts were noted in the ESI spectra for the longer analyte peptides containing several Arg residues. There appears to be an inverse relationship between the number of TFMSA's adducted and the charge state abundance for a particular adducted species, i.e., TFMSA adduction involves an ionic interaction with the basic group of Arg and prevents it from becoming a proton acceptor.

4. ELISA analysis of the synthetic peptides have permitted a tentative definition of the minimum epitope containing domain on RV C64-89 for mAb C9 as

```
   65          70          74
----N R G R G Q R R D W-->?
```

The involvement of N^{65} to G^{69}, the N-terminal side of the epitope is clear from both the standard and competitive ELISA results. The involvement of the residues from Q^{70} to W^{74} is less certain. This C-terminal side of the minimal limit is inferred from the +106 amu chemical modification that arose upon the micro-deformylation/TFMSA treatment of the assembly products at and after the addition of Asp^{73}. Confirmation of the location and the nature of this chemical change that renders the peptide non reactive with the antibody awaits further analysis.

5. The micro-chemistry described here generates analytc levels of peptides that are suitable for ELISA testing. This methodology coupled with mass spectrometric confirmation of structure permits a means to perform precise mapping studies on synthetic peptides having contiguous epitopes.

Acknowledgment

This work has been supported by grants from the National Institute of Health (grant #5RO1 GM43783-02 awarded to Richard M. Caprioli, and grant #A126943 awarded to Jerry S. Wolinsky).

References

1. Bowden, D.S., and Westaway, E.G. (1984). *J. Gen. Virol.,* **65**, 933-943.
2. Dominguez, G., Wang, C.-Y., and Frey, T.K. (1990). *Virology,* **177**, 225-238.
3. Wolinsky, J.S. (1990). *In* "Virology" (B.N. Fields, D.M. Knipe, R.M. Chanock, J.L. Melnick, B. Roizman and R.E. Shope, eds.) Rubella Virus, p. 815-840, Raven Press, New York.
4. Wolinsky, J.S., McCarthy, M., Allen-Cannady, O., Moore, W.T., Jin, R., Cao, S-N., Lovett, A. and Simmons, D. (1991). *J. Virol.,* **65**, In Press.
5. Karounas, D.G., Wolinsky, J.S., and Thomas, J.W. (1991). *J. Immunol.* (submitted).
6. Karounos, D.G., and Thomas, J.W. (1990). *Diabetes,* **39**, 1085-1090.
7. Oldstone, M.B.A. (1989). *Curr. Top. Microbiol. Immunol.,* **145**, 127-135.
8. Moore, W.T., and Caprioli, R.M. (1991). *In* "Techniques in Protein Chemistry II", (J. Villafranca, ed.), p. 511-527, Academic Press, San Diego, CA.
9. Fenn, J.B., Mann, M. Meng, C.K., Wong, S.F., and Whitehouse, C.M. (1989). *Science,* **246**, 64-71.
10. Smith, R.D., Loo, J.A., Edmonds, C.G., Barinaga, C.J., and Udseth, H.R. (1990). *Anal. Chem.,* **62**, 882-899.
11. Merrifield, R.B., Vizioli, L.D., and Boman, H.G. (1982). *Biochemistry,* **21**, 5020-5031.
12. Waxham, M.N., and Wolinsky, J.S. (1985). *Virology,* **143**, 153-165.
13. Geysen, H.M., Rodda, S.J., Mason, T.J., Tribbick, G., and Schoofs, P.G. (1987). *J. of Immunol. Meth.,* **102**, 259-274.
14. M. Bodanszky (1988). *In* "Peptide Chemistry A Practical Textbook", pp. 104-108, Spinger-Verlag.

TFMSA / TFA Cleavage in t-Boc Peptide Synthesis

Farzin Gharahdaghi, Jeffrey Mathers and Sheenah M. Mische

The Rockefeller University Protein Sequencing / HHMI Biopolymer Facility

New York, New York 10021

I. INTRODUCTION

The complete removal of the protecting groups employed in SPPS[1] is of great importance in order to obtain the desired peptide in satisfactory yield and high degree of purity. Most protecting group strategies in peptide synthesis are based on differential susceptibility to acids. HF is the most widely used method, but requires a highly specialized hood and vent, and is extremely dangerous. Our facility has space limitations which preclude HF cleavage. Therefore our peptides are cleaved by a commercial firm, which adds both time and expense to our analyses. It was for this reason that we began cleaving and deprotecting peptides with TFMSA/TFA, initially pioneered in solution phase synthesis by Yajima and coworkers (1-3). The main advantage of this method is that TFMSA, although the strongest of all organic acids, can be easily handled in conventional laboratory apparatus. The main disadvantages are 1) TFMSA is a viscous liquid, requiring peptides to be precipitated and 2) organic salts have a propensity for synthetic peptides cleaved with TFMSA, and therefore require desalting after cleavage.

---

[1] Abbreviations Used: ABI, Applied Biosystems Inc.; BOC, t-butyloxycarbonyl; Bom, benzyloxymethyl; Bzl, benzyl; CH_3CN, acetonitrile; ClZ, 2-chlorobenzyloxycarbonyl; DCC, dicyclohexylcarbodiimide; DCM, dichloromethane; NMP, N-methylpyrrolidone; DIEA, diisopropylethylamine; DMF, N,N-dimethylformamide; DMSO, dimethyl sulfoxide; HF, hydrogen fluoride; HOBt, 1-hydroxylbenzotriazole; MBzl, 4-methylbenzyl; Mts, N^G-mesitylene-2-sulfonyl; OBzl, benzyl ester; PAM resin, phenylacetamidomethyl resin; SPPS, solid phase peptide synthesis; TEA, triethanolamine; TFA, trifluoroacetic acid; TOS, 4-toluenesulfonyl; EDT, 1,2-ethanedithiol.

TECHNIQUES IN PROTEIN CHEMISTRY III

Our synthetic strategy involved the preparation of protected peptides on either PAM or BHA resin using a modified Merrifield SPPS method with BOC protected amino acids. N^G-mesitylene-2-sulfonyl-arginine (Arg(Mts)) was used as an alternative to Arg(Tos). Here we present our method for TFMSA/TFA cleavage and deprotection which we have used to synthesize peptides ranging from 8 to 32 amino acids in length. Data comparing TFMSA and HF cleaved peptides is presented, demonstrating that TFMSA is a feasible alternative to HF cleavage for the majority of peptides that have been synthesized in our facility. The application of TFMSA/TFA cleavage for the synthesis of phosphopeptides is also presented.

II. MATERIALS AND METHODS

Materials: NMP and CH_3CN were purchased from Burdick and Jackson, methanol and ethyl ether were obtained from Fischer; DMS, m-cresol, and EDT were from Aldrich. All other reagents and solvents and Boc protected amino acids were purchased from ABI, except for His(Bom), which was purchased from Bachem, and Tyr-O-[PO_3Bzl_2], purchased from Peninsula Laboratories. Preparative (21.4 mm ID x 25 cm) and analytical (4.6 mm ID x 25 cm) C8 Dynamax columns were obtained from Rainin. All other reagents and solvents were of reagent grade and used as supplied.

Peptide Synthesis: All oligopeptides were synthesized using an Applied Biosystems Model 430A peptide synthesizer using HOBt ester intermediates with NMP solid phase chemistry employing either PAM or BHA resins (as noted). All of the amino acids were single coupled for the first 20 cycles. Amino acids with significant charge differences and all amino acids beyond 20 amino acids were double coupled. The efficiencies of the individual couplings were monitored by standard ninhydrin assays (4), and ranged from 99.6 to 99.9%.

TFMSA/TFA Cleavage: Simultaneous side chain deprotections and peptide cleavages were performed with TFMSA/TFA using a modification of procedures published by ABI and Tam and coworkers (5,6). Table I outlines the methods used for both standard and low/high TFMSA/TFA cleavage.

HF Cleavage: Peptides (including peptides containing Arg(Mts)) were cleaved in liquid HF at -5°C in the presence of m-cresol, except where noted. All HF cleavages were performed by ABI using published methods (5).

TABLE I Standard and low/high TFMSA/TFA methods

A. Standard TFMSA/TFA Cleavage

Reagent	Volume	Glassware	Time	Note
1. Cleavage				
A. Thioanisole:EDT (2:1)	0.15 ml	100 ml round bottom flask	10 min	stir gently
B. TFA	1.00 ml	add to 1A	10 min	stir gently
C. TFMSA	0.10 ml (DROPWISE)	add to 1B	Pam resin 1 h BHA resin 1.5 - 2 h	slow addition of TFMSA; stir gently
2. Precipitation				
ethyl ether	100 ml	add to 1C	5 min	transfer to fritted glass funnel
3. Wash				
ethyl ether	200 ml	vacuum flask, fine frit glass funnel	dry for 20 min	
4. Extraction				
ethyl ether	3 x 1 ml	vacuum flask	dry for 30 min	flask contains 200 ml cold ethyl ether
5. Wash				
ethyl ether	300 ml	vacuum flask, ultra-fine fritted glass funnel	dry for 30 min	
6. Lyophilization	200-300 ml		12 - 15 h	dissolve peptide

B. Low - High TFMSA/TFA Cleavage

Reagent	Volume	Glassware	Time	Note
1. Low Condition				
A. Scavengers				
Met(O) m-cresol : DMS (1:3)	0.4 ml	100 ml round bottom flask	2 - 3 min	stir gently
Trp(CHO) EDT: m-cresol (1:4)	0.1 ml	add to 1A	2 - 3 min	stir gently
B. TFA	0.5 ml	add to 1B	2 - 3 min	stir gently
C. TFMSA	0.1 ml	add to 1C	3 h	stir gently -5 C - 0 C
2. Filtration				
ethyl ether	100 - 200 ml	vacuum flask, fine fritted glass funnel	20 min	under vacuum
3. High Condition: refer to Table 1A				

HPLC Analysis and Desalting: A Waters HPLC equipped with two model 600 pumps, a model 490 variable detector and Rheodyne injectors (2 ml and 5 ml sample volume) was used for both the analytical and preparative chromatography of synthetic peptides. Crude lyophilized peptide was solubilized in 0.1% TFA/CH_3CN and chromatographed at a flow rate of 1 ml/min on a Dynamax C8 analytical column (4.6 mm ID x 25 cm) with 0.1% TFA (buffer A) and 0.1% TFA/CH_3CN (buffer B) in a gradient of 0% to 60% buffer B over 60 minutes. Preparative desalting was performed on a Dynamax C8 preparative column (21.4 mm ID x 25 cm) with the same buffers and same gradient at a flow rate of 21.6 ml/min. Prior to desalting, crude solubilized peptide was neutralized (pH 8.0) with a solution of 10% NH_4HCO_3.

Mass Spectrometry: Mass spectra were recorded on a Bio-Ion 20 ^{252}Cf plasma desorption time-of-flight mass analyzer (Bio-Ion AB Uppsala, Sweden). Peptides were dissolved in 0.1% TFA and 5 ul of each analyzed using nitrocellulose targets. Spectra for each peptide were accumulated for 5-10 minutes and were performed by ABI (Foster City, Ca). Electrospray mass spectrometry was performed by The Rockefeller University Mass Spectrometry Resource Center.

Amino Acid Analysis: PTC amino acid analysis was performed using a Waters Pico-Tag Workstation and HPLC two pump gradient system equipped with a Model 490 mutiwavelength detector, WISP, and model 820 or 840 data station. Peptides were vapor hydrolyzed at 150°C for 1h, or 110°C for 22h, followed by derivatization and analysis, as described by Atherton (7).

Peptide Sequencing: Sequence analysis was performed on either an ABI 470A gas phase sequencer equipped with a 900A data acquisition system or a model 477A pulsed liquid sequencer, both equipped with on-line 120 PTH analyzers.

Calculations

$$\text{Theor. weight of crude peptide (\% CP)} = \frac{\text{Theor. weight peptide}}{\text{Theor. weight peptide/resin}} \times 100$$

$$\text{Expected recovery} = (\text{\%CP})(\text{submitted amount for cleavage})$$

$$\text{Percent Recovery} = \frac{\text{Peptide (mg) recovered post-cleavage}}{\text{Expected Recovery}} \times 100$$

$$\text{Percent Yield} = \frac{\text{Actual quantity calculated by AAA}}{\text{Submitted amount for AAA}} \times 100$$

TABLE II Summary of peptides cleaved and deprotected with TFMSA/TFA

PEPTIDE SEQUENCE	LENGTH	RESIN TYPE	CLEAVAGE	% RECOVERY
MLKVPNIS	8	PAM-SER	MET L/H	17%
SLKTTVPEV	9	BHA	1.5 HR	29%
KYGLAWERY	9	BHA	TRP L/H[a]	11%
CTENQENGSF	10	PAM-PHE	1 HR	69%
LPALESLWETK	11	BHA	TRP L/H[a]	38%
YPYDVPDYASLC	12	BHA	1.5 HR	45%
CPQPTITKQEVI	12	PAM-ILE	25 MIN	81%
CASVQGYMEGASK	13	PAM-LYS	MET L/H[b]	34%
SFKQRKSQSSSSS	13	BHA	1.5 HR	31%
KSSTETPPSYNQLN	14	PAM-ASN	1 HR	32%
FHLRAHAVDVNGNQC	15	PAM-CYS[c]	1.5 HR	51%
SRISLPLPNFSSLNLK	16	PAM-LYS	1.5 HR	28%
CNHFRFLKGYWSKFTDK	17	PAM-LYS	TRP L/H[a]	13%
CFIKLFGMAKIDNLLQE	17	PAM-GLU	MET L/H[b]	53%
KSKFLALGSKFRYSGRT	17	PAM-THR	1.5 HR	33%
MKKKAGPYSSVPPPSVSLC	19	PAM-CYS	MET L/H[b]	53%
SGQLQANGALIASLWRKVA	19	PAM-ALA	2.0 HR	36%
MSAPTPLAEASQIPTIPALS	20	PAM-SER	MET L/H[b]	49%
DNLFAQGNIDNDLSVAGRLNY	21	PAM-TYR	2.0 HR	50%
LELPHENELTVSERDSVMLG EISPHHDYYDSK	32	PAM-LYS	MET L/H[b]	68%

Protecting groups used are C(MOB), R(Mts), Y(Br-Z), D(OBzl), E(OBzl), H(Bom), K(ClZ), W(CHO), N(OBzl), M(O), S(Bzl)

[a] a Trp L/H: low-high cleavage method for Trp(CHO).

[b] Met L/H: low-high cleavage method for Met(O).

[c] PAM-Cys(Mb)

III. RESULTS AND DISCUSSION

Within the past year we have used TFMSA/TFA as an alternate cleavage method for peptides synthesized in our Facility. Under the conditions described in Table IA, standard TFMSA cleavage and deprotection proceeds by an SN1 mechanism, and is comparable to HF cleavage, whereas low/high procedures for both TFMSA or HF are both SN1 and SN2. The TFMSA low/high procedure (outlined in Table IB) allows the smooth conversion of Met(O) to Met and deformylation of Trp, as documented here. It should be noted that peptides synthesized on PAM resins are cleaved for one hour; this is in contrast to the 25 minutes recommended by ABI (5).

Table II summarizes peptides synthesized by t-Boc chemistry and cleaved with TFMSA/TFA. Peptides cleaved by this method require desalting, as TFMSA is quite viscous and therefore not as easily removed as HF. It should be noted that the lower recoveries of some of the smaller peptides could be due to the additional precipitation and filtration steps (Table II). However, the majority of peptides are recovered in ample quantity, and are comparable in quality to HF peptides even prior to desalting. An example of a direct comparison of peptide recovered after cleavage with TFMSA and HF is shown in Figure 1. As can be seen, both the crude and desalted peptides are comparable in quality and quantity, as determined by sequence analysis, amino acid analysis and mass spectrometry.

The use of MTS as a protecting group for Arginine has expanded the repertoire of peptides that are cleavable with TFMSA. 4-toluenesulfonyl (Tos) is the most common protecting group used for Arginine, yet it requires extended cleavage time with HF. Results presented here demonstrate that Arg(MTS) can be used successfully with either TFMSA or standard HF methods (Table III), and the presence of Arg(MTS) can be easily identified by sequence analysis, as shown in Figure 2.

The use of TFMSA in the synthesis of phosphopeptides is emerging as an important application of this cleavage method for t-Boc synthesis, as HF is precluded from use due to the hydrolysis of the phosphate group from the phosphoamino acid (8). The example presented here demonstrates that TFMSA can be used successfully for the cleavage and deprotection of phosphopeptides. A eight residue peptide was synthesized using Tyr-O-[PO_3Bzl_2]. The phosphoamino acid was double coupled, and the peptide was cleaved as described in Table 1B, with the following modifications: m-cresol:DMS:TFA:TFMSA (10:30:50:10 (v/v)) was used and the

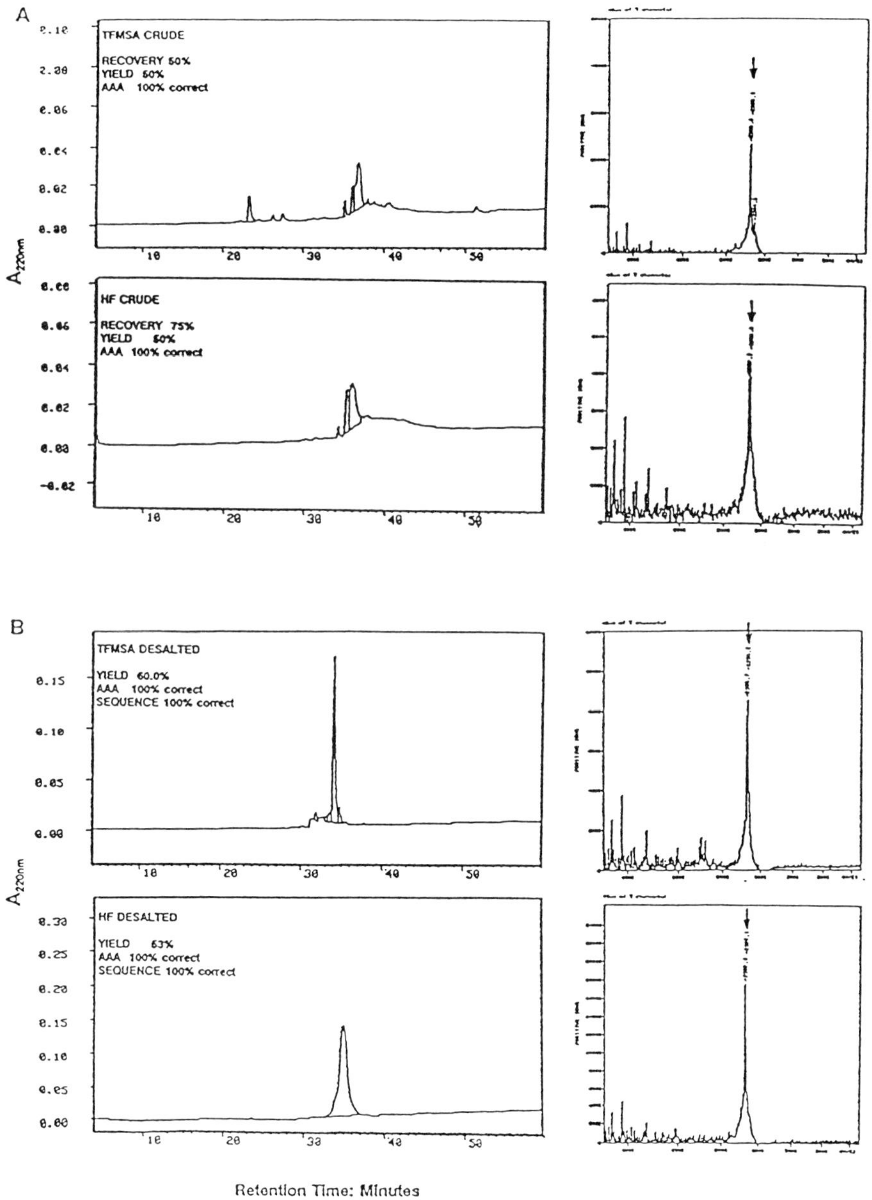

Figure 1 Analytical comparison of DNLFAQGNIDNDLSVAGRLNY cleaved with TFMSA and HF: After synthesis, 50% of the peptide was cleaved with TFMSA/TFA and the other 50% with HF. A) Crude peptide from each cleavage was analyzed by RPHPLC and AAA to determine the composition and weight of peptide, and PDMS. Note that % recovery and % yield was comparable for both methods. B) Each crude peptide was desalted as described and analyzed by HPLC, AAA, and PDMS.

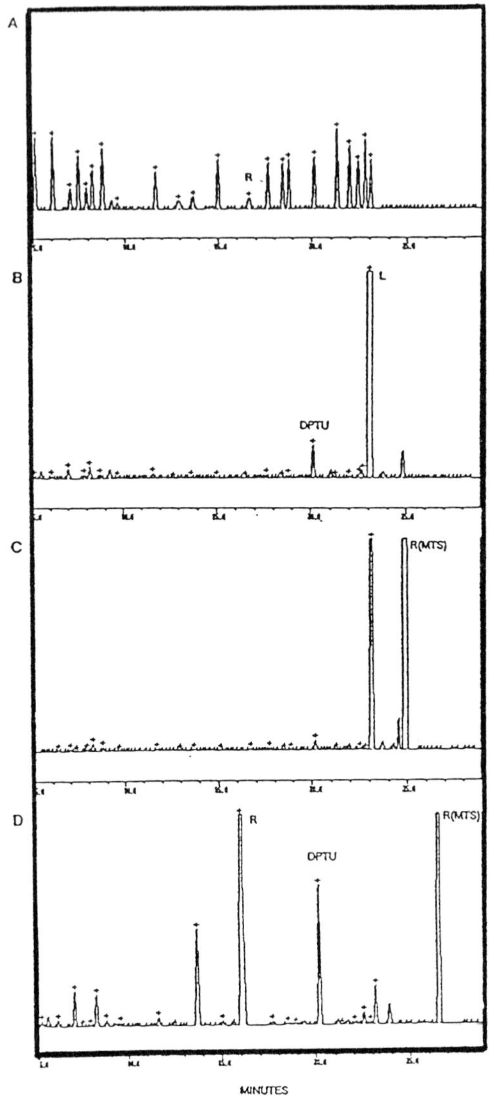

Figure 2 Sequence analysis of Arg(Mts): peptide-resin L-R(Mts) was sequenced in order to determine the identification of Arg containing Mts. Approximately 1 mg of peptide-resin was suspended in 50% 0.1% TFA/CH_3CN and 10 ul of the slurry applied to a blank GFC filter. A) PTH amino acid standard from on-line ABI 120 HPLC and 477A pulsed liquid sequencer. B) First cycle Leu; C) Second cycle Arg(Mts), with Leu carryover. Arg(Mts) elutes approximately 2.5 min after leucine under conditions described in *Materials and Methods*; D) Second residue of peptide SRISLPLPNFSSLNLK. Approximately 40 pmole of desalted peptide sequenced using an ABI 470A sequencer. Note that both Arg and Arg(Mts) are present, denoting incomplete deprotection.

TABLE III Sequence analysis of peptides synthesized with Arg(Mts)

PEPTIDE SEQUENCE	LENGTH	CLEAVAGE	EXTENT OF DEPROTECTION (%)*	CONDITION
KYGLAWERY	9	TFMSA	97	LOW/HIGH
FHLRAHAVDVNGNQC	15	TFMSA	100	STANDARD
SRISLPLPNFSSLNLK	16	HF	40	94 %
		TFMSA	74	STANDARD
CNHFRFLKGYWSKFTDK	17	TFMSA	100	LOW/HIGH
SGQLQANGALIASLWRKVA	19	HF	100	90 %
		TFMSA	100	LOW/HIGH
DNLFAQGNIDNDLSVAGRLNY	21	HF	100	90 %
		TFMSA	100	STANDARD

* Percent extent of deprotection = $\frac{\text{(peak area of Arg(Mts))}}{\text{(peak area Arg)(peak area Arg(Mts)}} \times 100$

peptide cleaved at 0°C for 4 h. Figure 3 represents chromatograms of both the crude (A) and desalted (B) phosphopeptide. The percent recovery was 25%, and electrospray mass spectral analysis (m/z 942.55; predicted value is 941.55) and absorbance spectra (phosphotyrosine $A_{max} = 265_{nm}$ @ pH 3.0) confirmed the presence of the phosphotyrosine residue.

In summary, TFMSA represents a good alternative to HF cleavage for peptide synthesis laboratories that do not have access to an HF apparatus. The health risks and costs associated with TFMSA are considerably less than those associated with HF, and TFMSA produces peptides equal in quality to those cleaved by HF.

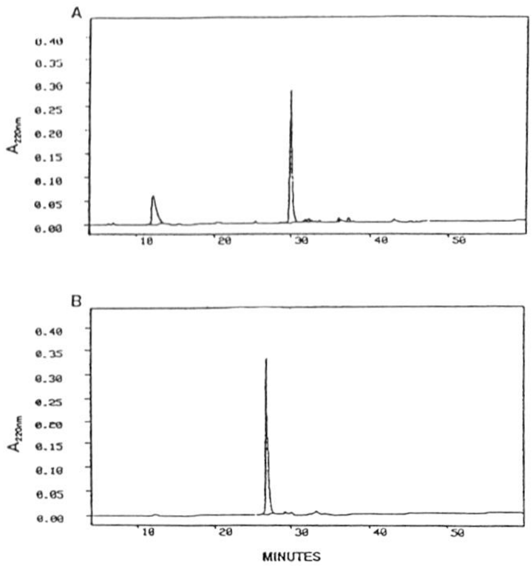

Figure 3 TFMSA/TFA cleavage for phosphopeptide synthesis: A eight residue peptide containing phosphotyrosine was synthesized with Boc-Tyr-O-$[PO_3Bzl_2]$. The phosphoamino acid was double coupled, the peptide cleaved and extracted using the TFMSA/TFA procedure outlined in *Materials and Methods*. HPLC profiles of A) crude phosphopeptide and B) desalted phosphopeptide.

IV. ACKNOWLEDGEMENTS

This work was supported in part by NIH Shared Instrumentation Grant (BRSIG). We gratefully acknowledge the PDMS data which was generated by Pau Yuan and Ling Chen of ABI, Inc.

V. REFERENCES

1. Yajima, H., Funakoshi, S., and Akaji,K. (1985) Int. J. Pep. Res. 26:337-361
2. Yajima, H., Akaji, K., Mitani, K., Fujii, N., Funakoshi, S., Adachi, H., Oishi, M., and Akazawa, Y. (1979) *Int J. Pep. Prot. Res.* 14:169-176
3. Yajima, H., Takayama, M., Kanaki, J., and Mitani, K. (1978) *J. Chem. Soc. Chem. Comm.* 482-483
4. Sarin, V.K., Kent, S.B.H, Tam, J.P. and Merrifield, R.B. (1981) *Anal. Biochem.* 17:147-157
5. ABI, Inc., In Introduction to Cleavage Techniques. ABI, Inc., Foster City, CA.
6. Tam, J.P., Heath, W.F., and Merrifield, R.B., (1986) *J. Am. Chem. Soc.* 108:5242- 5251
7. Atherton, D. *In* Techniques in Protein Chemistry (ed. T. Hugli) Academic Press, N.Y., N.Y. p 273-283
8. Kitas, E.A., Perich, J.W., Johns, R.B. and Tregear, G.W. (1988) *Tet. Lett.* 29:3591-3592

Synergy of Semisynthesis and Site-Directed Mutagenesis: Manipulating Genes to Create Sites for Autocatalytic Protein Fragment Religation

Carmichael J.A. Wallace
Department of Biochemistry
Dalhousie University
Halifax, N.S. Canada B3H 4H7

J. Guy Guillemette
Michael Smith
Department of Biochemistry
University of British Columbia
Vancouver, BC, Canada V6T 1W5

Yuko Hibiya
Department of Biochemistry
Keio University, Tokyo 160, Japan

I. Introduction

Semisynthesis is a means to protein engineering that relies, as the name implies, on the use of a large proportion of the natural protein in the assembly of the synthetic product. In a typical fragment condensation semisynthesis (1) the protein under study is side-chain protected, and cleaved, chemically or enzymatically, into a limited number of fragments which will then be isolated and purified. The fragment that contains the residue(s) of interest will be modified to introduce the desired substitution. This may involve total chemical synthesis of the fragment (1,2), its replacement by a corresponding peptide from another species (1), sequential degradation and resynthesis of the natural fragment (3), or fragment-specific chemical modification (4). Other peptides may be further protected or modified to facilitate the next step, fragment condensation. Product isolation will be followed by deprotection and purification.

The two main practical problems associated with such schemes are a requirement for relatively large amounts of protein, and the difficulty of the fragment condensation step. Conventional chemical approaches to coupling require harsh conditions (and therefore many additional protection and deprotection steps) and are inefficient. Nevertheless, semisynthesis has been employed with a number of proteins to yield informative analogues for structure-function studies (5).

TECHNIQUES IN PROTEIN CHEMISTRY III

Genetic manipulation could be used to assist the semisynthetic process overcome both these problems. Firstly, the cloning and overexpression of a particular gene may provide substantial amounts of an otherwise scarce protein for semisynthesis. Secondly, site-directed mutagenesis of such a cloned gene might be used to build into the protein structure cleavage and recombination sites for autocatalytic fragment religation (AFR).

AFR depends on the natural propensity of large fragments, in many different proteins, to mutually reassociate in complexes of native-like conformation. Thus the termini at the breakpoint(s) are brought into proximity and usually the correct orientation. If one of those termini is mildly activated to overcome the thermodynamic barrier to peptide bond formation, the reaction is substantially catalysed by these conformational factors. It can occur (but not reliably) in complexes of CNBr fragments, since this cleavage leaves the C-terminus at the breakpoint in the form of homoserine lactone, which provides weak activation (1,2). It has also been observed in the ternary complexes of fragments and proteolytic enzymes, when the presence of high concentrations of organic cosolvent modify the thermodynamics in favour of peptide bond formation by reducing water concentration and shifting carboxyl pK_as (6). Both types of AFR are limited in their scope, yet have been exploited in the semisynthesis of many protein analogues (2,7).

We have been elaborating a truly general approach to AFR by using elements of both these phenomena. Reverse proteolysis is used to add an activating element to a fragment C-terminus. Then that fragment is complexed with a contiguous fragment and the conditions adjusted to promote aminolysis of the activating element with peptide bond formation (8,9). We have demonstrated that the process will operate at many different breakpoints incorporating a wide range of C-and N-terminal amino acids (9) and with proteases of varied specificity (10).

Up to now we have been dependent on the natural disposition of suitable breakpoints, but it should be possible to combine semisynthesis and directed mutagenesis to create idealised sites. As the first example of this synergism we have made a mutant of yeast iso-I cytochrome c (S65M) with a novel site for cleavage and activation by cyanogen bromide.

In addition, a methionine residue naturally occurring at the variable position 64 that is cleaved by CNBr but does not function as a site for AFR was replaced by leucine. The product has been examined with two questions in mind.

(i) Does the introduced mutation have any deleterious effect on the function of the cytochrome?

(ii) Does the introduction of a site at this point lead to efficient religation?

II. Materials and Methods

A: *Mutant Preparation*

Oligonucleotide mutagenesis (11) was carried out essentially by the method of Kunkel (12) in which single-stranded DNA for mutagenesis is prepared from a strain of E.coli deficient in dUTPase and uracil-N-glycosylase (strain RZ1032, dut^-ung^-). The resulting DNA contains a proportion of uridine residues instead of thymidine and thus the wild type strand of the duplex produced after the mutagenic reaction is at a selective disadvantage during subsequent replication in dut^+ung^+ bacterial cells (strain JM101).

The parent cloned gene differed from the wild type iso-1-cytochrome c gene in that the cysteine codon at position 102 had been mutated to that of threonine. Previous studies had indicated that this substitution produces a cytochrome c protein which, unlike the wild-type version, does not dimerize during purification or autoreduce and so is more satisfactory for structural and functional studies (13). For the sake of simplicity, this form of the gene will be described henceforth as CYC1.

The shuttle plasmid vector (pEMBLYe30) has been described which has all the elements necessary for selection and replication in yeast and in E.coli (Leu2, 2μ ori, Amp^r, E.coli ori) but which can also be produced in a single stranded form (14). Thus since single stranded DNA is a convenient target for efficient in vitro mutagenesis, this vector may be used directly for manipulation of cloned sequences and for expression of these sequences in yeast cells. The yeast iso-1-cytochrome c expression vector pING 4 (fig. 1) was generated by ligating the complete yeast iso-1-cytochrome c gene together with its transcriptional regulatory elements into the multiple cloning site of pEMBL Ye30 (Inglis et al., 1991). Furthermore, the plasmid may be propagated in bacteria, and, since it contains an origin of replication for phage f1, can be converted to single stranded DNA by superinfection of plasmid carrying cells with phage f1 (14); single-stranded plasmid DNA is packaged and released along with helper DNA, and can be recovered easily by standard methods. This single-stranded DNA contains the coding strand of the iso-1-cytochrome c fragment, and may be used for nucleotide sequence analysis or for site-directed mutagenesis, using complementary synthetic oligonucleotide primers (11).

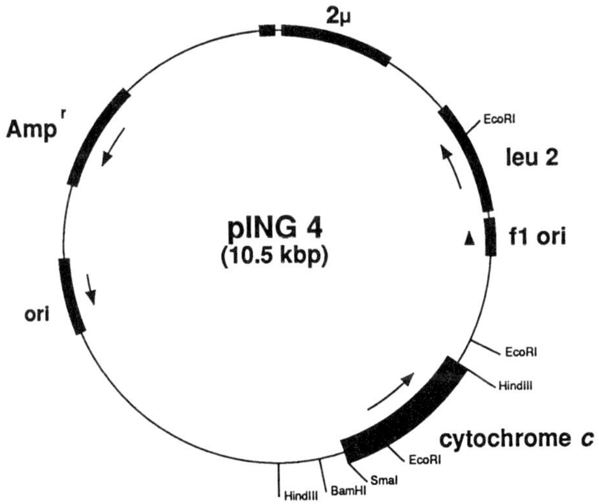

Figure 1: Yeast iso-1-cytochrome c expression vector pING 4. The plasmid contains the coding region and upstream activation sequences of yeast iso-1-cytochrome c which were cloned into the multiple cloning site of pEMBL Ye 30 (Baldari and Cesareni, 1985).

Met64 was replaced by Leu, and Ser65 replaced by Met using the mutagenic primer 5'GTACTCCATCAAGTTATTTT3' that was chemically phosphorylated at the 5' end to give what will be called [Met65]CYC1. The oligonucleotide was phosphorylated at the 5' end by chemical means using the reagent (2-cyanoethoxy)-2-(2'-O-4,4'dimethoxytrityloxyethylsulphonyl)-N, N diisopropylaminophosphine (15) as a 0.1M solution in anhydrous CH_3CN. Using the procedure of Kunkel (12) and chemically phosphorylated oligonucleotides, we have found that the efficiency of mutagenesis is generally high, between 30-80% and does not lead to unwanted mutations.

Positions 64, 65, and 102 are all naturally variable residues, so the S-M65 mutation is the significant change. The phagemid yeast shuttle vector containing the fully mutagenised CYC1 gene was inserted into yeast and propagated through the presence of the Leu2 selectable marker and the origin of replication of the yeast "2μ" circle. In a strain of yeast deficient for both isotypes of cytochrome c, GM3C-2 (16), the plasmid also confers the ability to utilize glycerol as the sole carbon source through expression of the cloned cytochrome c gene. In order to confirm the identity and integrity of the mutant genes after their introduction into yeast, plasmid DNA was re-isolated from yeast cells and re-introduced into bacteria for nucleotide sequencing as previously described (17).

B: Expression of the mutant and parent protein

Mother cultures of transformed yeast were grown up in YPG (3% glycerol) medium and used to inoculate 4 litre cultures. The cultures were shaken at 200 rpm for three days at 30°, then boosted with 1% lactate, and harvested 24h later by centrifugation at 5000 rpm in 500 ml bottles. Cells were lysed with 0.5M KCl with the addition of 25% Ethyl acetate overnight, the suspension clarified by centrifugation, made up to 50% $(NH_4)_2SO_4$ saturation and then centrifuged once more. The clear pink solution was dialysed against 12.5mM potassium phosphate buffer pH7 containing 2mM 2-mercaptoethanol and 1mM Na EDTA. Lysis and dialysis solutions contained 0.2mM PMSF, since proteolysis of product is a frequent problem. Purification of cytochromes employed cation- exchange chromatography on Trisacyl SP using a gradient of 0.08-0.24 potassium phosphate buffer, pH7. Purity was checked by UV-visible spectroscopy and gel-exclusion chromatography.

C: Functional tests of cytochromes c

The UV-Visible absorption spectra of mutant and parent cytochromes were recorded in the range 750 250nm, and compared. The absorbances of the 695nm charge transfer bands were titrated with rising pH, and pK_as for the loss of Methionine 80 sulphur-ferric iron coordination were calculated (2). The midpoint reduction potential was determined by titration with the ferri-/ferrocyanide couple (1), and electron-transfer activity was measured in the depleted-mitochondria succinate oxidase assay (18). Cation-exchange chromatography on a Waters 600 HPLC system used an SP5PW column and a potassium phosphate buffer, pH7, gradient.

D: Semisynthesis

Yeast cytochromes (2.5 mg) were digested for 16h in 88% HCOOH, using a 7.5:1 ratio of CNBr to protein. The digests were fractionated on sephadex G50 in 7% HCOOH and fragments corresponding to residues -5-64 and 65-103 (parent protein) or -5-65 and 66-103 (mutant) were isolated and freeze dried. Appropriate pairs of fragments were recombined in 1mL of 50mM potassium phosphate buffer and the solution degassed. The concentration of complex was approximately 0.1 mM. 1mg solid sodium dithionite was added to these mixtures that were then immediately sealed in glass-walled syringes to keep the solution anaerobic. After 24h, solutions were either loaded onto sephadex G50 columns for gel-exclusion chromatography in 7% formic acid, or buffer-exchanged on short G25 columns in 50mM phosphate, to remove excess dithionite. Religation yields were determined by comparisons of peak areas in elution profiles of gel-exclusion chromatography or IEHPLC, as well as by the extent of ascorbate reducibility. Resynthesised cytochromes are fully ascorbate reducible, but unreacted haem fragments are not. Horse cytochrome c was digested with a 3:1 ratio of CNBr protein and fragments (1-65)H and (66-104) isolated as above. Religation of -5-65 from yeast and horse 66-104 was attempted in the same manner as for the all-yeast complex.

II. Results and Discussion

A: Mutagenesis and expression of the iso-1 cytochrome c gene

Nucleotide sequencing of plasmid DNA re-isolated from transformed yeast confirmed that the desired mutation had been introduced into the CYC1 gene. Using the Kunkel method to generate scores of mutants, no unexpected changes were observed. We conclude that this method (12) does not lead to unwanted mutations. Yeast containing the plasmid bearing the mutant [Met^{65}] gene grew just as strongly as those containing the parent iso-1 cytochrome c gene. In both cases approximately 120g wet weight of cells was harvested from 4L cultures, although the yield of cytochrome from these cells differed. The parent gave a total of 14mg highly purified protein, but the mutant only 4mg. Since the phagemid used and the transformation procedures were identical for both, it seems unlikely that a variation of copy number caused this difference. It is possible that the mutation affects the rate of synthesis, the lability to cytoplasmic degradation or the efficiency of transport into the mitochondrion of the apoprotein, or even the addition of haem to the apoprotein by cytochrome c-haem lyase.

B: Functional properties of the mutant cytochrome

No significant difference was detected between the retention times on cation exchange HPLC of the mutant and parent proteins (table I). Although no change in global charge should attend the substitutions we have introduced, the elution position of cytochromes on ion-exchange chromatography is sensitive also to

Table I Retention times of cytochromes c on cation exchange HPLC. The column was equilibrated with, and samples loaded in, 0.04M potassium phosphate buffer, pH7, then developed with a gradient of buffer to 0.4m over 30 minutes.

Cytochrome	Retention time	
	Fe^{2+} form	Fe^{3+} form
CYC 1	15.20	16.01
[Met65] CYC 1	15.15	16.00

small changes in dipole moment resulting from the surface distribution of that charge (19). If a surface conformational change had occured, it would thus be signalled by changes in retention times. Internal conformational changes can also have detectable effects on the protein's properties, including the UV-visible spectrum and the pK_a of the loss of the 695 nm band that is sensitive to changes in haem crevice stability. Both the spectrum and pK_{695} of the analogue were identical to those of the parent protein.

Redox potential may also be affected by, among many factors, internal conformation. We measured an E'_m for [Met65] iso-1 cytochrome c of 265±5mV compared to 270-272mV determined for the native protein (1,13).

The final proof of unimpaired functionality lies in electron transfer activity. We measured this in an assay system in which the limiting step is the cytochrome c reductase-cytochrome c transfer (18). The specific activity of the analogue is 100% of that of the native protein (figure 2), and significantly higher than that of horse cytochrome c, even though the mitochondria are obtained from mammalian liver. This difference is related to redox potential differences (18).

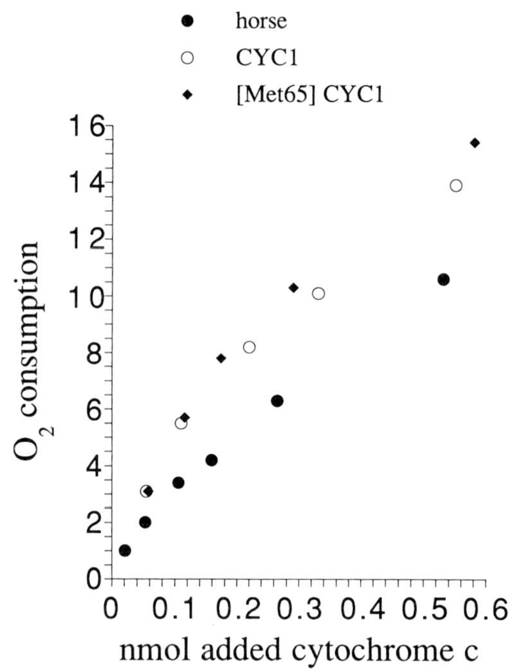

Figure 2: Assays of electron transfer activities of parent and mutant yeast, and horse cytochromes in cytochrome c-depleted rat liver mitochondria. Mitochondria are suspended in buffer including succinate as electron donor and O_2 as ultimate electron acceptor, in the chamber of a Clark-type O_2 electrode. Sequential additions of cytochrome c are made and O_2 consumption measured potentiometrically.

C: Autocatalytic fragment religation

CNBr cleavage gave fragments (-5-64)H and (65-103) or (-5-65)H and (66-104) in approximately 25% yield in either case, in line with expectation (1). When the fragments from the parent protein were recombined, exposed to AFR conditions and gel-filtered, 10% of the haem colour eluted at the holocytochrome position, but only 10% of this material was ascorbate reducible, implying a religation yield of approximately 1%. Previous work had failed to show religation (1). The other high molecular weight materials are probably polymers of the haem peptide. Under the same conditions, the mutant protein fragments recombined in greater than 50% yield, as judged by both ascorbate reducibility and peak areas on ion-exchange HPLC. The religated product elutes from this system with identical retention time as the uncleaved starting material. Thus the desired product is obtained in yield comparable to that seen in other AFR systems (10) and we have clearly succeeded in our objective of creating a novel and efficient AFR site. We also obtained >50% yield if the non-haem peptide was 66-104 from horse, demonstrating that the system will also function well in the creation of entirely novel structures.

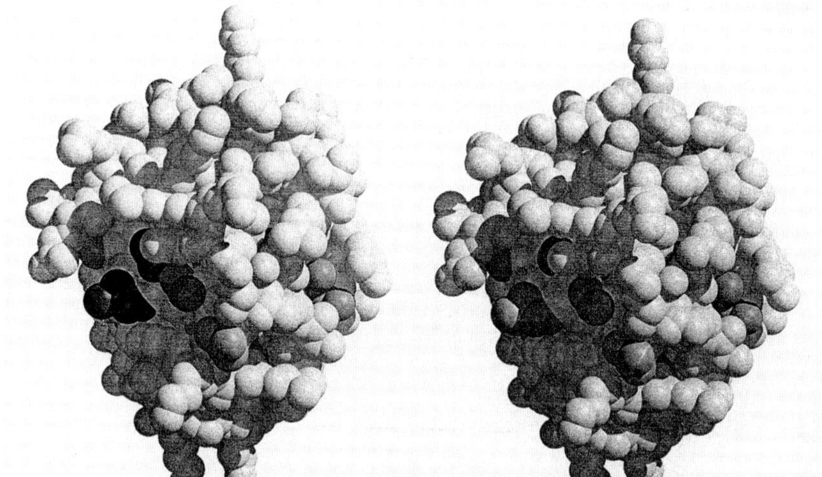

Figure 3: Space-fitting models of CYC1 and [Met65] CYC1 proteins generated by RASTER software. In each case the methionine residue has been replaced by homoserine lactone, and the 64-65 or 65-66 peptide bond is missing. The homoserine lactone ring and the side chain of the subsequent residue (either Ser^{65} or Glu^{66}) are emphasised in black. In the CYC1 model only the serine hydroxyl is visible. In [Met^{65}] CYC1 both the lactone and the Glu^{66} side chain, with the α-amino group of Glu^{66} between them, are clearly exposed at the molecular surface.

Why does the nearby naturally occuring methionine residue in the native protein not support AFR? We proposed, on inspection of this region of the protein native structure projected in helical wheel form, that to retain the conformation in the complex derived from parent protein the reactive elements, hydrophilic lactone and α-amino groups, would need to be buried in the hydrophobic interior. This thermodynamically unfavourable situation would inhibit approach of these groups (1). We have now modelled the structure of this complex with minimum deviation from the protein structure, and the structure of the (-5-65H): (66-103) complex, using Biosym Insight software and they are graphically represented, with RASTER 3D software, in figure 3. These models make the point that while the potential orientation of the termini is as described above for the non-productive complex, in that derived from the new mutant both α-amino group of Glu66 and homoserine lactone at position 65 are exposed in proximity at the molecular surface.

The new approach we have developed has already been used to produce a novel analogue, a chimera of yeast and horse cytochrome c structures. Functional studies of this material are reported elsewhere (20). The way is now open for the use of this site to generate analogues of yeast cytochrome c containing non-coded amino acids for functional studies (2) and crystal structure determinations (21). We will be applying the technology to alternative sites in this, and other proteins.

IV. Conclusions

1. We have successfully introduced into the iso-1-cytochrome c gene a mutation designed to provide a new site for cleavage and religation of the protein, and expressed the mutant protein in satisfactory yield.

2. The introduced mutation appears to cause no change in surface or internal conformation, or to the primary biochemical characteristics of the protein, reduction potential and electron transfer efficiency.

3. The mutant protein is appropriately cleaved by CNBr. Isolated fragments (-5-65)H and (66-103) can be complexed, and undergo Autocatalytic Fragment Religation (AFR) in greater than 50% yield. Thus site-directed mutagenesis is shown capable of facilitating semisynthesis, and this synergy will be exploited to create previously unattainable informative protein analogues.

References

1. Wallace, C.J.A., Marchiori, F., Borin, G., and Corradin, G. (1986) Biopolymers 25, 2121 2132.
2. Wallace, C.J.A., Mascagni, P., Chait, B.T., Collawn, J.F., Paterson, Y., Proudfoot, A.E.I. and Kent, S.B.H. (1989) J. Biol. Chem. 264, 15199-15209.
3. Wallace, C.J.A. and Corthesy, B.E. (1986) Protein Eng. 1, 23-27.
4. Wallace, C.J.A., and Rose, K. (1983) Biochem. J. 215, 651-658.
5. Offord, R.E. (1987) Protein Eng. 1, 151-157.
6. Proudfoot, A.E.I., Offord, R.E., Rose, K., Schmidt, M. and Wallace, C.J.A. (1984) Biochem. J. 221, 325-331.
7. Chaiken, I.M., Komoriya, A., Ohno, M. and Widmer, F. (1982) App. Biochem. Biotechnol. 7, 385-399.
8. Rose, K., Herrero, C., Proudfoot, A.E.I., Offord, R.E. and Wallace, C.J.A. (1988) Biochem. J. 249, 83-88.

9. Proudfoot, A.E.I., Rose, K. and Wallace, C.J.A. (1989) J. Biol. Chem. 264, 8764-8770.
10. Wallace, C.J.A. (1991) Proc. Eur. Pept. Symp. 21, 260-263.
11. Zoller, M.I. and Smith, M. (1982) Nucl. Acids. Res. 10, 6487-6500.
12. Kunkel, T.A. (1985) Proc. Natl. Acad. Sci. U.S.A. 82, 488-492.
13. Cutler, R.L., Pielak, G.J., Mauk, A.G. and Smith, M. (1987) Protein Eng. 1, 95-99.
14. Baldari, C. and Cesareni, G. (1985) Gene 35, 27-32.
15. Horn, T. and Urdea, M.S. (1986) Tetrahedron Letters 27, 4705-4708.
16. Faye, G., Leung, D.W., Tatchell, K., Hall, B.D. and Smith, M. (1981) Proc. Natl. Acad. Sci. U.S.A. 78, 2258-2262.
17. Inglis, S.C., Guillemette, J.G., Johnson, J.A. and Smith, M. (1991) Protein Eng., in press.
18. Wallace, C.J.A. and Proudfoot, A.E.I. (1987) Biochem. J. 245, 773-779.
19. Smith, H.T., Staudenmeyer, N. and Millet, F. (1977) Biochemistry 16, 4971-1977.
20. Wallace, C.J.A., Guillemette, J.G., Hibiya, Y. and Smith, M. (1991) submitted for publication.
21. Louie, G.V., Pielak, G.J., Smith, M. and Brayer, G. (1988) Biochemistry 27, 7870-7876.

Acknowledgements

We wish to thank NSERC and MRC of Canada for financial support, Angela Brigley-Cole for technical assistance, and Terry Lo and Gary Brayer for assistance with and use of the molecular modelling system.

State-of-the-Art Peptide Synthesis: Comparative Characterization of a 16-mer Synthesized in 31 Different Laboratories

A. J. Smith[1], J. D. Young[2], S. A. Carr[3], D. R. Marshak[4], L. C. Williams[5], and K. R. Williams[6]

[1]Beckman Center, Stanford University, Stanford, CA 94305
[2]Biological Chemistry, UCLA Medical School, Los Angeles, CA 90024
[3]Physical & Structural Chemistry, SmithKline Beecham Pharm., King of Prussia, PA 19406
[4]Cold Spring Harbor Laboratory, Cold Spring Harbor Laboratory, New York, NY 11724
[5]Norris Cancer Center, University of Southern California, Los Angeles, CA 90033
[6]Howard Hughes Medical Institute, Yale University, New Haven, CT 06510

I. Introduction

The Association of Biomolecular Core Facilities (ABRF) has 184 core facilities engaged in the synthesis and/or structural analysis of proteins and/or nucleic acids. These facilities operate in research environments and are located in academic, government and research institutions as well as in private industry. The ABRF recently formed a peptide synthesis/mass spectrometry subcommittee to evaluate the various aspects of peptide synthesis that are currently performed in its member laboratories. Similar ABRF research committees have been studying and reporting on the amino acid analysis and protein sequencing capabilities of core laboratories for the past three years (1-7). Peptide synthesis is a combination of several functions including synthesis, cleavage, purification and characterization of the final product. Each of these functions pose specific problems which impact the quality of the final product.

Our mechanism for evaluating the interrelationship of these various functions was to ask that our member laboratories synthesize a 16 residue peptide of predetermined sequence by their preferred chemistry. We requested that each core laboratory submit 10 mg of the crude, desalted peptide to the subcommittee for analysis. In addition, each laboratory had the option of also submitting a mg of the purified peptide. Characterization of the crude product should provide information about the effectiveness of the synthesis and cleavage chemistries. Analysis of the purified product should provide a monitor of purification techniques. Each of the 61 samples that was received in time for this study were subjected to a battery of

TECHNIQUES IN PROTEIN CHEMISTRY III

analytical techniques including amino acid analysis, HPLC, capillary electrophoresis (CE), mass spectrometry, and biological assay as the peptide was designed to be a substrate for at least two different protein kinases. It was anticipated that these analyses would help to identify optimal methods for characterizing synthetic peptides as well as provide data regarding common side-products and reactions which may occur during peptide synthesis. The data that are summarized in the following report indicate that while a variety of different synthesis strategies yielded the desired product in near homogenous form, the Fmoc/t-butyl strategy succeeded more often than the tBoc/benzyl approach within this group of core laboratories.

Since peptide synthesis is carried out in only 106 of ABRF's current member laboratories, the 36 laboratories that submitted samples in time for this study and the 4 laboratories that submitted samples after this deadline together represent nearly a 40% response rate. In view of the considerable time and effort that is required to synthesize and purify a 16 residue synthetic peptide, such a high response is indeed indicative of the critical need for more data concerning "state-of-the-art" peptide synthesis technology and capabilities. To our knowledge, this study is unique in that no other project has been attempted in which the same peptide has been simultaneously prepared in such a large number of different laboratories and then been so extensively characterized.

II. Materials and Methods

Participating laboratories were asked to synthesize the following peptide by the method most commonly used in their facility:

NH_2-Val-Lys-Lys-Arg-Cys-Ser-Met-Trp-Ile-Ile-Pro-Thr-Asp-Asp-Glu-Ala-COOH

To increase the possibility of obtaining significant variability between different synthetic approaches, the sequence contained a few problematical couplings such as the Trp-Ile-Ile sequence and potential sites of post-synthesis modification such as the Trp, Cys and Met residues. In addition, phosphorylation sites were included for cyclic AMP-dependent protein kinase (PKA, Ser #6) and for casein kinase II (CK II, Thr #12) so that the extent of nearby side chain modifications could be correlated with biological activity. Participants were asked to provide 10 mg of crude, salt free product and were given the option of also providing an additional mg of purified material. The samples were provided to the committee in coded form via a third party to maintain participant anonymity but yet allow the participants to identify data sets resulting from their samples. Each product was then subjected to the following analyses.

A. Amino Acid Analysis

These analyses were carried out on aliquots of 0.5% of each crude and 5% of each purified sample following hydrolysis for 16 hrs at 115°C in 100 μl 6 N HCl, 0.2% phenol. The analyzer was a Beckman 6300 with quantitation provided by a Nelson Analytical Model 4400 Data System.

B. Analytical HPLC/Capillary Electrophoresis

Samples were analyzed on a Perkin Elmer series 4 HPLC using a BioRad HIPore rp304 column, 250 mm x 4.6 mm. The linear gradient extended from 0.1% TFA to 70% acetonitrile, 0.09% TFA over a period of 33 min. The flow rate was 2 ml/min and the absorbance was monitored at 210 nm using a Perkin Elmer LC 95 detector. Capillary electrophoresis was carried out on an Applied Biosystems 270A instrument. The buffer was 20 mM sodium phosphate, pH 2.5. The voltage was 30 kV and electrophoresis was monitored at 220 nm.

C. Mass Spectrometry

Electrospray mass spectrometry (ESMS) was accomplished using a Sciex API-III triple quadrupole mass spectrometer with an atmospheric pressure ionization source (8). Samples were dissolved in 0.05% TFA at 0.5 μg/μl and 2 μl were introduced into the mass spectrometer via a Beckman model 507 autosampler. Two sequential 5 sec scans obtained from the eluted sample peak were summed for presentation. Plasma desorption mass spectrometry (PDMS) was carried out on an Applied Biosystems BioIon plasma desorption mass spectrometer (9-10). Samples were solubilized in water at a concentration of 11 mg/ml. Aliquots of 11.3 μg were electrosprayed onto nitrocellulose covered targets. The targets were incubated for 2 min at 20°C then dried by spinning under a stream of N_2. Targets were then washed with 50 μl of 0.1% TFA and dried under N_2 prior to PDMS. Spectra were collected at an accelerating voltage of 16 kV for 1 hr. ESMS data were recorded in both the positive and negative ion modes; PDMS was obtained in positive ion mode only.

D. Biological Assays

Aliquots were assayed as substrates for casein kinase II (CK II) and cyclic AMP-dependent protein kinase (PKA). [^{32}P]Phosphate incorporation into the peptide substrates was quantitated after incubation with each of the kinases at 30°C for 10 min. The CK II buffer was 50 mM MOPS, pH 7.0, 10 mM $MgCl_2$, 125 mM NaCl. The PKA buffer was 25 mM Tris-HCl, pH 7.4, 6 mM $MgCl_2$.

E. Quantitation of Results

In some instances, samples were grouped based on a semi-quantitative estimate of the overall % of the correct product as determined by HPLC, PDMS or ESMS. The following scale was used: 0 = no detectible correct product; 1 = <25% correct product; 2 = 25-75% correct product; 3 = >75% correct product. In the case of CE, the % referred only to the relative abundance of the largest 220 nm absorbance peak.

III. Results

A. Approaches and Instrumentation Used to Synthesize the ABRF Peptide

As indicated in Table I, a total of 36 crude and 25 purified samples were received in time to be included in this study. Half of these peptides were synthesized via tBoc and the other half via Fmoc chemistry. Applied Biosystems machines were used for 23 of the syntheses and Milligen/Biosearch instruments for 11 of the syntheses. One peptide was prepared on a Vega instrument and one was prepared manually using a Dupont RAMPS shaker. The majority (17/18) of the tBoc peptides had been made with a PAM resin while most (16/18) of the Fmoc peptides had been made on either a Wang or a PepSyn-KA resin. The most common cleavage methods relied on HF for tBoc peptides and TFA for Fmoc peptides. In all instances reversed phase HPLC was used for purification with buffer A containing 0.05 - 0.1% TFA. With the exception of one facility which used isopropanol and one which used an isopropanol/acetonitrile mixture, buffer B contained 60-100% acetonitrile with 0.05 - 0.1% TFA. C-18 Supports were used for 12 of the purifications while 7 facilities used C-8 and 6 used C-4 columns. The separations were performed at flow rates of 0.5 - 30 ml/min. All separations were at room temperature. The column dimensions varied from 0.46 - 2.5 cm in diameter and 21 of the 24 columns used were 22 - 30 cm long. It should be mentioned that one of the most homogenous tBoc peptide samples was classified as a crude preparation even though it had been subjected to gel filtration to remove salts and reagents.

B. Amino Acid Analysis

While acid hydrolysis/amino acid analysis is clearly the best approach for quantitating the total amount of peptide present, in this study amino acid analysis proved to be a poor predictor of peptide purity. The latter resulted from the fact that most problems (see below) occurred during cleavage and deprotection as opposed to during synthesis. Since no significant differences were seen in any of the 61 amino acid compositions that were obtained, none of these samples were contaminated by a substantial fraction of any

Table I. Approaches and Instrumentation Used to Synthesize the ABRF Test Peptide

	#	%		#	%
Number of samples received			Resin		
Crude	36	59	Pam	17	46
Purified	25	41	Merrifield	1	3
			Wang	9	24
Chemistry			PepSyn-KA	7	22
Boc/Benzyl	18	50	PAC	1	3
Fmoc/tButyl	18	50	Expansin	1	3
Instrumentation			Cleavage Type		
Applied Biosytems			HF	11	23
430A	19	51	TFMSA	4	11
431	4	11	Low HF/High HF	3	8
Milligen/Biosearch			TFA	18	49
9050	8	22	NaOH/Isopropanol	1	3
9500	2	5			
9600	2	5			

deletion or truncation products. Surprisingly, the overall average Trp value obtained using 6 N HCl, 0.2% phenol was as accurate as that for stable amino acids such as Ala or Asp:

Ala	(1)	1.07(SD $\pm$0.10)
Asp	(2)	2.26(SD $\pm$0.15)
Trp	(1)	0.98(SD $\pm$0.13)

C. HPLC/ESMS

As shown by the examples in Figure 1, these two techniques were in overall agreement regarding the relative abundance of the desired product in the crude samples. Hence, the overall average content of the desired product for 36 crude samples was 44% as determined by HPLC compared to 53% as determined by ESMS. In only 5 of the 36 crudes examined was there more than a 20% difference in the estimated degree of purity as determined by these two different approaches. In all of these instances, ESMS gave a higher estimate of the content of the desired product. One possible explanation for this slight discrepancy might be the presence of 210 nm absorbing, reagent peaks which would result in non-peptide peaks being included in the estimation of total material present in the HPLC analyses but not in the ESMS analyses. Although HPLC retention times are sufficiently reproducible to permit identification of the correct peak, this option is normally not available to the custom synthesis lab due to the lack of an authentic reference standard. However, in every instance in this study where HPLC indicated that there was a major peak that accounted for >60% of the absorbance detected, ESMS confirmed that the sample did indeed contain 65% or more of the desired product. Hence, while observing

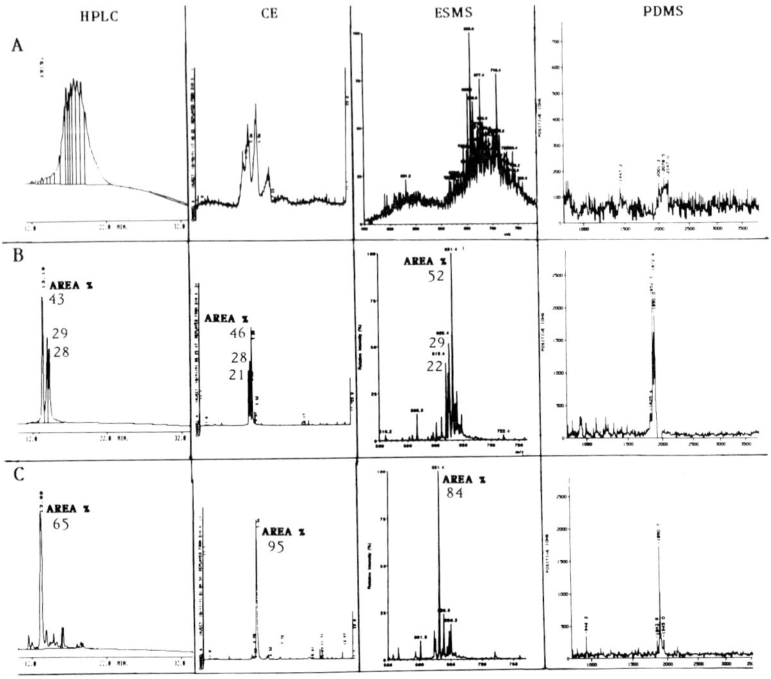

Figure 1. Examples of the variability in the abundance of the desired product in submitted samples. A, no desired product was detected in this sample. B, the relative abundance of the desired product was 43% by HPLC, 52% by ESMS, 35% by PDMS while CZE gave a 46% abundance for the major product. C, the relative abundance of the correct product was 65% by HPLC, 84% by ESMS, >95% by PDMS while CZE indicated a 95% abundance for the major product. The $(M + 3H)^{3+}$ region of the ESMS data is shown.

a single major peak in a "crude" HPLC profile certainly does not prove that this peak corresponds to a peptide with the correct structure, at least in this instance, it did seem to provide a good initial indicator.

The ESMS analyses permitted tentative identification of many of the side products that were major contaminants in the submitted samples. Hence, ESMS analysis of the tBoc sample shown in Fig. 1B revealed intense peaks at 1873 and 1855 which probably correspond to single and double dehydrations respectively. The latter may have resulted from succinimide formation within the Asp-Asp-Glu stretch in the peptide. The only contaminating peak detected in the Fmoc sample shown in Fig. 1C had a mass (1947) which would correspond to that of a t-butyl adduct. Table II summarizes the most common structural modifications that were tentatively identified by ESMS in the submitted samples. Since only one tBoc and three Fmoc crudes contained deletion products in >10% abundance (with the most extreme instance being a 30% Arg deletion) the coupling of the residues was only a minor cause of heterogeneity. The major problems appear to have occurred during the cleavage of the peptide from the resin and concurrent removal of the protecting groups from the side chains of the amino acids rather than during the actual synthesis. Dehydration, oxidation,

Table II. Common M_r Observed in ABRF Peptides by ESMS and Their Possible Structures

M_r Mono-isotopic	Mass Difference	Possible Product	Crude Peptides With >10% Contaminant			
			tBoc		Fmoc	
			#	%	#	%
1704.9	-186	Trp Deletion	1	6	2	11
1734.8	-156	Arg Deletion	-	-	1	6
1854.9	-2 x 18	Double Dehydration	2	11	-	-
1872.9	-18	Single Dehydration	4	22	-	-
1906.9	+16	Single Oxidation	4	22	-	-
1918.9	+28	Formyl-Trp	1	6	-	-
1935.0	+44	Met sulfoxide + Formyl-Trp	1	6	-	-
1947.1	+56	t-Butyl Adduct	3	17	11	67
1981.0	+90	O-Benzyl Adduct	3	17	-	-
2103.3	+212	Mtr-Arg	-	-	2	11
2113.3	+222	Fmoc Adduct	-	-	1	6

t-butyl and O-benzyl products were common contaminants in the tBoc synthesized peptides with t-butyl and Mtr adducts being frequently found in the Fmoc synthesized peptides (Table II). In general, the Fmoc peptides contained a greater percentage of the desired product than the tBoc syntheses. As shown in Table III, 100% of the Fmoc peptides contained >25% of the desired product (based on ESMS), while only 56% of the tBoc peptides fell into the same category. In fact, no detectable correct product was found in 6 of the 18 tBoc crude samples that were examined by ESMS.

D. Capillary Electrophoresis

In general, there was good agreement between the results of the CE and HPLC analyses. Hence, 29 of the 34 crudes that were assayed by both these approaches were given either the same quality score (see Materials and Methods for an explanation of quality score) by both these approaches or the quality score based on CE was one above that based on HPLC. Since the reproducibility of CE without an internal standard was too variable to allow the correct product to be identified by its electrophoretic mobility, the CE quality scores referred only to the relative complexity of the sample not, as in the case of HPLC, to the relative content of the desired product. Hence, it is not surprising that the quality scores based on CE would be somewhat higher than those based on HPLC. In no instance did CE appear

Table III. Mass Spectrometric Characterization of the ABRF Test Peptide

	%Desired Product	ESMS			PDMS
		%tBoc[a]	%Fmoc[b]	%Total	%Total[c]
Crude Peptides	0	33	0	17	22
	<25	11	0	6	33
	25-75	39	61	50	39
	>75	17	39	28	6
Purified Peptides	0	42	9	26	28
	<25	0	0	0	28
	25-75	8	9	9	16
	>75	50	82	65	28

[a]Number of tBoc peptides characterized by ESMS; crude = 18, purified = 12.
[b]Number of Fmoc peptides characterized by ESMS; crude = 18, purified = 11.
[c]Number of peptides characterized by PDMS; crude = 36, purified = 25.

to indicate the presence of a larger number of major species than was detected by HPLC. Both HPLC and CE gave a closely spaced doublet for a crude sample which by ESMS appeared to contain 73% of the desired product and 22% of the t-Butyl derivative. Further studies are underway to characterize the individual peaks within this doublet.

E. Plasma Desorption Mass Spectrometry

Based on the data that are summarized in the last two columns in Table III and the generally good agreement among the ESMS/HPLC/CE analyses, it appears that PDMS selectively ionized protected peptides in the presence of unprotected peptides and hence led to an underestimate of the content of the desired product. Whereas ESMS indicated that 78% of the crude samples contained >25% of the desired product, PDMS suggested that only 45% of the submitted samples would fall into this same category. In this sense, PDMS was actually more sensitive than ESMS for detecting some structural modifications. The potential problem with relying solely on PDMS is illustrated by the fact that PDMS did not detect any desired product in 6 crude samples that, based on ESMS, contained from 33 - 69% correct product. Two likely explanations can be given for these observations. A relatively large amount (~11 μg) of sample was applied to the nitrocellulose-coated foil, and the foils were washed following sample adsorption. There may have been selective adsorption of the more hydrophobic components of the peptide mixtures, which are usually the incompletely deprotected species. Second, the incompletely deprotected products may suppress the ionization of the desired peptide under fission-fragment bombardment in PDMS. Typically for PDMS, a series of dilutions of sample should be applied, and they should be tested with and without rinsing, either with dilute aqueous acid or pure water. Inherently, the peak areas that appear in PDMS spectra of mixtures do not quantitatively reflect the proportion of the components. This limitation is common to all the mass spectrometric techniques employed. Qualitatively, however, the ESMS results on these and other samples closely paralleled the HPLC data. ESMS appeared, therefore, to provide a more realistic appraisal than PDMS did of the purity of the crude peptides. Analysis time is also considerably shorter in ESMS (seconds) compared to PDMS ($\geq$ 1 hr).

F. Biological Assays

Enzymatic assays provided insight into the consequences of the varied structural alterations that have been found in this study. Hence, oxidation of Met or Cys lowered PKA activity, as predicted. Dehydration or incomplete deprotection of Asp and/or Glu lowered the CKII substrate activity. Incomplete deprotection of Arg, seen in several of the Fmoc syntheses, lowered PKA substrate activity with less effect on CKII. Formylation or deletion of Trp were observed, and both modifications had some effect on both activities. As peptides were found that contained no detectable desired product (based on ESMS) but yet retained full CKII activity, biological activity by itself did not seem to represent a reliable assay for synthetic peptide purity. Particularly when applied to crude samples, biological assays can also be easily influenced by the presence of non-specific reagent activators and/or inhibitors.

IV. Conclusions

Since 6 of the 36 crude samples and 6 of the 23 purified samples that were received did not contain detectable amounts of the desired product, custom peptide synthesis has clearly not yet advanced sufficiently to be considered a routine service. One surprising finding is that while it is clearly possible to synthesize the ABRF standard peptide via either the tBoc or Fmoc approach (3 tBoc and 7 Fmoc crude samples contained greater than 75% of the correct product as estimated by ESMS, see Table III), the core laboratories that participated in this study more often succeeded with the latter approach. Hence, 100% of the Fmoc as compared to only 56% of the tBoc crude samples contained at least 25% of the desired peptide. The nature of the modifications that have been detected suggest that most problems with the ABRF standard peptide occurred during cleavage as opposed to during the actual synthesis itself. We suspect therefore that the relatively mild conditions required for Fmoc, as compared to tBoc cleavage, account for the generally better results that were obtained with the Fmoc approach. Although insufficient data are available to draw firm conclusions, the three tBoc synthesized crude peptides that contained >75% correct product, used formyl protection for tryptophan and cyclohexyl protection for glutamic and aspartic acid. While all three of these cleavages were carried out at -2° or -5°, each was done in the presence of different scavengers. Additional studies are clearly needed to identify the variable(s) that account for the wide range in the quality of the tBoc crude samples that were submitted in this study.

Based on the results that have been described, a quantitative separation technique coupled with a mass spectrometric technique (HPLC and ESMS in this study) provides the most accurate approach for examining the purity and authenticity of synthetic peptides. Although based on ESMS, 20% of the submitted samples did not contain detectable amounts of the intended structure, these samples nonetheless all had the correct amino acid compositions and some had full CKII or PKA activity. The latter finding makes it difficult to estimate the possible impact on biological research of altered synthetic peptides. As illustrated by the CKII and PKA assays, the impact will depend not only on the exact modification but also on the intended application for which the peptide was synthesized. To avoid this uncertainty a far greater proportion of core facilities must have routine access to ESMS or to a comparable mass spectrometric technique. Another interesting finding that bears on this topic is that in less than half of the 23 instances, where ESMS data is available on both the crude and corresponding purified samples, did HPLC purification result in any significant increase in the relative abundance of the intended product. In 5 instances HPLC purification actually led to a significant decrease in purity. The latter appeared to result either from oxidation during purification or more commonly, from purifying the wrong absorbance peak, presumably because these laboratories lack the mass spectrometric capabilities that are essential to identify common structural alterations found in synthetic peptides. These data demonstrate that without an effective method (ie., bioassay or mass spectrometry) to identify the absorbance peak that is enriched for the correct product, HPLC fractionation may be of little value. Since a recent ABRF survey of 124 core laboratories demonstrated that less than 9% of the core facilities that offer peptide synthesis are equipped for mass spectrometry (11), a considerable capital investment is urgently needed to improve the quality control capabilities of peptide synthesis facilities.

Acknowledgements

We especially thank all the ABRF core facilities that participated in this study.

References

1. Niece, R. L., Williams, K. R., Wadsworth, C. L., Elliott, J., Stone, K. L., McMurray, W. J., Fowler, A., Atherton, D., Kutny, R., and Smith, A. J. (1989). *In* "Techniques in Protein Chemistry" (Hugli, T. E., ed) pp 89-101, Academic Press, New York
2. Speicher, D., Grant, G., Niece, R., Blacher, R., Fowler, A., and Williams, A. (1990). *In* "Current Research in Protein Chemistry: Techniques, Structure, and Function" (Villafranca, J. J., ed) pp 159-166, Academic Press, New York
3. Crabb, J. W., Ericsson, L., Atherton, D., Smith, A. J. and Kutny, R. (1990). *In* "Current Research in Protein Chemistry: Techniques, Structure, and Function" (Villafranca, J. J., ed) pp 49-61, Academic Press, New York
4. Ericsson, L. H., Atherton, D., Kutny, R., Smith, A. J., and Crabb, J. W. (1991). *In* "Methods in Protein Sequence Analysis" (Jornvall, H., Hoog, J.-O. and Gustavsson, A.-M., eds) pp 143-150, Birkhauser Verlag, Basel, Switzerland
5. Niece, R., Ericsson, L., Fowler, A., Smith, A., Speicher, D., Crabb, J., and Williams, K. (1991). *In* "Methods in Protein Sequence Analysis" (Jornvall, H., Hoog, J.-O. and Gustavsson, A.-M., eds) pp 133-141, Birkhauser Verlag, Basel, Switzerland
6. Tarr. G., Paxton. R., Pan, Y.-C., Ericsson, L., and Crabb, J. (1991). *In* "Techniques in Protein Chemistry II" (Villafranca, J.J., ed) pp 139-150, Academic Press, New York
7. Yuksel, K., Grant, G. Mende-Mueller, L., Niece, R., Williams, K., and Speicher, D. (1991). *In* "Techniques in Protein Chemistry II" (Villafranca, J.J., ed) pp 151-162, Academic Press, New York
8. Edmonds, C. C. and Smith, R. D. (1990). *In* "Methods in Enzymology" (McCloskey, J. A. vol ed) Vol 193, pp 412-431.
9. Macfarlane, R. D. (1990). *In* "Methods in Enzymology" (McCloskey, J. A. vol ed) Vol 193, pp 263-279.
10. Roepstorff, P. (1990). *In* "Methods in Enzymology" (McCloskey, J. A. vol ed) Vol 193, pp 432-440.
11. Niece, R., Beach, C., Cook, R., Hathaway, G., and Williams, K. (1991) *FASEB J.*, in press

SECTION V

Amino Acid Analysis

APPLICATIONS OF AUTOMATIC PTC AMINO ACID ANALYSIS

Karen A. West and John W. Crabb
Protein Chemistry Facility
W. Alton Jones Cell Science Center, Inc.
Lake Placid, New York 12946

I. INTRODUCTION

In addition to the 1972 Nobel Prize in Chemistry, the importance of amino acid analysis in protein chemistry is recognized annually by the Stein/Moore Award of the Protein Society. Thanks to these and other investigators, amino acid analysis has become a deceptively simple methodology provided sufficient sample is available. However, micro-amino acid analysis remains one of the most difficult analytical procedures in protein chemistry. Amino acids vary widely in their stability/lability to acid hydrolysis, oxidation and modification. Minor variability in hydrolysis and derivatization conditions as well as instrument performance contribute a statistical dimension to the measurement while background contributions (contamination) complicate the analysis further. We are interested in improving high sensitivity amino acid analysis technology and have participated in the evaluation of a commercially available automatic phenylthiocarbamyl (PTC) amino acid analyzer capable of on line, automatic hydrolysis (West and Crabb, 1989; West and Crabb, 1990). We report here applications of the automatic PTC methodology to polyvinylidene difluoride (PVDF) immobilized samples and to tryptophan and cysteic acid analysis.

II. EXPERIMENTAL PROCEDURES

Instrumentation. PTC amino acid analysis was performed with an Applied Biosystems Model 420H hydrolyzer/derivatizer, Model 130 analyzer and Model 920 computer. The Model 420H was plumbed for diisopropylethylamine (DIEA) vapor delivery to the transfer flask according to the manufacturer's specifications. Manual vapor phase HCl hydrolysis was performed at 150°C for 1 h in the presence of phenol. Methanol in bottle #4 contained 1 mg/ml EDTA to enhance recovery of acidic and basic residues and methanol in bottle #6 contained 30% methylene chloride (Burdick and Jackson, HPLC grade) for improved cleaning of sample frits.

TECHNIQUES IN PROTEIN CHEMISTRY III

Analysis of PVDF-immobilized Samples. Samples were dissolved in 25 mM N-ethylmorpholine acetate pH 8.5 and adsorbed to PVDF (Millipore) by incubating 200-300 μl aliquots containing 1 μg protein with 5.5 mm PVDF discs overnight at 4°C with gentle shaking. Each disc was removed with forceps, washed 3x with water and vacuum dried. Amino acid analysis of protein remaining in solution showed that adsorption was 90-100% efficient. Manual hydrolysis was performed at 150°C for 1 h. Manual extraction of the PVDF protein hydrolysate was with 30% methanol (2X 50 μl) with vortexing prior to PTC-derivatization. Automatic hydrolysis derivatization and extraction were performed with the Model 420 hydrolyzer/derivatizer. For automatic hydrolysis, perforated teflon discs (~8 mm diameter) supplied by Applied Biosystems were placed over the PVDF to hold the membrane in place on the sample frit. Before applying samples, teflon discs and sample frits were washed using standard automatic hydrolysis and derivatization precycles. Automatic hydrolysis of PVDF samples was performed with the manufacturer's recommended PVDF hydrolysis cycle. Automatic PTC derivatization and extraction of PVDF samples was performed with a modified derivatization cycle (Table II).

Tryptophan Analysis. A 40 μl aliquot of 40% dodecanethiol (Fluka) in heptane was applied to the sample frit 10-15 minutes before sample application and autohydrolysis performed with the standard hydrolysis cycle as described by Bozzini et al. (1989). Response factors for tryptophan were determined from autohydrolysis and analysis of standard proteins containing 1-6 mole tryptophan per mole protein (Table III).

Cysteic Acid Analysis. Oxidation of cysteine/cystine to cysteic acid was performed by manual vapor phase HCl hydrolysis at 150°C for 1 h in the presence of 2% (v/v) dimethylsulfoxide (DMSO) (Aldrich, HPLC grade) as a modification of the method originally described by Spencer and Wold (1969). Response factors for cysteic acid were determined from oxidative hydrolysis and analysis of standard proteins containing 1-8 mole Cys per mole protein (Table V). The derivatization cycle and Model 420H were modified as recommended by the manufacturer to bubble DIEA vapor through the derivatized sample in the transfer flask. This step neutralizes the negative charge on cysteic acid by ion pairing and consequently increases its RP-HPLC retention time and enhances quantification.

Calculations. Percent error and average % error were calculated as described previously (West and Crabb, 1990):

% Error = 100 X (Experimental - True Residue Value)/True;

Average % Error = Sum (Absolute % Error)/16.

III. RESULTS AND DISCUSSION

A. Analysis of PVDF-Immobilized Samples

The best average recovery (~70%) we have obtained from protein hydrolysates immobilized on PVDF has been by manual vapor phase hydrolysis followed by manual extraction (2X 50 μl 30% methanol with vortexing) and automatic PTC derivatization (Table I). Although compositions with less than 20% error have been obtained, the average error from multiple analyses performed in this manner have tended to be greater than 20% (Table I). Comparable accuracy with significantly reduced recovery was obtained from automatic derivatization/extraction of manually hydrolyzed samples using the standard Model 420 derivatization cycle (data not shown). To enhance recovery using automatic extraction,

TABLE I

PTC AMINO ACID ANALYSIS ON PVDF
COMPARISON OF HYDROLYSIS/EXTRACTION METHODS

Sample/ Method	Compositional Error (%)	Recovery (%)	n
Manual Hydrolysis/Manual Extraction			
ß-Lactoglobulin	25	65	15
Cytochrome C	24	89	13
Myoglobin	24	60	6
	$\bar{x}$ ~ 24 ± 18 %SD	$\bar{x}$ ~ 71 ± 25 %SD	n = 34
Manual Hydrolysis/Automatic Extraction			
ß-Lactoglobulin	25	52	9
Cytochrome C	25	60	11
Myoglobin	23	44	6
	$\bar{x}$ ~ 24 ± 29 %SD	$\bar{x}$ ~ 52 ± 24 %SD	n = 26
Automatic Hydrolysis/ Automatic Extraction			
ß-Lactoglobulin	15	37	9
Cytochrome C	14	62	9
Myoglobin	19	30	9
	$\bar{x}$ ~ 16 ± 30%SD	$\bar{x}$ ~ 43 ± 21 %SD	n = 27

SD, standard deviation; n, number of analyses

TABLE II

MODIFIED MODEL 420 PTC AMINO ACID ANALYSIS PROGRAM FOR PVDF-SAMPLES

STEP #	FXN #	FUNCTION NAME	TIME/TEMP
1 - 18		STANDARDPROGRAM	unmodified
18	8	PRESS #5, #6	30
insert 1	31	HEATER	35
" 2	29	#6 TO TF	1
" 3	22	ARGON TO TF	3
" 4	17	DRV WAIT	30
" 5	22	ARGON TO TF	120
" 6	23	DRY TF	600
" 7	31	TF HEATER	15
19	21	DSOLV SMPLE	2
19 - 43		STANDARD PROGRAM	unmodified

Cycle Time: 43 mins

we modified the derivatization cycle as shown in Table II to include a 100% methanol extraction which is dried in the transfer flask prior to the normal sodium acetate buffer extraction. With the modified derivatization cycle, recovery was increased by about one third over the standard cycle to about 52% (Table I). Automatic hydrolysis/derivatization/extraction yielded less average error (~16%) than manual hydrolysis but with lower recovery (~43%) for the proteins analyzed (Table I). Notably, the perforated teflon discs recently supplied by the manufacturer reliably hold the PVDF samples in place during hydrolysis/analysis. Following either manual or automatic hydrolysis of the protein solutions (1μg) used for generating Table I data, analyses prior to PVDF adsorption yielded average compositional errors in the 7-12 % range. This implicates the hydrolysis/extraction process on PVDF as a significant source of error and an area where further methods development research is warranted.

Amino acid analysis of samples on PVDF is more problematic than those in solution. In 1990 the Association of Biomolecular Resource Facilities found that 29 core facility laboratories using manual hydrolysis and various extraction procedures obtained an average ~27% error and ~52% recovery from triplicate hydrolyses/analyses of PVDF-immobilized myoglobin (0.9 μg) (Tarr et al., 1991). In contrast, participants in the study obtained an average 13.5% error and 98% recovery from replicate hydrolyses (1.33 μg) of myoglobin provided dry and salt-free in the absence of PVDF. The higher error associated with PVDF samples did not appear to be due to poor recovery of a few select residues but rather increased difficulties with all the residues; no superior extraction methodology was evident (Tarr et al., 1991). Relative to these results, the data in Table I indicates that current automatic hydrolysis technology can provide improved accuracy of PVDF analyses with somewhat reduced recovery.

B. Tryptophan Analysis

The potential for determining tryptophan by autohydrolysis in the presence of dodecanethiol was first demonstrated by Bozzini et al. (1989) and described more recently in the February 1991 Applied Biosystems Research News. In these communications, a linear response was shown in the 0.1-6 μg range for Trp recovery from autohydrolysis of an 18 residue test peptide containing 1 mole Trp per mole peptide (M_r~2154). However, partial and variable destruction of Trp (20-45%) was also observed dependent upon the amount of peptide hydrolyzed. To compensate for partial destruction without using correction factors, we evaluated Trp recovery from standard proteins containing 1-6 mole Trp using the autohydrolysis methodology. Reproducible Trp response factors (Table III) that agreed with each other were obtained from cytochrome C, α-lactalbumin and lysozyme over the mass range hydrolyzed (about 0.3-1.3 μg). The response factors determined from the protein hydrolysates were about 40% less than that determined from a nonhydrolyzed 300 pmol Trp amino acid standard and about 20% lower relative to 300 pmol Trp hydrolyzed in the presence of dodecanethiol. The average response factor determined from the protein hydrolysates (Table III) was used to calculate the Trp content of other proteins (Table IV). The compositional data in Table IV support the efficacy of the Trp analysis as well as demonstrate that good overall accuracy can also be obtained with the methodology.

TABLE III

TRYPTOPHAN RESPONSE FACTORS FROM HYDROLYSIS/ANALYSIS OF STANDARD PROTEINS

Sample	TRP Response factor	Amount Hydrolyzed(μg)	n
Cytochrome C	403	0.3	3
1 Trp/mole	357	0.5	3
	499	0.9	3
	$\bar{x}$~ 420 ± 14% SD		n = 9
α-Lactalbumin	419	0.5	3
4 Trp/mole	396	0.7	3
	417	1.3	3
	$\bar{x}$~ 411 ± 2.5% SD		n = 9
Lysozyme	413	0.4	3
6 Trp/mole	425	0.7	3
	410	1.3	3
	$\bar{x}$~ 416 ± 1.5% SD		n = 9
Overall Average	415 ± 8.5% SD		n = 27

Trp response factors were determined following automatic hydrolysis in the presence of dodecanethiol as described in Experimental Procedures. Response factor units are area/pmol. SD, standard deviation; n, number of analyses.

Autohydrolysis in the presence of dodecanethiol provides a facile method for routine Trp analysis provided the instrument has been calibrated properly. Calibration for Trp using protein hydrolysates has consistently yielded more accurate results in our laboratory than using a Trp amino acid standard in conjunction with correction factors. This calibration method is still subject to variability due to the condition of the column, sample frits, solvents and reagents therefore concurrent analysis of Trp control samples with unknowns should always be performed to verify accurate calibration.

C. Cysteic Acid Analysis

Cysteic acid analysis was the most frequent method used by core facilities to determine the cysteine content in the 1989 and 1991 amino acid analysis studies sponsored by the Association of Biomolecular Resource Facilities (Crabb et al., 1990; Strydom et al., 1992). Although an incomplete compositional analysis results from oxidation of cysteine/cystine to cysteic acid (Cya) because His, Met, Tyr and Trp are also modified, the simplicity and efficacy of the method accounts largely for its popularity.

TABLE IV

TRYPTOPHAN AMINO ACID ANALYSES

Amino Acid	ß-Lactoglobin	Myoglobin	Trypsin	Chymotrypsin	BSA	BSA	Ovalbumin
Asx	16.8 (16)	11.2 (10)	21.8 (22)	21.1 (23)	53.3 (54)	58.6 (54)	30.4 (31)
Glx	26.5 (25)	20.6 (19)	14.9 (14)	15.9 (15)	79.6 (79)	81.4 (79)	50.4 (48)
Ser	6.8 (7)	5.4 (5)	31.9 (34)	26.2 (28)	27.1 (28)	28.5 (28)	37.1 (38)
Gly	4.0 (3)	16.3 (15)	27.2 (25)	29.4 (23)	19.4 (16)	31.9 (16)	23.9 (19)
His	1.8 (2)	9.6 (11)	3.0 (3)	2.0 (2)	16.6 (17)	15.5 (17)	6.2 (7)
Arg	3.1 (3)	2.3 (2)	2.2 (2)	4.5 (4)	24.9 (24)	26.5 (24)	16.7 (15)
Thr	7.6 (8)	6.7 (7)	9.6 (10)	21.2 (23)	33.2 (34)	31.3 (34)	15.3 (15)
Ala	15.1 (14)	15.9 (15)	14.2 (14)	24.3 (22)	48.0 (46)	50.1 (46)	36.7 (35)
Pro	7.9 (8)	3.9 (4)	7.8 (8)	9.5 (9)	27.4 (28)	26.6 (28)	14.2 (14)
Tyr	3.8 (4)	1.8 (2)	9.3 (10)	3.7 (4)	19.4 (19)	18.8 (19)	9.3 (10)
Val	9.4 (10)	6.2 (7)	12.9 (17)	20.2 (23)	32.9 (36)	32.6 (36)	27.4 (31)
Met	5.0 (4)	2.1 (2)	2.6 (2)	2.8 (2)	4.6 (4)	5.2 (4)	17.0 (16)
Ile	7.6 (10)	6.6 (9)	12.5 (15)	7.6 (10)	11.6 (14)	11.9 (14)	20.6 (25)
Leu	24.0 (22)	16.9 (17)	13.4 (14)	20.7 (19)	64.1 (61)	63.8 (61)	32.8 (32)
Phe	3.9 (4)	6.5 (7)	3.0 (3)	6.0 (6)	26.8 (27)	24.3 (27)	19.5 (20)
Lys	15.1 (15)	18.9 (19)	13.5 (14)	13.9 (14)	53.7 (59)	55.6 (59)	20.6 (20)
TRP	2.1 (2)	1.5 (2)	3.8 (4)	8.6 (8)	2.1 (2)	3.1 (2)	3.4 (3)
Amount hydrolyzed							
μg	0.8	0.5	1.0	0.5	1.4	0.5	0.5
pmol	41.1	28.7	44.7	19.7	20.5	7.5	12.5
Compositional Error (%)	7.2	5.0	7.8	10.7	6.3	14.0	7.7

Values are residues/mole from a single analysis and are uncorrected. Numbers in parenthesis are known values from sequence analysis. Trp was quantified using an average response factor derived from cytochrome C, a-lactalbumin and lysozyme (Table III). One half of the amount hydrolyzed was analyzed. Compositional error reflects the average error in all residues except Trp and Cys. Autohydrolysis in the presence of dodecanethiol and PTC amino acid analysis were as described in Experimental Procedures All samples were RP-HPLC purified. Analysis of BSA below ~1 μg protein consistently yielded inflated Trp values; this probably is due to non-covalent association of Trp with BSA (Peters, 1975).

Cysteic acid analysis is particularly suitable for micro-analysis because oxidation with volatile reagents such as performic acid (Hirs, 1956) or DMSO (Spencer and Wold, 1969) negates the necessity for desalting prior to analysis and thereby reduces sample losses. Recently, Beam and Speicher (1990) emphasized the utility of oxidative hydrolysis for the micro-analysis of cysteic acid. Incorporating sample oxidation into the requisite hydrolysis step (i.e., acid hydrolysis in the presence of DMSO) streamlines the analysis, reducing time and labor investments.

Variability, however, can be introduced into Cya analysis from several sources including incomplete oxidation, partial destruction during hydrolysis and inadequate chromatography. As an approach to limit variability and compensate for partial destruction, we evaluated the Cya recovery from oxidative hydrolysis of standard proteins containing 1-8 mole Cys. Proteins were subjected to manual HCl vapor phase hydrolysis (150°C/1 h) in the presence of 2% DMSO in 6x50 mm tubes all within the same 40 ml screw cap bottle. As shown in Table V, reproducible Cya response factors were obtained for cytochrome C, ß-lactaglobulin and lysozyme over the mass range hydrolyzed (0.2-1.0 µg). The response factors determined from the protein hydrolysates were about 30% lower than that obtained from a nonhydrolyzed Cya amino acid standard (120-300 pmol).

TABLE V

CYSTEIC ACID RESPONSE FACTORS
FROM HYDROLYSIS/ANALYSIS OF STANDARD PROTEINS

Sample	CYA Response Factor	Amount Hydrolyzed (µg)	n
Cytochrome C	649	0.2	2
2 Cys/mole	727	0.4	3
	767	0.7	3
	$\bar{x}$~ 714 ± 6.9% SD		n = 8
ß-Lactoglobulin	740	0.3	3
5 Cys/mole	711	0.5	3
	702	0.9	3
	$\bar{x}$~ 718 ± 2.3% SD		n = 9
Lysozyme	706	0.4	3
8 Cys/mole	736	0.5	3
	699	1.0	3
	$\bar{x}$~ 714 ± 2.3% SD		n = 9
Overall Average	715 ± 4.4% SD		n = 26

Cya response factors were determined following manual vapor phase HCl hydrolysis in the presence of DMSO as described in Experimental Procedures. Response factor units are area/pmol. SD, standard deviation; n, number of analyses.

As described above for the Trp analyses, the average Cya response factor (Table V) was used to calculate the Cya content of other proteins subjected to oxidative hydrolyses (Table VI). The compositional data in Table VI show that high quality cysteic acid analyses can be obtained using this methodology. It should be emphasized that the cysteic acid analysis data in Tables V and VI are from manual oxidative hydrolysis and that this procedure is not yet recommended for automatic hydrolysis. As with Trp analysis, concurrent analysis of control and unknown samples should always be performed to verify that the analysis system has been calibrated accurately.

TABLE VI

CYSTEIC ACID AMINO ACID ANALYSES

Amino Acid	BSA		RNAse		α-Lactalbumin		Trypsin		Chymotrypsin	
CYA	34.4	(35)	7.4	(8)	8.2	(9)	11.5	(12)	9.8	(10)
Asx	66.2	(54)	17.0	(15)	22.2	(24)	22.1	(22)	23.7	(23)
Glx	86.5	(79)	12.8	(12)	14.2	(13)	15.0	(14)	15.6	(15)
Ser	20.0	(28)	13.3	(15)	6.1	(7)	32.1	(34)	25.9	(28)
Gly	22.3	(16)	4.1	(3)	8.0	(6)	30.1	(25)	32.8	(23)
Arg	24.2	(24)	4.5	(4)	1.0	(1)	2.1	(2)	4.0	(4)
Thr	33.5	(34)	8.6	(10)	5.6	(7)	9.3	(10)	18.8	(23)
Ala	49.4	(46)	13.1	(12)	3.3	(3)	15.5	(14)	24.7	(22)
Pro	27.0	(28)	4.3	(4)	2.2	(2)	8.5	(8)	9.1	(9)
Val	33.2	(36)	9.1	(9)	5.1	(6)	13.9	(17)	20.5	(23)
Ile	11.8	(14)	2.2	(3)	6.2	(8)	12.7	(15)	8.0	(10)
Leu	60.7	(61)	2.3	(3)	12.3	(13)	13.2	(14)	18.2	(19)
Phe	23.6	(27)	2.7	(3)	2.9	(4)	2.3	(3)	5.0	(6)
Lys	62.7	(59)	10.8	(10)	11.9	(12)	14.9	(14)	16.3	(14)
Amount hydrolyzed										
μg	0.6		0.4		0.5		0.3		0.1	
pmol	8.9		26.7		33.6		12.1		3.6	
Amount analyzed										
μg	0.5		0.3		0.4		0.3		0.1	
pmol	8.1		24.1		30.2		10.9		3.2	
Compositional Error (%)	12.4		11.9		11.4		11.4		12.1	

Values are residues/mole from a single analysis and are uncorrected. Numbers in parentheses are known values from sequence analysis. Cysteic acid (CYA) was quantified using an average response factor derived from cytochrome C, ß-lactoglobin and lysozme (Table V). Compositional % error reflects the average error in all residues except His, Tyr, Met, Trp and CYA. Manual hydrolysis and PTC amino acid analysis were as described in Experimental Procedures. All samples were RP-HPLC purified.

IV. CONCLUSIONS

1. Automatic hydrolysis is our method of choice for amino acid analysis of most samples because of (i) reduced labor requirements and (ii) improved results relative to manual hydrolysis (i.e., better accuracy and recovery, West and Crabb, 1990). The automatic hydrolyzer has performed reliably in our laboratory for the last 17 months with very little downtime (<5%) due to malfunction.

2. Automatic hydrolysis of PVDF-immobilized samples can now be performed reliably with the Model 420H/130/920 system using perforated teflon discs to hold the samples in place. In general, we observe greater accuracy but lower recovery from automatic vs. manual hydrolysis/extraction of PVDF samples (Table I). A modified derivatization/extraction cycle (Table II) enhances recovery from PVDF samples on our instrument.

Despite the above advantages from on line automatic hydrolysis and the high sensitivity attainable with PTC-amino acids, precolumn derivatization methodology is susceptible to interference from buffers and salts (Tarr et al. 1991). Adsorption of samples to PVDF out of salt solutions (eg, Tris, urea, glycerol, SDS, etc.) offers a possible way around this problem. Potential benefits such as this justify further research to improve PVDF-based amino acid analysis technology.

3. Tryptophan analysis performed by autohydrolysis in the presence of dodecanethiol yields reproducible and reasonably accurate results (Tables III and IV) provided the analysis system has been calibrated properly. Trp response factors derived from autohydrolysis/analysis of standard proteins provides improved Trp quantification and eliminates the need for correction factors for partial destruction during hydrolysis. Concurrent analysis of control and unknown samples is recommended.

4. Cysteic acid analysis following oxidative hydrolysis in the presence of DMSO appears to be one of the easiest methods available for Cys quantification. Cya response factors derived from the analysis of standard proteins (Table V) provides improved quantification (Table VI). Concurrent analysis of control and unknown samples is recommended. Adaption of this methodology to automatic hydrolysis could further streamline and enhance the overall analysis process.

ACKNOWLEDGEMENTS

Supported in part by USPHS grants EY06603, CA37589, DK38639, NSF grant DMB-8516111 and by Applied Biosystems, Inc. We thank Marina LaDuke for assistance with word processing and Dr. Ken Williams for valuable discussions.

REFERENCES

Beam, K.S. and D.W. Speicher (1990) Poster T122, Microanalysis of Cysteine/Cystine in Liquid Samples or on PVDF Membranes using Vapor Phase Hydrolysis and PTC-Amino Acid Analysis, Fourth Symposium of the Protein Society, San Diego, CA, August 11-15, 1990.

Bozzini, M., R. Bello, A. Chui and D. Dupont (1989) Poster M201, Methods Development on an Automated Hydrolysis System, Third Symposium of the Protein Society, Seattle, WA, July 29-August 2, 1989.

Crabb, J.W., L. Ericsson, D. Atherton, A.J. Smith and K.L. Kutny (1990) In: Current Research in Protein Chemistry (J.J. Villafranca, ed.), Academic Press, San Diego, pp. 49-61.

Hirs, C.H.W. (1956) J. Biol. Chem. 219, 611-621.

Peters Jr., T. (1975) In: The Plasma Proteins, Vol. 2 (F. Putnam, ed.), Academic Press, New York, pp. 133-181.

Spencer, R.L. and F. Wold (1969) Anal. Biochem. 32, 185-190.

Strydom, D.J., G.E. Tarr, Y-C.E.Pan and R.J.Paxton (1992) In: Techniques in Protein Chemistry III (R.H. Angeletti, ed.), this volume.

Tarr, G.E., R.J. Paxton, Y.-C. Pan, L.H. Ericsson and J.W. Crabb (1991) In: Techniques in Protein Chemistry II (J.J. Villafranca, ed.), Academic Press, San Diego, pp. 139-150.

West, K.A. and J.W. Crabb (1989) In: Techniques in Protein Chemistry (T.E. Hugli, ed.), Academic Press, San Diego, pp. 295-304.

West, K.A. and J.W. Crabb (1990) In: Current Research in Protein Chemistry (J.J. Villafranca, ed.), Academic Press, San Diego, pp. 37-48.

LC-Determination of D-and L-Amino Acids in Peptide Hydrolysates by Pre-column Derivatization with a Chiral Reagent

Hans Brueckner
Matthias Langer
Institute of Food Technology, University of Hohenheim, Stuttgart, Germany

Herbert Godel
Hewlett-Packard GmbH, Waldbronn, Germany

Introduction

For the separation of mixtures of amino acids (AA) by HPLC, pre-column derivatization with OPA together with thiols has found wide acceptance (1,2). The failure of the OPA/thiol reagents to react with secondary AA such as proline or hydroxyproline has been overcome by an additional derivatization step with FMOC-Cl (3). It has also been shown that derivatization of DL-AA with OPA and various N-acyl-L-cysteines (ACLC) yields diastereomeric isoindole derivatives which are separable by HPLC (4,5,6). From investigations on the usability of various ACLC for the separation of mixtures of DL-AA (4) we assumed that it might be possible to resolve all protein L-AA from their corresponding D-enantiomers by fully automated pre-column derivatization with OPA and N-isobutyryl-L (or D-) Cysteine (IBLC and IBDC, respectively) employing a suitable liquid chromatograph,

TECHNIQUES IN PROTEIN CHEMISTRY III

commercially available stationary phases, and simple linear gradient elution conditions (7). We report here on the automated derivatization and HPLC separation of a standard mixture of DL-AA with OPA-IBLC and sensitive fluorescence detection of the diastereomers formed. Applications of the method are also shown on hydrolysates of peptide antibiotics and a synthetic Bradykinin analogue.

Materials and Methods

For HPLC an HP 1090 liquid chromatograph was used equipped with an autoinjector and autosampler and an HP 1046A fluorescence detector. The excitation wavelength was 230 nm and the emission wavelength was 445 nm. The column (250 * 4 mm) and guard-column (20 * 4 mm) used were filled with Hypersil ODS 5 µm (Shandon, Chesire, England).

For gradient elution two mobile phase systems were used: Mobile phase A consisted of 23 mM sodium acetate, adjusted pH 6.0, mobile phase B consisted of methanol/acetonitrile (600:50, v/v). The flow rate was set a 1 ml/min, the gradient was linear from 100% A TO 53.5% B in 75 min.

AA-standards (1 nmol/AA) and samples were automatically derivatized with 1 µl OPA/IBLC by using the derivatization device of the instrument. IBLC is obtainable on request from Novabiochem, Laeufelfinge, Switzerland. Samples were hydrolyzed in 6 N HCl at 110 C for 18 h.

Results and Discussion

The separation of 17 pairs of DL-AA, including DL-Asn, Dl-Gln, DL-Trp and Gly and L-homo-Arg (used as internal standard), derivatized with the OPA/IBLC-reagent is shown in Fig. 1a.

a

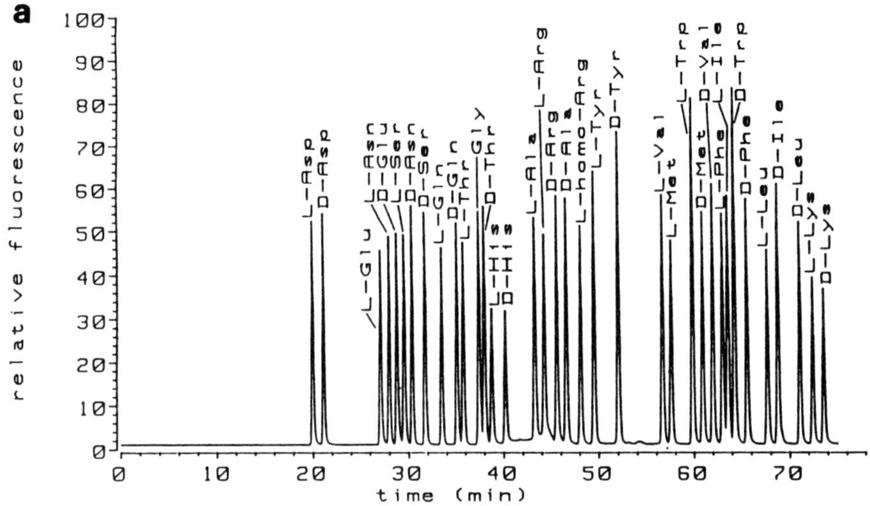

Fig 1a: HPLC-analysis of a DL-amino acid standard mixture derivatized with OPA/IBLC.

About 1 nmol of each amino acid enantiomer in the standard can be detected routinely and would allow the detection of approximately 0.1%-0.5% D-amino acid found in an excess of L-amino acid using fluorescence detection.

The precision of the method was evaluated by injection of a 36-component DL-amino acid standard using the chromatographic conditions described in materials and method by performing seven consecutive injections. The percent relative standard deviation values for all of the retention times were less that 0.6% and for the peak areas less the 1.4% except for histidine and lysine with 2.5%.

The linearity of the method (peak area versus concentration) over a range of 1 pmol to 1 nmol was tested for DL-amino acid standards containing 1, 5, 25, 100, 250, or 1000 pmol of each amino acid. The derivatizations were performed at each concentration level. All of the calculated correlation coefficients were 0.998 or better for all of the amino acids in the standard solutions.

Applications of the method are shown on total hydrolysates of natural Bacitracin (fig. 1b) and a Bradykinin analogue (Fig. 1c) synthesized by the solid phase procedure.

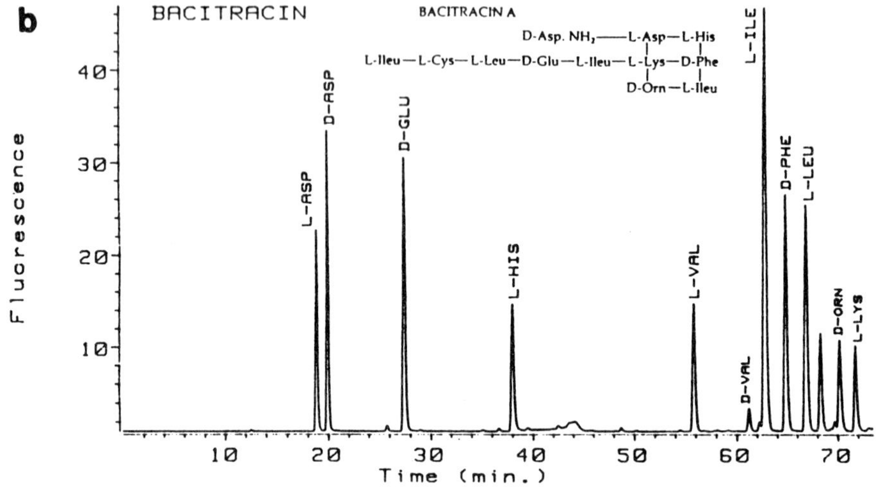

Fig. 1b: HPLC amino acid elution profile of hydrolyzed natural Bacitracin (2.6 mg hydrolyzed, 1.7 µl analyzed).

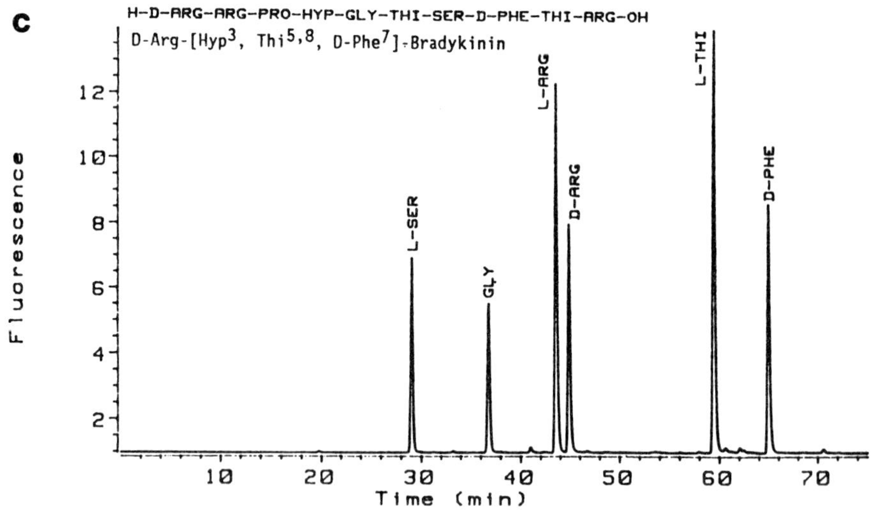

Fig 1c: HPLC amino acid elution profile of hydrolyzed synthetic Bradykinin analogue (1.9 mg hydrolyzed, 1.4 µg analyzed).

Bacitracin (Sigma), containing about 70% Bacitracin A, shows the expected L- and D-AA, but also contains significant amounts of L- and D-Val. This finding and the non-stoichiometry of D- and L-Asp are attributed to the natural microheterogeneity of Bacitracin . Amino acid analysis of the synthetic Bradykinin analogue (Fig. 1c) shows the following stoichiometry (D-Phe - 1.00); expected/found); D-Arg (1/1.12), L-Arg (2/1.81), Gly (1/0.78) L-Ser (1/0.85). From the results it is concluded that automated pre-column derivatization of DL-AA with OPA/IBLC and separation of the resulting diastereomers by HPLC is a valuable completion of existing LC-methods such as separation of DL-AA by chiral variants of Sanger's reagent (8), as well as established gas-chromatographic methods using chiral stationary phases (9). Advantages of the HPLC method are that, if present, acid sensitive amino acids such as DL-Asn, DL-Gln and DL-Trp, and also DL-Arg and DL-His (which require special care in GC) are easy to derivatize and resolve.

Disadvantages of the method are that secondary amino acids and cysteine are not yet to determine routinely.

References

1. Alvarez-Coque, M.C.G., Hernandex, M.J.M., Camanas, R.M.V., and Fernandez, C.M. (1989) Anal. Biochem. 178, 1-7.
2. Graser, T.A., Godel, H., Albers, S., Foeldi, P., and Fuerst, P. (1985) Anal. Biochem. 151, 142-152.
3. Schuster, R. (1988) J. Chromatogr. 431, 271-284.
4. Brueckner, H., Wittner, R., and Godel, H. (1989) J. Chromatogr. 476, 73-83.
5. Buck, R.H., and Krummen, K. (1987) J. Chromatogr. 387, 255-265.
6. Euerby, M.R., Partridge, L.Z., and Gibbons, W.A. (1989) J. Chromatogr. 483, 239-252.
7. Brueckner, H., Wittner, R., and Godel, H. (1991) Chromatographia, in press.
8. Brueckner, H., and Keller-Hoehl , C. (1990) Chromatographia, in press.
9. Frank, H., Nicholson, G.J., and Bayer, E. (1977) J. Chromatogr. Sci. 15, 174-176.

Amino Acid Analysis of PVDF Bound Proteins

Farzin Gharahdaghi
Donna Atherton
Michael DeMott
Sheenah M. Mische

The Rockefeller University
Protein Sequencing/HHMIBiopolymer Facility
New York, New York 10021

I. INTRODUCTION

In conjunction with sensitive analytical techniques, electroblotting of SDS PAGE separated proteins to PVDF membrane has greatly simplified the preparation of homogeneous proteins and peptides for protein chemistry. It has now become a routine technique for both N-terminal sequence analysis, as well as preparation for proteolytic digestion and subsequent internal peptide sequence determination. Accurate amino acid compositions of PVDF bound proteins and peptides is necessary for estimating both the quantity and amino acid content of the protein of interest, especially for assessing the suitability of a particular cleavage method for peptide generation and subsequent amino acid sequencing. As a result, the need for a reproducible, sensitive technique for amino acid analysis of PVDF bound proteins and peptides is critical. Amino acid analysis on PVDF has been performed with high protein concentrations (>2 ug), but accurate analyses of lower amounts have been more problematic (3,5-7). Here we describe a reproducible manual hydrolysis method for determining the amino acid composition of a PVDF bound protein of one microgram. Five standard proteins ranging in molecular weight from 14.3 kD to 77 kD were electroblotted onto PVDF from SDS PAGE and then hydrolyzed, quantitatively eluted and analyzed by PTC amino acid analysis. The accuracy, precision, and recovery at 0.5, 1.0 and 2.0 micrograms were determined. As presented here, an amino acid composition on

TECHNIQUES IN PROTEIN CHEMISTRY III

1 ug of electroblotted protein can be determined with 85% average accuracy, 5% average precision, and 88% average recovery. Lower levels (0.5 μg) can be determined with approximately 79% accuracy, 11% precision and 32% average recovery.

II. EXPERIMENTAL PROCEDURES

A. Reagents and Materials

PVDF membranes were from Millipore, phenylisothiocyanate (PITC), standard amino acid mixture H, constant boiling 6N HCl (Sequanal grade) and mercaptoethanol were purchased from Pierce. Acrylamide, N,N'-methylenebisacrylamide, sodium dodecyl sulfate, bromphenol blue, CAPS buffer and Ponceau S were purchased from Sigma Chemical Co. Triethylamine (TEA; Gold Label) was from Aldrich; redistilled phenol (Molecular Biology grade) from Boehringer Mannheim. Ethanol (100%) was from Industrial Chemical Co., acetonitrile (HPLC grade) from Burdick and Jackson, sodium acetate (HPLC grade) was purchased from Fisher and glacial acetic acid (aldehyde free) was from J.T. Baker. Pyrex disposable tubes were from Corning, glass storage vials were purchased from VWR Scientific, WISP glass and disposable limited volume inserts from Waters, and disposable 1.5 ml microfuge tubes from Sarstedt. RT20 Clean-Pak pipette tips were purchased from Rainin.

Water was distilled daily (the distillation apparatus used has no exposed teflon or polyethylene tubing, the water is collected in glass only, and the glassware used is dedicated and pyrolyzed) and filtered first through Millipore Norganic cartridges (replaced every 15 days) and finally through a 0.22 μm Millipore filter (GSWP 047). Prepurified nitrogen (99.998%) from Matheson gas products was filtered through a Millipore 0.22 μm Millistak-GS disposable filter (replaced every 3 months).

B. Proteins

Bovine serum albumin (BSA)(lot # 68F-9401), soybean trypsin inhibitor (STI)(lot # 36F-8530), human transferrin (HT) (lot # 50H-9316), egg albumin (OVALB) (lot # 68F-9407) and lysozyme (LZ) (lot # 115F-8540) were obtained from Sigma. Each protein was solubilized in 0.1%TFA (1 mg/ml, based on supplier information) and aliquots corresponding to 0.5 μg, 1.0 μg and 2.0 μg were hydrolyzed in triplicate and derivatized for amino acid analysis. Based on these analyses, 0.5, 1.0 and 2.0 μg of each protein was analyzed by SDS

PAGE electrophoresis followed by transfer to PVDF membrane. Amino acid analysis of standard protein solutions containing 0.5, 1.0 and 2.0 µg, as determined by our analysis were also used as controls for calculating the efficiency of transfer to, and elution from PVDF.

C. Electrophoresis and Electroblotting

Proteins were electrophoresed on a Hoefer Mighty Small II slab gel apparatus using a standard Laemmli SDS-PAGE protocol (2). After electrophoresis, gels were transferred to PVDF (including a second PVDF back-up sheet) using a Hoefer TE 22 Mini Transphor apparatus according to Matsudaira (4). After transfer, membranes were stained in 0.1% Ponceau-S/1% acetic acid, briefly washed with 5%-10% acetic acid to remove excess stain, rinsed 2-3 times in Milli-Q water, and then stored at 4° C until analysis was performed. An estimate of transfer efficiency (>90%) was based on the quantity of protein remaining in gels after staining with Coomassie blue or amido black, and NH_2-terminal sequence analysis.

D. Hydrolysis Methods

1. *Solution Samples*

Hydrolysis of standard proteins in acid vapor (6N HCl, 0.1% phenol, 22h, 110°C), derivatization of the free amino acids with PITC, and analysis by HPLC was performed as described (1).

2. *PVDF Samples*

Areas of PVDF membranes containing stained peptides were excised and transferred to pyrolyzed tubes. Blank strips corresponding to an empty lane for each gel were also hydrolyzed for background quantitation. The area derived from the blank analyses were subtracted from the hydrolyzate analyses. These corrected peak areas were then used for subsequent calculations. Table I outlines the method used for hydrolysis, extraction and derivatization of PVDF samples for amino acid analysis. All samples were analyzed immediately after derivatization.

E. Calibration

Each set of analyses was calibrated to Pierce standard amino acid mixture H (1 nmol / 10µl) on PVDF. Briefly, a prewetted PVDF strip (0.5 x 0.5 cm prewet with 1 µl of acetonitrile) was applied with five serial 2 µl aliquots of the standard mixture, with complete drying between applications. The strip was dried completely followed by hydrolysis, extraction and derivatization as described.

Table I Methods and Procedures

Method	Solution	Condition	Time
Hydrolysis	6N HCl, 0.1% phenol	vapor (Pico-Tag Workstation)	22 h 110 C
Extraction	50 ul Milli-Q H2O 100 ul CH3CN 20 ul 0.1% TFA	1. add sequentially 2. sonicate & vortex 3. remove the strip	10 min
Dry (Eluent)		Savant Speed-Vac	45 - 60 min
Redry	ethanol:Milli-Q H2O:triethylamine (2:2:1) 10 ul total volume	Savant Speed-Vac	20 min
Derivatization	ethanol:Milli-Q H2O:triethylamine:PITC [a] ((7:1:1:1) 20 ul total volume	RT	20 min
Dry		Savant Speed-Vac	30 min

[a] For samples less than 1.0 ug, dilute PITC was used (1:10)

F. Amino Acid Analysis

A Waters Pico-Tag Workstation and HPLC two pump gradient system equipped with a Model 490 multiwavelength detector, WISP, and either a Model 840 or 820 data station was used for analysis of the PTC amino acid hydrolyzates. Table II details the calculations used in this study. Precision (average % SD) is an estimate of the agreement among the various determinations, while accuracy (average % error) measures the agreement of the experimental result with the theoretical result. Both are expressed as percent to minimize magnitude effects. Percent error allows comparison among proteins, but exaggerates error for low frequency amino acids, especially low molecular weight size proteins.

III. RESULTS AND DISCUSSION

We have presented here a reproducible and reliable manual extraction and vapor hydrolysis method for determining amino acid compositions of PVDF bound proteins ranging in size from 14.3 kD to 77 kD. Three concentrations (0.5 µg, 1.0 µg and 2.0 µg) of each protein were electrophoresed by SDS PAGE, electroblotted to PVDF membrane, hydrolyzed and extracted from

Table II Definition of Terms and Calculations

Term	Definition
AA	Amino Acid
n	number of analyses
% mole AA	100 (pmol AA)/(total pmol of all AA)
Residue/mole protein	$\frac{\text{(\% mole AA)(true total of all AA)}^{a}}{100}$
% Error (per each AA residue)	$\frac{\text{((residue/mole AA) - (true total of all AA))}^{a}}{\text{true residue/mole protein}} \times 100$
% Avg. Error (per each AA residue)	$\frac{\text{sum \% error AA}}{n}$
% Avg. Compositional Error	$\frac{\text{sum \% avg error}}{16}$
Standard Deviation (SD)	$\sqrt{\frac{\text{sum of [(residue/mole) - (mean of residue/mole)]}^{2}}{n-1}}$
% SD	$\frac{100 \times \text{SD}}{\text{mean residue/mole}}$
Avg % SD	$\frac{\text{sum of (\% SD residue/mole of all AA)}}{16}$
Recovery (ug)	sum of ((pmol AA)(MW of AA))
% Recovery	$\frac{\text{recovery (ug)}}{\text{amount electroblotted (ug)}} \times 100$

[a] True total of all amino acids, based on amino acid sequence (excluding cys and trp).

PVDF. Samples were derivatized with PITC (1:1), and analyzed as described. Figure 1 shows a representative chromatogram of 100 pmol of standard H mixture after elution from PVDF, as well as a representative blank PVDF strip (Figure 1B). Each set of protein hydrolyzates were calibrated on the basis of standard mixture H hydrolyzed and eluted from PVDF. In order to verify the recovery values for the protein hydrolyzates, a comparison of the recovery and the precision of standard mixture H with and without hydrolysis and extraction from PVDF was performed. The results are shown in Table III. As can be seen, the recovery of the standard from PVDF is within 1% of the standard, and the ratio of the two is equal to 1, confirming that the standard used for the calibration of the proteins is accurate.

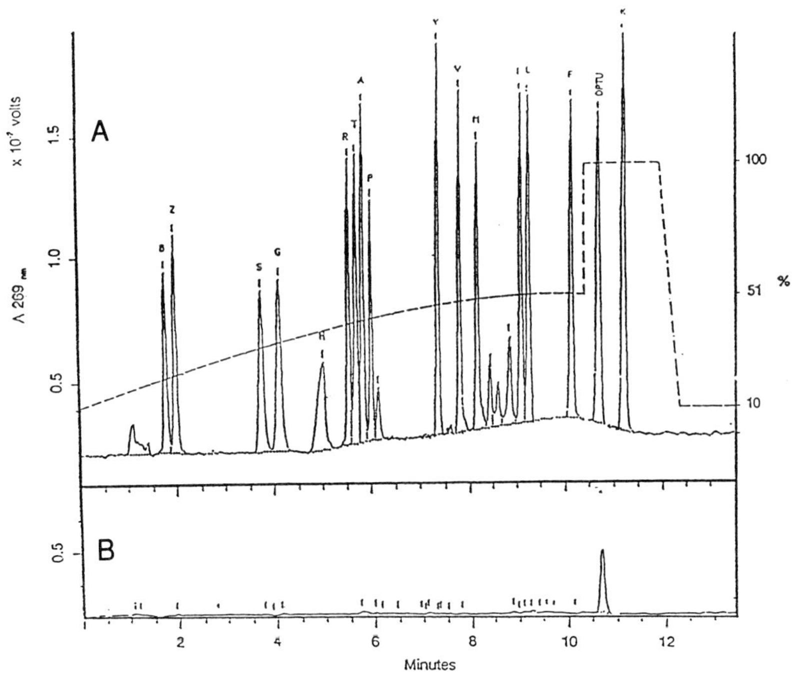

Figure 1: A) Chromatogram of analysis of 100 pmol of standard H eluted from PVDF. 1 nmol of standard H was applied to PVDF, eluted and derivatized, as described in *Experimental Procedures*. 10/100 µl was analyzed. PTC amino acids are identified by single letter code. B) Chromatogram of PVDF blank. 10/30 ul of the PVDF eluant was analyzed and normalized for 100 pmol scale for analysis. The total AA of the blank PVDF was 0.04 µg. The absorbance was measured at 269 nm. No corrections were made for hydrolytic destruction or amino acid losses.

Table III Comparison of Calibration Standard H with and without Hydrolysis and Extraction from PVDF

Amino Acid	Standard Mixture H on PVDF w/ hydrolysis & extraction		Standard Mixture H no PVDF, no hydrolysis		+PVDF [c] / - PVDF
	peak area[a]	%SD	peak area[b]	%SD	
B	41661	2.1	49298	3.6	0.8
Z	47464	2.6	47362	3.7	0.9
S	45839	1.4	46769	1.8	1.0
G	53309	1.3	47678	3.2	1.1
H	37690	1.8	41964	3.0	0.9
R	41351	2.2	49948	3.1	0.8
T	54077	1.9	46722	3.5	1.2
A	60754	1.8	55252	1.8	1.1
P	38763	3.4	40407	2.7	1.0
Y	63597	1.0	58781	3.8	1.1
V	57327	0.6	52488	4.1	1.1
M	54457	5.7	60241	3.5	0.9
I	60446	2.2	54851	3.9	1.1
L	68686	2.5	56752	2.7	1.2
F	67310	2.9	60165	0.6	1.1
K	82779	1.2	98759	3.8	0.8
TOTAL	*875510*		*867437*		
AVERAGE		*2.2*		*3.1*	*1.0*
n =	5		3		

[a] Peak area resulting from analysis of 10% of PVDF extract following hydrolysis (assuming 1 nmol hydrolyzed on PVDF)

[b] Peak area from 100 pmol standard H mixture without hydrolysis or PVDF

[c] Peak area ratio of a/b

As shown in Table IV, 1 microgram of PVDF bound protein resulted in an average of 88% sample recovery, as determined by amino acid analysis. Amino acid compositions as calculated gave an average error of 15% (range 11% - 16%) with an average percent standard deviation or precision of 5% (range 3.5% - 6.1%). Amino acid compositions of two micrograms of electroblotted protein yielded a decreased average error of 10%, 5.7% average standard deviation and an increased recovery of 90%. These results suggest that 1 µg or more of protein can be eluted from PVDF. However, for electroblotted proteins of 0.5 micrograms, an average of only 32% was recovered, with an average error of 21% for amino acid composition (range 16% - 24%) and an average precision of the data was 10.7% (range 9% - 17%).

Previous reports have demonstrated recoveries ranging from 28% to 47% for 2 micrograms or more of electroblotted proteins hydrolyzed and extracted with 30% methanol/0.1 N HCl, with low recoveries attributed to poor transfer to PVDF (5,7). In the study presented here, the efficiency of transfer was estimated to be >90%. Therefore, inefficient transfer is unlikely to cause the low recovery of 0.5 ug samples shown here. Also, it should be noted that in previous studies, acetonitrile proved less effective than methanol for extraction of the hydrolyzed amino acids in their laboratory (7). However, a key difference is that our PVDF is stained with Ponceau S, whereas earlier studies were all performed with Coomassie blue; the staining procedure used for the latter often results in proteins being fixed to membrane. Additional studies are in progress to regarding the problems of extraction from the membrane.

Table IV Accuracy, Precision, and Recovery of Electroblotted Proteins (14 kD - 77 kD)

Protein	Amt. Electroblotted (μg)	MW	n	Avg % Error	Avg % Recovery	Avg % SD
HT	0.5	77.0 kD	3	16.3	40	9.1
BSA	0.5	67.0 kD	3	23.0	20	9.3
Ovalb	0.5	45.0 kD	3	23.2	40	14.8
STI	0.5	20.1 kD	3	20.5	20	3.5
Lysozyme	0.5	14.3 kD	3	24.1	40	16.9
Average				***21.4***	***32***	***10.7***
HT	1.0	77.0 kD	3	11.2	70	3.5
BSA	1.0	67.0 kD	6	16.6	100	5.4
Ovalb	1.0	45.0 kD	3	14.0	100	3.6
STI	1.0	20.1 kD	3	16.2	90	6.1
Lysozyme	1.0	14.3 kD	3	15.0	80	5.8
Average				***14.6***	***88***	***4.9***
Ovalb	2.0	45.0 kD	3	9.8	90	6.2
STI	2.0	20.1 kD	3	10.1	90	5.1
Average				***10***	***90***	***5.7***

Representative amino acid compositions of two of the proteins analyzed in this study (egg albumin and soybean trypsin inhibitor) are presented in Table V. Residues per mole for each were calculated from triplicate analyses of each level and were compared to the known composition of each protein. As can be seen, this technique produces data that is quite reproducible for low level analyses of protein bound to PVDF. The precision, accuracy and recovery are presented as well, and in graph form in Figure 2A and 2B. As can be seen, His quantitation is the most problematic for this technique. For Met, error and variability is due in part to a coeluting PVDF/protein product (see Figure 1B), and Pro errors may be due to contributions from NH_3.

Figure 2C summarizes the precision and accuracy for all proteins analyzed at 0.5 and 1.0 µg levels. As can be seen, the basic amino acids as well as proline and methionine present the most difficulty. At the 0.5 µg level Arg, His, Lys have higher errors and lower recovery. This may represent the higher affinity of basic residues for PVDF as opposed to hydrophilic or acidic residues. Met and Pro appear to be due to low residue values and resultant integration problems. If the basic amino acids are disregarded and the average percent error for analyses performed at 0.5 microgram are recalculated, the average percent error decreases to 17.1% Further studies are being pursued to increase the recovery of samples of less than one microgram.

The method described here produces the highest recovery and best accuracy and precision based on the published literature for manual hydrolysis. As compared to the results summarized in the 1990 AAA ABRF report (6), this method compares favorably. The average performance of the facilities at one microgram is equivalent to our results at 0.5 microgram. Our results at 1 microgram equals the best of the ABRF labs for accuracy and precision of the composition. The average ABRF recovery of protein was 52% of the expected 0.9 micrograms *adsorbed* to PVDF. Our yield by this method was greater than 90% for one microgram *electroblotted* to PVDF.

In summary, this method provides a reproducible technique with excellent recovery and adequate accuracy and precision for determining an amino acid composition on samples submitted on PVDF membrane. We have implemented this technique in our Facility with good results for unknown proteins that cannot be analyzed by conventional methods. It is also ideal for proteins that cannot be desalted by other methods.

Table V Amino acid Compositions of Electroblotted Proteins

Amino Acid	Soybean Trypsin Inhibitor				Egg Albumin (Ovalb)			
	Known Composition	Residue/mole[a]			Known Composition	Residue/mole[a]		
		0.5	1.0	2.0[b,c]		0.5	1.0	2.0[b,c]
B	26	23.7	17.3	28.4	31	23.7	38.1	29.7
Z	18	21.6	17.6	20.5	48	46.7	54.7	48.6
S	11	9.2	10.9	11.1	38	44.4	30.7	34.3
G	16	12.9	14.2	15.7	19	18.4	15.9	21.4
H	2	0.0	1.1	2.3	7	0.0	6.3	6.6
R	9	6.9	10.6	8.8	15	9.5	16.6	16.2
T	7	7.0	7.2	6.7	15	15.9	14.7	16.9
A	8	6.2	8.8	8.6	35	39.2	28.8	34.1
P	10	10.0	11.2	11.1	14	14.1	14.5	16.4
Y	4	4.8	5.1	4.0	10	12.5	11.9	11.2
V	14	17.2	12.9	12.6	31	32.3	24.4	30.9
M	2	1.9	2.3	2.5	16	17.3	15.2	12.9
I	14	17.1	15.0	12.4	25	29.1	23.9	23.3
L	15	14.7	17.0	11.4	32	37.8	36.4	32.3
F	9	9.9	9.3	8.7	20	21.9	17.6	21.8
K	10	12.0	14.7	9.9	20	13.1	26.3	20.1
Average % SD		*3.5*	*6.1*	*6.2*		*14.6*	*3.6*	*6.2*
Average % Error		*20.5*	*16.2*	*9.8*		*23.2*	*14.0*	*9.8*
Average % Recovery		*20.0*	*90.0*	*90.0*		*40.0*	*90.0*	*100.0*

[a] Residue/mole =(%mole AA)(true total of all AA)/100

[b] micrograms

[c] n=3 for each microgram value

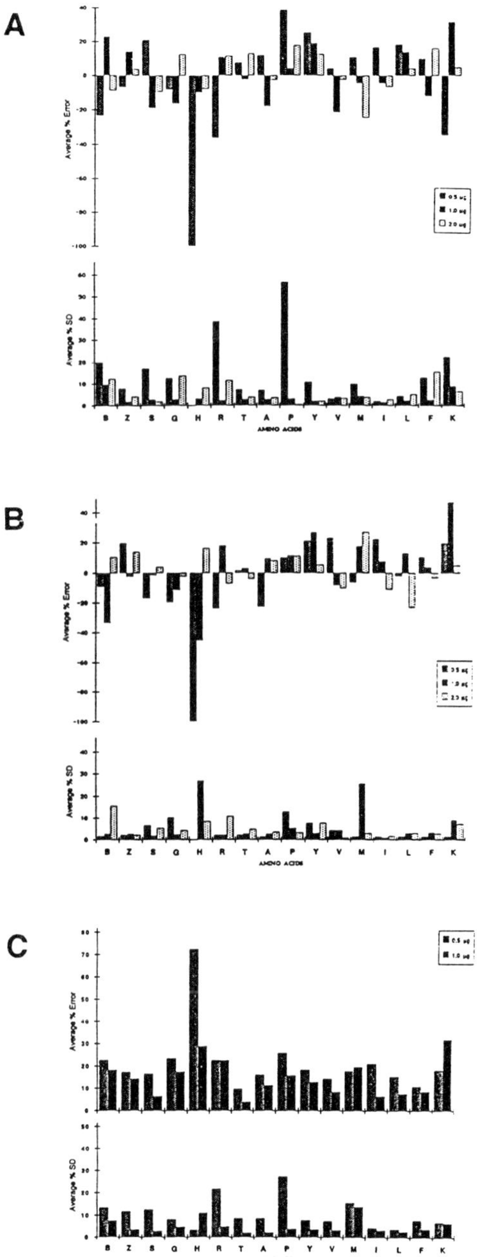

Figure 2: A.) Comparison of average error and precision per amino acid for egg albumin and B.) soybean trypsin inhibitor. n=3 for each concentration of protein. C.) Comparison of average error and precision for all amino acids for all proteins. n=3 for each of the five proteins analyzed with the exception of BSA, where n=6 at the 1.0 μg level.

REFERENCES

1. Atherton, D. (1989) Successful PTC amino acid analysis at the picomole level. In Techniques in Protein Chemistry (T.E. Hugli ed.) Academic Press. p 273-283
2. Laemmli, U.K. (1970) Cleavage of structural proteins during the assembly of the head of bacteriophagee T4. Nature 227:680-685
3. Nakagawa, S. and Fukuda, T. (1989) Direct amino acid analysis of proteins electroblotted onto PVDF membrane from SDS PAGE. Anal. Biochem. 181:75-78
4. Matsudaira, P. (1987) Sequence from picomole quantities of proteins electroblotted onto polyvinylidene difluoride membranes. J. Biol Chem. 262:10035-10038
5. Ploug, Michael, Jensen, A.L, and Barkholt, V. (1989) Determination of amino acid compositions and NH2-terminal sequences of peptides electroblotted onto PVDF membranes from Tricine SDS-PAGE: Application to peptide mapping of human complement component C3. Anal.Biochem. 181:33-39
6. Tarr, G.E., Paxton, R.J., Pan, Y.-C. E., Ericsson, L.H., and Crabb, J.W. (1991) Amino acid analysis 1990: the third collaborative study from the association of biomolecular resource facilities (ABRF). In Techniques in Protein Chemistry II (J.J. Villafranca, ed.) Academic Press 1991. p 139-150
7. Tous, G.I., Fausnaugh, J.L., Akinyosoye, O., Lackland H., Winter-Cash, P., Vitorica, F.J., and Stein, S. (1989) Amino acid analysis on polyvinylidene difluoride membranes. Anal. Biochem., 179:50-55

Collaborative Trial Analyses of ABRF-91AAA

Daniel J. Strydom
Center for Biochemical and Biophysical Sciences and Medicine
Harvard Medical School
Boston, Massachusetts 02115

George E. Tarr
Department of Cardiology
Children's Hospital
Boston, Massachusetts 02115

Yu-Ching E. Pan
Department of Protein Biochemistry
Hoffmann-La Roche, Inc.
Nutley, New Jersey 07110

Raymond J. Paxton
Department of Protein Chemistry
Immunex Corp.
Seattle, Washington 98101

I. Introduction

Collaborative trials for assessment of the reliability of amino acid analyses have been done since the early fifties. The first 30 years of trials have been summarized and discussed by Williams (1), and more recently the ABRF series of trials (2-5) have provided information on various aspects of amino acid analyses. As has been noted previously (1), amino acid analysis is not as simple and routine as the wide use of the technique may suggest. Consequently, a large measure of self-evaluation by facilities providing this service is very necessary, and continuing education in the techniques and alternatives remains essential. The objective of the ABRF, in this instance, is then to provide a mechanism for self-evaluation and improvement of amino acid analytical capabilities of core facilities, by coordinated analyses of standard samples.

The ABRF studies to date have examined the recovery, accuracy and precision of amino acid analysis using either synthetic peptides, purified proteins, or a protein hydrolysate. These studies have in part concentrated on the sensitivity of analysis, with hydrolyses and analyses being performed in the 0.1-5 μg range. Laboratories have also been asked to follow well defined protocols to simplify data reduction and subsequent comparisons. The accuracy of analyses has generally been in the 80-90% range. The present study was an attempt at investigating the quality of analyses

TECHNIQUES IN PROTEIN CHEMISTRY III

under conditions where laboratories might feel more comfortable, and where ample supply of sample was available. Three general aspects were explored. The first two were a realistic assessment of the accuracy of absolute quantitation and of the accuracy of relative yields of residues obtainable at present. The third aspect was the analysis of the problem amino acids tryptophan and cystine/cysteine, where we were aiming at discovering which methods are used, and how well they performed.

II. Methods and Materials

A. *Sample Preparation*

Samples were generously prepared by Dr. Sam Margolis, National Institute of Standards and Technology, Gaithersburg, Maryland. Glassware and ampules were cleaned in a pyrolyzing oven to minimize contamination. Bovine serum albumin (7% solution, Standard reference Material 927a) was diluted by absorbance to a concentration of 0.75 mg per ml and dispensed into cleaned ampules using a Micromedic dispensing pump (1 ml) set at 20% volume. The samples were then lyophilized and sealed under an atmosphere of air. The samples were sent by regular mail to ABRF facility directors. The samples were distributed without identification of the protein as BSA. A questionnaire asked for a general strategy of analysis, a single answer of how many nmoles of each amino acid residue was present in the vial (a ballpark figure of 0.1 mg was given as guide) and a number of questions relating to the methodology and calibration.

The results were received by the committee after the identifying cover sheets were removed by an independent collaborator, Dr. B. Holmquist, to insure anonymity.

B. *Data Reduction*

The data was coded into personal computer-based spreadsheets, and analyzed as follows. The amount of protein predicted by each reported residue was found by dividing by the known number per molecule for that residue. Individual amounts were compared with the average value for a given analysis and any that deviated by more than 15% were discarded; a mean amount was calculated from the remaining "relevant residues". Residues/mol were then determined by dividing the nanomoles of amino acid by this mean. Recoveries were calculated from the mean and the molecular weight of BSA and compared to the expected 0.15 mg to obtain a relative recovery (accuracy of quantitation).

The accuracy of each residue compared to the true value is expressed here as %Error. The overall accuracy of the composition is expressed as the Average %Error. Cystine/cysteine and tryptophan errors were not included in the average percent error calculation.

Equations:

$$\%\text{Error} = 100 \times (\text{experimental} - \text{true residue value}) / \text{true}$$

$$\text{Average } \%\text{Error} = \Sigma \text{ absolute } \%\text{Error for 16 amino acids} / 16$$

III. Results and Discussion

A. *Strategy and Instrumentation*

Unlike previous ABRF AAA studies, laboratories were asked to develop their own strategy for this year's sample, and sufficient sample was provided so that a number of analyses could be performed. Table I summarizes the instrumentation and the methodologies which were employed by the 51 laboratories which participated in the trial. Seven of the facilities used two different methodologies for this analysis, for a total of 58 "sites". As in previous studies, the number of pre- and post-column methods was similar, with PTC and ninhydrin chemistries being most popular.

Forty-six sites performed cystine/cysteine analyses (2 of these reported no result for Cys) and 21 sites attempted tryptophan analyses, although for the latter only 16 sites reported results with 3 of these being 0 nanomoles. The amount of sample used per hydrolysis ranged from a low of 0.6 μg to a high of 45 μg. Fifty-four sites (93%) combined data from more than one hydrolysis. Approximately 60% of the sites performed a single analysis of each hydrolysate, whereas 9% performed replicate analyses; the remaining sites didn't provide this information. Twenty sites performed time course hydrolysis studies, although in some cases it was unclear whether this data was used in reporting the results. This data was most often used to correct for serine and threonine destruction (values being extrapolated back to 0 hours), or to correct for decreased yields of isoleucine and valine (values taken from later time points). Eight sites performed multilevel hydrolyses, although in most cases the rationale was not explained. Asparagine and glutamine determinations were also suggested, but none of the participating facilities reported these values.

Table I. Instrumentation and Methods

Instrumentation[a]				Methodology			
Pre-column		Post-column		Pre-column		Post-column	
ABI 420/130	11	Beckman	17	PTC	30	ninhydrin	21
Waters	13	St. Johns	1	OPA	1	OPA	3
H/P 1090	2	LKB	1	OPA/FMOC	2	Fluram	1
Waters/Beckman	1	Dionex	1				
Beckman	1	Pharmacia	1				
ABI/Waters	1	Home-made	3				
Total	29		24		33		25

[a]More than one analysis was reported from some single instruments. Those instruments were only counted once.

B. *Recovery and Accuracy of the Analyses*

Previous collaborative trials had concluded that the excellent (<3% error) accuracy claimed by developers and manufacturers of amino acid analysis techniques (6), could be achieved by only a fraction of the participating laboratories. It is clear that such accuracy and precision may generally be achievable on standard mixtures, if at all, but not on hydrolysed samples. The present study has seen broadly the same trends as previously, with some improvement, probably due to the larger amount of sample.

Figure 1 summarizes the yields that were attained by all the sites, relative to the 150 μg theoretical value. Although some sites achieved close to the theoretical yield, most did not. Discarding 18 sites as outliers (differing more than 2 standard deviations from the mean), the remaining 40 sites had an average yield of 91.2 $\pm$ 6.1 %. This could signify an error in the "theoretical" yield, which is based among others on optical density measurements and amino acid analyses, or a pervasive bias. Such a bias could be due to pipetting errors (plastic-tipped dispensers), derivatizing inefficiencies or manipulative losses or to incorrect calibration, due perhaps even to incorrect standards. This study cannot address this question. The use of an internal standard may alleviate some of these potential problems. Sixteen sites used an internal standard, Nle (13 sites) being the predominant choice, with 2 sites choosing α-aminobutyric acid and one hydroxyproline. Since only one of these analyses differed by more than two standard deviations from the mean, the overall result was demonstrably better than without internal standards. Conversely, many of the sites not using internal standards achieved good quantitation. The outliers in the latter group were however much more egregious. Although these data say that accuracy in absolute quantitation may not be good, the precision at 6% is better than in past collaborative trials.

The average of the absolute percent error of each residue in each sample, in comparison to the value obtained by sequence, is summarized in Figure 2. As in previous studies, the range extends from very good, 2.8% for site 41, to very bad, over 25% for site 10. The average for all sites of 9.47% error is similar to that observed for high level hydrolyses (9% error) in the 1989 study (3) and better than the 13.45% error in the 1990 study (5). For the 21 sites with >10% average errors, the problem can be attributed in many cases to one or two bad residues, with

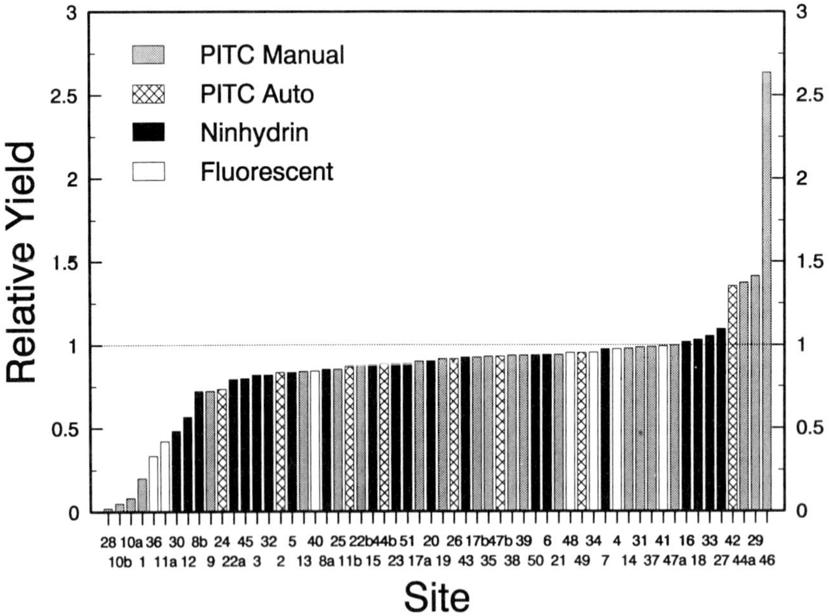

Figure 1. Yields of analyses, relative to the 150 μg dispensed to the sites.

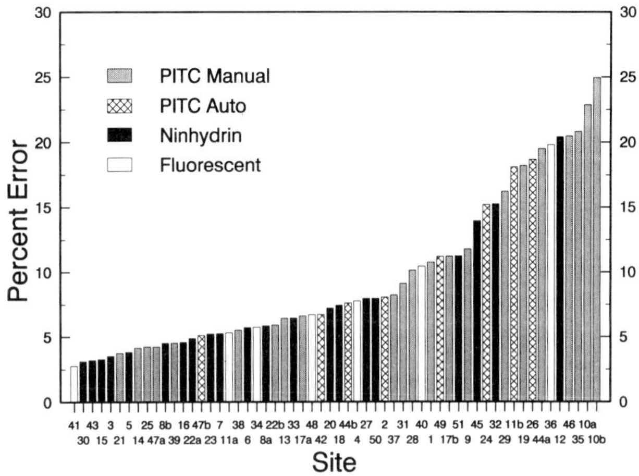

Figure 2. The average of the absolute percent error of each residue in each sample. These values were calculated relative to the sequence values.

methionine and glycine being the most common. By deleting the worst residue in the average %error calculation, the average %error of 12 of these sites decreases to 10% or less. Among the methods, the classical ninhydrin methodology generally afforded better data than the pre-column PITC methods, although all methods had individual analyses in the low %error range. This perhaps reflects the larger experience base of the classical methodology.

The relative errors for each residue are summarized in Figure 3. Note that no samples were removed from the set. The general trend is that tryptophan, cystine, and methionine remain problematic in many laboratories. A discussion of the various problems highlighted by these comparisons is as follows.

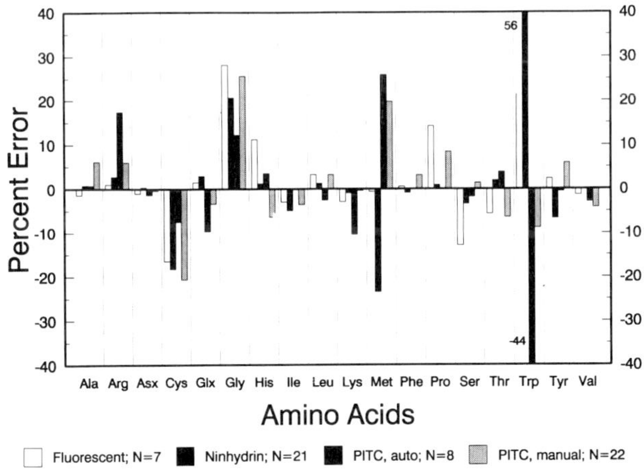

Figure 3. Relative errors for each amino acid residue. No samples were removed from the set.

Tryptophan. Most laboratories do not analyze for tryptophan, probably due to these difficulties. Trp is the least abundant residue in BSA at 2 residues per 582, which of course compounded the problems experienced. The raw values and methodology for tryptophan estimation are summarized in Table II. The methanesulfonic acid based method of hydrolysis accounted for half of the reported Trp analyses (8 sites), with only 3 of them within 20% of the sequence value. The excessively high value (162% or 220% error) found e.g. by one of the PITC and one of the OPA based analyses, is probably due to a chromatographic artifact, while difficulties in integrating small amounts of Trp in addition to hydrolytic losses probably underlie the low values found in other analyses. Three sites reported a total lack of tryptophan. No method can be statistically selected as superior, although the thioglycolic acid method (2 successful analyses out of two) may need some attention.

Table II. Analyses of Tryptophan in ABRF-91AAA

Hydrolysis Medium	Analysis Method	Percent Error of Individual Sites		
Methane Sulfonic Acid	Ninhydrin	-100.0	-35.2	
	ptc	-10.7	-69.8	6.7
	fluor	162.2	-100.0	3.6
Thioglycolic acid		-3.1	-13.1	
Dodecane thiol		9.3	-100.0	
HCl	As dihydro tryptophan	41.8		
Mercaptoethane sulfonic acid		72.8	26.0	
3% phenol		220.3		

Cystine/Cysteine was determined by 44/58 sites and raw error values are summarized in Table III. Oxidative preparation of samples remains the most popular, 23 samples being analyzed after such treatment. Direct analyses of HCl hydrolysates were done in 12 cases, and 4 sites used induced disulfide exchange during hydrolysis to form tractable derivatives. A further 4 sites reduced and alkylated the protein and then did a "normal" analysis. Disulfide-exchange and prior alkylation were the most successful methods.

If 25% error is tolerated, three quarters of the samples remain. The "direct" analyses fall into two clear groups, those with more than 45% losses and those with less than 18% error, perhaps pointing to the general utility, but also to considerable uncontrolled problems. The oxidative methods, presumably popular due to their historical use, average 18% low, with a range from +6.8 to -63.1%. Some of the

higher values were obtained by use of standard proteins (sites 25 and 39) to correct for both chemical and hydrolytic losses, but site 31 still had only a 50% yield.
Methionine: Large variations in methionine quantitation were found. The biggest factor is most probably the low level of methionine in BSA - only 4 residues. Figure 3 shows methionine to be analyzed in the most biased fashion by the two major methods. Ninhydrin, when in error, consistently underestimates Met, with values as low as a total lack; no site overestimated it. This could imply either a consistent hydrolysis problem (oxidation) or an integration problem on a small peak. The PITC methodology frequently overestimates Met, being up to 3 fold high in 2 cases. Low values were also seen, which is consistent with a hydrolysis problem as seen with ninhydrin. The high values are most probably due to erroneous identification of a common "reagent" peak. Overall, the error rate in Met correlates well with the average error rate of all residues, with only 2 of the best 28 and 7 of the remaining 30 analyses having a higher than 20% error in Met. There was no correlation with the additive used for hydrolysis.

Table III. Analyses of Cystine/Cysteine in ABRF-91AAA

Derivative	Analysis Method	N	average	std dev	range
performic acid	ninhydrin	8	-16.8	14.7	-36.5 - 4.2
	ptc	6	-14.6	17.0	-49.1 - 6.6
	fluor	2	-35.4	6.7	-42.1 - -28.7
alkylated		4	5.8	14.9	-7.4 - 29.4
None	ninhydrin	3	-40.5	40.6	-90.3 - 13.8
	ptc manual	7	-36.3	29.3	-81.1 - -3.6
	ptc auto	2	-17.4	2.1	-19.5 - -15.3
hcl/dithio propionic acid		3	4.5	10.2	-3.7 - 18.9
dmso/HCl		6	-18.8	25.1	-63.1 - 6.8
dithiodiglycol		1	-0.5	-	
"abdf" titration		1	-4.3	-	
periodate		1	-12	-	
Total		44			
Summary					
Oxidized		23	-18.2	18.7	-63.1 - 6.8
Alkylated		4	5.8	14.9	-7.4 - 29.4
Disulfide exchange		4	3.2	9.1	-3.7 - 18.9
Direct		12	-34.2	31.9	-90.3 - 13.8

Aspartic acid: The worst problems in Asp analyses were seen with the PITC methodology, where 3/8 automated analyses had negative errors larger than -10%, and 5/22 of the manual sites had similar errors. Two of these sites had large positive errors. These negative errors are probably due to derivatization or solubilization problems caused by leachates from the glassware used for hydrolysis. This may sometimes be alleviated by appropriate chemical procedures (7).

Glutamic acid: In the past it has been suggested that Glu values may be overestimated due to erroneous calibration with standards in which partial formation of pyroglutamic acid has taken place. More Glu values are low (or high) than Asp values are, but in the majority of cases they parallel errors in Asp. A few very large errors cause the large deviations seen in Figure 3. The slightly positive average error seen for ninhydrin may reflect the expected calibration errors but are inconsequential in the context of the generally high errors seen in this study as a whole.

Histidine: The fluorescent methods overestimate His, while the manual PITC methods underestimate it. Thus, 4 of the 7 fluorescent sites overestimated His by at least 2 residues out of 17. Contrariwise 6 of the 22 manual PITC sites underestimated His by at least 2 residues, with 2 of those yielding only half the theoretical value. Calibration values were similarly low (see below, Section D). As noted before, this may be due to chemical problems in the derivatization, and/or chromatographic problems. The ninhydrin sites fared better with only one underestimating (and that at 40%) and two overestimating.

Tyrosine: BSA may have 20 tyrosyl residues (8), as opposed to the 19 residues we have assumed for these calculations (9). The marked negative bias in the ninhydrin-based analyses of Tyr may therefore be 50% larger than shown in Figure 3 and is typically associated with hydrolysis conditions. The magnitude of these errors precludes their being only due to integration errors of the usually broad Tyr peaks. The positive bias of the PITC-based analyses may similarly be decreased to a small positive value. Since hydrolysis conditions should be independent of analysis method, this positive bias in the PTC analyses is probably due to "reagent" peaks which elute in close proximity to Tyr in most chromatography systems.

Lysine: Manual PITC methodology occasionally lead to low yields of Lys (4/22 worse than -10% error), while the automated sites showed markedly negatively biases (4/8 worse than -10% error). This would suggest a pervasive problem which may be due to solubility properties of the PTC-amino acids or the quality of the PITC.

Arginine: Only 2 of the 8 auto-PITC sites could analyze Arg with less than 10% error, mostly positively biased. A small number of the manual PITC sites likewise were positively biased. Such bias has been seen in unpublished studies to be due, among others, to contaminated methanol used during derivatization.

Proline: 2 of the 3 sites which did analysis of Pro by a fluorescent derivative had large positive errors, emphasizing that extra care is still needed for analysis of secondary amino acids. Ninhydrin sites had few problems with Pro, in contrast to conventional wisdom. A strong correlation of overall error with more than 10% error in Pro, is seen with the PITC methods, where 7 of the 8 erroneous Pro values fall in the 12 analyses with the worst average percent error. Proline reacts very easily with PITC to form the most stable PTC-derivative, so that a bad proline value may be taken as a predictor of erroneous analyses in general. Standards with markedly variant relative color yields and especially high Pro color yields will most facilely alert operators to this condition in a set of samples.

Glycine: Very large errors were seen for glycine, probably due to adventitious contamination, some perhaps already in the original samples (about 5%). There are however some samples which are very much in error - 5 are higher than 50% in error, ranging to 200%.

To summarize the error-level of the whole set of analyses, they can be categorized in three sets; <5% average error (n=14), 5-10% average error (n=23), and >10% error (n=21). The notable differences between the best sites and the average sites was in the ability to accurately quantitate methionine and histidine. The worst sites, on the other hand, had great difficulty in analyzing methionine and possibly had contamination problems as evidenced by high glycine values.

C. The Influence of Hydrolysis Conditions

Hydrolysis conditions were varied, although nearly all of the 58 (effective) sites partaking in the study used 6 *N* HCl as the digestion acid of choice. One site used 2:1 HCl:trifluoroacetic acid, and one site used this in addition to 6 *N* HCl. Gas-phase versus liquid-phase hydrolyses are equally popular with 26 sites doing the former and 24 the latter. No difference is obvious in the quality of the analyses done by these two procedures. Interestingly, no participant used micro-wave assisted hydrolysis.

The early work in this field strongly suggested the use of a good vacuum during hydrolyses (*e.g.* 10), and many sites seem to still adhere to this. Thus 13 sites used the traditional <0.1 mm Hg pressure, while 15 sites used between 0.1 and 1 mm Hg. A further 9 sites used higher pressures, 12 did not report their vacuum conditions and one site used no vacuum. The 8 automated hydrolyses made use of vacuum as required by that instrument. No correlation was seen of quality of analysis and vacuum conditions.

The commonly used 110°C as hydrolysis temperature is still the most popular with 38 sites using temperatures ranging from 106° to 120°, and hydrolyses times from 18 hrs to 96 hrs. Temperatures of 147° to 170° were used by 12 sites for manual hydrolyses and by the 8 automated sites. In the latter cases hydrolyses times were from 30 min to 3 hrs. The higher temperatures lead to both excellent and very bad analyses, pointing perhaps again to the importance of operator experience, rather than specific techniques. The use of an inert gas to assist in displacing oxygen from hydrolysis tubes was seen at 41 sites; 29 used Nitrogen, 11 Argon and one Helium. No difference in analysis quality correlated with this.

Extrapolation of the analytical values for Ser/Thr and Ile/Val determined from multiple hydrolysis times, were no more successful in yielding "correct" values, than values taken from single time period hydrolyses.

The only hydrolysis conditions that seemed to correlate weakly with quality, was the use of additives, such as phenol. The use of mercaptans or disulfide-containing reagents coincided with none of the lowest ranking 22 analyses, while only four phenol-treated samples fell in the lowest 15. It is not clear whether the automated analyses used additives. Additives are not however absolutely necessary, as demonstrated by sites 7, 25, 38 and 47.

D. Calibration

Accurate amino acid analyses require accurate calibration, an obvious fact that tends to be discounted. In the case of precolumn analysis with PITC, the accuracy of calibration can be judged from the extinction coefficients of the PTC-amino acids in the analytical solvent at the analytical wavelength. These constants are known (11).

Table IV

	CALIBRATION % ERROR					BSA % ERROR			
	Raw	Absolute				Raw	Absolute		
	Ave	Ave	H-non	>1-1	Berr	Ave	Ave	H-non	>1-1
Ala	0.4	3.6	1.5	1.0	-0.4	2.9	5.6	6.2	-4.3
Arg	5.4	6.7	0.3	4.9	3.7	3.0	7.3	5.2	-4.1
Asx	2.4	5.8	-3.5	-1.3	-1.8	-0.5	5.5	-0.6	2.3
Glx	-2.0	3.5	3.0	-0.0	1.8	-2.9	6.4	6.1	-2.4
Gly	-3.4	4.4	-4.3	-1.6	-1.4	13.3	14.2	-2.3	-6.0
His	-11.2	11.2	1.8	3.3	-6.6	-7.3	10.8	-3.7	-3.2
Ile	1.0	5.4	1.6	2.0	2.0	-4.8	7.9	3.6	-4.2
Leu	1.6	3.8	0.5	-0.7	-0.3	0.7	6.1	6.0	-8.0
Lys	-0.5	3.4	-3.0	-2.1	0.5	-0.4	7.5	2.0	-8.9
Met	-2.3	9.2	7.8	-0.4	1.6	7.1	16.5	8.5	-0.5
Phe	3.0	4.6	-2.1	0.3	-0.4	0.7	5.6	3.8	-7.2
Pro	-3.0	4.9	0.9	0.7	0.7	5.5	7.4	4.9	-3.8
Ser	-8.5	10.0	8.7	-3.4	3.5	-3.8	10.9	0.7	-7.1
Thr	7.4	9.2	-2.6	-1.1	4.6	-7.4	10.7	5.8	-4.3
Tyr	-1.1	7.2	4.8	-0.9	1.1	1.2	6.3	5.5	-4.8
Val	-1.3	4.1	1.5	-1.3	-2.5	-3.2	5.6	2.7	-0.8
av-WC	-0.8	6.1	1.1	-0.0	0.4	0.2	8.4	3.1	-4.2

Calibration "errors" for PTC-amino acid analyses were computed from areas per nmol for each site's standard minus the extinctions at 254 nm, with data sets normalized to the non-problematic amino acids E A P L & F. Column 1 shows the crude ave % errors, with negative values meaning lower recovery values than expected, while column 2 shows the average absolute % errors. Sites (n=15) were divided three different ways into roughly equal bins; columns 3-5 show the results of subtracting in each pair the errors of one from the other. H-non: hydrolyzed standards - non-hydrolyzed. >1-1: calibration using the average of >one standard - only one. Berr: sites reporting <5% ave error (ignoring W & C) in their analysis of BSA - those with >6% ave error. The BSA columns have a similar meaning. One calibration error each for G & Y and one BSA error were excluded from the calculation because of extreme values.

Some of the results are shown in Table IV for the 15 sites submitting calibration data based on areas at 254 nm.

The His calibration error was the largest and was recovered consistently low (note equal magnitude of "raw" and absolute % errors). Moreover, it was the strongest predictor of problems in the analysis of BSA: the large negative number in column 5 indicates that the sites with better analyses did markedly better calibrating His. The calibration error is sufficiently great to account for the large error in His values for BSA (column 7), while similarly high errors for Gly, Met, Ser & Thr are better explained as contamination and hydrolysis losses.

The expectation that sites which hydrolyzed their standard(s) would show a high "error" for Ser was fulfilled (column 3), as Ser is partially destroyed by hydrolysis. Did they therefore do better determining Ser in BSA? Column 8 shows they did not. Perhaps contaminating Ser offset the error necessarily experienced by the sites which did not compensate for that loss. Thr is also partially destroyed by hydrolysis, but is rarely a contaminant, so some other process must be inflating its apparent recovery: note that Met is high in column 3, meaning it is lost on hydrolysis. The most likely product is Met sulfoxide, which is known to coelute with Thr in many PTC separations. The high error for Thr in column 8 shows that this "compensation" does not work; however, sites with the better analyses (column 5) did show lowered recovery of Thr (& Ser), suggesting a beneficial compensation in some cases. But in general, the values in columns 3 & 8 suggest a disadvantage in hydrolyzing one's standard.

The benefits of averaging calibration standards are far more clear: all but one of the values in column 9 are negative, indicating lower errors in the BSA analysis. The calibration is distinctly more accurate as well, if one ignores the aberrant values for Arg & His (column 4). But the general impression of this data is noise: the absolute "raw" averages (columns 1 & 6) are often much less than the average absolute values (columns 2 & 7), indicating the cancellation of fluctuations about a mean; and the average difference is small between hydrolyzing versus not, averaging many vs one, or even doing in general a good analysis of BSA vs doing a poor one.

Recommendations, especially to those doing PTC analyses: average several calibration standards and pay attention to response factors that are far from the usual or a priori expected values, especially for His. Unusual values imply problems in sampling (contamination), chemistry (incomplete reaction, loss), chromatography (coeluting garbage peak, baseline irregularity), or integration.

E. The "Best" Analyses

The quality of many of the analyses reported in this trial was good and Table V reports an example of a few such analyses. These represent the "best" analyses, including Trp and Cys, as acquired by each of the different types of method.

Table V. Examples of Best Total Analyses of ABRF-91AAA by Each Method[a]

Site #	41	6	33	2	39	47a	Seq[b]
	Fluor	Ninhydrin		Auto	PITC		
Ala	46.2	47.4	44.8	46.3	47.2	45.9	46
Arg	22.9	23.0	22.2	26.0	23.8	26.4	23
Asx	53.3	55.7	52.8	59.0	47.6	57.5	54
Cys	34.4	30.0	22.2	37.4	37.3	33.7	35
Glx	82.4	84.0	78.8	81.8	81.2	80.2	79
Gly	17.2	19.4	16.8	20.8	17.7	15.8	16
His	16.5	16.9	17.6	15.3	16.4	17.2	17
Ile	12.6	11.8	13.8	11.8	13.7	12.1	14
Leu	62.5	60.2	61.6	61.6	62.9	61.4	61
Lys	57.8	60.6	58.3	56.4	58.7	59.5	59
Met	3.9	3.5	2.3	4.6	3.8	4.1	4
Phe	27.2	26.6	26.4	26.3	27.5	26.5	27
Pro	29.2	29.9	29.8	27.6	29.5	28.6	28
Ser	28.2	27.1	21.8	25.9	26.3	27.4	28
Thr	33.4	34.4	35.2	31.1	32.8	35.4	34
Trp	2.1	1.3	1.9	2.2	2.1	1.7	2
Tyr	19.3	19.7	18.0	19.0	20.2	18.3	19
Val	35.8	33.5	35.6	33.5	34.9	32.0	36

[a]Values are given as residues per mol of protein.
[b]from ref. 9.

F. Comparison with previous analyses of BSA

Table VI summarizes a few earlier analyses of BSA (12-15) together with the current analyses. The analyses of 1946 (microbiological) were good by 1991 standards, with a better average error than most of thc present analyses. The chromatographic analyses of 1949, 1964 and 1970, all done before the sequence was known, are better in average error than nearly all of this year's analyses. Although the quantities used in those studies were considerably larger than the present, these data demonstrate an ideal for present methodology which is surely inherently attainable. Operator experience is part of the answer, but if the chromatography of those analyses are compared to current chromatograms, the better resolution of the earlier analyses are apparent, suggesting that some improvement may be attained through improved resolution of the fast present-day separations. The existence of many baseline disturbances in the fast analyses likewise compare unfavorably with the steady baselines of analyses at the >100 nmol level. Site 41, using OPA postcolumn methodology, had the lowest average %error in this trial and compares very favorably with the older analyses. The average of all the analyses and of the best 14 analyses, both are very near the true analysis, suggesting that the large errors are mostly random and therefore average out. The higher average error values for Cys and Gly in the pooled data of all 58 sites, but not in the selected group, signifies specific problems with those residues which could be alleviated with proper technique.

Table VI. Comparison of Analyses of ABRF-91AAA with Literature Values

	Percent Error						
	all 1991	top 14 1991	site #41 1991	1946[a]	1949[b]	1964[c]	1970[d]
Ala	1.1	0.3	0.4	nd	4.1	0.0	-1.4
Arg	4.5	2.6	-0.6	2.1	0.4	-2.7	3.2
Asx	-1.5	-0.6	-1.4	-3.2	3.6	0.4	2.4
Cys	-18.1	-8.2	-1.7	0.6	4.5	2.0	3.5
Glx	-2.4	1.8	4.3	-5.1	-3.1	-1.9	4.5
Gly	20.6	7.9	7.3	0.3	3.1	-0.3	-3.9
His	-1.2	-0.3	-2.7	-4.4	3.5	-1.0	0.8
Ile	-4.7	-5.2	-10.3	3.2	-3.2	-0.1	1.7
Leu	0.4	0.5	2.5	11.8	4.5	0.7	-2.4
Lys	-3.5	0.3	-2.0	-5.6	1.4	4.5	0.7
Met	0.2	-2.5	-2.5	-12.2	-7.8	-5.8	-4.6
Phe	-0.3	0.6	0.6	-9.3	0.7	-1.3	-2.5
Pro	3.1	2.9	4.4	13.2	0.3	6.7	2.5
Ser	-3.5	-3.5	0.8	-0.8	-1.8	-7.2	-3.4
Thr	-2.9	-2.3	-1.9	4.7	-1.8	0.7	-2.3
Trp	6.9	9.5	3.6	-9.2	-2.4	1.5	3.8
Tyr	-1.3	-0.3	1.8	4.5	0.3	6.4	5.4
Val	-3.1	-1.0	-0.7	0.9	-4.3	1.1	-1.1
abs av	3.4	2.0	2.8	5.4	2.7	2.5	2.7

[a]ref. 12 [b]ref. 13 [c]ref. 14 [d]ref. 15

Acknowledgements

This work was partially supported by NSF grant DIR 9003100 to John Crabb (W.Alton Jones Cell Science Center) on behalf of the ABRF. We would like to thank the anonymous ABRF member facilities that participated in the study, Dr. Sam Margolis for his interest in the study and for preparing the samples, and Dr. Barton Holmquist for receiving and handling the voluminous results and enabling the maintenance of anonymity of the collaborating facilities.

References

1. Williams, A.P. (1981) *in* "Amino Acid Analysis", J.M. Rattenbury, Editor, Ellis Horwood Ltd., pp 138-152.
2. Niece, R.L., Williams, K.R., Wadsworth, C.L., Elliot, J., Stone, K.L., McMurray, W.J., Fowler, A., Atherton, A., Kutny, R. and Smith, A. (1989) *in* "Techniques in Protein Chemistry (T.E. Hugli, ed), Academic Press, San Diego, pp 89-101.
3. Crabb, J.W., Ericsson, L.H., Atherton, D., Smith, A.J. and Kutny, R. (1990) *in* "Current Research in Protein Chemistry (J.J. Villafranca, ed) Academic Press, San Diego, pp 49-61.
4. Ericsson, L.H., Atherton, D., Kutny, R., Smith, A.J. and Crabb, J.W. (1991) *in* "Methods of Protein Sequence Analysis 1990 (H.Jörnvall and J.-O.Höög, eds), Birkhauser Verlag, Basel.
5. Tarr, G.E., Paxton, R.J., Pan, Y.-C.E., Ericsson, L.H., & Crabb, J.W. (1991) *in* "Techniques in Protein Chemistry II", Academic Press, pp 139-150.
6. Spackman, D.H., Stein, W.H. & Moore, S. (1958) Anal. Chem. 30, 1190.
7. Mora, R., Berndt, K.D., Tsai, H., & Meredith, S.C. (1988) Anal. Biochem. **172**, 368.
8. Hirayama, K., Akashi, S., Furuya, M., & Fukuhara, K. (1990) Biochem. Biophys. Res. Commun. **173**, 639.
9. Brown, J.R. (1975) Fed. Proc. Fed. Amer. Soc. Exp. Biol. **34**, 591; Reed, R.G., Putnam, F.W., & Peters, T.,Jr. (1980) Biochem. J. **191**, 867; and Protein Identification Resource Entry : ABBOS.
10. Moore, S., and Stein, W.H. (1963) Methods Enzymol. **6**, 819.
11. Tarr, G.E. (1986) *in* "Methods of Protein Microcharacterization (J.E.Shively, ed) Humana Press, Clifton NJ, pp. 155-194.
12. Brand, E. (1946) Ann. NY Acad. Sci. **47**, 187.
13. Stein, W.H. & Moore, S. (1949) J. Biol. Chem. **178**, 79.
14. Spahr, P.F. & Edsall, J.T. (1964) J. Biol. Chem. **239**, 850.
15. King, T.P. & Spencer, M. (1970) J. Biol. Chem. **245**, 6134.

SECTION VI

Structural Analysis of Glycoproteins

Identification of O-Glycosylation Sites with a Gas Phase Sequencer

John L. Abernethy, Yang Wang, Allen E. Eckhardt, Robert L. Hill

Department of Biochemistry
Duke University Medical Center
Durham, North Carolina 27710

I. Introduction

Glycoproteins generally contain carbohydrate structures attached via either of two linkages (1). They may be N-linked to the amide nitrogen of asparagine (2,3), or they may be O-linked glycosidically to the hydroxyl groups of threonine or serine residues (4). O-linked oligosaccharides are characteristically found on mucins or cell surface mucin-like molecules, which are often heavily glycosylated, containing up to 65-85% carbohydrate by weight (4,5). A number of problems are frequently encountered in structural studies of heavily glycosylated glycoproteins. Although it is desirable to analyze intact glycopeptides, such glycoproteins are frequently resistant to proteolytic cleavage, and high yield purification of glycopeptides may be hampered by microheterogeneity of the extent of glycosylation at each serine or threonine residue. Attempts to remove the carbohydrate and label the attachment sites by β-elimination reactions are not ideal, since this reaction is neither totally specific nor complete, particularly in the case of threonine residues (6,7,8). Partial chemical deglycosylation procedures, such as trifluoromethanesulfonic acid:anisole (2:1) (9) may be an alternative, but carbohydrate structural information is lost, and the procedure still leaves one sugar residue attached.

Sequencing methods using standard gas phase reagents and polybrene carrier for peptide immobilization are generally unsatisfactory for glycopeptides. Sequencer cycle "blanks" are usually observed at the attachment site of the large N-linked carbohydrate units; however the interpretation of HPLC chromatograms of O-glycosylated amino acid phenylthiohydantoin (PTH) derivatives may be much more deceptive. Inefficient extraction of potentially unstable glycosylated amino acid 2-anilino-5-thiazolinone (ATZ) derivatives into the usual butyl chloride washes may cause the derivative, or its breakdown products, to be washed out over several cycles. Thus, the precise localization of the O-glycosylation site may be obscured.

This report describes a methodology for sequencing glycopeptides and identifying O-glycosylated threonine residues.

TECHNIQUES IN PROTEIN CHEMISTRY III

II. Methods

A. *Peptide Synthesis*

Peptides were synthesized on PAC resins using FMOC-BOP chemistry in a Milligen/Biosearch 9600 Peptide Synthesizer (10,11). The peptides were cleaved from the resin with TFA/dimethylsulfide/dichloromethane (14:1:5) for two hours at room temperature, then purified by gel filtration on Sephadex G-25 (2.4 x 131 cm) in 10% acetic acid, followed by HPLC (see below).

B. *Glycosylation of Synthetic Peptides*

The UDP-GalNAc: polypeptide N-acetylgalactosaminyltransferase has been purified using 5-Hg-UDP-GalNAc-thiopropyl Sepharose 6B (Wang et al, unpublished). Peptides, up to 50 nmols, were incubated in 0.1 M imidazole-HCl pH 7.2, 10 mM $MnCl_2$, 0.1% Triton-X-100, 0.02% sodium azide, with enzyme (2.3 m U total in four additions over four days; 1 U = 1 µmol/min incorporation into apomucin from ovine submaxillary glands) and UDP-[^{14}C]GalNAc (500 nmol total; specific activity 4 x 10^3 cpm/nmol) at 37°C for up to four days in 100-500 µl.

The enzymatically glycosylated peptides were initially fractionated on a Spice C18 cartridge (loaded in water to separate reaction byproducts, and eluted with 60% methanol to separate the peptides from Triton-X-100). Then the individual glycosylated peptide species were separated by HPLC on a Vydac C18 (218TP54) column, eluted with 0.1% TFA and an acetonitrile gradient, and monitored at 214 nm. Peaks were collected manually, and aliquots were taken for scintillation counting. Peptide peaks were quantitated by taking aliquots for amino acid analysis (see below).

C. *Sequence Analysis*

A portion of each peptide peak was attached via the carboxyl terminus to Sequelon AA™ membranes (Milligen) using the N-(dimethylaminopropyl)-N'-ethylcarbodiimide (EDC) coupling method outlined by the manufacturer (12).

The membrane disk was immersed in 0.5 ml of 30% acetonitrile for several minutes, then removed, allowed to dry, cut into three pieces and sandwiched between two cartridge seals in the cartridge of an Applied Biosystems 470A sequencer. A modified sequencer program (Table I) was utilized which eliminated the post-cleavage drying step, added a post-cleavage ethyl acetate extraction, and incorporated multiple pulsed butyl chloride deliveries, following in part the suggestions of Speicher (13) and Tempst and Riviere (14). For peptides with larger carbohydrate groups attached, methanol many be used as the extractant instead of ethyl acetate (see discussion below). Aliquots of collected sequencer fractions were analyzed for 1) PTH-amino acids, 2) radioactivity and 3) galactosamine content. PTH-amino acids were dissolved in 5% aqueous acetonitrile containing dithiothreitol (1 mg/100 ml) and chromatographed on an Applied Biosystems reverse-phase phenylthiohydantoin column (2.1 mm x 22 cm) in a Hewlett-Packard 1090 HPLC system, with the sodium acetate:tetrahydrofuran (5%) buffer (A) and the acetonitrile (B) gradient recommended by the manufacturer. Dithiothreitol (10 mg/liter) was the reductant used in the HPLC buffers and solvents.

Table I. Modified Sequencer Program[a]

STEP	CARTRIDGE	FLASK	TIME	STEP	CARTRIDGE	FLASK	TIME
1	DEL R2	AR DRY	120	35	DEL R3	PAUSE	177
2	PREP R1	AR DRY	6	36	DEL R3	AR DRY	3
3	DEL R1	AR DRY	3	37	DEL R3	PAUSE	177
4	AR DRY	AR DRY	3	38	DEL R3	AR DRY	3
5	PAUSE	AR DRY	37	39	DEL R3	PAUSE	180
6	AR DRY	AR DRY	10	40	DEL R3	COLLECT	20
7	DEL R2	AR DRY	60	41	DEL R3	DEL S4	3
8	DEL R2	DEL R4	9	42	DEL R3	AR DRY	8
9	DEL R2	AR DRY	4	43	DEL R3	COLLECT	30
10	DEL R2	PAUSE	330	44	DEL R3	DEL S4	20
11	DEL R1	PAUSE	3	45	DEL R3	AR DRY	5
12	AR DRY	PAUSE	3	46	AR DRY	FC STEP	2
13	PAUSE	PAUSE	37	47	DEL S2	EMPTY	8
14	AR DRY	PAUSE	10	48	PAUSE	EMPTY	32
15	DEL R2	PAUSE	400	49	TRANS W/ AR	PAUSE	30
16	DEL R1	PAUSE	3	50	PREP TRANS	PAUSE	6
17	AR DRY	PAUSE	3	51	TRANS W/ S3	PAUSE	15
18	PAUSE	PAUSE	37	52	PAUSE	DEL R4	9
19	AR DRY	PAUSE	10	53	PAUSE	AR DRY	4
20	DEL R2	PAUSE	400	54	PAUSE	PAUSE	10
21	AR DRY	PAUSE	60	55	TRANS W/ S3	PAUSE	15
22	DEL S2	PAUSE	3	56	PAUSE	AR DRY	10
23	DEL S1	AR DRY	30	57	TRANS W/ AR	PAUSE	5
24	DEL S2	AR DRY	35	58	TRANS W/ S3	PAUSE	18
25	AR DRY	AR DRY	2	59	PAUSE	AR DRY	10
26	DEL S2	AR DRY	35	60	TRANS W/ AR	PAUSE	5
27	AR DRY	AR DRY	2	61	TRANS W/ S3	PAUSE	15
28	DEL S2	AR DRY	35	62	PAUSE	AR DRY	10
29	AR DRY	AR DRY	60	63	TRANS W/ AR	PAUSE	5
30	PAUSE	AR DRY	200	64	TRANS W/ S3	PAUSE	15
31	AR DRY	AR DRY	60	65	PAUSE	AR DRY	10
32	DEL R3	AR DRY	100	66	TRANS W/ AR	PAUSE	30
33	DEL R3	DEL S4	8	67	DEL S3	AR DRY	30
34	DEL R3	AR DRY	8	68	AR DRY	AR DRY	120

[a]Reagents and Solvents are from Applied Biosystems. R1, 5% PITC in n-heptane; R2, 12.5% trimethylamine; R3, TFA; R4, 25% TFA in water; S1, S2, and S3 are n-heptane, ethyl acetate, and l-chlorobutane, respectively, with added DTT (0.4 mg/200 ml); S4, acetonitrile with added DTT (10 mg/200 ml).

Aliquots of sequencer cycle fractions were hydrolyzed in constant boiling HCl in a vacuum dessicator (15) for four hours at 105°. After drying and dissolution in water, portions were injected into a Dionex Carbopac PA1 (4 x 250) column eluted in 0.015 N sodium hydroxide for determination of galactosamine by pulsed amperometric detection (E1 = 0.05V, E2 = 0.6V, E3 = -0.6V, T1 = 5, T2 = 2, T3 = 1) (16). A Hewlett-Packard Series 300 workstation was used for data analysis and integration of both the PTH-amino acids and galactosamine from the pulsed amperometric detector.

Table IIa. Sequencer Yields for Peptides A-E

		A + B (IY = 40%, RY = 96%)[a]				C (IY = 37%, RY = 95%)				D (IY = 38%, RY = 96%)				E (IY = 17%, RY = 97%)		
Cycle	AA	PTH[b] Pmole	(X/T)[b]	CPM	GalN Pmole	PTH Pmole	(X/T)	CPM	GalN Pmole	PTH Pmole	(X/T)	CPM	GalN Pmole	PTH Pmole	(X/T)	CPM
1	VAL	2940		38		1070		34		1530		32		214		32
2	LYS	3120		80		336		190		827		41		51		125
3	THR	780	(0.23)	421	349	279	(0.81)	698	603	196	(0)	49	ND[c]	136	(0.76)	189
4	GLU	33		310	177	34		211	140	16		135		8		96
5	ALA	2940		195		877		122		1120		217		277		141
6	THR	576	(0.12)	329	250	205	(0.30)	219	161	219	(0.81)	748	774	169	(0.86)	224
7	THR	855	(0.66)	1230	1010	211	(0.85)	490	288	359	(0.88)	867	934	141	(0.90)	216
8	PHE	2590		514	263	722		201	67	1060		485	339	196		109
9	ILE	2350		188		701		88		1100		254		168		75

[a]Initial yields (IY) include covalent coupling yields plus the initial sequencer yields. RY = Repetitive yield.
[b]PTH yields for threonine are quantitated for the PTH-threonine peak alone. X/T values are the ratios of peak heights for the anomalous peak X (correlating with PTH-Thr-GalNAc) with respect to the PTH-Thr peaks. Yields of PTH-amino acids, but not galactosamine, are corrected for carry-over.
[c]ND, not detected.

Table IIb. Sequencer Yields for Peptides F and G

		F (IY = 57%, RY = 94%)[a]				G (IY = 41%, RY = 95%)			
Cycle	AA	PTH[b] Pmole	(X/T)[b]	CPM	GalN Pmole	PTH Pmole	(X/T)	CPM	GalN Pmole
1	VAL	3180		66		3350		125	ND[c]
2	LEU	2400		48		3890		115	
3	GLY	2460		69		2900		88	ND[c]
4	THR	439	(0)	67	ND[c]	786	(0.86)	2231	1640
5	THR	177	(0.7)	294	538	1090	(0.88)	2520	1730
6	ALA	2570		381	271	3030		783	540
7	VAL	2240		119		2860		308	

[a]See Footnote *a* in Table IIa.
[b]See Footnote *b* in Table IIa.
[c]See Footnote *c* in Table IIa.

D. Amino Acid Analysis

Aliquots were hydrolyzed *in vacuo* in constant boiling HCl at 110°C for varying times (15). After drying, OPA and FMOC derivatives were made and chromatographed on a Hewlett-Packard Aminoquant system (17).

III. Results and Discussion

The gas phase sequencer methodology was modified in several ways to optimize the direct identification of glycosylation sites when sequencing glycopeptides. First, the peptide is covalently coupled via its carboxyl groups (carboxyl terminus and aspartic and glutamic acid side chains) to arylamine groups on a modified PVDF membrane--Sequelon AA™ (Milligen/Biosearch). This coupling was accomplished with the water soluble N-(dimethylaminopropyl)-N'-ethylcarbodiimide hydrochloride (EDC) in MES buffer at pH 5.0 (12). This solid phase attachment procedure was used to prevent potential washout of the peptide during the extraction steps, which otherwise might contaminate collected sequencer cycles with carbohydrate-containing peptide material. The solid phase attachment also permits the use of more stringent washing conditions or solvents, as necessary, for extraction of the carbohydrate-containing 2-anilino-5-thiazolinone derivatives from the peptide immobilization support. In this vein, the post-cleavage drying step of the sequencer program (Table I, steps 46-48) is omitted to allow residual TFA to enhance the extraction, and an initial ethyl acetate extraction is included in addition to the pulsed butyl chloride deliveries (13) for extraction from the PVDF membranes.

The results of the sequencing experiments, using the modifications described above, on two synthetic peptides O-glycosylated on threonine residues with a purified UDP-GalNAc:polypeptide N-acetylgalactosaminyltransferase are summarized in Tables II and III and Fig. 2. UDP-[^{14}C]GalNAc was used to facilitate identification and aid in peptide purification. Mixtures of glycopeptide species were obtained and separated by HPLC. They were then coupled to the membrane and subjected to sequence analysis by the program listed in Table I.

Table III. Glycosylation of Synthetic Substrates by GalNAc Transferase[a]

Substrate		VAL-LYS-THR-GLU-ALA-THR-THR-PHE-ILE	
Products	1)	VAL-LYS-THR-GLU-ALA-THR-THR-PHE-ILE	A
	2)	VAL-LYS-THR-GLU-ALA-THR-THR-PHE-ILE	B
	3)	VAL-LYS-THR-GLU-ALA-THR-THR-PHE-ILE	C
	4)	VAL-LYS-THR-GLU-ALA-THR-THR-PHE-ILE	D
	5)	VAL-LYS-THR-GLU-ALA-THR-THR-PHE-ILE	E
Substrate		VAL-LEU-GLY-THR-THR-ALA-VAL	
Products	1)	VAL-LEU-GLY-THR-THR-ALA-VAL	F
	2)	VAL-LEU-GLY-THR-THR-ALA-VAL	G

[a]Glycosylation sites are underlined.

Except for peptides A and B (Table III), which were sequenced as a mixture, the other peptides were homogeneous, since the HPLC system easily resolved the mono-, di-, and triglycosylated species. In the case of peptides C and D, two different diglycosylated species were resolved. The sequencer cycles were collected in the fraction collector, dried in a vacuum dessicator, and reconstituted with a known volume of 5% acetonitrile containing DTT. Twenty per cent aliquots were removed for scintillation counting before PTH HPLC analysis. After injection, the rest of the cycle was hydrolyzed for analysis of galactosamine. This procedure resulted in some delay before injecting the PTH amino acids, as compared to an "on-line" PTH injection system. However, this combined analysis procedure was designed to identify definitively the amino sugar as quantitatively as possible. Table II(a and b) lists the stepwise yields of PTH-amino acids, radioactivity, and recovered galactosamine. The identificaton of the glycosylated PTH derivative of threonine on HPLC was surprisingly deceptive. Fig. 1 shows a typical sequencer cycle for a nonglycosylated and a glycosylated (with GalNAc) threonine, comparing cycle 4 from peptides F and G. The results do not represent partial glycosylation at a specific threonine residue, since the mono- and diglycosylated peptides were resolved on HPLC. Cycle 4 from peptide F shows a typical result for unmodified threonine--a peak in the position of standard PTH-threonine and multiple peaks corresponding to the β-elimination product, α-aminocrotonic acid, and its DTT adducts (18-20). Surprisingly, the corresponding glycosylated threonine from cycle 4, peptide G, shows a relatively large peak for PTH-threonine, in addition to an anomalous peak "X" eluting just before PTH-serine. Few if any, of the β-elimination products are observed. This may suggest that an apparently unstable PTH-Thr-GalNAc derivative breaks down via a hydrolysis route rather than via β-elimination. The time delay in injecting the PTH derivative onto the HPLC may also be a factor in this destruction. Consequently, as long as the extraction of the glycosylated ATZ derivative is effective, it is better to rely on a direct identification of the sugar after hydrolysis by ion exchange chromatography-pulsed amperometric detection (16), amino acid analysis of aminosugars, or by one of several HPLC techniques (21-23).

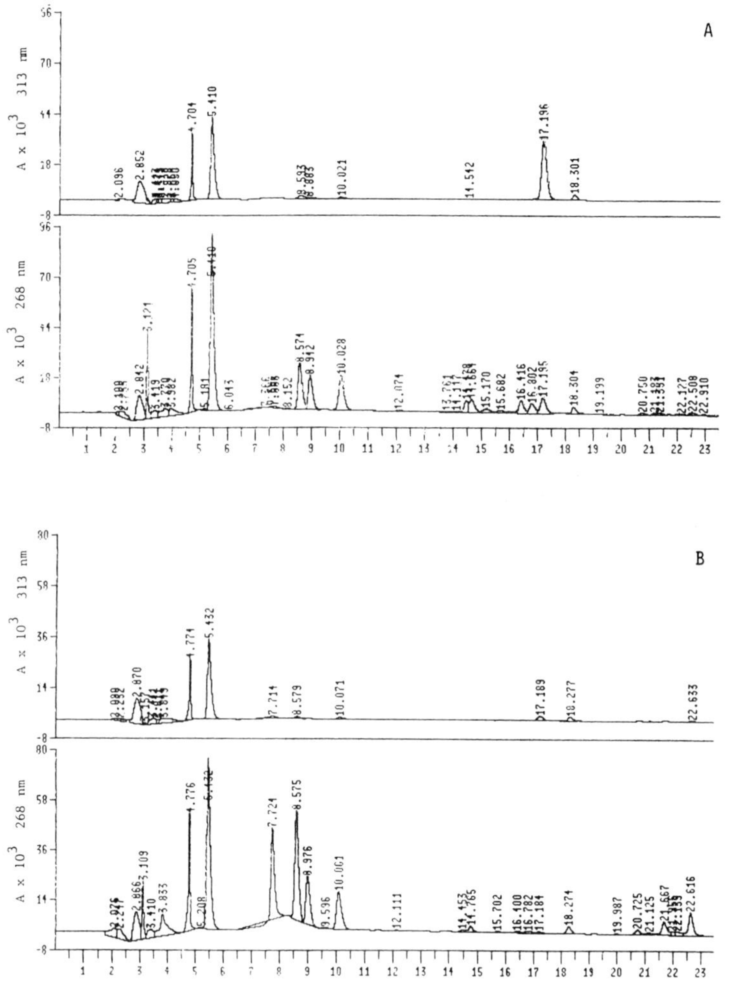

Figure 1. Comparison of sequencer cycles #4 for peptides F (Panel A, nonglycosylated PTH-Thr) and G (Panel B, glycosylated PTH-Thr). Retention times for peaks shown are as follows: "X", the anomalous peak correlating with PTH-Thr-GalNAc, 7.7 min; PTH-Thr, 8.57 min; PTH-Gly, 8.9 min; DMPTU, 10.0 min; PTH-dehydrothreonine, 17.196 min; DTT adducts of PTH-dehydrothreonine, 14.5-16.8 min.

Figure 2 shows plots of radioactivity from sequencer cycles of five glycosylated products of one peptide.

These changes have been successfully used on a gas phase sequencer--Applied Biosystems 470A. Thus far they have not been tried on a "pulsed liquid" sequencer, although one might expect results as good or better on that instrument, since much more TFA is available for extracting the glycosylated ATZ derivatives. In addition, for the larger carbohydrate groups such as the N-linked oligosaccharides, one might anticipate using stronger extraction solvents, such as methanol, as has been used for solid phase sequencing previously (12,24-26).

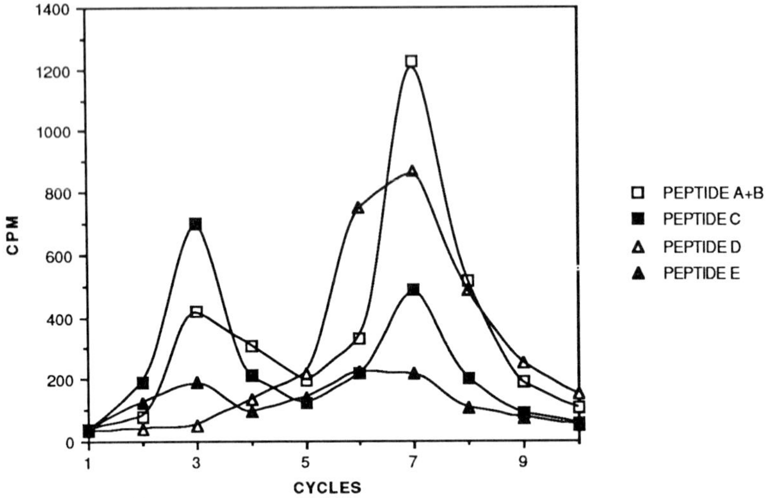

Figure 2. Plots of [^{14}C]GalNAc recovered in sequencing cycles of peptides A-E. (Peptides A and B are both monoglycosylated species and were inseparable on HPLC. They were sequenced as a mixture.) Twenty percent aliquots were removed for scintillation counting.

In general, standard gas phase sequencing methods used on an oligosaccharide group larger than one sugar residue yield only a "blank" sequencer cycle at the attachment site amino acid residue. Consequently, the carbohydrate site has to be inferred indirectly. For example, in the case of N-linked sugars, one may know that the peptide contains a potential asparagine from compositional analysis, or may be able to correlate peptide sequencing with cDNA sequencing to find Asn-X-Ser/Thr consensus sequences (27). A few investigators have been able to identify directly amino acid residues attached to one, but generally not to two or more, sugar residues. Paxton et al. (28,29) identified an anomalous HPLC peak corresponding to the PTH derivative of asparaginyl-linked GlcNAc when sequencing a peptide from carcinoembryonic antigen, which had been partially chemically deglycosylated with TFMSA:anisole (10). They commented on the poor extraction of this monoglycosylated derivative using butyl chloride alone, and observed no peak at all for the diglycosylated derivative--Asn-GlcNAc-GlcNAc. Robb et al. (30,31) have observed an aberrant PTH derivative corresponding to Thr-GalNAc, and possibly one corresponding to threonine with attached N-acetylgalactosamine, galactose, and N-acetylneuraminic acid. In addition, Gooley et al. (32) have observed a peak corresponoding to a glycosylated threonine, and Talbo et al. (33) have observed two unidentified peaks eluting earlier than PTH-Thr and -Ser, apparently corresponding to PTH-Thr-GalNAc.

This sequence methodology of using a solid phase support in a gas phase sequencer exhibited no significant incompatibilities. The solid phase attachment procedure is relatively quick and simple, and allows adjustment of washing/extraction conditions and solvents without suffering peptide washout. An increase in carry-over was observed in the last few cycles, but this did not affect sequence interpretation, and the carboxyl terminal amino acid (attached to the membrane) was identified in high yield. Interestingly, a slight amount of preview of the following cycle could be observed at internal glutamic acid attachment site cycles, but this did not interfere with sequence assignment.

Another demonstration of these glycosylation-sequence studies is that radiolabeling of the carbohydrate makes identification of the glycosylation sites comparatively easy (Table II and Fig. 2). In this case, [^{14}C]GalNAc was incorporated onto potential threonine acceptor sites by a purified UDP-GalNAc: polypeptide N-acetylgalactosaminyltransferase. Scintillation counting of sequencer cycle aliquots easily revealed the sites of glycosylation. In principle, other methods could also be used, such as *in vitro* biosynthetic labeling in cell or tissue culture systems, or terminal galactose or N-acetylgalactosamine labeling by galactose oxidase and reduction with tritiated borohydride (34).

In the past, β-elimination of O-linked oligosaccharide groups in alkaline solution, followed by modification of the resultant double bonds by reduction with tritiated borohydride or by addition of sulfite, has been used to label putative carbohydrate attachment sites (6-8). (Other nucleophiles, such as methylamine, ethanethiol, 2-mercaptoethanol, and methylmercaptan, have been used to label the β-elimination products of phosphoserine and phosphothreonine residues, α-aminoacrylic and α-aminocrotonic acid, respectively (35,36)). However, the β-elimination reaction is neither totally specific nor complete, particularly in the case of threonine residues (6,35). In addition, the alkaline solutions employed may cause non-specific peptide bond cleavage, as well as untoward chemical modifications of amino acid residues (37-39).

A most exciting future prospect is the use of glycosyl transferases of highly restricted specificity, in conjunction with glycosidases, to label specific oligosaccharide structures, so that individual structures can be localized to unique attachment sites. A number of the more common transferases have already been purified and/or cloned. As more of these enzymes are cloned and expressed, particularly in soluble form, they will undoubtedly have a great impact on the analysis and localization of carbohydrate structures by both chemical and mass spectroscopic methods (1,40).

In sum, a solid phase attachment procedure using Sequelon AA™ membranes has been successfully used in a gas phase sequencer, along with a modified extraction program, for the direct identification of O-glycosylation sites, without any prior modification of these sites.

References

1. Beyer, T.A., Sadler, J.E., Rearick, J.I., Paulson, J.C., and Hill, R.L. (1981) Adv. in Enzymology **52,** 23-175.
2. Kornfeld, R., and Kornfeld, S. (1976) Ann. Rev. Biochem. **45,** 217-237.
3. Kornfeld, R., and Kornfeld, S. (1985) Ann. Rev. Biochem. **54,** 631-664.
4. Jentoft, N. (1990) Trends Biochem. Sci. **15,** 291-294.
5. Carlstedt, I., Sheehan, J.K., Corfield, A.P., and Gallagher, J.T. (1985) Esays Biochem. **20,** 40-76.
6. Simpson, D.L., Hranisavljevic, J. and Davidson, E.A. (1972) Biochem. **11,** 1849-1856.
7. Downs, F., Peterson, C., Murty, V.L.N., and Pigman, W. (1977) Int. J. Peptide Protein Res. **10,** 315-322.
8. Downs, F., Herp, A., Moschera, J., and Pigman, W. (1973) Biochim. Biophys. Acta **328,** 182-192.
9. Edge, A.S.B., Faltynek, C.R., Hof, L., Reichert, L.E., Weber, P. (1981) Anal. Biochem. **118,** 131-137.
10. Hudson, D. (1988) J. Org. Chem. **53,** 617-624.
11. Castro, B., Dormoy, J.R., Evin, G., and Selve, C. (1975) Tetrahedron Lett. **14,** 1219-1222.

12. Pappin, D.J.C., Coull, J.M., and Koester, H. (1990) *In* "Current Research in Protein Chemistry: Techniques, Structure and Function" (J.J. Villafranca, ed.), pp. 191-202. Academic Press, New York.
13. Speicher, D.W. (1989) *In* "Techniques in Protein Chemistry" (T.E. Hugli, ed.), pp. 24-35. Academic Press, New York.
14. Tempst, P., and Riviere, L. (1989) Anal. Biochem. **183,** 290-300.
15. Meltzer, N.M., Tous, G.I., Gruber, S., and Stein, S. (1987) Anal. Biochem. **160,** 356-361.
16. Hardy, M.R., Townsend, R.R., and Lee, Y.C. (1988) Anal. Biochem. **170,** 54-62.
17. Blankenship, D.T., Krivanek, M.A., Ackermann, B.L., and Cardin, A.D. (1989) Anal. Biochem. **178,** 227-232.
18. Edman, P., and Henschen, A. (1975) *In* "Protein Sequence Determination", 2nd Ed. (S.B. Needleman, ed.), pp. 232-279. Springer-Verlag, New York.
19. Margolies, M.N., and Brauer, A. (1978) J. Chromatog. **148,** 429-439.
20. Hunkapiller, M.W. (1985) Applied Biosystems User Bulletin, #14.
21. Honda, S., Akao, E., Suzuki, S., Okuda, M., Kakehi, K., and Nakamura, J. (1989) Anal. Biochem. **180,** 351-357.
22. Jentoft, N. (1985) Anal. Biochem. **148,** 424-433.
23. Hawke, D.H., Hsi, K.-L., Chen, L., and Yuan, P.-M.; Fifth Symposium of the Protein Society, Baltimore, MD, June 22-26, 1991.
24. Yuen, S.W., Sheer, D.G., Yuan, P.-M., Mattaliano, R.J.; Fourth Symposium of the Protein Society, San Diego, CA, August 11-15, 1990.
25. Sheer, D.G., Yuen, S.W., Yamane, D.K., Mattaliano, R.J., and Yuan, P.-M.; Fourth Symposium of the Protein Society, San Diego, CA, August 11-15, 1990.
26. Yuen, S.W., Tu, H.-M., Nika, H., and Mattaliano, R.J.; Fifth Symposium of the Protein Society, Baltimore, MD, June 22-26, 1991.
27. Marshall, R.D. (1972) Ann. Rev. Biochem. **41,** 673-702.
28. Paxton, R.J., Mooser, G., Pande, H., Lee, T.P., and Shively, J.E. (1987) Proc. Natl. Acad. Sci. USA **84,** 920-924.
29. Hefta, S.A., Paxton, R.J., Hefta, L.J.F., Lee, T.D., and Shively, J.E. (1989) *In* "Techniques in Protein Chemistry" (T.E. Hugli, ed.), pp. 195-206. Academic Press, New York.
30. Robb, R.J., Kutny, R.M., Panico, M., Morris, H.R., and Chowdhry, V. (1984) Proc. Natl. Acad. Sci. USA **81,** 6486-6490.
31. Robb, R.J., Kutny, R.M., Panico, M., Morris, H., DeGrado, W.F., and Chowdhry, V. (1983) Biochem. Biophys. Res. Commun. **116,** 1049-1055.
32. Gooley, A.A., Classon, B.J., Marschalek, R., and Williams, K.L.; Fifth Symposium of the Protein Society, Baltimore, MD, June 22-26, 1991.
33. Talbo, G., Højrup, P., Rahbek-Nielsen, H., Andersen, S.O., and Roepstorff, P. (1991) Eur. J. Biochem. **195,** 495-504.
34. Morell, A.G., and Ashwell, G. (1972) Methods Enzymol. **28,** 205-208.
35. Mega, T., Nakamura, N., and Ikenaka, T. (1990) J. Biochem. (Tokyo) **107,** 68-72.
36. Meyer, H.E., Hoffmann-Posorske, E., Korte, H., and Heilmeyer, L.M.G. (1986) FEBS Lett. **204,** 61-66.
37. Bohak, Z. (1964) J. Biol. Chem. **239,** 2878-2887.
38. Jarboe, C.J., Noll, B.W., Hass, L.F. (1971) Biochem. Biophys. Res. Commun. **43,** 1029-1034.
39. Donovan, J.W. (1967) Biochem. Biophys. Res. Commun. **29,** 734-740.
40. Paulson, J.C., and Colley, K.J. (1989) J. Biol. Chem. **264,** 17615-17618.

GLYCOFORM, GLYCOTYPE LINKAGE AND BRANCHING DETAIL OF THE CD2 ADHESION DOMAIN BY DESOLVATION MASS SPECTROMETRY: SFCI-MS AND ESI-MS

Bruce B. Reinhold[1], Ellis L. Reinherz[2,4], and Vernon N. Reinhold[1,3]

[1]Department of Nutrition, Harvard University School of Public Health, Departments of [2]Medicine and [3]Biological Chemistry and Molecular Pharmacology, Harvard Medical School and the [4]Laboratory of Immunobiology, Dana Farber Cancer Institute, Boston, MA 02115

Maria H. Knoppers and Michael A. Recny

Procept Inc., Cambridge, MA 02139

I. INTRODUCTION

With increasing frequency, glycoprotein N-linked and O-linked carbohydrate residues are being recognized as important structural entities mediating critical physiological processes (1,2,). Earlier indirect evidence of glycan involvement in biological functions is now being replaced with more definitive proof by studying cellular events altered by the modification or deletion of carbohydrate residues. Three developments, and their subsequent applications have made major contributions to this understanding: (*i*) the techniques of site-directed mutagenicity; (*ii*) the isolation, purification, and characterization of specific enzymes that bring about deglycosylation; and, (*iii*) the discovery, synthesis, and utilization of a series of specific inhibitors that block N-linked glycosylation and glycan processing within the cell. Cellular adherence and lymphocyte homing have been reported to be partially mediated by specific carbohydrate sequences on various cell-surface receptors (3-6). For example, ICAM-1 is an endothelial cell-surface glycoprotein that is a counterreceptor for the leukocyte integrins LFA-1 and MAC-1 (7-10), and ICAM-1 is partially responsible for lymphocyte, monocyte and neutrophil adhesion to endothelial cells (11-13). Recent evidence suggests that the extent of glycosylation on ICAM-1 may regulate adhesion to LFA-1 and MAC-1 *in vivo* (14). Carbohydrates may also serve as direct ligands for cell-cell interactions, as has been shown recently for the selectin family of adhesion proteins (reviewed in ref. 15).

TECHNIQUES IN PROTEIN CHEMISTRY III

The T-lymphocyte transmembrane glycoprotein CD2 (T11) also plays an important role in facilitating adhesion between T-lymphocytes and their cognate cellular partners (reviewed in ref. 16). CD2 functions by binding to the surface antigen LFA-3 on antigen presenting cells, which leads to recognition of the antigen:major histocompatibility complex (MHC) by the CD3-Ti antigen receptor (17) and subsequent up regulation of the binding affinity between the LFA-1:ICAM-1 congener pair (18). Therefore, CD2 plays a fundamental role in mediating the immune response by both promoting and regulating cell-cell contact. Functional analysis of soluble, two-domain CD2 receptor ($T11_{ex2}$) previously established that the structural basis for CD2-mediated adhesion lies within the first extracellular domain of the CD2 polypeptide (19,20). The CD2 adhesion domain contains a single N-linked glycosylation site at Asn^{65}, which always is glycosylated when $T11_{ex2}$ is expressed in eucaryotic cells. Interestingly, this glycosylation site is flanked by two regions within the polypeptide sequence implicated in both activation and adhesion functions (21). Preliminary evidence suggests that the N-linked glycans at Asn^{65} are required for functional binding of CD2 and LFA-3 (22). Therefore, the N-linked glycans present within the adhesion domain were analyzed for carbohydrate structural detail by desolvation mass spectrometry.

This report utilizes two recently developed techniques in mass spectrometry for N-glycan characterization. In contrast to the ballistic-ionization approaches utilized in desorption mass spectrometry, (e.g., Fast atom bombardment, liquid desorption or plasma desorption), the current "state-of-the-art" technology employs direct liquid injection of the solvent and dissolved analytes using various techniques of aerosol formation. The unifying principles here are (*i*) the spatial dispersion of the solvated analyte by generating an aerosol from the bulk liquid and (*ii*) the subsequent desolvation, and hence isolation in the gas phase of single molecules of the analyte from the microscopic solvent/solute clusters. Currently, three mass spectrometry (MS) ion sources are dependent on the principles of desolvation and aerosol formation; thermal spray (TS-MS), electrospray ionization (ESI-MS), and supercritical fluid/chemical ionization (SFCI-MS). Of the techniques currently available, ESI-MS and SFCI-MS probably represent the least energetic, or "softest" molecular ionization approach that also encompasses remarkable sensitivity. These two techniques can be used independently to characterize separate components of carbohydrate structure. ESI-MS, in a more non-evasive manner, can be used to profile glycan glycoform distribution from isolated glycopeptides, while SFCI-MS is most effective in characterizing the details of carbohydrate structure such as linkage, branching, and primary sequence. Using this latter approach, glycoproteins (or glycopeptides) are first deglycosylated and labeled at the reducing-end (to enhance sensitivity), chemically modified (to maximize molecular specificity), and derivatized (for SFC mobile phase miscibility).

II. EXPERIMENTAL PROCEDURES

Purification of $T11_{ex2}$ from CHO cells. $T11_{ex2}$ was purified from CHO-cell conditioned media by passage over an anti:CD2 immunoaffinity column prepared by crosslinking 1.0 gram anti:$T11_2$ monoclonal antibody (3PT2H9) to 150 ml Protein A-sepharose (Repligen). Nonspecifically bound protein was washed off the column with PBS and $T11_{ex2}$ was eluted with 50 mM Na-citrate buffer pH 3.0. Fractions containing $T11_{ex2}$ were pooled, diluted with an equal volume of 0.1% trifluoroacetic acid/water (Solvent A) and loaded onto a preparative C4 reverse phase HPLC column (2.1 x 250 cm; Vydac) equilibrated in 90% Solvent A/10% Solvent B (0.1% trifluoroacetic acid/90% acetonitrile/10% water). The column was developed with a linear gradient from 25%-50% Solvent B over 50 minutes at a flow rate of 10 ml/min. Fractions containing $T11_{ex2}$ were pooled, diluted with an equal volume of 50 mM Na-citrate buffer, pH 5.5 plus three volumes distilled water and loaded onto a column containing 50 ml CM-Sepharose FastFlow media (Pharmacia) equilibrated in the same buffer. $T11_{ex2}$ was eluted by stepwise adjustment of the buffer to 500 mM NaCl, concentrated to 5 mg/ml and stored at -70 °C for subsequent studies. Purity was monitored by SDS-PAGE under reducing conditions (23). Protein concentrations were determined using a quantitative coomassie dye-binding assay (BioRad), and $T11_{ex2}$ antigen was quantitated using a sandwich ELISA employing two different anti:CD2 mAbs as previously described (20).

Clostripain digestion of $T11_{ex2}$ and purification of $T11_{105}$. A 50 mg aliquot of $T11_{ex2}$ (200 ug/ml), was dialyzed into PBS containing 1 mM $CaCl_2$. Freshly prepared dithiothreitol (DTT) was added to a final concentration of 2 mM and the sample was digested with 1.0 mg clostripain (Promega: 280 units/mg) for 40 min at 37°C. The enzyme was inactivated by adding freshly prepared 20 mM iodoacetamide, incubated for 30 min in the dark at 4°C, dialyzed against 50 mM NaOAc buffer, pH 5.5 and concentrated to 2 mg/ml in Centriprep concentrators (Amicon; 10,000 MWCO). The digestion mix was applied to a 500 ml Sepharose S-100HR (Pharmacia) column (2.5 x 100 cm) equilibrated in 50 mM NaOAc buffer, pH 5.5 and eluted at 2.5 ml/min. Aliquots containing the $T11_{105}$ adhesion domain (residues 1-105) were pooled based on SDSPAGE and TSK 3000SW HPLC analysis.

Instrumentation (ESI-MS). The instrument used in this study was a Finnigan-MAT TSQ-700 instrument, (Finnigan-MAT Corp., San Jose, CA), equipped with an electrospray ion source and an VG ZAB-2SE-FPD high performance mass spectrometer (VG FISONS, Manchester, UK) equipped with an SFCI-MS (24) and FAB-MS ion sources. For ESI-MS samples were dissolved in water or acetonitrile solutions and analyzed by syringe pump, flow injection at a rate between 5 and 20 ul/min directly into the electrospray chamber through a stainless steel hypodermic needle. Raw data were deconvoluted by an algorithm designed to extract molecular weight information from the peak spacings exhibited by such multiply-charged ions. This computational approach (pattern recognition by a relative or level-2 entropy algorithm) was written to detect such patterns embedded in a complex spectrum, while avoiding the complications of contaminating components, overlapping peaks, and erratic baselines (25). Treatment of the data in this manner suppresses artifacts associated with the earlier deconvolution approach (26), and improves spectral dynamic range and confidence in peak detection. This data handling technique has proven to be most useful in resolving the demanding problems of glycoprotein profiles and other post-translational modifications.

III. RESULTS AND DISCUSSION

Characterization of recombinant, CHO-derived $T11_{ex2}$ and $T11_{105}$. A cDNA construct was previously designed to produce a soluble, two-domain CD2 receptor ($T11_{ex2}$) encoding the first 182 amino acids of the molecule (20). This cDNA was subsequently reengineered into a new vector for high level expression of $T11_{ex2}$ in Chinese Hamster Ovary (CHO) cells. $T11_{ex2}$ is secreted from CHO cells as a mixture of monomeric glycoforms due to heterogeneity within the carbohydrate chains occupying each of the three N-linked glycosylation sites (Asn^{65}, Asn^{117} and Asn^{126}). SDS-PAGE analysis of $T11_{ex2}$ purified from CHO cells illustrates a broad band composed of several protein species with apparent molecular weights between 34 kDa and 38 kDa (Fig. 1A). Isoelectric focusing (IEF) analysis demonstrates that these bands represent multiple isoelectric glycoforms (Fig. 1B). Previously, we showed that digestion of baculovirus expressed $T11_{ex2}$ with papain produced a stable, amino-terminal fragment ($T11_{pap}$) that contained all of the immunoadhesion properties of two-domain CD2 (20). Digestion of CHO-derived $T11_{ex2}$ with clostripain produces a similar sized, 105 amino acid residue adhesion domain ($T11_{105}$) from specific cleavage of $T11_{ex2}$ at Arg^{105} (Fig. 1A). Functional analysis of $T11_{105}$ in a sandwich ELISA using two conformationally dependent anti-CD2 mAbs indicates that $T11_{105}$ is biologically equivalent to $T11_{pap}$.

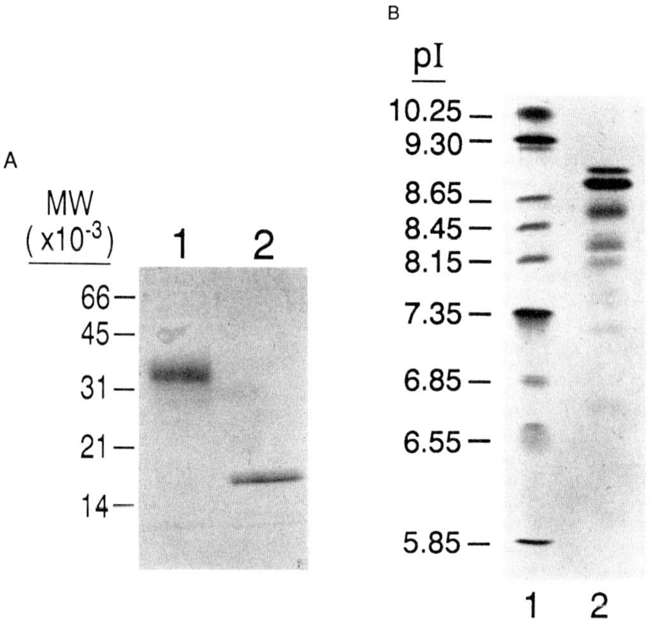

FIGURE 1: *Panel A:* SDS-PAGE analysis of CHO $T11_{ex2}$ (lane 1) and purified $T11_{105}$ (lane 2)
Panel B: Nondenaturing IEF analysis of CHO $T11_{ex2}$ (lane 2) and IEF standards (lane 1)

The adhesion domain of human CD2 contains one consensus N-linked glycosylation site at Asn^{65}. This glycosylation site is flanked by two polypeptide regions within the adhesion domain previously implicated in CD2-LFA3 binding (21), and preliminary evidence gathered from analysis of CD2 mutants lacking this N-glycosylation site suggests that occupancy of the Asn[65] site with N-linked glycans is required for expression of functional CD2 receptors on the cell surface (22). Therefore, we were interested in determining the primary structure of the N-linked glycans within the CD2 adhesion domain.

Glycoform profiling (ESI-MS). Clostripain digestion of $T11_{ex2}$ and HPLC purification of the subsequent mixture provided three polypeptides; Domain I (residues 1-105), Domain IIa (residues 106-146), and Domain IIb (residues 147-182). Each peptide fraction was lyophilized, redissolved in the injection solvent and analyzed directly by ESI-MS (Figs. 2a, 2b, and 2c). The voltage differential between the needle tip and source electrode (-3.5 KV) induces an electrohydrodynamic instability at the liquid surface, resulting in the expulsion of charged droplets from the surface and directed toward the source electrode. These droplets transverse a nitrogen gas bath (approx. 1000 torr) where multiple collisions cause the droplets to rapidly evaporate. As the droplet radius decreases, the electric field on the droplet increases due to a constatn net charge until the liquid surface is once again unstable and expels smaller charged droplets. This process repeats itself until other physical events at different length scales intervene. The exact nature of these events and their relevant scales is not yet clear but probably involves mechanisms of declustering and ion formation when the surface properties of the droplet/cluster are likely to be quite different from the bulk. A unique feature of ESI-MS is the generation of ions from a single molecular species with a range of multiply charged states. The important advantage of this technique is that multiple charging brings the effective mass (m/z) of large peptides or small proteins into a range easily handled by small mass spectrometers. However, the complication resulting from multiple charging is to accurately predict the component molecular weight from a large series of multiply charged states.

The ESI spectrum of CD2 domain I (Fig. 2a) illustrates an envelope of multiply charged ions, centered around 800 Da, covering a range between +10 to +20 charges per peak. Associated with each major peak were a series of satellite peaks attributed to attached carbohydrate residues. Deconvolution of this data by a pattern recognition algorithm (25) provided the mass spectrum shown in the insert (Fig. 2a). These results indicated a prominent ion (base ion) equivalent to the predicted polypeptide mass of $T11_{105}$ (12,438 Da) plus an additional mass of 1,206 Da. This additional mass is equal to a single N-linked high mannose glycan, $(Man)_nGN_2$, where n = 5. Consistent with this interpretation are the incriminating ions at 162 Da intervals (m/z 13,806 and 13,968) which we interpret as being representative of one and two additional mannose increments, $(Man)_nGN_2$, n = 6 & 7, respectively. The unlabeled peaks associated with these ions appear to be salt adducts that are frequently observed in ESI-MS data. The spectrum obtained from CD2 domain IIa is presented in Figure 2a (raw data) accompanied by the deconvolution results (insert). Although we will not provide an interpretation of this spectrum, preliminary indications are that both of the two predicted N-linked sites are glycosylated, and the N-glycans are of the complex glycotype. This latter information can be derived from the mass intervals within the deconvoluted spectrum which are representative of neuraminyl and aminolactosyl residues; hexose intervals were not observed. These intervals may reflect glycoform differences resulting from either of the two N-linked sites. Thus, it remains impossible to definitively assign either structure. CD2 domain IIb provided the ESI spectrum in Figure 2c which, upon deconvolution, provided the simple spectrum (insert) that would be expected for a nonglycosylated peptide with molecular weight-related ion m/z=3907.7 (residues 147-182).

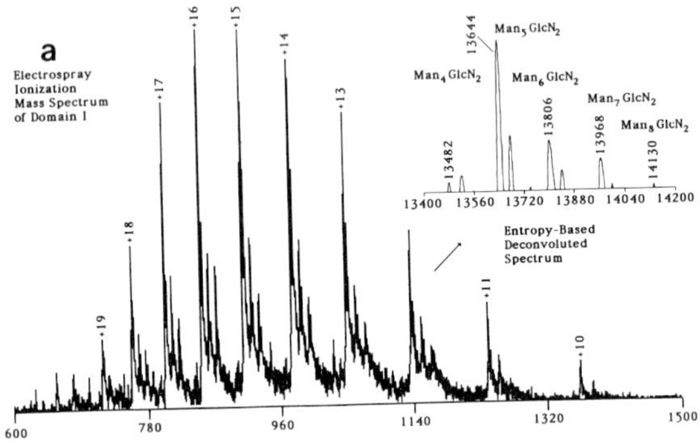

Fig. 2a

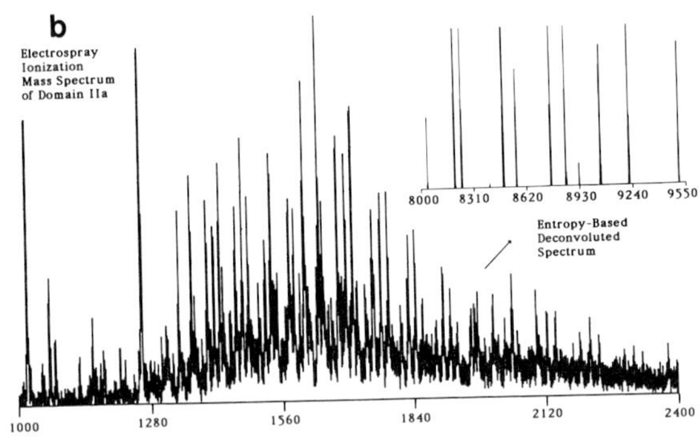

Fig. 2b

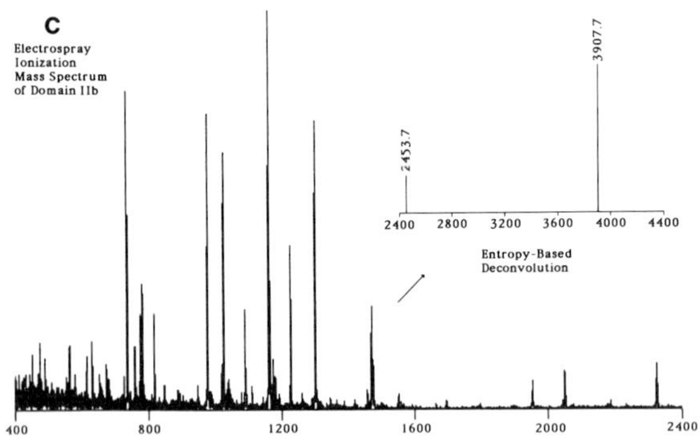

Fig. 2c

FIGURE 2: Electrospray ionization mass spectra of CD2 Domain I (Fig.2a), Domain IIa (Fig. 2b) and Domain IIb (Fig. 2c), and insets showing the entropy-deconvoluted spectra for each domain. The charge states for the Man_5 glycoform for Domain I are labeled.

Linkage, branching & sequence (SFCI-MS, FAB-MS). To characterize the details of N-glycan structure, the CD2 adhesion domain polypeptide was deglycosylated, dialyzed, lyophilized, periodate oxidized and derivatized by methoxime-peracetylation. Pre-analysis chemistry imparts improved molecular specificity that allows differentiation from other isomeric components. The susceptibility to periodate oxidation is dependent on the specific glycotype, glycan linkages and branching patterns. Trapping the product aldehydes can be accomplished with a number of reagents, of which, we have found methoximes to be most satisfactory. As a final step, the remaining hydroxyl groups are derivatized by acetylation to enhance supercritical fluid miscibility. Following this approach, all high mannose glycans are modified, and many isomers are converted to unique products. For these latter compounds, a molecular weight determination would be adequate to define linkage, branching, and sequence. Some samples, unfortunately, give rise to structures that are still isobaric and subsequent collision induced dissociation must be performed to confirm the structure. From earlier studies, it is anticipated that collision induced dissociation (CID) will provide a fragmentation pattern sufficiently unique for structural assignment. However, for the CD2 domain I glycopeptide, we were able to detect a unique molecular ion $(MH)^+$ at m/z=2184 by FAB-MS following the chemical derivatization of the peptide. This mass ion could only be accommodated by one of two $(Man)_5GN_2$-PFBAB structures illustrated in Figure 3. Under CID analysis, the 3-linked antenna have proven to be most labile and a comparable study of these structures would readily differentiate them. The full characterization of these details by CID is currently under investigation.

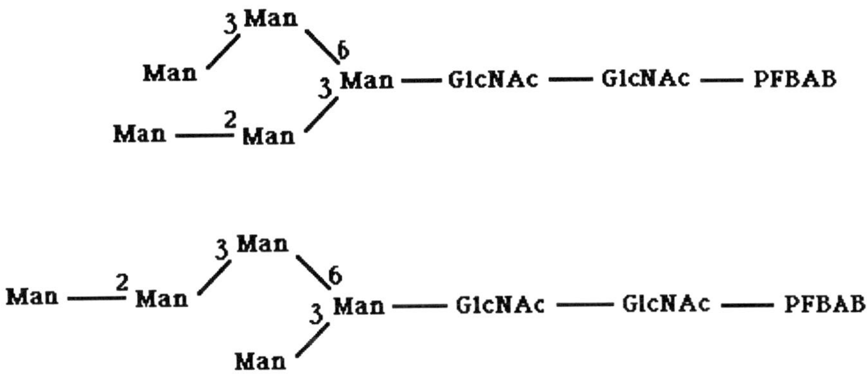

FIGURE 3: Putative CD2 Domain I $(Man)_5GN_2$-PFBAB glycoconjugates.

ACKNOWLEDGEMENTS

We thank Dr. Michael Concino for development of the $T11_{ex2}$ expressing CHO cell line, Kristin Gordon, Theresa Sampo, and Kyle Wallace for expert technical assistance, and Dr. Michael Luther, Dr. Sanjay Khandekar and Edith Neidhardt for critical discussions.

REFERENCES

1. Rademacher T.W., Parekh R.B., and Dwek R.A. (1988) *Ann. Rev. Biochem.* **57**, 785-838.
2. Cumming, D. (1991) *Glycobiol.*, **1**, 115-130.
3. Parish C.R., McPhun, V., and Warren H.S. (1988) *J. Immunol.* **141**, 3498-3503.
4. Coombe, D.R., and Rider, C.C. (1989) *Immunol. Today* **10**, 289.
5. Paulson , J.C. (1989) *Trends Biochem. Sci.* **14**, 272274.
6. Yagita, M., and Saksela, E. (1990) *Scand. J. Immunol.* **31**, 275-284.
7. Rothlein, R., Dustin, M.L., Marlin, S.D., and Springer, T.A. (1986) *J. Immunol.* **137**, 1270-1274.
8. Marlin, S.D., and Springer, T.A. (1987) *Cell* **51**, 813-819.
9. Dustin M.L, Rothlein, R., Bhan, A.K., Dinarello, C.A., and Springer, T.A. (1986) *J. Immunol.* **137**, 245-254.
10. Staunton, D.E., Marlin S.D., Stratowa, C., Dustin, M.L., and Springer T.A. (1988) *Cell*, **52**, 925-933.
11. Diamond, M.S., Staunton, D.E., deFougerolles, A.R., Stacker, S.A., Garcia-Aguilar, J., Hibbs, M.C., ad Springer, T.A. (1990) *J. Cell. Bio.* **111**, 3129-3139.
12. Dustin, M.L. and Springer, TR.A. (1988) *J. Cell. Bio.* **107**, 321-331.
13. Smith, C.W., Marlin, S.D., Rothlein, R., Toman, C., and Anderson, D.C. (1988) *J. Clin. Invest.* **83**, 2008-2017.
14. Diamond, M.S., Staunton, D.E., Marlin, S.D., and Springer, T.A. (1991) *Cell* **95**, 961-971.
15. Springer, T.A., and Lasky, L.A. (1991) *Nature* **349**, 196-197.
16. Moingeon, P.E., Chang, H-C., Sayre, P.H., Clayton, L.K., Alcover, A., Gardner, P., and Reinherz, E.L. (1989) *Immunol. Rev.* **111**, 111-144.
17. Moingeon, P.,E., Chang, H-C., Waller, B.P., Stebbins, C.C., Frey A.Z. and Reinherz, E.L. (1989) *Nature* **339**, 312-314.
18. Moingeon, P.E, Lucich, J.L., Stebbins, C.C., Recny, M.A., Waller, B.P., Koyasu, S., and Reinherz, E.L. (1991) *Eur J Immunol.* **21**, 605-610.
19. Sayre, P.H., Hussey, R.E., Chang, H-C., Ciardelli, T.L., and Reinherz E.L. (1989) *J. Exp. Med.* **169**, 995-1009.
20. Recny, M.A., Neidhardt, E.A., Sayre, P.H., Ciardelli, T.L. and Reinherz E.L. (1990) *J. Biol. Chem.* **265**, 8542-8549.
21. Peterson, A., and Seed, B. (1987) *Nature* **329**, 842-846.
22. Luther, M.A., et. al. (1991) *J. Cell. Biochem.* **15G**, 175.
23. Laemmli, U.K. (1970) *Nature* **227**, 680-685.
24. Reinhold, V.N., Sheely, D.M., Kuei, J., and Her, G.-R. (1988) *Anal. Chem.* **60**, 2719-2722.
25. Reinhold, B.B., and Reinhold, V.N. (1991) *J. Am. Soc. Mass. Spectrom.* **2**, (in press).
26. Mann, M., Meng, C.K., and Fenn, J.B. (1989) *Anal. Chem.* **61**, 1702-1708.

Characterization of the Glycosylation on Recombinant Human Low-Affinity Nerve Growth Factor Receptor

Christine A. Settineri[1], Iris Leung[2], Lawrence S. Cousens[2], Barbara Chapman[2], Frank R. Masiarz[1,2] and Alma L. Burlingame[1]

[1]Department of Pharmaceutical Chemistry, Mass Spectrometry Facility, University of California, San Francisco, CA 94143
and
[2]Chiron Corporation, Emeryville, CA 94608

I. INTRODUCTION

Nerve growth factor (NGF) is a neurotrophic factor that promotes the survival and function of sympathetic and neural crest-derived sensory neurons subsequent to specific interactions with receptors present on the plasma membrane (Banarjee et al., 1973; Frazier et al., 1974). In addition, specific nerve growth factor receptors (NGF-R) are found on cholinergic neurons in the brain (Schwab et al., 1979; Honneger et al., 1982; Taniuchi et al., 1986), as well as on non-neuronal derivatives of the neural crest, including pheochromocytoma cells (Herrup and Thoenen, 1979), melanoma cells (Fabricant et al., 1977), Schwann cells (Zimmerman and Sutter, 1983) and neurofibroma cells (Ross et al., 1984).

There are two types or states of the NGF-R, described as low- and high-affinity, as determined by binding studies (Sutter, et al., 1979). Recent evidence has shown that the high-affinity receptor is composed of members of the *trk* family of proto-oncogene products (Nebreda et al., 1991), 140-kDa proteins which are a members of the tyrosine kinase family of transmembrane receptors. The high affinity receptor is believed to be either made up of members of the *trk* family alone, or in combination with the low-affinity receptor, a 75-kDa protein which is not a tyrosine kinase (Nebreda et al., 1991). Sequencing of the cloned human NGF-R (low affinity) cDNA in 1986 (12) showed that this receptor is a member of the family of membrane glycoproteins containing multiple cysteine-rich domains (such as epidermal growth factor receptor and insulin receptor (Ullrich et al, 1984; Ullrich et al., 1985)). Therefore, to fully understand the binding and subsequent action of NGF, a complete structural analysis of the NGF-R is critical .

In order to study the structure of the NGF binding region in this glycoprotein, a recombinant form (rhNGF-R) was prepared which contains the N-terminal 222 amino acids ending just before the transmembrane region of the native full length receptor. rhNGF-R is a soluble form of the receptor that retains its ability to bind NGF. This protein contains one consensus site for *N*-linked glycosylation at Asn-32 and a Ser, Thr and Pro-rich *C*-terminal domain which

TECHNIQUES IN PROTEIN CHEMISTRY III

may contain *O*-linked glycosylation. We have used enzymatic deglycosylation combined with reductive amination derivatization strategies previously developed in this laboratory (Webb et al., 1988) to determine the structure and branching of the major *N*-linked carbohydrates on rhNGF-R.

II. METHODS

Expression and Purification. Recombinant NGF-R produced in Chinese hamster ovary (CHO) cells (Chapman et al., 1991) was purified from conditioned medium using ultrafiltration and diafiltration, followed by chromatography on Q-Sepharose, phenyl-Sepharose and Sephacryl S-200, and was estimated to be >90% pure from analysis on SDS-PAGE. The resulting product bound NGF with a K_d of approximately 1 nM. Purified rhNGF-R was reduced and carboxymethylated (RCM) at protein concentrations of 1-4 mg/ml in 200 mM N-ethylmorpholine acetate (NEMOAc), 6 M guanidine-HCl and 3 mM ethylenediamine-tetraacetic acid (EDTA) (pH 8.5). The protein was reduced with 2-mercaptoethanol for 1.5 hours at 60°C, followed by alkylation with NaIOAc for 1 hour at 25°C in the dark.

Enzymatic Digestions. Endoproteinase glu-C (*Staphylococcus aureus* V8) digestions were performed in 50 mM ammonium bicarbonate (pH 7.8) or 50 mM ammonium acetate (pH 4.0) at a weight ratio of 8% (2 x 4%) for 24-48 hours at 37°C. Trypsin/chymotrypsin digestions were performed using sequencing grade trypsin and chymotrypsin at a weight ratio of 5% in 100 mM ammonium bicarbonate buffer (pH 8.5) with 0.1 mM calcium chloride. Digestions were performed at 37 °C for 5 hours with trypsin, followed by 18 hours with chymotrypsin. N-glycanase (PNGase F) digestions were performed at 37°C in 100 mM ammonium bicarbonate (pH 8.0), using 4-5 units of enzyme over 4-5 days. Neuraminidase (from *Arthrobacter ureafaciens*) digestions were performed in 50 mM sodium acetate (pH 5.0) at 37°C for 40 hours. Aliquots of 160 mUnits were added at 0 and 22 hours. O-glycanase (Endo-α-N-acetylgalactosaminidase) digestions were performed in 50 mM sodium phosphate (pH 7.6) at 37°C for 46 hours. 5 mUnits were added at 0 and 22 hours.

Reverse phase chromatography was performed on a Beckman 114M HPLC system using an Applied Biosystems 757 detector and Vydac reverse phase columns. Carbohydrate analysis was performed using a Dionex AI-450 Model II high performance anion exchange chromatography system with pulsed amperometric detection (Hardy et al., 1988).

Mass Spectrometry. Liquid secondary ion mass spectrometric (LSIMS) analysis of peptides and oligosaccharides was carried out in the positive and negative modes using a Kratos MS50S double focusing mass spectrometer equipped with a high field magnet and a cesium ion source (Aberth et al., 1982) and gun at 10 keV (Falick, et al., 1986). A matrix of 1:1 thioglycerol:glycerol (with 1% TFA in the positive ion mode only) was used for LSIMS analysis.

ABOE Derivatization. *n*-Octyl *p*-aminobenzoate (ABOE) derivatization (Poulter et al., 1991) was performed using a reagent stock solution made up of 500 µl MeOH, 5 µl glacial acetic acid, 2 mg *n*-octyl *p*-aminobenzoate and 1 mg $NaBH_3CN$ (final pH 4.5) at a concentration of 30 µl per nmol of oligosaccharide. The reaction was carried out in an oven at 85°C for 1-2 hours. After cooling, the mixture was dissolved in 20 µl MeOH and excess reagent extracted twice with 150 µl $H_2O/CHCl_3$ (1:1).

III. RESULTS AND DISCUSSION

Treatment of the purified extracellular domain of recombinant human NGF-R (rhNGF-R) with N-glycanase, neuraminidase and O-glycanase all affected the migration on analytical SDS-PAGE relative to the migration of untreated rhNGF-R, indicating the presence of *N*- and *O*-linked glycosylation (not shown). In order to determine the composition and relative quantities of carbohydrates on this protein, intact rhNGF-R was hydrolyzed using different conditions specific for quantitation of neutral sugars, amino sugars and sialic acid (Hardy et al., 1988). Carbohydrate composition analysis was performed using high pH anion-exchange chromatography with pulsed amperometric detection (Hardy et al., 1988) and the results are shown in Table I. The presence of the monosaccharides fucose, galactosamine (from N-acetylgalactosamine, GalNAc), glucosamine (from N-acetylglucosamine, GlcNAc), galactose (Gal), glucose (Glc) and mannose (Man) confirms the presence of *N*- and *O*-linked carbohydrates. Therefore, protease digestion followed by mass spectrometry was used to try to map the protein and possibly find *O*-linked glycopeptides (Settineri et al., 1990). In addition, the intact, reduced and alkylated protein was treated with *N*-glycanase in order to isolate the *N*-linked oligosaccharide(s) attached to Asn-32 for derivatization and sequencing by mass spectrometry.

rhNGF-R was reduced and carboxymethylated before digestion with either endoproteinase glu-C or a combination of trypsin followed by chymotrypsin, and the peptides were separated by reverse-phase HPLC. Each fraction was analyzed by LSIMS to generate a peptide map. However, two portions of the protein were not mapped. The first portion (residues 24-33) contains the consensus sequence for *N*-linked glycosylation (Bause and Hettkamp, 1979) at Asn-32. Since the nonglycosylated peptide was not found, this suggests that Asn-32 is glycosylated. If so, proteolytic digestion would yield a glycopeptide which was either too hydrophilic or too large (>3500 Da) to be seen under our routine LSIMS mapping conditions. The second unmapped portion of the molecule was the Ser, Thr and Pro-rich *C*-terminus (residues 187-222). This could mean that the full length extracellular domain was not secreted, or that this portion of the protein contains extensive *O*-linked glycosylation, yielding glycopeptides not seen with routine LSIMS mapping. While the LSIMS analysis yielded some molecular ions that did not match predicted cleavage products, no *O*-linked glycopeptides were positively identified from peptide mapping of this protein.

In order to isolate the *N*-linked carbohydrates, 5 nmoles of reduced and carboxymethylated rhNGF-R were treated with N-glycanase, which cleaves the β-aspartylglycosylamine linkage of all known types of Asn-linked sugars and

Table I. Carbohydrate Analysis Results for Intact NGF-R RCM (mol/mol hydrolyzed NGF-R)

	Neutral Sugars	Amino Sugars	Sialic Acid
Sialic acid	--	--	1.5
Fucose	1.1	--	--
Galactosamine	2.9	9.4	--
Glucosamine	2.1	5.8	--
Galactose	5.4	4.7	--
Glucose	0.9	3.4	--
Mannose	1.9	0.5	--

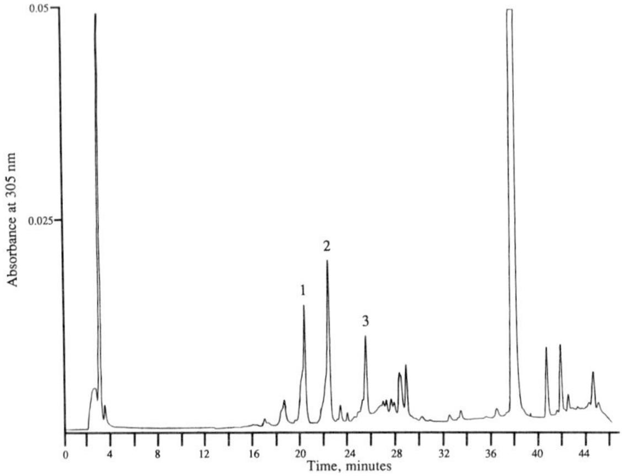

Figure 1. Reverse phase HPLC of ABOE derivatized N-linked oligosaccharides released from rhNGF-R RCM. Derivatized oligosaccharides were eluted on an analytical C_{18} column with water and acetonitrile (no TFA). UV detection of the chromophoric ABOE group was at 305 nm. Peaks eluting after fraction 3 are due to residual ABOE reaction components.

converts the attachment site Asn to Asp (Tarentino et al., 1985). After N-glycanase treatment, the free carbohydrates were separated from the intact protein by analytical HPLC on a C4 column. The deglycosylated protein was recovered and the water-eluted carbohydrates were derivatized with the hydrophobic chromophore, *n*-octyl *p*-aminobenzoate, to form the ABOE derivative (Poulter et al., 1991).

Reverse phase HPLC of the resulting derivatized oligosaccharides indicated three major species labelled 1 through 3 in Figure 1. Since only one chromophore is present per molecule of oligosaccharide, the relative peak intensities of the derivatized oligosaccharides on reverse phase HPLC indicates the relative quantities of each oligosaccharide species present on the glycoprotein. Negative ion LSIMS of peak 3 yielded an $M\text{-}H^-$ = 2019.4 and fragment ions corresponding to successive losses of hexose (Hex), N-acetylhexosamine (HexNAc) and Hex (Figure 2). Using the known sugars present from the carbohydrate analysis, this molecular ion could only correspond to a monosaccharide composition of $Hex_5HexNAc_4fucose_1ABOE_1$. Negative ion LSIMS of the largest peak, fraction 2, yielded an $M\text{-}H^-$ = 2310.0 and fragment ions corresponding to successive losses of sialic acid (NeuAc) , Hex, HexNAc and Hex, as well as a weaker series of ions showing successive losses of Hex, HexNAc and Hex, indicating the presence of two branches of different structures. Negative ion LSIMS of peak 1 yielded an $M\text{-}H^-$ = 2601.4 and fragment ions corresponding to successive losses of NeuAc and HexHexNAc. Figure 1 contains two smaller peaks eluting before peak 1, which could be oligosaccharide species with additional sialic acid residues. However, we have been unable obtain LSIMS spectra from these fractions.

From these data, the major *N*-linked structures of rhNGF-R were determined to be fucosylated diantennary complex oligosaccharides as shown in Figure 3. Of these three species, approximately 45% are monosialyl, 35% are disialyl and 20% are asialyl (from the UV absorbance of the ABOE derivatives).

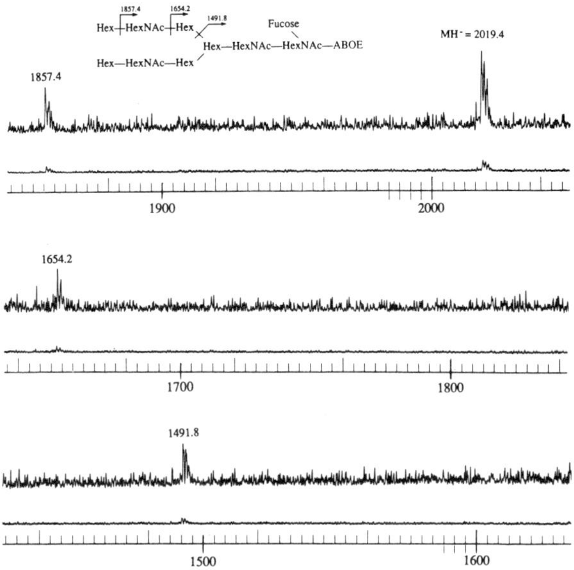

Figure 2. Negative ion LSIMS of peak 3 from figure 1. This fraction yielded an MH^- = 2019.4 and fragment ions corresponding to successive losses of Hex, HexNAc and Hex. Using the known sugars present from the carbohydrate analysis, this molecular ion could only correspond to a monosaccharide composition of $Hex_5HexNAc_4fucose_1ABOE_1$.

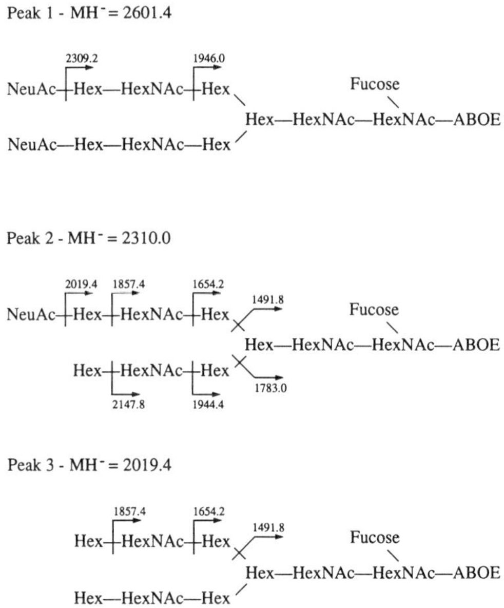

Figure 3. Major N-linked carbohydrate structures from Asn-32 of rhNGF-R. The carbohydrate structures were determined to be fucosylated diantennary N-acetyllactosamine (complex) type. (Hex = hexose; HexNAc = N-acetylhexosamine; NeuAc = neuraminic acid; ABOE = *n*-octyl *p*-aminobenzoate) The fucose is assumed to be on the innermost HexNAc residue from previous literature on the specificity of fucosyltransferases (Wilson et al., 1976; Kobata, 1984).

However, it is possible for some of the sialic acid residues to be hydrolyzed during the derivatization procedure, so the exact ratios of sialylated species may not be accurate (unpublished results from this laboratory). It is interesting to note that the size and branching of the structures determined in this study are identical to the major carbohydrate structures determined for the two *N*-linked sites on soluble CD4 receptor expressed in CHO cells (Carr et al., 1989). The CD4 structures were determined using positive ion fast atom bombardment (FAB) MS of permethylated oligosaccharides released from PNGase F treated glycopeptides. However, unlike derivatization with ABOE, permethylation does not allow separation or relative quantitation of the different carbohydrate species attached to a single site, because no UV-absorbing chromophore is present after permethylation of an oligosaccharide mixture.

IV. CONCLUSIONS

We have used carbohydrate analysis using high pH anion exchange chromatography with pulsed amperometric detection and enzymatic deglycosylation, combined with reductive amination derivatization strategies to determine the structure and branching of the major *N*-linked carbohydrates on rhNGF-R. The structures were determined to be fucosylated diantennary complex oligosaccharides, shown in Figure 3. From the UV absorbance of the ABOE derivatives, approximately 45% are monosialyl, 35% are disialyl and 20% are asialyl. However, since it is possible that a small percentage of the sialic acid groups may be cleaved from the oligosaccharides during the ABOE derivatization procedure, these are only rough estimates of the relative amounts of sialylated species. Experiments are in progress to determine the linkages of the *N*-linked structure as well as the structure and attachment sites of the *O*-linked carbohydrates on this protein.

ACKNOWLEDGEMENTS

This work was supported by National Institutes of Health grant DRR01614 and National Science Foundation Biological Instrumentation Program DIR8700766. The authors would like to thank Gena Reza, Mark Wininger and Peter Brown of Chiron Corporation for the production of conditioned medium containing rhNGF-R. We also thank Susan Kaufman and Gregory Stauber for quantitative analysis of NGF binding to the receptor glycoprotein.

REFERENCES

Aberth, W., Straub, K. M., Walls, F. C., and Burlingame, A. L. (1982) *Anal. Chem.* **54**, 2029-2034.

Banarjee, S. P., Synder, S. H., Cuatrecasas, P. and Greene, L. A. (1973) *Proc. Natl. Acad. Sci. USA.* **70**, 2519-2523.

Bause, E., and Hettkamp, H. (1979) *FEBS Letters.* **108**, 341-344.

Carr, S. A., Hemling, M. E., Folena-Wasserman, G., Street, R. W., Anumula, K., Barr, J. R., Huddleston, M. J., and Taylor, P. (1989) *J. Biol. Chem.* **264**, 21286-21295.

Chapman, B. S., Reza, G. C., Brown, P. C., Wininger, M. T., Cousens, L. T., Leung, I., Scandella, C. J., Stauber, G. B., Brown, S. A., and Kaufman, S. E. (1991) manuscript in preparation.
Fabricant, R. N., DeLarco, J. E., and Todaro, G. J. (1977) *Proc. Natl. Acad. Sci. USA.* **74**, 565-569.
Falick, A. M., Wang, G. H., and Walls, F. C. (1986) *Anal. Chem.* **58**, 1308-1311.
Frazier, W. A., Boyd, L. F., and Bradshaw, R. A. (1974) *J. Biol. Chem.* **249**, 5513-5519.
Hardy, M. R., Townsend, R. R., and Lee, Y. C. (1988) *Anal. Biochem.* **170**. 54-62.
Herrup, K., and Thoenen, H. (1979) *Exp. Cell Res.* **121**, 71-78.
Honneger, P., and Lenoir, D. (1982) *Dev. Brain Res.* **3**, 229-238.
Johnson, D., Lanahan, A., Buck, C. R., Sehgal, A., Morgan, C., Mercer, E., Bothwell, M., and Chao, M. (1986) *Cell* **47**, 545-554.
Kobata, A. (1984) *in* Biology of Carbohydrates, vol. 2, V. Ginsburg and P. Robbins, Eds., John Wiley, New York, 87-131.
Nebreda, A. R., Martin-Zanca, D., Kaplan, D. R., Parada, L. F., and Santos, E. (1991)*Science* **252**, 558-561.
Poulter, L., Karrer, R., and Burlingame, A. L. (1991) *Anal. Biochem.* **195**, 1-13.
Ross, A., Grob, P., Bothwell, M., Elder, D., Ernst, C., Marano, N. Ghrist, B., Slemp, C., Herlyn, M., Atkinson, B., and Koprowski, H. (1984) *Proc. Natl. Acad. Sci. USA.* **81**, 6681-6685.
Schwab, M. E., Otten, U., Agid, Y., and Thoenen, H. (1979) *Brain Res.* **168**, 473-483.
Settineri, C. A. , Medzihradszky, K. F., Chu, C., George-Nascimento, C., Masiarz, F. R., and Burlingame, A. L. (1990) *Biomed. Environ. Mass Spectrom.* **19**, 665-676.
Sutter, A., Riopelle, R. J., Harris-Warrick, R. M. and Shooter, E. M. (1979) *J. Biol. Chem.* **254**, 5972-5982.
Taniuchi, M., Schweitzer, J. B., and Johnson, E. M., Jr. (1986) *Proc. Natl. Acad. Sci. USA.* **83**, 1950-1954.
Tarentino, A. L., Gomez, C. M., and Plummer, T. H., Jr. (1985) *Biochemistry* **24**, 4665-4671.
Ullrich, A., Bell, J. R., Chen, E. Y., Herrera, R., Petruzzelli, L. M., Dull, T. J., Gray, A., Coussens, L., Hayflick, J. S., Liao, Y.-C., Tsubokawa, M., Mason, Seeburg, P. H., Grunfeld, C., Rosen, O. M., and Ramachandran, J. (1985) *Nature* **313**, 756-761.
Ullrich, A., Coussens, L., Hayflick, J. S., Dull, T. J., Gray, A., Tam, A. W., Lee, J., Yarden, Y., Libermann, T. A., Schlessinger, J. Downward, J., Mayes, E. L. V., Whittle, N., Waterfield, M. D., and Seeburg, P. H. (1984) *Nature* **309**, 418-425.
Unpublished results from this laboratory.
Webb, J. W., Jiang, K., Gillece-Castro, B. L., Tarentino, A. L., Plummer, T. H., Byrd, J. C., Fisher, S. J., and Burlingame, A. L. (1988) *Anal. Biochem.* **169**, 337-349.
Wilson, J. R., Williams, D., Schachter, H. (1976) *Biochem. Biophys. Res. Commun.* **72**, 909-916.
Zimmerman, A., and Sutter, A. (1983) *EMBO J.* **2**, 879-885.

NONENZYMIC GLYCATION OF HEMOGLOBIN
STRUCTURAL DETERMINANTS OF SITE SELECTIVITY

Parimala Nacharaju and A. Seetharama Acharya
Division of Hematology, Department of Medicine
Albert Einstein College of Medicine
Bronx, New York, NY 10461

I. INTRODUCTION

The nonenzymic glycation, a post-translational protein modification reaction, involves the reaction of glucose with amino groups of proteins. A number of body proteins such as hemoglobin (1), albumin (2), collagen (3), lens crystallin (4) and membrane proteins (5,6) are known to undergo nonenzymic glycation. The nonenzymic addition of glucose to the proteins exhibits very high site selectivity. Only a limited number of lysine residues of the protein are glycated and even with these, the reactive groups are derivatized with different rates. This reaction involves the formation of a labile Schiff base (aldimine) as the first step, which is subsequently rearranged (Amadori rearrangement) to a stable ketoamine (Fig. 1). Since the Schiff base formation requires unprotonated amino group, α-amino groups of the proteins, with their pK_a (around 7) lower than that of ε-amino groups (around 10), are expected to be the preferred sites of glycation under the physiological conditions (pH 7.4, 37°C). However this is not true, except with HbA. Even here, only α-amino group of β-chain is reactive and that of α-chain is not. The high selectivity of only few ε-amino groups as the major glycation sites in most of the proteins (7,8) indicates that the structural aspects of the protein that facilitate the Amadori rearrangement rather than the pK_a of the amino groups of glycation hot spots, dictate the site selectivity of nonenzymic glycation.

There have been considerable efforts made in recent years to identify the structural factors of the proteins that could influence the site selectivity of nonenzymic glycation. Shapiro et al. (1) suggested that the carboxyl groups that are oriented properly at the glycation sites may catalyse the

TECHNIQUES IN PROTEIN CHEMISTRY III

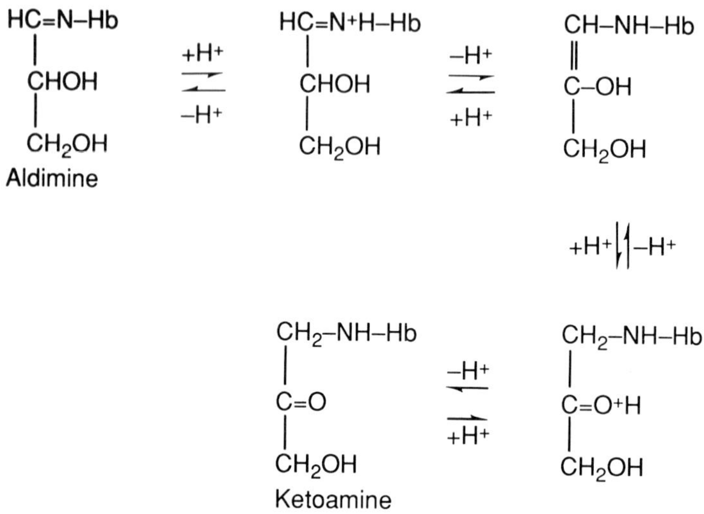

Figure 1. Schematic representation of the isomerization of aldimine (protein bound aldose) to ketoamine (protein bound ketose), i.e., glucose isomerase like activity of nonenzymic glycation sites.

Amadori rearrangement. Lysine residues lying in a Lys-Lys or Lys-His sequence have been hypothesized to be the major glycation sites in proteins (8). Watkins et al. (7) proposed that lysines located in or near the phosphate binding region of proteins are the facile sites for this reaction. Comparing the reactivity of Val-1(α) and Val-1(β) of Hb A for nonenzymic glycation, we had previously invoked a role for His-2(β) in the site selectivity of Val-1(β) for glycation (9). More recently, Shilton and Walton invoked the concept of the catalysis of glycation by histidine residues which are present close to the reaction site (10). In principle, this concept is similar to that of Iberg and Fluckiger (8), except that this incorporates the conformational aspect of the site.

Recently, we have estimated the aldimine to ketoamine isomerization potential of the microenvironment of the glycation sites of Hb A (11). The *in vitro* nonenzymic glycation sites of Hb A with glyceraldehyde are Lys-16(α), Val-1(β), Lys-66(β), Lys-61(α), Lys-82(β) and Val-1(α), in the decreasing order of reactivity. Since the formation of Schiff base is the first step in both nonreductive (in absence of $NaCNBH_3$) and reductive (in presence of $NaCNBH_3$) glycation, the ratio of extent of nonreductive and reductive modification of a given site is expected to reflect the aldimine to ketoamine isomerization potential of the microenvironment at that site. Lys-16(α), the most reactive site in nonenzymic glycation of Hb A, has an isomerization potential nearly three orders of magnitude higher than that of Val-1(α), the least reactive nonenzymic glycation site (11). From the given distinct catalytic power of the individual glycation sites, it is clear that the structural aspects of protein around the glycation site dictates the potential for the Amadori rearrangement.

The structural factors, that endow the microenvironment of the glycation site the potential to catalyse the Amadori rearrangement, may be arising from either the amino acid sequence (primary structure) or conformational (tertiary/quaternary) aspects of the protein. A general similarity in the structural aspects of the protein of the various nonenzymic glycation sites of Hb A is not readily apparent from the crystal structure of Hb A (Fig. 2). Lys-16(α) is a surface residue at the carboxy terminal end of the A-helix of the α-chain. Val-1(β) and Val-1(α) are the non-helical amino terminal residues of β- and α-chains respectively. Lys-66(β) and Lys-61(α) are in the

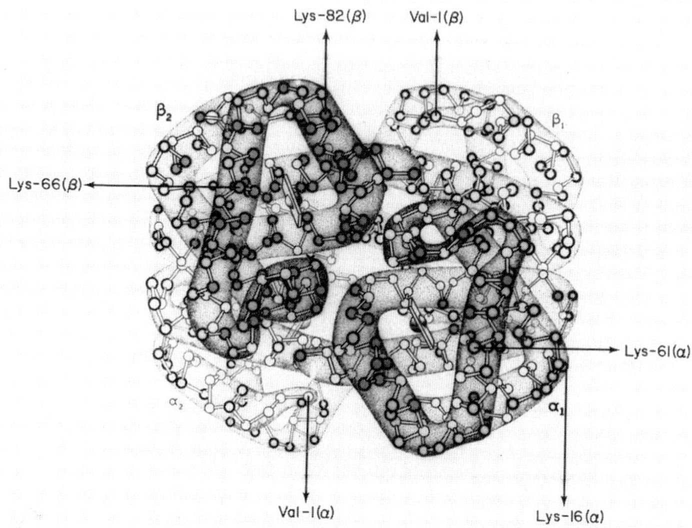

Figure 2. Nonenzymic glycation sites of Hb A.
Adapted from the figures of Dickerson and Geis (12).

middle of the E-helix (E10) of the β- and α-chains respectively. Lys-82(β) is at the EF corner, and it is a part of DPG-binding pocket of Hb A, formed by the positively charged residues of two β-chains. Val-1(β) also forms a part of the same pocket. This makes one to speculate that the factors that endow a aldimine to ketoamine isomerization potential to the glycation hot spots are originating from the tertiary/quaternary structural factors of the protein. In an attempt to establish the contribution of such structural factors, the nonenzymic glycation of Lys-16(α) and Val-1(α) has been now investigated in the α-chain and α(1-30). The choice of these two as nonenzymic glycation substrates is based on the fact that both of them contain the two nonenzymic glycation sites of Hb A, but α-chain is lacking the quaternary structural elements of Hb A, and α(1-30) has lost both the quaternary and tertiary structural elements of the respective microenvironment. In all the three substrates the primary structure around the respective glycation sites remains the same. The results of these investigations are presented here.

II. MATERIALS AND METHODS

A. Preparation of α(1-30) Peptide

The globin, separated from the purified hemoglobin A by acid acetone precipitation, was taken in 10 mM ammonium acetate, pH 4.0, and digested with S. aureus V_8 protease at 37°C for 2 hr and then lyophilized. An enzyme to globin ratio of 1:200 (w/w) was employed for this digestion. The lyophilized sample was redissolved in 0.1% TFA and concentrated using Amicon concentrator, equipped with YM-10 membrane. Most of the undigested globin and larger peptides remained in the concentrator. The filtrate, containing the smaller peptides like α(1-30) and β(1-22), was lyophilized. α(1-30) was purified by RPHPLC, using a semi preparative Aquapore RP-300 column. A linear gradient of 5-50% acetonitrile containing 0.1% TFA was used to elute the peptides.

B. Preparation of α-chain

The purified Hb A (0.1 mM) in 20 mM phosphate buffer, pH 6.0, containing 0.2 M NaCl, was reacted with p-hydroxymercuribenzoate (1 mM), for overnight at 4°C. After the incubation the reaction mixture was centrifuged to eliminate the small amount of precipitate that formed during the incubation. The clarified sample was dialysed against 10 mM phosphate buffer, pH 6.0, and loaded on CM-52 column (2.5 x 50 cm), pre-euilibrated with the same buffer. The PMB-chains of Hb A were eluted with a linear gradient of 10 mM phosphate buffer, pH 6.0, to 15 mM phosphate buffer, pH 8.0 (1000 ml each). The fractions, containing α-chain, were pooled and concentrated. This sample was dialysed against PBS, pH 7.4, and saturated with carbonmonoxide.

C. Kinetics of Nonenzymic Glycation of Hb A

Purified Hb A (1 mM) was incubated with 10 mM [^{14}C]glyceraldehyde in PBS, pH 7.4, at 24^{0}C. Aliquots were drawn at different time intervals and gel filtered on a column of Sephadex G-25, equilibrated and eluted with PBS. The amount of glyceraldehyde incorporated into the protein was determined from the ^{14}C-lable associated with the protein.

D. Nonenzymic Glycation of α-chain and Hb A

Carbonmonoxy forms of α-chain (2 mM) and Hb A (1 mM) were incubated separately with 10 mM [^{14}C]glyceraldehyde in PBS, pH 7.4, at 24^{0}C for 3 hrs. After incubation the reaction mixture was clarified from any precipitate that might have formed during incubation, and gel filtered on a column of Sephadex G-25 to separate the modified protein from the excess of reagent. The extent of nonenzymic glycation was determined as described above.

E. Reconstitution of Tetramer from Glycated α-chain

To avoid the precipitation of nonreductively modified α-chain during borohydrid reduction, to convert the ketoamine derivative to 2,3-dihydroxy propyl derivative, it was necessary to reconstitute the modified chain into tetramer by hybridizing with the complementary β-chain before reduction. The glycated PMB α-chain and unglycated PMB β-chain were taken in a ratio of 1:1.1 in 50 mM phosphate buffer, pH 7.4, and incubated with 100 mM 2-mercaptoethanol for 2 hrs at 4^{0}C. The hybridized sample was concentrated and gel filtered on a column of Sephadex G-25, equilibrated and eluted with PBS.

F. Nonenzymic Glycation of α(1-30)

α(1-30), at a concentration of 2 mM, was incubated with 10 mM [^{14}C]glyceraldehyde in PBS, pH 7.4, for 90 minutes at 37^{0}C. After the incubation the peptide was separated from the excess reagent by RPHPLC. The extent of modification was determined as discussed above, the peptide being quantitated by amino acid analysis.

G. *Tryptic Digestion of Glycated Samples*

The reconstituted tetramer and nonreductively glycated Hb A were reduced with $NaBH_4$, as described previously (13). The heme free globins were prepared from these samples by acid acetone precipitation. These globin samples (1mg/ml) were digested with trypsin in 0.1 M NH_4HCO_3 at pH 7.8 and 37°C for 16 hrs. The glycated α(1-30) was also reduced with $NaBH_4$, and triptic digestion was carried out in 50 mM phosphate buffer, pH 7.8, at 24°C for only 5 hrs. The ratio of enzyme to substrate was 1:100 (w/w). After digestion the samples were lyophilized to get salt free tryptic digests.

H. *Tryptic Peptide Map*

Tryptic digests of all the samples were subjected to RPHPLC, on a semi preparative Aquapore RP-300 column and the peptides were eluted with a linear gradient of 5-50% acetonitrile, containing 0.1% TFA.

III. RESULTS AND DISCUSSION

A. *Influence of Temperature on the Site Selectivity of Nonenzymic Glycation of Hb A*

In order to minimize the denaturation of α-chain during incubation, the glycation was carried out at room temperature (24°C), instead of 37°C, the temperature usually used for all our previous studies. For comparison, the glycation of Hb A was also studied at 24°C. Kinetics of nonenzymic glycation of Hb A at 24°C was studied to verify the linearity of the reaction at this temperature. Under the experimental conditions (1 mM Hb A and 10 mM glyceraldehyde) the incorporation of glyceraldehyde into the protein was linear, atleast during the initial stages of the reaction (Fig. 3) as at 37°C (11). However the rate of reaction is reduced significantly as the temperature is reduced from 37 to 24°C.

The lower reactivity of Hb A at 24°C could either be the consequence of changes in the site selectivity of nonenzymic glycation or a slower rate of overall reaction at the same sites or it could be a combination of both. Accordingly, the tryptic peptide map of the protein, nonenzymically glycated for 3 hrs, was generated. The relative distribution of glycation on various sites is nearly the same at both temperatures (Table I). Thus relative influence of lowering the temperature on reactivity of the individual glycation sites is the same. The overall conformational aspects of Hb A are changed very little on lowering the temperature from 37 to 24°C. Thus the lower rate

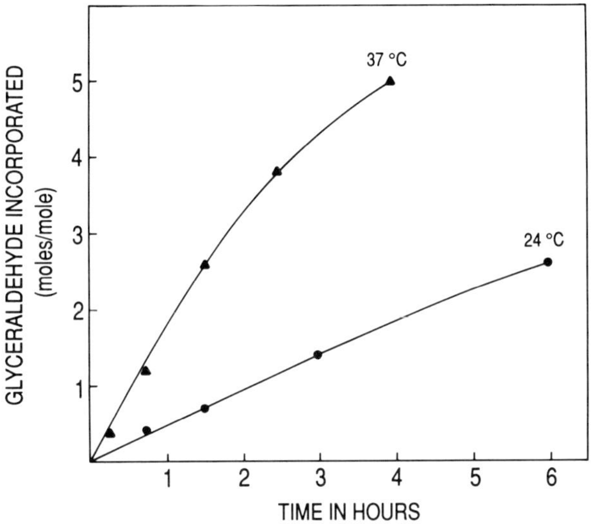

Figure 3. Influence of temperature on the nonenzymic glycation of Hb A.

simply reflects the reduction in the number of fruitful collisions of the reaction, i.e., the extent of formation of aldimine.

Table I. Influence of Temperature on the Site Selectivity of Nonenzymic Glycation of Hb A[1]

Glycation Site	% of total modification[2] at 24°C	% of total modification[2] at 37°C
Val-1(α)	2	2
Val-1(β)	22	20
Lys-16(α)	33	38
Lys-66(β)	18	16
Lys-61(α)	12	12
Lys-82(β)	12	12

[1]The nonenzymic glycation of Hb A (1 mM) was carried out with 10 mM glyceraldehyde in PBS, pH 7.4. The incubation period was 3 hrs at 24°C and 90 min at 37°C.
[2]The total modification of six sites is taken as 100%.

B. Nonenzymic Glycation of α-chain

The α-chain contains the most and least reactive nonenzymic glycation sites of Hb A, Lys-16(α) and Val-1(α), respectively, in addition to a third reactive site, Lys-61(α). The nonenzymic glycation of isolated α-chain at 24°C and pH 7.4 for three 3 hrs, incorporated about 0.35 moles of glyceraldehyde per mole of chain. Under identical conditions, 0.7 moles of glyceraldehyde were incorporated per αβ dimer of Hb A, nearly 50% of the reaction being on α-chain. This accounts for an incorporation of 0.35 moles of the reagent per mole of α-chain in Hb A. Thus the overall extent of nonenzymic glycation in the isolated α-chain is nearly the same as that when it is present in the tetramer. However, the relative ratio of the reagent to amino groups in the case of isolated chain is twice that of Hb A (β-chain is absent in the nonenzymic glycation of isolated α-chain). Neverthless, the extent of nonenzymic glycation is still comparable to that in tetramer suggesting an overall decrease in the rate of nonenzymic glycation on the loss of quaternary structural elements. In order to determine the site selectivity of nonenzymic glycation in the isolated α-chain the tryptic peptide map of the glycated α-chain has been established and compared with that of Hb A (Table II). The reactivity of Lys-16(α) reduced to about 40% in the isolated chain when compared to that in the intact Hb A, where as Val-1(α) of the chain is nearly 10 times more reactive than of tetramer. The reactivity of Lys-61(α) is also reduced by about 50%, compared to that in Hb A. Thus the site selectivity in the isolated α-chain is distinct from that of the α-chain in the tetramer. The decrease in the reactivity of lysine residues clearly reflects the contribution of the quaternary structure of Hb A in endowing the catalytic power to the microenvironment of the respective glycation sites to isomerize the aldimine to ketoamine. On the other hand, the reactivity of Val-1(α) increases in the absence of the quaternary structure of the protein demonstrating the down modulating potential of the quaternary structure of this region. The α-amino terminus of α-chain in the tetramer is proximal to

Table II. Influence of Quaternary Structure of Hb A on the Reactivity of Nonenzymic Glycation Sites

Glycation Site	Moles of aldose per mole of residue Hb A	α-chain
Val-1(α)	0.006	0.063
Lys-16(α)	0.098	0.037
Lys-61(α)	0.032	0.015

The nonenzymic glycation of Hb A (1 mM)/ α-chain (2 mM) was carried out with 10 mM glyceraldehyde in PBS, pH 7.4, at 24°C. The incubation period was 3 hrs.

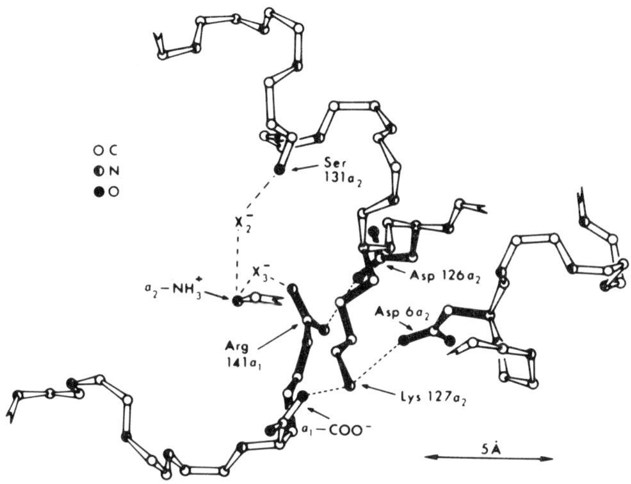

Figure 4. The anion binding sites of Hb A. Adapted from the figures of O'Donnell et al. (14).

the α-carboxyl of the second α-chain, and the quaternary structure of this region generates a chloride ion binding pocket of Hb A (Fig. 4). Such a quaternary structural arrangement has apparently contributed to a decrease in the Amadori rearrangement potential of the microenvironment of Val-1(α), since the propensity of Val-1(α) to form aldimine with glyceraldehyde is nearly the same as that of Val-1(β) in the tetramer (9).

C. Nonenzymic Glycation of α(1-30)

The segment 1 to 30 of α-globin also contains the most and least reactive nonenzymic glycation sites of Hb A, Lys-16(α) and Val-1(α), respectively. This peptide exhibited a very low reactivity towards nonenzymic glycation at 24°C, the temperature used for α-chain. Accordingly, this reaction of the peptide was carried out at 37°C and compared with that of Hb A at the same temperature. In this reaction, only 0.01 moles of glyceraldehyde were incorporated per mole of peptide. under similar conditions, 2.6 moles of glyceraldehyde reacted with each mole of Hb A, accounting for 0.24 groups modification of Lys-16 and 0.01 groups modification of Val-1, in each of α-chain (Table III). With the peptide, almost all the reaction was with Val-1(α), and Lys-16(α) was nearly unreactive. Thus the reactivity of Val-1(α) in the peptide and Hb A is comparable. Even

Table III. Influence of Tertiary/Quaternary Structure of Hb A on the Reactivity of nonenzymic Glycation Sites

Glycation Site	Moles of aldose per mole of residue	
	Hb A	α(1-30)
Val-1(α)	0.013	0.010
Lys-16(α)	0.240	almost unreactive

The nonenzymic glycation of Hb A (1 mM)/ α(1-30) at a concentration of 2 mM, was carried out with 10 mM glyceraldehyde in PBS, pH 7.4 at 37°C. The incubation period was 90 min.

though the molar ratio of the aldose to amino groups is much more higher with the peptide (8 mM amino groups vs 10 mM glyceraldehyde) as compared to that with Hb A (48 mM amino groups vs 10 mM glyceraldehyde), the reactivity of Val-1(α) in both the substrates is nearly same indicating the reduction in the reactivity of this residue on loss of tertiary/quaternary structure of the substrate. We have seen earlier that the loss of quaternary structure alone (as in the case of isolated α-chain) leads to an increase in the nonenzymic glycation of this residue by about 10 fold. This implies that the truncation of α-chain to generate α(1-30), i.e., loss of tertiary interactions of the chain results in a 10-fold reduction in the nonenzymic glycation potential of Val-1(α). As pointed out earlier, the reactivity of ε-amino groups in the peptide is undetectable under the present conditions. Thus these studies clearly establish that the isomerization potential at Lys-16(α) in Hb A, a glycation site with highest isomerization potential, is endowed to it by a synergy of the tertiary and quaternary structural aspects of the protein.

IV. CONCLUSIONS

From the results presented here, it is clear that the site selectivity of nonenzymic glycation is modulated considerably by the tertiary and quaternary structure of the protein. A comparison of the reactivity of Val-1(α) and Lys-16(α), the residues with lowest and highest reactivity for nonenzymic glycation of Hb A, in three different forms of substrate, isolated peptide segment (30 residues), isolated α-chain (141 residues) and Hb A (574 residues), demonstrates that the nonenzymic glycation potential of the amino groups of protein is not a reflection of the primary structure around the glycation susceptible sites. Conformational aspects of the glycation site endows the microenvironment of the respective sites with distinct aldimine to ketoamine isomerization potential. Nonenzymic glycation studies with mutant hemoglobins which perturb the stereochemical aspects of the individual glycation sites in very precise terms, should help us to establish the presence or absence of unique structural elements of the region, such as the presence of catalytic residues that could facilitate the Amadori rearrangement. As the generality of the molecular aspects that facilitate the nonenzymic glycation of proteins emerges, one could visualize the translation of these structural concepts into the design and construction of useful nonenzymic glycation inhibitors.

ACKNOWLEDGEMENTS

This work is partly supported by NIH Grant HL-38665, Feasibility Grant from American Diabetes Association and a Grant-in-Aid from the Diabetes Research Training Center of Albert Einstein College of Medicine.

REFERENCES

1. Shapiro, R., Mc Manus, M.J., Zalut, C. and Bunn, H.F. (1980) *J. Biol. Chem.*, **255**, 3120-3127.
2. Day, J.F., Thorpe, S.R. and Baynes, J.W. (1979) *J. Biol. Chem.*, **254**, 595-597.
3. Robins, S.P. and Bailey, A.J. (1972) *Biochem. Biophys. Res. Commun.*, **48**, 76-84.
4. Steves, V.J. Rouzer, C.A., Monnier, V.M. and Cerami, A. (1978) *Proc. Natl. Acad. Sci. USA*, **75**, 2918-2922.
5. Bailey, A.J., Robins, S.P. and Tanner, M.J.A. (1976) *Biochim. Biophys. Acta*, **434**, 51- 57.
6. Miller, J.A., Gravellese, E. and Bunn, H.F. (1980) *J. Clin. Invest.*, **65**, 896-901.
7. Watkins, N.G., Thorpe, S.R. and Baynes, J.W. (1985) *J. Biol. Chem.*, **260**, 10629-10636.
8. Iberg, N. and Fluckiger, R. (1986) *J. Biol. Chem.*, **261**, 13542-13545.
9. Acharya, A.S., Sussman, L.G. and Manning, J.M. (1983) *J. Biol. Chem.*, **258**, 2296-2302.
10. Shilton, B.H. and Walton, D.J. (1991) *J. Biol. Chem.*, **266**, 5587-5592.
11. Acharya, A.S., Roy, R.P. and Dorai, B. (1991) *J. Protein Chem.*, (In Press).
12. Dickerson, R.E. and Geis, I. (1983) *Hemoglobin: Structure, Function, Evolution and Pathology*, The Benzamin/ Cummings Publishing Company, Inc., California, p116.
13. Acharya, A.S. and Manning, J.M. (1980) *J. Biol. Chem.*, **255**, 1406-1412.
14. O'Donnell, S., Mandaro, R., Schuster, T.M. and Arnone, A. (1979) *J. Biol. Chem.*, **254**, 12204-12208.

A Unified Approach to Glycoprotein Primary Structure Analysis:
Identification, Isolation, and Characterization of both Peptide and Pendant Carbohydrate of Glycopeptides

David H. Hawke, Kuo-Liang Hsi, Lynn R. Zieske,
Ling Chen and Pau-Miau Yuan
Applied Biosystems Inc., Foster City, CA 94404

I. Introduction

As protein chemists, we find the primary structure determination of proteins is complicated by the presence of a variety of posttranslational modifications, especially aminoterminal blockage, phosphorylation, and glycosylation. Based on a survey of the NBRF-PIR protein sequence database we were surprised to realize that the occurrence of glycosylation in proteins is in fact much higher than either aminoterminal acetylation or phosphorylation. Admittedly, the database is not an unbiased selection of naturally occurring proteins, but it is indeed remarkable that >50% of the total sequence entries (release #25) exhibit the N-linked consensus sequence N-X-S/T, indicating a potential N-linked site. For this and other reasons we decided to investigate glycoprotein chemistry in more detail.

Our strategy, as outlined in Scheme 1, is:

1. Screen for the presence of glycosylation by lectin blotting;
2. Fragment the glycoprotein;
3. Locate the glycopeptides
   a. Peptide maps may be screened for glycopeptides, as above (1.), and further purified as required, or
   b. Glycopeptides may be isolated directly by affinity methods based on information derived above;
4. Once isolated the intact glycopeptide may be subjected to:

TECHNIQUES IN PROTEIN CHEMISTRY III

a. Peptide sequencing to identify the site(s),
b. Carbohydrate analyses to identify those structures as well,
c. Mass analysis to obtain molecular weight information;
d. The peptide and oligosaccharide may be cleaved apart and the separate portions analysed.

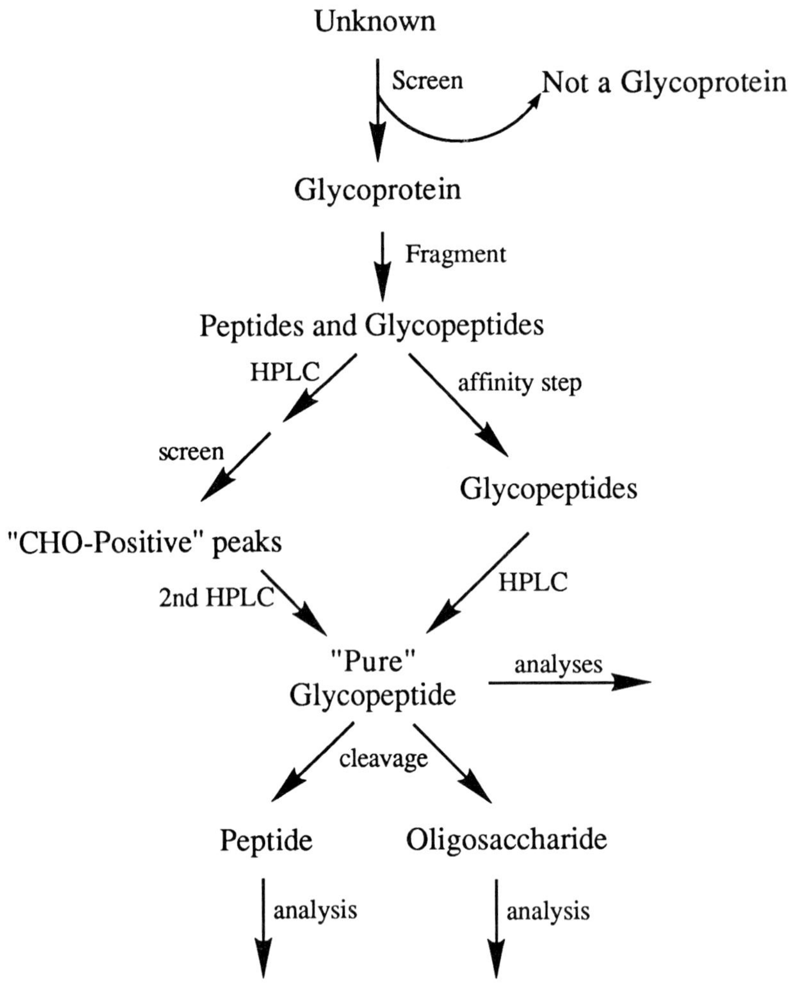

Scheme 1. Overall Strategy for Glycoprotein Primary Structure Analysis.

We have been interested in seryl-carboxypeptidases for some time, because of their usefulness as analytical reagents for protein chemists. Two of these of particular utility, CPY (carboxypeptidase Y from *Sacch. cer.*) and CPP (carboxypeptidase P from *Pen. janth.*) are also glycosylated. The structure of CPY is fairly well known, while relatively little has been published about CPP. We therefore chose them as models to study glycoprotein/glycopeptide analysis.

II. Materials and Methods

General. Chemicals were either reagent grade or HPLC grade, as appropriate. 1-methyl-3-phenyl pyrazolin-5-one (PMP, abbreviation as per Honda, reference 1), was purchased from Aldrich and recrystallized from methanol. Deuterated solvents were also from Aldrich. Carboxypeptidase P, carboxypeptidase Y, and endoproteinase Asp N were purchased from Boehringer-Mannheim. Trypsin was from CalBiochem. Biotinylated concanavalin A, concanavalin A-sepharose 4B resin, diaminobenzidine hydrochloride, hydrogen peroxide (30% solution), polyvinylpyrrolidone 360, and Triton X-100 were from Sigma. Avidin-biotinylated horseradish peroxidase reagent (ABC kit) and biotinylated lectin kits were purchased from Vector Laboratories. PNGase F and Endo H were from Genzyme. Oligosaccharides were obtained from Oxford Glycosystems. ProBlott™ is a product of Applied Biosystems Inc (ABI). Other reagents and solvents were from Applied Biosystems.

Sequence analysis was performed on an ABI Model 477A Pulsed Liquid Protein Sequencer using standard Normal-2 cycles. Cf-252 plasma desorption time-of-flight mass spectra were collected on a BioIon 20™ biopolymer mass analyzer (ABI). Typically, samples were applied to a nitrocellulose-coated target (as supplied) in a small volume (2-5 µl) of ethanol-water. Spectra were acquired for 1-10 million fission events with an accelerating potential of between 14 and 17 kV.

Peptides were separated on an ABI 140 B Solvent Delivery System, with a 785A Programmable Absorbance Detector (214 nm) and 112A Injector, using an Aquapore® RP-300 column (2.1x100 mm, ABI) in TFA/Water/MeCN (see figure legends). A second dimension HPLC employed 15 mM NH_4OAc (pH 7.0) instead of TFA. PMP derivatives (254 nm) were separated on a Model 120A using a 2.1x100 mm Aquapore® OD-300 column (both ABI) at 50°. Injections were made manually using a Rheodyne 7125, or automatically by a Gilson autoinjector (Model 231/401). Data was integrated with a Hewlett-

Packard 3396A. A simple TFA/Water/MeCN solvent system was used (see figure legends).

Lectin Blotting. This procedure is described in detail elsewhere (2). Briefly, samples were spotted onto a ProBlott membrane, air dried, and blocked with 2% PVP-360 in 50 mM Tris, 0.5 M NaCl pH 7.5 (TN) buffer for 1 hour at RT. The solution was replaced with biotinylated lectin (10 μg/ml in blocking buffer) for 1-2 hours, rinsed 3 times with TN/0.1% Triton X-100, then treated with avidin-biotinylated HRP complex for 1 hour (pre-formed ABC reagent mixture: one drop each of reagents "A" and "B" in 5 ml TN, incubated at RT for 30 minutes, then diluted to 15 ml). After another 3-fold rinse, the membrane was developed with DAB/H_2O_2 (0.05% DAB-HCl, 0.03% of 30% H_2O_2 in TN) 1-3 minutes until sufficient brownish color developed, rinsed (H_2O) and air dried.

Protease Digestion. Glycoprotein (0.5-1nmol) was dissolved in 50 μl of digestion buffer (200 mM NH_4HCO_3, pH 7.8). Endoproteinase Asp N was added (E/S ratio of 1:100), incubated at 37° for 12 hours, then terminated by adding TFA to adjust to pH 2. Tryptic digestion was performed under similar conditions (E/S ratio 1:50).

Micro-batch Affinity Method. About 0.1 ml of Con A-sepharose 4B resin was washed with binding buffer (3 times 200 μl of 5 mM NaOAc 100 mM NaCl, 1 mM Ca^{2+} and 1 mM Mn^{2+}, pH 5.5) in a microcentrifuge tube, removing supernatant by centrifugation. The resin was then suspended with 200 μl of digestion mixture (1-10 nmol) in binding buffer and allowed to stand 15 min with occasional shaking. The supernatant was removed and the resin washed again with binding buffer (3 times 200 μl). Bound peptides were released by suspending the resin in elution buffer (3 times 150 μl TN, pH 6.0 made up to 1N acetic acid, 0.5 M α-methyl mannoside). The pooled supernatant was vacuum dried and the eluate separated by reverse phase HPLC.

Hydrolysis. Oligosaccharide containing materials were hydrolyzed in 2 M TFA for either 4 hours at 100° or 2 hours at 120°. Microcentrifuge tubes, Pierce microvials, and plastic autosampler vials have all been used, with varying degrees of convenience.

Enzymatic Deglycosylation. Endo H deglycosylation was carried out in 100 μl of 100 mM sodium citrate buffer pH 5.5 using 5 nanomoles of glycopeptide substrate and 50 mU enzyme. After 15 hours at 37°, the reaction was stopped by drying and direct PMP derivatization.

PMP Derivatization. Samples for derivatization were dried into the bottom of either small diameter plastic microcentrifuge tubes, or directly into 300 μl plastic autosampler vials. If desired, a volume containing internal standard (rhamnose) was added and dried also. Typically 10 µl each of 0.5 M PMP in methanol and 0.3 M NaOH were added to the tubes and incubated at 70° (heating block) for 30 minutes. Then 3 µl 1 N HCl was added and the sample was evaporated to dryness (Speed Vac). The sample was reconstituted with 20 µl each of water and chloroform, agitated, allowed to separate, and the aqueous phase retained for analysis.

III. Results and Discussion

A classical approach of carbohydrate chemists to glycoprotein analysis is to completely deglycosylate the glycoprotein (e.g. using hydrazinolysis), isolate and purify the resulting oligosaccharides, then do structure determinations. Disadvantages to this scheme are that site information is lost, and the peptide bonds are usually completely destroyed as well. Large amounts of material were required, but improvements in sensitivity have been made for example by reductive tritium labelling at the reducing end. Site information, however, can be retained by peptide fragmentation, followed by gel-permeation chromatography for glycopeptide enrichment, as the carbohydrate-bearing peptides tend to be of higher molecular weight. More recently comparative mapping strategies, e.g. +/- peptide N-glycosidase F maps (on HPLC), followed by analysis, perhaps MS, have been used by some groups to identify glycosylation sites (3,4). While conceptually straightforward, comparative maps often contain large numbers of peptides, and are further complicated by the problems of incomplete digestion, as well as incomplete resolution of fragments.

The approach described herein was intended to simplify the situations outlined above by judicious use of affinity methods, noting that the decision of which affinity material to use could come from the initial glycoprotein screening using lectins (5,6), or possibly antibodies. In brief, with research unknowns the question "is it glycosylated?" may be answered with high sensitivity by lectin screening. The kind of carbohydrate structures present may be at least partly inferred by the specificity of the lectins which are found to bind the glycoprotein. The question of which sites are glycosylated may be addressed by mapping and screening, comparative mapping, or affinity isolation. The primary advantage of the affinity approach is that early in the separation scheme the selectivity is based on the presence of carbohydrate, overcoming the problem of co-purification of non-glycosylated fragments. Subsequent purification steps are therefore much cleaner and more readily interpreted.

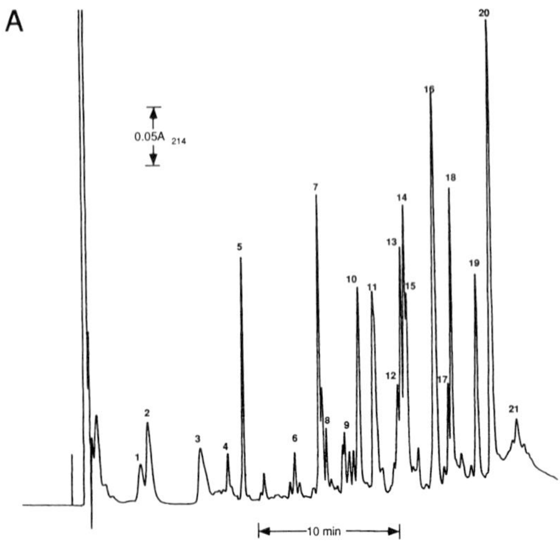

Figure 1A. Reverse Phase Chromatography of the CPP Tryptic Map (500 pmol). Peak 11 gave a positive stain using Con-A. Peptides were separated at a flow rate of 0.3 ml/min with a linear gradient from 0-70% acetonitrile containing 0.1% TFA (balance water). The OD-300 column was used at ambient temperature.

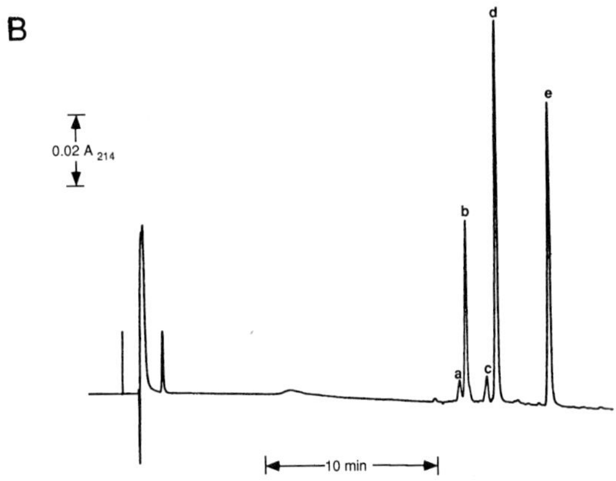

Figure 1B. Rechromatography of the Carbohydrate-Positive Peak from 1A (250 pmol). Only peak d was carbohydrate-positive. The separation was carried out using 15 mM NH_4OAc (pH 7.0) instead of 0.1% TFA in the same acetonitrile solvent on RP-300 Detection was at 214 nm.

A. Application to CPP

We applied both mapping/screening, and affinity approaches to characterize the carbohydrate site in CPP (7), a glycoprotein of about 51 kD. Peptide maps were produced using both trypsin and endoproteinase Asp-N, each of which contained only one Con-A positive peak. A second chromatographic pass under neutral conditions was required in each case to produce a fraction "pure" in peptide sequence. Figure 1 illustrates these two steps for the tryptic map. Alternatively, the crude digest was bound to immobilized Con-A, eluted with α-methyl-mannoside, and fractionated by HPLC. This result is shown in Figure 2. The single eluted peak past the breakthrough was >95% pure in peptide by sequencing. The sequence found was AYIDFSYXNTYQQLIDDSQR, and on sequencing deglycosylated material D was found (instead of X) at residue 8, as expected for the glycosylation site.

PDMS of both the intact and deglycosylated forms (8) indicated that the carbohydrate could be a $HexNAc_2Hex_5$ structure, more specifically $GlcNAc_2Man_5$ since it bound to Con-A and occurs at an N-linked site. Further, the intact glycopeptide may be recovered after

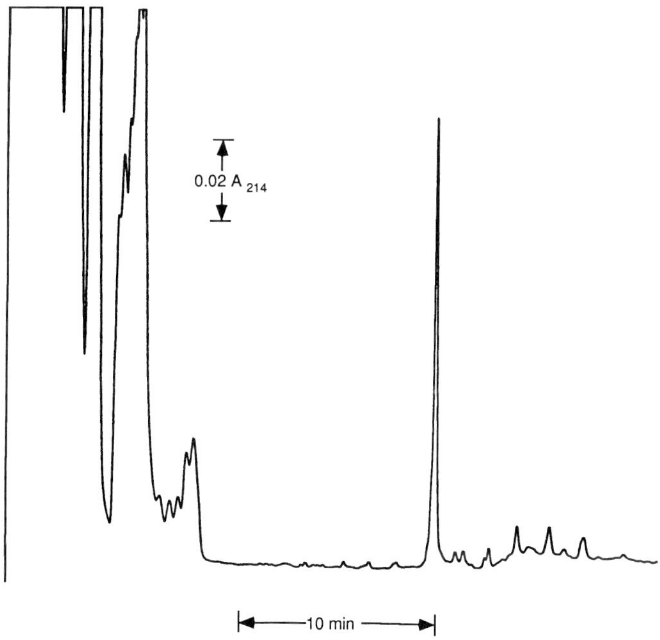

Figure 2. Reverse Phase Chromatography of the Affinity-Purified Tryptic Digest (250 pmol). Chromatography as in Figure 1A.

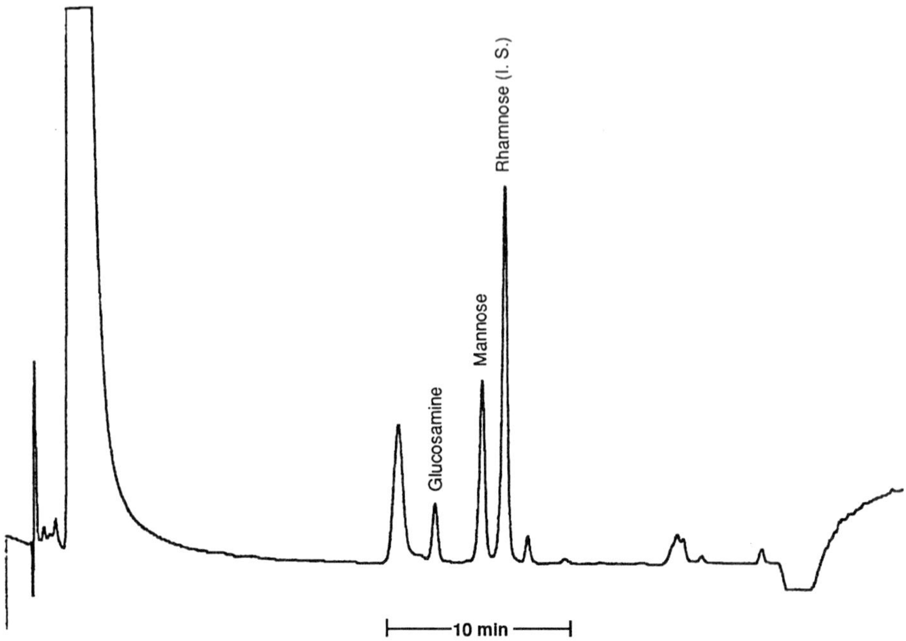

Figure 3. HPLC of the PMP-Derivatized Hydrolysate from the CPP Glycopeptide (300 pmol). Separation was performed on an OD-300 column at 50C and 200 ml/min. Gradient program: 5 min hold at 5%B, linear gradient to 30%B over 35 min, followed by a 90%B flush. "A" solvent was 0.1% TFA/water, "B" was 0.1% TFA in 90% MeCN/water, with detection at 254 nm.

PDMS for further manipulations (8). Carbohydrate composition analysis using PMP as a label for reducing sugars was performed (1), and gave the profile shown in Figure 3. Since significant amounts of only glucosamine and mannose were found (ratio 30:70), the assignment must be $GlcNAc_2Man_5$. In addition, after deglycosylation with PNGase F or Endo H, the crude reaction mixture was subjected to PMP derivatization, and HPLC. The resulting purified PMP-oligosaccharides were analysed by PDMS to give MH^+ ions at 1568.0 (theory: 1568.6) and 1363.7 (theory: 1363.4) respectively. The results of Endo H experiment are shown in Figure 4. The major peak (Figure 4A) was the PMP-oligosaccharide (PMP-$GlcNAc_1Man_5$), and the mass spectrum is shown in Figure 4B.

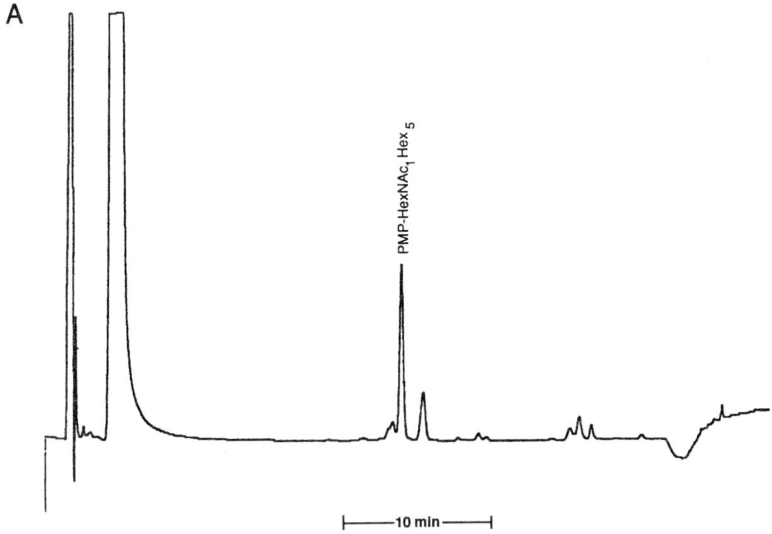

Figure 4A. HPLC of PMP-Derivatized Endo-H Product from CPP (5nmol). After Endo-H digestion of the CPP glycopeptide, the products were derivatized and separated under the same conditions as Figure 3.

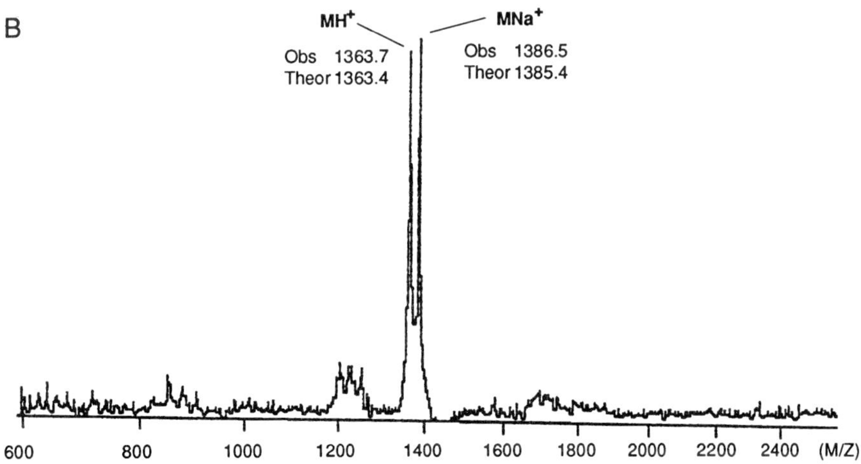

Figure 4B. PDMS of the Purified Endo-H Derived PMP Oligosaccharide of CPP. The main peak from Figure 4A was collected and mass analysed. The observed mass is in good agreement with that expected for PMP-$GlcNAc_1Man_5$.

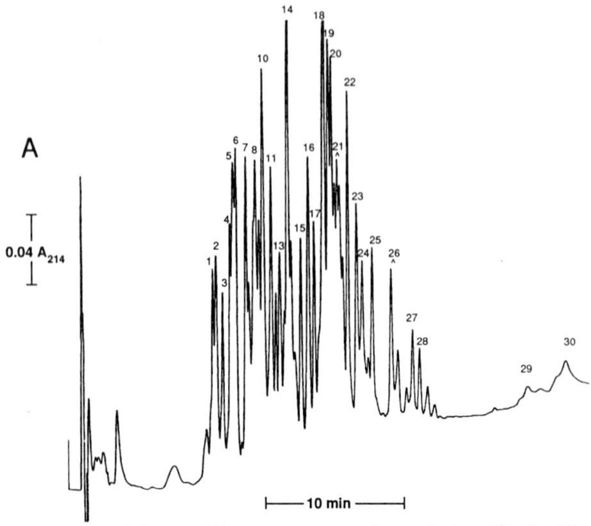

Figure 5A. Reverse Phase Chromatography of the CPY Chymotryptic Map (500 pmol). Chromatography was performed as in Figure 1A. The protein was reduced and alkylated with 4-vinylpyridine.

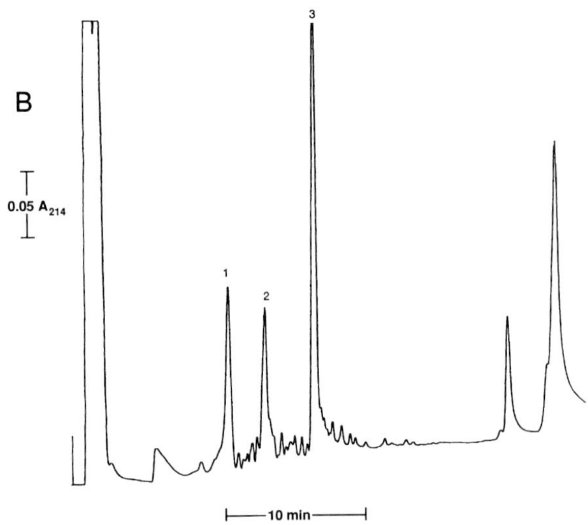

Figure 5B. Reverse Phase Chromatography of the Affinity-Purified Chymotryptic Digest of CPY (500 pmol). Chromatography was performed as in Figure 1A.

B. Application to CPY

CPY is a glycoprotein of about 61kD (9) containing four N-linked carbohydrate attachment sites, all occupied by high-mannose type oligosaccharides. In addition, many of these chains are phosphorylated (10), but the site-specific information has not been reported. Initial application of both screening and affinity approaches to the isolation of these sites using trypsin or Asp-N gave curious results. In each case only three of the four known sites were found, but not the same three! Using trypsin, the N368 site was not recovered, while with Asp-N we could not confirm the N87 site. Reexamination of the protein sequence showed that in each case the resistant site was in a rather large peptide fragment, making it difficult to isolate.

We therefore tried chymotrypsin, which typically yields many smaller fragments as compared to either enzyme above. A reverse phase chymotryptic map of reduced, alkylated CPY is shown in Figure 5a. For comparison, a second aliquot of crude digest was affinity purified on Con-A, and the eluate was chromatographed under the same RP conditions. This result is shown in Figure 5b. Peptide sequencing revealed that peaks 1 and 3 contained essentially pure peptides corresponding to sites N368 and N87 respectively. Peak 2 contained mainly 2 peptides, corresponding to sites N13 and N168, which were easily separated using a second RP-HPLC scheme. Further results from this glycoprotein will be reported elsewhere. A key observation derived from this study so far is that the choice of fragmentation enzyme is very important. If too many peptides are generated site information may be lost, but if too few are produced they may be more difficult to isolate.

IV. Conclusions

We have outlined our general strategy for the analysis of glycoproteins, utilizing lectin staining, available peptide separation and sequencing techniques, a simple micro-batch affinity method, PDMS, and PMP labeling of carbohydrates. Advantages are that most protein laboratories have the necessary instrumentation, and both peptide and carbohydrate structural information can be acquired at levels commensurate with the demands of modern protein chemistry.

References

1. Honda, S., Akao, E., Suzuki, S., Okuda, M., Kakehi, K. and Nakamura, J. (1989) Analyt. Biochem. 180, 351-357.
2. Hsi, K-L., Chen, L., Hawke, D.H., Zieske, L.R. and Yuan, P-M. (1991) Analyt. Biochem, in press.
3. Carr, S.A. and Roberts, G.D., (1986) Analyt. Biochem. 157, 396-406.
4. L'Italien, J.J., (1986) J. Chrom. 359, 213-220.
5. Merkle, R.K. and Cummings, R.D. (1987) Meth. Enz. 138, 232-259.
6. Osawa, T. and Tsuji, T. (1987) Ann. Rev. Biochem. 56, 21-42.
7. Yokoyama, S., Oabayashi, A., Tanabe, O., Sugarwara, S., Araki, E. and Ichishima, E. (1977) Appl. Microbiol. 27, 953-960.
8. Chen, L., Kochersperger, M., Bozzini, M. and Yuan, P-M , abstract #THRP63, 39th ASMS, Nashville TN, May 19-24, 1991.
9. Hasilik, A. and Tanner, W. (1978) Eur. J. Biochem. 91, 567-575.
10. Ballou, L., Hernandez, L.M., Alvarado, E. and Ballou, C.E. (1990) Proc. Nat. Acad. Sci. USA 87, 3368-3372.

SECTION VII

Protein Folding and Conformation

Refolding of human recombinant insulin-like growth factor II (IGF-II) in vitro, using a solubilizing affinity handle

Göran Forsberg, Elisabet Samuelsson[1], Henrik Wadensten, Tomas Moks[1] and Maris Hartmanis

Kabi Pharmacia KabiGen, S-112 87 Stockholm, Sweden
and
[1]Department of Biochemistry and Biotechnology, Royal Institute of Technology, S-100 44 Stockholm, Sweden

I. Introduction

Recombinant proteins or peptides are presently being produced in a variety of expression systems, using several different host cells. Expression in *Escherichia coli* is frequently used, since very high expression levels can be obtained. However, recombinant proteins produced intracellularly in *E. coli* tend to precipitate, forming inclusion bodies. To dissolve these, strong denaturing agents must be used. The native product may then be obtained using a properly designed refolding procedure. Proteins secreted from *E. coli* normally accumulate in the periplasmic space, where they can aggregate. Recently, an *E. coli* system where proteins are secreted to the medium, was described (Moks *et al*, 1987; Nilsson *et al*, 1991). In this system, the product is initially expressed as a fusion protein, where a 14 kDa IgG-binding peptide derived from staphylococcal protein A is used as fusion partner. After site-specific cleavage and separation of the fusion partner, a correctly folded protein can be obtained.

The insulin-like growth factors are homologous to insulin and promote growth in a variety of cells. For a review of the insulin-like growth factors and their actions, see Humbel, 1990. So far, IGF-I is the best studied polypeptide among the IGF-s. It has been produced in several heterologous organisms such as yeast (Elliot *et al*, 1990) and *E. coli* (Moks *et al*, 1987). However, even when producing IGF-I as a secreted product from these organisms, most of the IGF-I is found in biologically inactive forms with mismatched disulfides. Both incorrect monomeric as well as multimeric forms have been identified. The

TECHNIQUES IN PROTEIN CHEMISTRY III

misfolded forms can be converted into the native form using simple redox systems (Forsberg *et al*, 1990). However, even though the protein concentration in these systems has been low, 0.1 mg/ml or less, precipitation and aggregation can be observed during the redox procedures. Recently, it was shown that such undesired effects can be less pronounced, when a hydrophilic fusion partner is used, which increases the solubility of IGF-I during the redox procedure (Samuelsson *et al*, 1991).

Insulin-like growth factor II, IGF-II, consists of 67 amino acid residues in a single chain. In analogy with IGF-I, there are 6 cysteine residues forming 3 disulfide bonds. In this study, the recombinant IGF-II was initially produced as a secreted fusion protein, ZZ-IGF-II, in *E. coli*, as described by Wadensten *et al*, 1991. However, most full-length fusion protein was found in multimeric aggregates held together with disulfide bonds. A redox system, using the fusion partner as a solubilizer, was developed to transform multimeric, misfolded variants into the monomeric, correctly folded molecule. A comparison between IGF-II refolding as a fusion protein and as a native molecule, is also shown.

II. Materials and methods

Expression and purification of ZZ-IGF-II

In the expression system used here, IGF-II was initially produced as a 22 kDa fusion protein (Wadensten *et al*, 1991). The fusion partner was a synthetic IgG-binding polypeptide derived from *Staphylococcus aureus* protein A (Nilsson *et al*, 1987). After fermentation, the cells were removed by centrifugation and the supernatant was applied to an IgG-Sepharose column (Pharmacia LKB Biotechnology, Sweden) as described by Forsberg *et al*, 1989. The fusion protein was eluted with 0.2 M acetic acid, pH 3.2 and was subsequently lyophilized.

Protein separation and characterization using gel filtration

The material eluted from IgG-Sepharose was characterized using gel filtration on an analytical Superdex 75 column (Pharmacia LKB Biotechnology, Sweden). The column was equilibrated and eluted with 150 mM ammonium acetate, pH 8.0, at a flow rate of 0.5 ml/min. For preparative studies, a preparative Superdex 75 (26x600 mm) was used.

Protein solubility determinations

The protein was dissolved in 50 mM NH_4HCO_3, pH 8.0, to various concentrations between 5 µM and 500 µM, as determined with amino acid analysis. To each sample, 10 mM DTT was added and the samples were incubated for 30 min at 37 °C. The samples were centrifuged and the protein concentration in the supernatants was determined with amino acid analysis, as described by Forsberg *et al*, 1990.

Refolding

The protein was dissolved in a solution containing 26 mM NH_4HCO_3, pH 8.0, and 140 mM 2-mercaptoethanol, to protein concentrations between 5 μM and 1 mM. The samples were reduced for 30 min at 37 °C and then reoxidized by dialysis against 26 mM NH_4HCO_3, pH 8.0, at ambient temperature. The dialysis buffer was changed twice and the samples were analysed on reverse phase HPLC and on the Superdex 75 column.

Reverse phase HPLC analysis

The samples were generally analysed with reverse phase HPLC. The column used was a Kromasil C_8 (4.6x250 mm) with 5 μm particles and 100 Å pore diameter. The gradient used was from 35 to 44 % acetonitrile in 0.25 % pentafluoropropionic acid over 43 min at a temperature of 40 °C and a flow rate of 1 ml/min.

III. Results and discussion

Expression

Following fermentation, the culture medium was clarified and the fusion protein was recovered using IgG-Sepharose. Using this expression system, yields of approximately 900 mg fusion protein per liter were obtained (Wadensten *et al*, 1991). A site specific cleavage of a methionine linker in the fusion protein with cyanogen bromide (Gross and Witkop, 1962), was used to liberate IGF-II. A final yield of 5-7 mg/l was obtained after the fusion partner and the multimeric forms had been removed by chromatographic methods.

However, a size exclusion chromatography analysis of the affinity purified material from the IgG-Sepharose column, revealed that two thirds of the full-length protein consisted of multimeric aggregates (Figure 1). These multimeric forms are held together with disulfide bonds (data not shown). Since the fusion partner lacks cysteine residues, the aggregates must be a result of incorrect disulfide bonds between the IGF-II moieties in the fusion protein.

Solubility studies

In order to increase the yield of IGF-II, a refolding of the multimers using a redox procedure was studied. Since the solubility of reduced IGF-II is very low, the conditions must be chosen carefully to avoid protein precipitation and aggregation. The solubility might be increased, using denaturating agents or organic solvents in the redox buffers. However, in this study these possibilities were not examined, since an alternative strategy was used. In the redox procedure, the object was to take advantage of the high solubility of the protein A domain in aqueous solutions. Instead of refolding multimeric forms of IGF-II after cleavage, the entire multimeric fusion protein was refolded. Here, the hydrophilic protein A domain keeps the more hydrophobic IGF-II moiety soluble throughout the redox procedure. Such an approach was recently used when designing a redox procedure for IGF-I, in which the protein A domain solubilized the growth factor (Samuelsson *et al*, 1991). Using this procedure, correctly folded IGF-I could be recovered from the multimeric fusion protein. The refolding was performed at a concentration of 1-2 mg/ml, without use of any solubilizing agents.

To investigate the solubilizing effects of the fusion partner in the IGF-II system, the following experiment was performed: Different concentrations, between 5 μM and 500 μM, of reduced IGF-II, with or without fusion partner, were incubated in 50 mM NH_4HCO_3, pH 8.0 at 37 °C for 30 min. The samples were centrifuged and the protein concentration in the supernatant was determined using amino acid analysis. The results showed that reduced IGF-II precipitates at concentrations above approximately 15 μM, equivalent to 0.1 mg/ml, while the IGF-II fusion stays soluble even at 10 mg/ml, a thirtyfold higher molar concentration.

Purification and refolding procedures

A well defined batch consisting of monomeric and multimeric fusion protein was initially prepared. Fusion protein from the IgG-Sepharose column was loaded on a S Sepharose ion exchange column (Pharmacia LKB Biotechnology). The column was washed with 20 mM ammonium acetate, pH 5.3, which

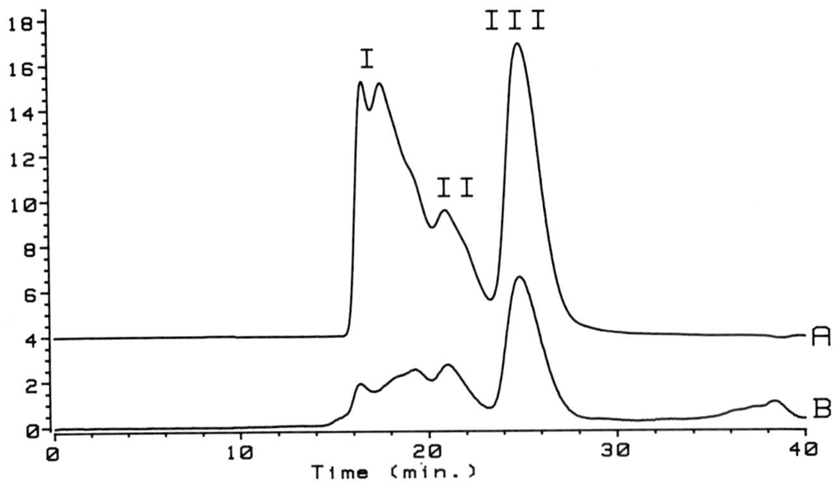

Figure 1) Characterization of the fused IGF-II, ZZ-IGF-II, using gel filtration, as described in materials and methods. The material analysed in A) was obtained from IgG-Sepharose and in B) obtained after subjecting the multimers to a redox procedure. The following peaks were identified: I) Multimeric forms of ZZ-IGF-II, II) Dimeric form of ZZ-IGF-II and III) Monomeric form of ZZ-IGF-II. The analytical conditions are described in materials and methods. The absorbance was measured at 280 nm.

removed degraded forms of the fusion protein, followed by elution of the full-length fusion protein with 500 mM ammonium acetate, pH 6.0. The material obtained was subsequently applied to a Superdex 75 column (Pharmacia LKB Biotechnology), which separated monomeric fusion protein from multimeric forms (Figure 1). The multimeric forms constituted 70 % of the total full-length fusion protein. The fraction containing multimers was lyophilized and redissolved in a buffer containing 50 mM NH_4HCO_3, pH 8.0, to a final concentration of 11 μM, 0.25 mg/ml. After addition of 140 mM 2-mercaptoethanol, the sample was reduced at 37 °C for 30 min. The reduced protein was then dialysed against 50 mM NH_4HCO_3, pH 8.0, containing 0.1 mM EDTA. Here, the reducing agent was removed, allowing the protein to reoxidize in air.

The reoxidized protein was analyzed by size exclusion chromatography (Figure 1b). This refolded material consisted of 60 % monomers. The peaks containing monomers and multimers were collected. Reverse phase HPLC analysis of these peaks showed that the monomeric fraction consisted mainly of fused IGF-II with a correct disulfide pairing (Figure 2).

However, if the refolding procedure was performed at higher protein concentrations, the amount of monomeric fusion protein obtained decreased (Table I). It is possible that an optimized refolding performed using other pH-values or other buffers, would result in a higher degree of monomers at higher protein concentrations.

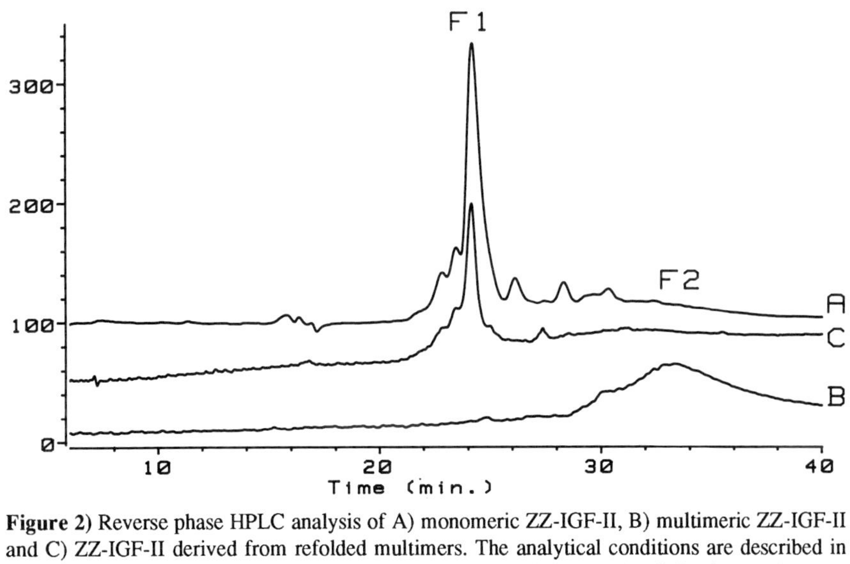

Figure 2) Reverse phase HPLC analysis of A) monomeric ZZ-IGF-II, B) multimeric ZZ-IGF-II and C) ZZ-IGF-II derived from refolded multimers. The analytical conditions are described in materials and methods. The absorbance was measured at 220 nm. The following peaks were identified: F1) Monomeric, correctly folded fusion protein and F2) Multimeric forms.

Table I Data from the redox experiments

	ZZ-IGF-II			IGF-II	
Protein conc. (mg/ml)	0.25	0.75	1.25	0.03	0.1
Percent correctly folded monomers after redox[a]	60	40	20	25	5

[a] As determined by size exclusion chromatography and reverse phase HPLC.

The fusion protein obtained after refolding was further characterized. The material was cleaved with cyanogen bromide and analysed with reverse phase HPLC. Figure 3 shows a HPLC-chromatogram of refolded and subsequently cleaved protein. Peak C1 indicates native IGF-II, which is present only after cleavage of monomeric fusion protein, formed during fermentation or after refolding. The other identified peaks are C2: fusion partner and C3: uncleaved fusion protein. More than 80 % of the IGF-II present in the monomeric fusion protein was recovered after chemical cleavage. In contrast, virtually no IGF-II with a correct structure is obtained when the pooled multimers (peaks I and II in Figure 1) were cleaved and analyzed (data not shown). However, the

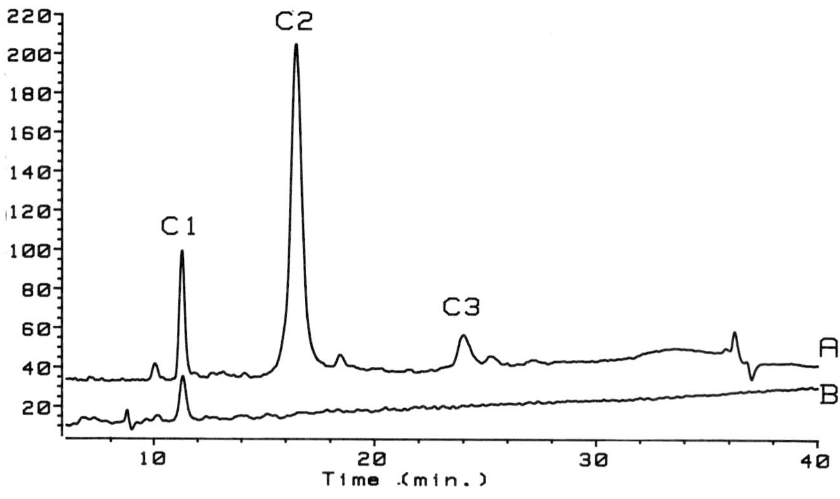

Figure 3) Reverse phase HPLC analysis of A) multimeric ZZ-IGF-II subjected to a redox procedure and subsequently cleaved with cyanogen bromide and B) unfused IGF-II after the redox procedure. The following peaks were identified: C1) Native IGF-II, C2) Fusion partner, ZZ, and C3) Uncleaved fusion protein. The analytical conditions are described in materials and methods. The absorbance was measured at 220 nm.

monomeric fusion protein formed after refolding of the multimeric forms yielded correctly folded IGF-II after cleavage, as seen in the HPLC-chromatogram. The major IGF-II containing peak, obtained after cleavage of monomers formed during fermentation, represents the native IGF-II structure, as described by Wadensten *et al*, 1991. In addition, using this redox procedure no monomeric IGF-II with mispaired disulfides was observed.

Refolding of IGF-II without using a fusion partner

Purified and correctly folded IGF-II was reduced and reoxidized using the same procedure as for the fusion protein. The protein concentration was 4 µM, 0.03 mg/ml, corresponding to a three-fold lower molarity, than used in the fusion protein experiment. The reoxidized protein was analyzed by size exclusion chromatography (data not shown) and reverse phase HPLC (Figure 3), revealing that correctly folded IGF-II was formed. The yield of correctly folded IGF-II was approximately 25 %. This yield should be compared to the 60 % obtained when refolding fused IGF-II at a threefold higher molar concentration (Table I). The redox procedure was also carried out with unfused IGF-II at 13 µM, 0.1 mg/ml, at the found solubility limit. However, at this concentration, only 5 % of the material was recovered as a correctly folded product.

The refolding of unfused IGF-II resulted in protein precipitation, probably due to protein aggregation during the refolding procedure. The amount precipitated was not determined. In contrast, no precipitation was observed when refolding the fused IGF-II even at a concentration of 2.5 mg/ml.

Even though the refolding procedure for unfused IGF-II may be optimized, the advantages of refolding IGF-II as a fusion protein are obvious. A recent study has shown that the same technique is advantageous to use also for refolding of IGF-I. The concept presented here may be useful in general, when proteins with low solubilities have to be refolded. Since fusion partners are often used as affinity handles during the purification procedure, it is advantageous if they also can be used as solubilizers. A specific problem for the expression of recombinant proteins in *E. coli* is the intracellular formation of inclusion bodies. As these inclusion bodies have to be dissolved using strong denaturing agents, the proteins must subsequently be refolded. We believe that the concept described here, can be beneficial also for refolding of recombinant proteins derived from inclusion bodies.

IV. Conclusions

Here we have shown that a hydrophilic fusion partner may act as a solubilizer of insulin-like growth factor II during refolding. The fusion partner apparently does not interfere with the refolding. Instead it seems to keep the more hydrophobic IGF-II moiety soluble, making denaturating agents or organic solvents unnecessary in the refolding buffers. The fusion partner can then subsequently be removed using site specific cleavage. This concept can be used to refold proteins which normally precipitate during refolding.

Acknowledgements

We are grateful to Kristina Zachrisson for amino acid analysis and Christina Kalderén for fermentations. We also thank Björn Nilsson for valuable discussions.

References

Elliott, S., Fagin, K.D., Narhi, L.O., Miller, J.A., Jones, M., Koski, R., Peters, M., Hsieh, P., Sachdev, R., Rosenfeld, R., Rohde, M.F. and Arakawa, T. (1990) *J. Prot. Chem.* **9**, 95-104.

Forsberg, G., Baastrup, B., Brobjer, M., Lake, M., Jörnvall, H. and Hartmanis, M. (1989) *BioFactors* **2**, 105-112.

Forsberg, G., Palm, G., Ekebacke, A., Josephson, S. and Hartmanis, M. (1990) *Biochem. J.* **271**, 357-363.

Gross, E. and Witkop, B. (1962) *J. Biol. Chem.* **237**, 1856-1860.

Humbel, R.E. (1990) *Eur. J. Biochem.* **190**, 445-462.

Moks, T., Abrahmsén, L., Holmgren, E., Bilich, M., Olsson, A., Uhlén, M., Pohl, G., Sterky, C., Hultberg, H., Josephson, S., Holmgren, A., Jörnvall, H. and Nilsson, B. (1987) *Biochemistry* **26**, 5239-5244.

Nilsson, B., Forsberg, G. and Hartmanis, M. (1991) *Methods Enzymol.* **198**, 3-16.

Nilsson, B., Moks, T., Jansson, B., Abrahmsén, L., Elmblad, A., Holmgren, E., Henrichson, C., Jones, A. and Uhlén, M. (1987) *Protein Eng.* **1**, 107-113.

Samuelsson, E., Wadensten, H., Hartmanis, M., Moks, T. and Uhlén, M. (1991) *Bio/Technology* **9**, 363-366.

Wadensten, H., Ekebacke, A., Hammarberg, B., Holmgren, E., Kalderén, C., Tally, M., Moks, T., Uhlén, M., Josephson, S. and Hartmanis, M. (1991) *Biotechnol. Appl. Biochem.* **13**, 412-421.

Conformational Stability of the Molten Globule of Cytochrome *c* : Role of Electrostatic Repulsion

Yuji Goto and Shin Nishikiori
Department of Biology, Faculty of Science,
Osaka University, Toyonaka, Osaka 560, Japan

I. Introduction

The molten globule structure, a compact structure with a significant amount of secondary structures, but showing a largely disordered tertiary structure, has been detected for several proteins and studied extensively as a candidate for the major intermediate of protein folding (1-4). The acidic molten globule of ferricytochrome *c* (cytochrome *c*) was the first example for which the term molten globule was used (1). While several acid-denatured proteins, including cytochrome *c*, are maximally unfolded at pH 2 by HCl in the absence of salt, they undergo cooperative transformation to a molten globule structure upon addition of anions from either salts or acids (5-8). Recently, the location of secondary structure in the molten globule of cytochrome *c* was examined by NMR using the H-D exchange of amido protons (9). This revealed that a significant amount of α-helical structure, similar to that in the native structure, is present in the molten globule. On the other hand, thermodynamic details of how the molten globules are stabilized are unknown.

A series of studies carried out by Goto and Fink (5-8) suggested the importance of charge repulsion in the stability of acidic molten globules. Goto *et al.* (8) examined the salt- and acid-induced formation of molten globules and proposed that the conformation of acid-denatured proteins is determined by a balance of charge repulsions between positive groups, which favor the extended

TECHNIQUES IN PROTEIN CHEMISTRY III

conformation, and mainly hydrophobic interactions, which stabilize the molten globule, and that shielding of the former by anion binding manifests the latter.

If electrostatic repulsions determine the conformation of acid-denatured proteins, removal of the positive charges should result in stabilization of the molten globule. Acetylation with acetic anhydride is a reaction which removes the charge of an amino group ($R\text{-}NH_3^+$):

$$R\text{-}NH_3^+ + (CH_3CO)_2O \rightarrow R\text{-}NH\text{-}COCH_3 + CH_3COOH + H^+ \quad (1)$$

Horse cytochrome *c* has 19 lysyl amino groups and the N-terminal α-amino group is acetylated during protein synthesis. To confirm this suggestion, we prepared a series of acetylated cytochrome *c* species with various degrees of acetylation. The conformation of acetylated cytochrome *c* species showed quantitatively that the charge repulsions do, in fact, unfold the compact molten globule of cytochrome *c*.

II. Materials and Methods

A. *Acetylated Cytochrome c*

Acetylated cytochrome *c* species with different degrees of acetylation were purified by preparative electrofocusing. Horse cytochrome *c* (Sigma, type IV) at 10 mg/ml in 0.4 M phosphate buffer (pH 7.0) was acetylated with acetic anhydride for 2 h at room temperature. The concentration ratio of amino groups to reagent was varied from 1:0.1 to 1:10 to obtain the various degrees of acetylation. The pH was kept constant, when necessary, by adding a small volume of 1 N NaOH. The protein solution was dialyzed against deionized water and then applied to a preparative flat-bed electrofocusing system with the LKB 2117 Multiphor. The Sephadex G-75 superfine gel bed contained 7 M urea in addition to carrier ampholite (pH 3-10) to remove the difference in mobility of the species with the same number, but with different sites, of modification. One hundred milligrams of the acetylated cytochrome *c* was loaded at the anode side of the gel bed and then electrofocusing was carried out for 10 h at 8 W constant power. The acetylated cytochrome *c* molecules were

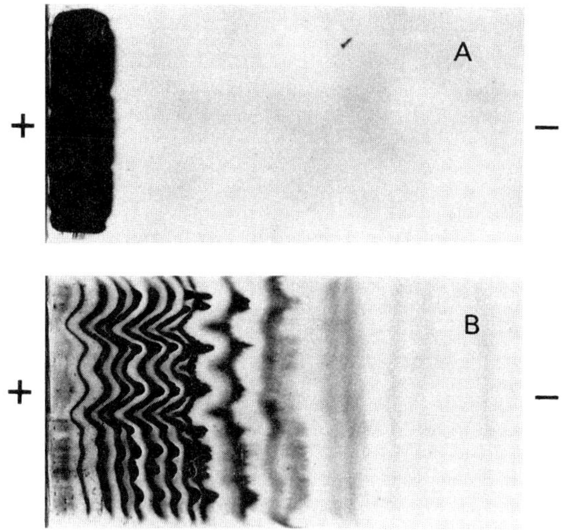

Figure 1. Separation of horse cytochrome *c* species with various extents of acetylation by preparative electrofocusing in the presence of 8 M urea. (A), before electrofocusing; (B), after electrofocusing.

separated as many red stripes (Fig. 1). The fractions of modified cytochrome *c* were cut out and the protein was eluted with deionized water. The protein solution was dialyzed against deionized water to remove the carrier ampholite and urea, and then concentrated.

Analytical electrofocusing was carried out with the LKB 2117 Multiphor on polyacrylamide gels 2 mm thick containing carrier ampholite covering a pH range of 3-10 and 8 M urea. The protein samples were loaded at the anode side and electrofocusing was carried out for 1 h at 24 W. The protein bands were stained with Coomassie brilliant blue R250.

B. Ionspray Ionization Mass Spectrometry

Ionspray ionization mass spectrometry was carried out using a Parkin-Elmer Sciex API-III triple quadrupole mass spectrometer equipped with an ionspary ion source, which was calibrated with ammonium adducts of polypropylglycol. Ionspray, interplate and orifice voltages were +5000, +650, and +90V, respectively. Protein samples of 1 μg, dissolved in 2 μl of 50% methanol and 1% formic

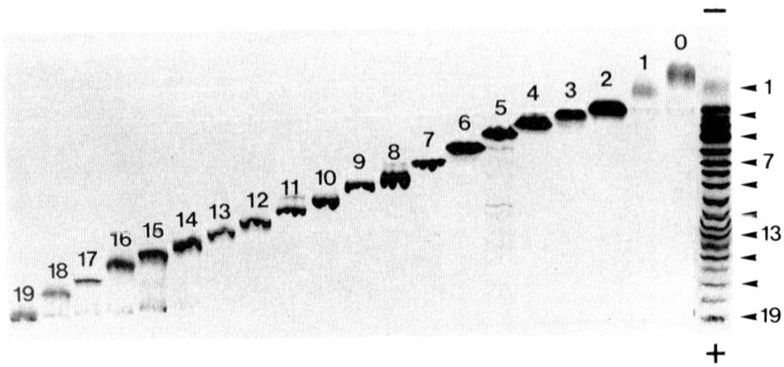

Figure 2. Analytical electrofocusing of acetylated horse cytochrome *c* molecules in the presence of 8 M urea. Figures indicate the numbers of acetylated amino groups. The right lane shows a mixture of species with various extents of acetylation.

acid, were applied through a HPLC injector (Waters 625 LC system) at flow rate of 40 µl/min.

C. Circular Dichroism Measurements

Circular dichroism (CD) spectra were measured with a Jasco spectropolarimeter, model J-500A. The CD spectra were measured at protein concentrations of 8 µM with a 2-mm cell from 250 to 195 nm.

The concentration of intact cytochrome *c* was determined from the extinction coefficient of 1.06 x 10^5 M^{-1} cm^{-1} at 410 nm at pH 7. Those of the modified species were determined assuming that the extinction coefficient of pyridine hemochrome, 29,000 M^{-1} cm^{-1} at 550 nm, is independent of acetylation (10).

III. Results

A. Preparation of Acetylated Cytochrome c Species

Figure 2 shows the analytical electrofocusing of the purified acetylated cytochrome *c* species. Taking advantage of the visible

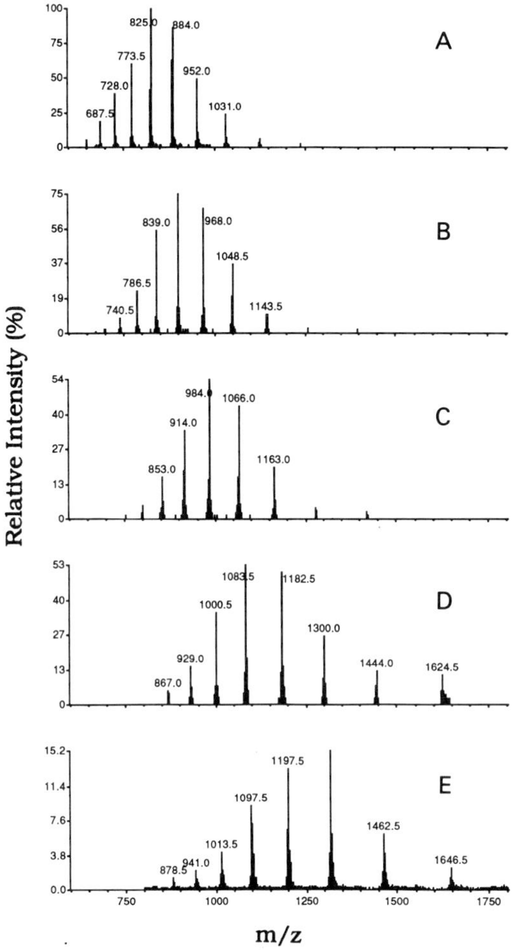

Figure 3. Ionspray ionization mass spectra of the purified acetylated cytochrome *c* species. Numbers of acetylated lysyl amino groups were 0 (A), 5 (B), 10 (C), 15 (D), and 19 (E).

color of cytochrome *c*, we were able to purify all of the 19 species with different extents of acetylation. Since the pI of intact cytochrome *c* is 10.1, the pI is expected to decrease to as low as 4 for the 19-acetylated species on the basis of the intrinsic pK_a values (11), although we did not determine the pI values. It should be mentioned that the purified species can be a mixture of various molecules with the same number of acetylations, but at different sites.

Acetylation of an amino group increases the molecular mass by 42 Da. To confirm the number of acetylations, we measured the molecular mass by ionspray ionization mass spectrometry (12,13), which has an accuracy in the range of ±0.01 to 0.05%. Figure 3 shows the ionspray ionization mass spectra of cytochrome *c* with various

Table 1. Identification of the number of acetylated amino groups by electrofocusing and ionspray ionization mass spectrometry[a]

Estimated number of acetylations	Estimated M.M.	Measured M.M. ± S.D.	ΔM.M.	Most Probable Charge
0	12360.1	12360.0 ± 2.0	-0.1	15
1	12402.1	12403.8 ± 3.1	1.7	15
2	12444.2	12442.8 ± 3.2	-1.4	15
3	12486.2	12487.0 ± 0.9	0.8	14
4	12528.2	12527.7 ± 1.9	-0.6	15
5	12570.3	12570.7 ± 1.6	0.4	14
6	12612.3	12611.3 ± 3.7	-1.0	14
7	12654.4	12652.6 ± 3.2	-1.8	13
8	12696.4	12695.9 ± 1.6	-0.5	13
9	12738.4	12737.7 ± 2.2	-0.8	13
10	12780.5	12778.6 ± 2.8	-1.9	13
11	12822.5	12821.5 ± 3.0	-1.0	13
12	12864.5	12863.7 ± 1.5	-0.8	12
13	12906.6	12906.7 ± 2.8	0.1	12
14	12948.6	12946.7 ± 2.9	-2.0	12
15	12990.7	12990.8 ± 3.1	0.1	12
16	13032.7	13031.1 ± 3.9	-1.6	11
17	13074.7	13074.5 ± 4.7	-0.3	11
18	13116.8	13117.6 ± 5.1	0.9	11
19	13158.8	13160.1 ± 3.1	1.3	10

[a]M.M. and S.D. indicate the molecular mass and standard deviation, respectively. ΔM.M. indicates the difference between estimated and measured M.M. The numbers of acetylations were first estimated by electrofocusing, and then confirmed by agreement of the estimated molecular mass with the average molecular mass determined from the mass spectra.

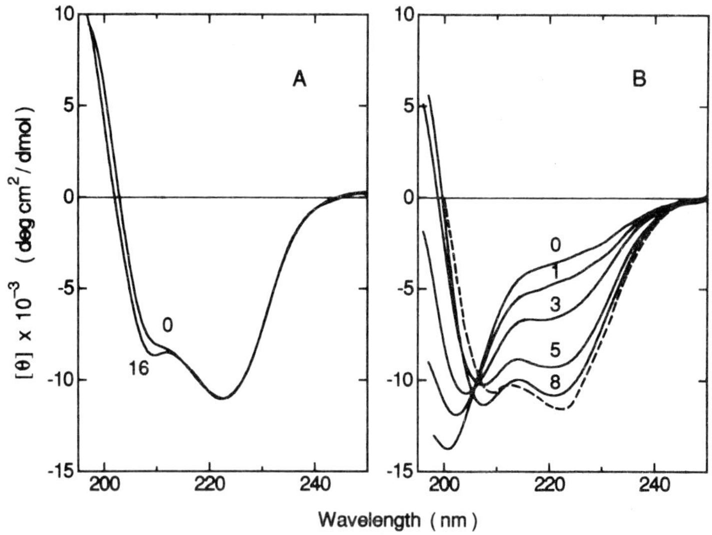

Figure 4. Far-UV CD spectra of the intact and acetylated cytochrome *c* in 10 mM phosphate buffer (pH 7.0) (A) and in 18 mM HCl (pH 1.8) (B) at 20°C. The figures indicate the numbers of acetylated amino groups. Dotted line in B shows the CD spectrum of the molten globule of intact cytochrome *c* in 18 mM HCl (pH 1.8) and 0.5 M NaCl.

extents of acetylation. The maximum in charge distribution shifted from m/z 825 for the intact species to m/z 1300 for the 19-acetylated species. The increase in the most probable m/z value was as expected because the number of amino groups, being one of the probable protonation sites, decreases with the progress of acetylation. As shown in Table 1, most probable charge state shifted from +15 for the intact species to +10 for the 19-acetylated species. The measured values of molecular mass agreed well with the estimated values based on electrofocusing.

B. Conformation of Acetylated Species

Figure 4 shows the far-UV CD spectra of the acetylated cytochrome *c* species. The far-UV CD spectrum at pH 7 was not affected by acetylation, indicating that the secondary structure was independent of acetylation (Fig. 4a). The sharp peaks at around 280

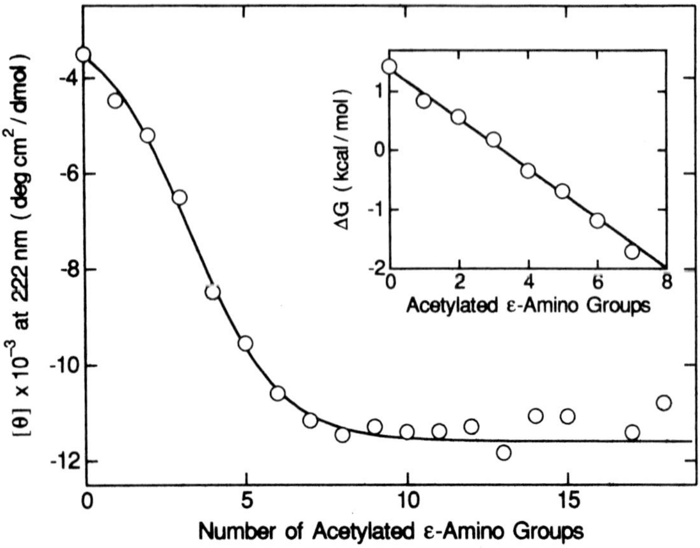

Figure 5. Dependence on the degree of acetylation of the ellipticity at 222 nm in 18 mM HCl (pH 1.8) at 20°C. The inset shows the dependence on the degree of acetylation of ΔG_F (see the text). The curve was calculated on the basis of the linear relationship shown in the inset.

nm of the intact cytochrome *c* were also observed for the 16-acetylated species (not shown). Thus, acetylation does not affect the global structure of cytochrome *c* at pH 7.

On the other hand, the secondary structure in 18 mM HCl (pH 1.8) is critically dependent on acetylation (Fig. 4B). While the CD spectrum of the intact protein showed an absence of any significant ordered structure, the spectrum changed with the progress of acetylation to one with minima at 222 nm and 208 nm, representing the formation of α-helix. The spectra at different degrees of acetylation showed an isodichroic point at 205 nm, consistent with a two-state folding mechanism. The fine aromatic CD was not observed for the acetylated species (not shown). The far-UV CD spectrum of the 16-acetylated cytochrome *c* species was similar to that of the molten globule induced by 0.5 M NaCl at pH 2 (Fig. 4B).

Figure 5 shows the dependence of ellipticity at 222 nm on the extent of acetylation at 20°C. The acetylation induced a cooperative folding transition. The apparent midpoint of the transition was at 3.5-acetylations.

IV. Discussion

We have shown that, whereas intact cytochrome *c* is unfolded largely at pH 2 in the absence of salt, acetylation of amino groups induces the refolding transition to the molten globule state. These results can be explained simply if we assume that the acid-denatured cytochrome *c* consists of two conformational states, *i.e.* the fully unfolded (U) and molten globule (MG) states, and that the transition between the two states is determined by a balance of electrostatic repulsions and opposing forces. Although the present results indicate nothing about the opposing forces, we believe that hydrophobic interactions are the major forces involved (5,8).

The free energy change (ΔG_F) of formation of the molten globule can be calculated from the acetylation-induced folding transition measured by the ellipticity at 222 nm by

$$\Delta G_F = -RT\, ln \frac{[MG]}{[U]} = -RT\, ln \frac{[\theta]_U - [\theta]}{[\theta] - [\theta]_M} \qquad (2)$$

where $[\theta]_U$, $[\theta]_M$ are the ellipticities of the unfolded and molten globule states, respectively, $[\theta]$ is the ellipticity of the modified species, R is the gas constant, and T is the temperature in Kelvin. The inset of Fig. 5 shows the dependence on the degree of acetylation of the ΔG_F. The dependence can be approximated by a linear relationship. While the intercept on the left axis gives the ΔG_F of the intact cytochrome *c*, the slope indicates the contribution of a unit charge to ΔG_F. For the intact cytochrome *c*, the molten globule state is less stable than the unfolded state by 1.4 kcal/mol at pH 2 and 20°C. It was clarified that a positive charge destabilizes the molten globule by 400 cal/mol at pH 2 and 20°C.

Although charge repulsion is one of the important factors determining the conformational stability of proteins, there is little evidence to suggest that such repulsion is sufficient to produce protein unfolding. The present results provide a clear example of charge repulsions producing unfolding of the compact protein structure. Because a similar mechanism determines the stability of other acid-denatured proteins (5-8) , the Lys, Leu model peptide (14), and melittin (manuscript in preparation), charge repulsion appears to be of critical importance when the stability of a folded state is highly marginal.

V. Acknowledgement

We thank Dr. Y. Umeda, Takara Shuzo Co. Ltd. (Otsu, Japan) for carring out ionspray ionization mass spectrometry.

VI. References

1. Ohgushi, M. & Wada, A. (1983) *FEBS Letters* **164**, 21-24.
2. Ptitsyn, O. B. (1987) *J. Prot. Chem.* **6**, 273-293.
3. Kuwajima, K. (1989) *Proteins* **6**, 87-103.
4. Kim, P. S. & Baldwin, R. L. (1989) *Ann. Rev. Biochem.* **59**, 631-660.
5. Goto, Y. & Fink, A. L. (1989) *Biochemistry* **28**, 945-952.
6. Goto, Y. & Fink, A. L. (1990) *J. Mol. Biol.* **214**, 803-805.
7. Goto, Y., Calciano, L. J. & Fink, A. L. (1990) *Proc. Natl. Acad. Sci. U.S.A.* **87**, 573-577.
8. Goto, Y., Takahashi, N. & Fink, A. L. (1990) *Biochemistry* **29**, 3480-3488.
9. Jeng, M.-F., Englander, S. W., Elove, G. A., Wand, A. J. & Roder, H. (1990) *Biochemistry* **29**, 10433-10437.
10. Wada, K. & Okunuki, K. (1968) *J. Biochem.* **64**, 667-681.
11. Tanford, C. (1968) *Adv. Protein. Chem.* **23**, 121-283.
12. Bruins, A. P., Covey, T. R. and Henion, J. D. (1987) *Anal. Chem.* **59**, 2642-2646.
13. Smith, R. D., Loo, J. A., Barnaga, C. J., Edmonds, C. G. & Udseth, H. R. (1989) *J. Chromatogr.* **480**, 211-232.
14. Goto, Y & Aimoto, S. (1991) *J. Mol. Biol.* **218**, 387-396.

Calcium-Induced Folding of Troponin-C: Formation of Homodimeric and Heterodimeric Two-Site Domains From Synthetic Peptides

Gary S. Shaw
Robert S. Hodges
Brian D. Sykes
Department of Biochemistry and MRC Group in Protein Structure and Function
University of Alberta
Edmonton, Alberta Canada

I. Introduction

The muscle protein troponin-C (TnC) contains four calcium-binding sites arranged as a pair of two-site domains. In the absence of calcium the C-terminal domain, containing calcium-binding sites III and IV, is mostly unfolded. However, upon calcium binding this domain undergoes a significant conformational change to form a pair of pseudosymmetric calcium-binding sites, each arranged in a helix-loop-helix motif. In this calcium-bound state, x-ray crystallographic studies have revealed two common features of TnC and many other calcium-binding proteins; a short three-residue β-strand and numerous hydrophobic interactions at the interface of the two calcium-binding sites (Strynadka and James, 1989). These structural features are likely to have a strong influence on the calcium binding properties of the protein and the overall stability of the two-site domain.

One approach to study the calcium-induced conformational change in TnC is to utilize synthetic peptides which encompass a single helix-loop-helix calcium-binding site (Reid *et al.*, 1981). In particular, we have recently shown, using two-dimensional nuclear magnetic resonance (2D NMR) spectroscopy that calcium-binding to a 34-residue peptide analogue of site III in TnC (SCIII) results in a significant conformational change to form the helix-loop-helix motif and a further peptide association to yield a symmetric two-site SCIII homodimer similar in structure to the C-terminal domain of TnC (Shaw *et al.*, 1990). A comparison of the interactions which occur at the interface of the two helix-loop-helix "subunits" in the SCIII homodimer revealed that many of these contacts were hydrophobic and analogous to those observed in the x-ray structures of TnC (Herzberg and James, 1988; Satyshur *et al.*, 1988). These observations suggest that the use of synthetic calcium-binding peptides represent a unique and practical method for studying the calcium-induced folding in calcium-binding proteins such as TnC. Further, by altering the amino acid composition of the peptides synthetically the importance of the site-site interactions which occur at the peptide-

TECHNIQUES IN PROTEIN CHEMISTRY III

peptide interface can be determined and utilized for the design of a synthetic calcium-binding protein with modified calcium affinity (Monera *et al.*, 1990). To accomplish this, we have expanded our initial studies of the site III TnC peptide (SCIII) to incorporate calcium binding to a mixture of SCIII and a 34-residue peptide representing site IV of TnC (SCIV). In this work we show, primarily by 2D NMR techniques, that calcium binding to this peptide mixture induces a conformational change in both peptides and a specific association to form a SCIII/SCIV heterodimer similar to the site-site association in TnC.

II. Experimental

A. Peptidc Synthesis and Purification. Two 34-residue peptides encompassing sites III and IV of chicken skeletal troponin-C were synthesized by stepwise solid-phase techniques (Hodges *et al.*, 1988) and purified by reverse phase HPLC. A detailed account of the synthesis and purification protocals is described elsewhere (Shaw *et al.*, 1991a). The sequences of the two peptides are given below:

```
     93        100       105       110       115       120       125
Ac-KSEEELANAFRIFDKNADGYIDIEELGEILRATG amide        SCIII
    129        135       140       145       150       155       160
Ac-VTEEDIEDLMKDSDKNNDGRIDFDEFLKMMEGVQ amide        SCIV
```

B. Preparation of Samples for NMR Experiments. Peptide solutions of SCIII for 2D NMR experiments were about 4 mM in 600 μl of either D_2O or H_2O, using 30 mM imidazole buffer and 50 mM KCl at pH 7.3. In the case of the SCIII/SCIV peptide mixture, 500 μl solutions of each peptide were made up in D_2O containing 50 mM KCl and diluted to 1 ml with D_2O. From each of these samples, 2 μl aliquots were removed and diluted to 25 μl with H_2O. A composite sample containing 10 μl of each of these solutions was analyzed by reverse-phase HPLC in order to determine the relative concentrations of SCIII and SCIV. Based on these results, the appropriate volumes of each peptide solution were combined, lyophilized and then redissolved in 85% H_2O/15 % D_2O to 500 μl giving a final concentration of each peptide of about 4 mM. The sample was made 10 mM in calcium and pH adjusted to 7.2. A similar SCIII/SCIV peptide solution was used for 1D NMR experiments but in 500 μl D_2O, 30 mM imidazole, 50 mM KCl and pH 7.30. In addition, this sample was initially made calcium-free using a heated chelex treatment (Shaw *et al.*, 1991a).

C. ^1H NMR Experiments. All 1D ^1H NMR experiments were performed at 30°C on a Varian VXR-500. Spectra were recorded using a spectral width of 6000 Hz and 2 sec acquisition time for 256 transients. A 2 sec presaturation pulse was used to suppress the HDO signal. 2D NMR experiments were acquired at 15°C and 30°C on a Varian VXR-500 or Varian Unity 600 spectrometer. For double quantum filtered correlated spectroscopy (DQF-COSY) experiments 64 transients for 256 t_1 increments were collected for SCIII and 32 transients for 320 increments were collected for SCIII/SCIV. Nuclear Overhauser effect spectroscopy (NOESY) were acquired in the phase sensitive mode (States *et al.* 1982) using a 20 ms random delay to suppress zero-quantum coherences. A total of 64 transients for each of 256 t_1 increments were acquired for SCIII and 32 transients for 320 t_1 increments were acquired for SCIII/SCIV each using 300 ms mixing period.

DQF-COSY spectra were processed using a sine bell weighing function for 4096 points in ω_1 and ω_2. NOESY spectra were processed using a

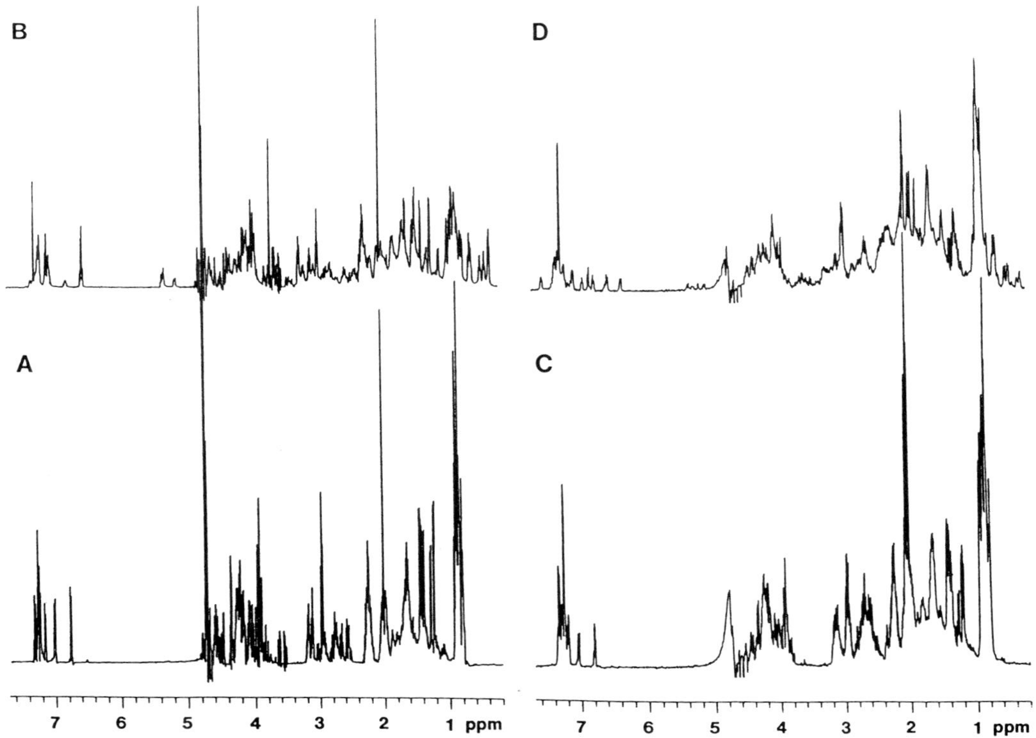

Figure 1. 500 MHz ^{1}H NMR spectra of (A) 430 μM apo-SCIII; (B) SCIII + Ca; (C) 160 μM apo-SCIII/SCIV and (D) SCIII/SCIV + Ca in D_2O, 50 mM KCl and 30 mM imidazole-d_4. The vertical scales of the apo-peptides (A, C) have been adjusted to give the same apparent peptide concentration. In each case the vertical scale for the Ca-peptides (B, D) is the same as for its apo-peptide (A, C).

combination of gaussian multiplication and apodization to ensure the fids decayed to zero at the end of the acquisition period.

III. Results and Discussion

The two 34-residue peptides of this study, SCIII and SCIV were shown to be homogeneous by a variety of analytical methods including reverse-phase HPLC , amino acid composition and mass spectroscopy. The sequence of SCIII (residues 93-126) has two amino acid substitutions from chicken skeletal TnC; position 101 where Ala has replaced Cys to prevent complications from oxidation and position 112 where Tyr has replaced Phe. This latter change was made so that Y112, which coordinates the calcium ion via its backbone carbonyl oxygen, could be positively identified in 1D NMR spectra. Peptide SCIV is identical to the native sequence of residues 129-162 in TnC.

In order to determine the calcium-induced conformational changes of SCIII and SCIII/SCIV a series of 1D ^{1}H NMR experiments in combination with 2D NMR experiments were used. The 1D ^{1}H NMR spectra of SCIII and SCIII/SCIV in the absence and presence of calcium are shown in Figure 1. Only the spectra for SCIII/SCIV will be discussed here as a detailed account for SCIII has been reported elsewhere (Shaw *et al.*, 1990; Shaw *et al.*, 1991b). In the absence of calcium the ^{1}H resonances for SCIII/SCIV are narrow and non-dispersive. Examination of the aromatic region of this peptide mixture reveals that the resonances for the four phenylalamine residues (F102, F105, F151 and F154) are centered near 7.30 ppm and the single tyrosine residue Y112 forms two doublets at 6.80 ppm and 7.05 ppm. These chemical shifts, as well as the congested methyl resonances centered at 0.80 ppm are consistent with the SCIII/SCIV peptide mixture adopting a "random coil" conformation. Further we have compared the ^{1}H NMR spectrum of apo-SCIII/SCIV with that of apo-SCIII and apo-SCIV and have found that the apo-SCIII/SCIV spectrum is a simply a composite of the individual SCIII and SCIV spectra. This suggests that little interaction occurs between the SCIII and SCIV peptides in solution in the absence of calcium.

The addition of calcium to each peptide solution results in significant changes throughout the ^{1}H NMR spectra of SCIII and SCIII/SCIV. In the case of SCIII we have previously shown that these arise from calcium binding, induction of a significant conformational change and the association of two peptides to form a two-site symmetric homodimer. Examination of the ^{1}H NMR spectrum of calcium-saturated SCIII/SCIV (Ca-SCIII/SCIV) shows that although this spectrum is significantly different from apo-SCIII/SCIV, it does not resemble that of calcium-saturated SCIII (Ca-SCIII). In particular the dispersive nature of the aromatic resonances and the αCH resonances in Ca-SCIII/SCIV are markedly different from Ca-SCIII. In analogy with the peptide association observed for SCIII (Shaw *et al.*, 1990; Shaw *et al.*, 1991b) we now provide evidence from 2D NMR experiments which show that the SCIII/SCIV peptide mixture also undergoes a calcium-induced conformational change and peptide association to form a two-site heterodimer.

In order to show the formation of the Ca-SCIII/SCIV heterodimer, specific interactions between residues at the interface of the two peptide subunits must be identified. This requires the complete sequential assignment of all resonances in the ^{1}H NMR spectra of Ca-SCIII and Ca-SCIII/SCIV. In this work the differences between the DQF-COSY and NOESY spectra of these two dimeric species were used to determine the interactions between sites. Specifically, three different regions of the spectra were used corresponding to interactions between aromatic residues, aromatics and aliphatics and those residues comprising a short β-strand between the sites.

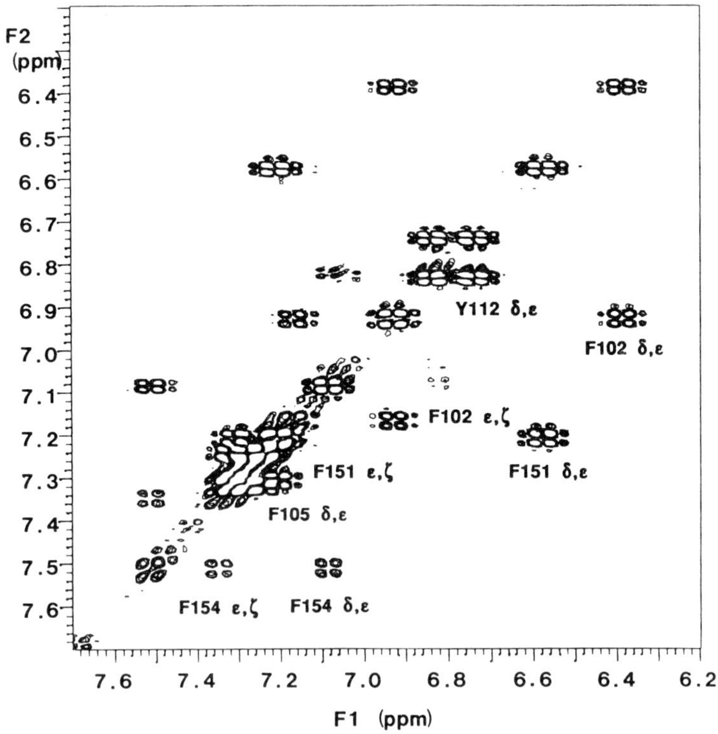

Figure 2. Aromatic region of DQF-COSY spectrum for Ca-SCIII/SCIV acquired at 600 MHz in 85% H_2O, 50 mM KCl, pH 7.2 and 15°C. The crosspeaks indicated have been derived from the sequence specific assignment of Ca-SCIII/SCIV.

The aromatic region of the DQF-COSY spectra of Ca-SCIII/SCIV is shown in Figure 2. This spectrum shows well resolved crosspeaks which have been assigned to F102 δ,ε; F151 δ,ε; F102 ε,ζ; F154 δ,ε; F154 ζ,ε and Y112 δ,ε. In addition, two crosspeaks near the diagonal, corresponding to F105 δ, ε and F151 ε, ζ, have been identified. A comparison of the chemical shifts for these residues to those determined for TnC (Tsuda *et al.*, 1988; Drabikowski *et al.*, 1985) reveals an excellent agreement (Shaw *et al.*, 1991c). This suggests that in principle calcium addition to the SCIII/SCIV peptide mixture induces a conformational change in the peptides and the formation of a SCIII/SCIV heterodimer.

Further evidence for the formation of a SCIII/SCIV heterodimer can be derived from the aromatic region of the NOESY spectrum shown in Figure 3. For each aromatic residue crosspeaks (nOes) representing δ,ε and ε,ζ interactions are present, analogous to these found in the DQF-COSY spectrum in Figure 2. However in the NOESY spectrum many other crosspeaks are also present. These arise from short "through-space" interactions corresponding to short internuclear distances between protons on neighboring amino acids, usually the result of the fold of the protein. In this case, several nOes are present between F102, F151 and F154. In particular interactions are found for F102 δ and F151 δ, F151 ε, F154 δ, and F154 ε; F102 ε and F154 δ and F151 δ; F151 δ and F154 δ; and F151 ζ and F154 ε. These observations show that these three aromatic residues

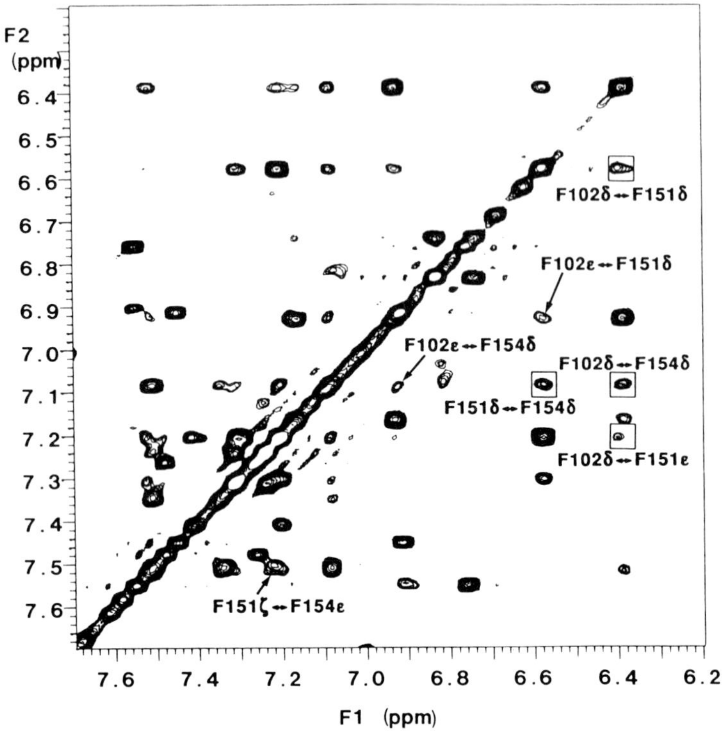

Figure 3. Aromatic region of NOESY spectrum for Ca-SCIII/SCIV acquired at 600 MHz in 85% H_2O, 50 mM KCl, pH 7.2 and 15°C. Inter-residue crosspeaks are either boxed or indicated with an arrow. Intra--residue crosspeaks are not marked but are easily identified by comparison with Figure 2.

are all close in space even though F102 resides in SCIII and F151 and F154 are in SCIV. A comparison of these interactions with those observed by Drabikowski *et al.* (1985) for the C-terminal domain of TnC also shows a remarkable similarity. These results are as expected for a calcium-induced association of SCIII/SCIV to form a heterodimer similar in structure as the C-terminal domain of TnC. In the x-ray structures of TnC (Herzberg and James, 1988; Satyshur *et al.*, 1988) F102, F151 and F154 are integral members of a hydrophobic core at the interface of sites III and IV. In this arrangement F102 packs against F154 while F151 stacks against these two residues such that the distances between the centre of the F102 aromatic ring and the F151 and F154 rings are less than 6Å. The observation of nOes between the aromatic rings of these residues suggests a similar packing of aromatic rings in SCIII/SCIV is also possible.

Calcium binding to the TnC peptides SCIII and SCIV results in the specific formation of a SCIII/SCIV heterodimer with little or no formation of the SCIII or SCIV homodimeric species. This result is similar to that obtained for tropomyosin coiled coils (Lehrer *et al.*, 1989; Lehrer and Qian, 1990) and the Fos and Jun leucine zipper peptides (O'Shea *et al.*, 1989). In both of these cases preferential heterodimer formation results from the relative instability of one of the homodimeric species compared to the heterodimer or the second homodimer. This is likely the case for SCIII and SCIV as preliminary studies have shown that the SCIV homodimer is quite unstable compared to either the SCIII homodimer or the SCIII/SCIV heterodimer. As with the tropomyosin and Fos/Jun dimer studies this thermodynamically suggests that the SCIII/SCIV heterodimer will be preferentially formed if $\Delta G_{SCIII/SCIV} - [\Delta G_{SCIII/SCIII} + \Delta G_{SCIV/SCIV}] < 0$. This hypothesis is currently being investigated through guanidine denaturation studies of the three calcium saturated peptides dimers

Acknowledgements

We would like to thank Paul Semchuk for his help in the synthesis and purification of the peptides. We are also grateful to Sue Smith for typing this manuscript. This study was supported by a grant from the Medical Research Council of Canada and an Alberta Heritage Foundation for Medical Research Fellowship (G.S.S.).

References

Drabikowski, W., Dalgarno, D. C., Levine, B. A., Gergely, J., Grabarek, Z. and Leavis, P. C. (1985) *Eur. J. Biochem.* **151,** 17.

Herzberg, O. and James, M. N. G. (1988) *J. Mol. Biol.* **203,** 761.

Hodges, R. S., Semchuk, P. D., Taneja, A. K., Kay, C. M., Parker, J. M. R. and Mant, C. T. (1988) *Peptide Research* **1,** 19.

Lehrer, S. S., Qian, Y. and Hvidt, S. (1989) *Science* **246,** 926.

Lehrer, S. S. and Qian, Y. (1990) *J. Biol. Chem.* **265,** 1134.

Monera, O., Semchuk, P. S., Oikawa, K., Shaw, G. S., Strynadka, N. J., Kay, C., Sykes, B. D. and Hodges, R. S. (1991). *In* "Peptides 1990" (E.Giralt and D. Andreu, eds.) p 570, ESCOM Science Publ., Leiden, The Netherlands.

O'Shea, E. K, Rutkowski, R., Stafford, W. F. and Kim, P. S. (1989) *Science* **245,** 646.

Reid, R. E., Gariépy, J., Saund, A. K. and Hodges, R. S. (1981) *J. Biol. Chem.* **256,** 2742.

Satyshur, K. A., Rao, S. T., Pyzalska, D., Drendel, W., Greaser, M. and Sundaralingam, M. (1988) *J. Biol. Chem.* **263,** 1628.

Shaw, G. S., Hodges, R. S. and Sykes, B. D. (1990) *Science* **249,** 280.

Shaw, G. S., Hodges, R. S. and Sykes, B. D. (1991a) *Biochemistry*, in press.

Shaw, G. S., Golden, L. F., Hodges, R. S. and Sykes, B. D. (1991b) *J. Am. Chem. Soc.*, in press.

Shaw, G. S., Hodges, R. S. and Sykes, B. D. (1991c). *In* "Peptides : Chemistry and Biology" (J. A. Smith and J. E. Rivier, eds.) ESCOM Science Publ., Leiden, The Netherlands, in press.

States, D. J., Haberkorn, R. A. and Rubin, D. J. (1982) *J. Mag. Res.* **48,** 286.

Strynadka, N. C. J. and James, M. N. G. (1989) *Ann. Rev. Biochem.* **58,** 951.

Tsuda, S., Hasegawa, Y., Yoshida, M., Yagi, K. and Hikichi, K. (1988) *Biochemistry* **27,** 4120.

The Role of Aqueous Solvation in Protein Folding: Insights from Aqueous-Alkane Partitioning

Jonathan M. Keske, J. Malcolm Bruce[1], and P. Leslie Dutton
Dept. of Biochem. and Biophys., Univ. of Pennsylvania, Philadelphia, PA 19104
and
[1]Dept. of Chem., Univ. of Manchester, Manchester M13 9PL United Kingdom

I. Abstract

The interaction of a protein with water can be considered to be similar in nature to that of a multi-substituted alkane. The energies involved in the movement out of water of such compounds have been estimated by a number of methods [1-15], but without regard to the individual lone pair basicities, electrostatic and hydrophobicity contributions. Further refinement of the solvation analysis methodology developed previously [16], provides, we believe, a potentiality for new insights into the energetics of protein folding and protein-cofactor/ substrate/ drug interactions.

II. Introduction

Transfer free energies have been utilized to determine the energetic cost of removal of protein sidechains from water to the protein interior, which is commonly ascribed as being similar to n-octanol in its ability to solvate the sidechain [3-5]. Although recent mutational studies on protein stability [17-19] seem to provide support for an alcoholic protein interior, the octanol-water transfer process is to date poorly understood. One may consider protein folding as occurring in two chemically dissimilar regions namely, the protein-water interface and the bulk protein interior. Previous researchers have analyzed the statistical distribution of protein sidechains within these two regions [10,11and 20] and not surprisingly, found a correlation with the "hydrophobic" character of the sidechain. Thus, it is our purpose in this manuscript to further characterize the octanol-water transfer process and to quantitatively highlight the interfacial behavior of protein sidechains.

The decrease in accessible surface area accompanying protein folding is believed to contribute substantially to the overall stability of the folded structure. Although this is almost certainly the case, methodologies based solely upon changes in surface area fail to recognize the uniqueness of the

TECHNIQUES IN PROTEIN CHEMISTRY III

folded structure since the free energy gained from removal of a portion of the protein surface area does not necessarily dictate any specific three-dimensional structure. The energetic gain upon removal of protein surface area from an aqueous environment is dependent on the initial interfacial tension of the region buried and not on the bulk transfer process of passing from an aqueous to a alkane (alkanol) environment. Assuming a near uniform character of the protein surface in its unfolded state it is possible to derive a reasonable estimate of its interfacial tension with water, which according to initial investigation appears to be on the order of 40-50 cal mol^{-1} A^{-2}, this value being significantly less then the interfacial tension of water and alkane (72 cal mol^{-1} A^{-2}) [21, 22]. The ability of the backbone peptide groups and the hydrophilic sidechains to lower the interfacial tension between the aqueous environment and the protein cavity has the effect of reducing the stability of the folded state to an extent necessary to ensure a characteristic three-dimensional fold, as well as, providing for suitable solubility in water. It is fortuitous that previous researchers chose octanol as the reference state, although this does not necessitate that the protein interior be like octanol in its ability to interact specifically with protein residues.

Our previous work clearly demonstrates that the strengths of hydrogen bonds cannot be represented by a potential function which simply treats the interactions as occurring between oxygen-oxygen, nitrogen-oxygen and nitrogen-nitrogen pairs but must also incorporate the differential proton affinity of the two groups [23-25]. Secondly, the high dielectric of water induces a large image contribution to the electrostatic stabilization of ions and dipoles near the protein-water interface [26]. Therefore from both an electrostatic and H-bonding perspective it is a logical extension to segregate the protein into the two distinct regions of solvation discussed above.

Typically, the characteristics of the protein interior are approximated by selectively choosing an immiscible secondary phase from which to derive a transfer free energy (ΔG_{tr}) suitable for estimation of the free energy of stabiliziation (ΔG_s). It is our contention that a distinction must be made between the energetics of movement of a residue out of water (ΔG_{tr}) and its contribution to the overall stability of the protein if protein stability is to be understood at the atomic level.

III. Methodology

The current approach is a further refinement of the solvation analysis methodology published previously [16]. These refinements are as follows:

a) Cavity: The use of a Lee and Richards type procedure to estimate the free energy of cavity formation (ΔG_{cav}), from correlation of water-hexane ΔG_{tr} for a series of n-alkanes as a funtion of their solvent accessible surface area as shown in **Figure 1A**, from which a slope of 28.7 cal mol^{-1} A^{-2} was obtained.

b) Electrostatics: The excess electrostatic contribution (ΔG_{ele}) is derived from the water -hexane ΔG_{tr} of a select series of non-hydrogen bonding compounds rather than approximated from a pseudo chloroform (diethyl ether)-hexane transfer process, which appears to underestimate the electrostatic

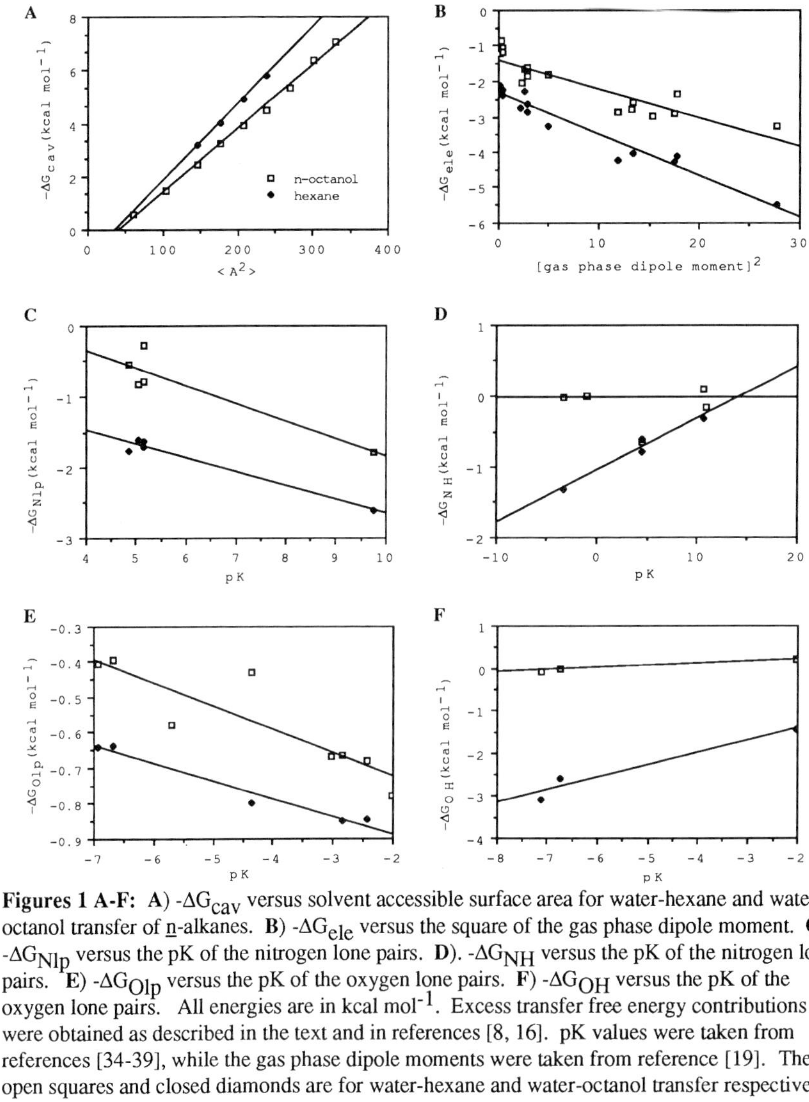

Figures 1 A-F: **A**) $-\Delta G_{cav}$ versus solvent accessible surface area for water-hexane and water-octanol transfer of n-alkanes. **B**) $-\Delta G_{ele}$ versus the square of the gas phase dipole moment. **C**) $-\Delta G_{Nlp}$ versus the pK of the nitrogen lone pairs. **D**). $-\Delta G_{NH}$ versus the pK of the nitrogen lone pairs. **E**) $-\Delta G_{Olp}$ versus the pK of the oxygen lone pairs. **F**) $-\Delta G_{OH}$ versus the pK of the oxygen lone pairs. All energies are in kcal mol^{-1}. Excess transfer free energy contributions were obtained as described in the text and in references [8, 16]. pK values were taken from references [34-39], while the gas phase dipole moments were taken from reference [19]. The open squares and closed diamonds are for water-hexane and water-octanol transfer respectively.

contribution to the water-hexane transfer. Reaction field theory [28-30] implies that the dipolar interaction strength should have a quadratic dependence on the dipole moment; hence, ΔG_{ele} was plotted against the square of the gas phase dipole moment as illustrated in **Figure 1B** (m = -0.12 kcal mol^{-1} D^{-2}, b = -2.38 kcal mol^{-1} and $R^2 = 0.96$) . Further refinements on this work are currently under way so as to separately address the electrostatic contributions arising from polarizability and dipolar interactions.

c) **Hydrogen bonding**: It was necessary to reevaluate, as a result of the changes emanating from refinements **a**) and **b**), the functional dependence of the hydrogen bond strength on lone pair basicity. The current status of the hydrogen bond strength (ΔG_{Olp}, ΔG_{Nlp}, ΔG_{OH}, and ΔG_{NH}) versus lone pair basicity (pK) correlations are as follows: The oxygen and nitrogen lone-pair interactions with the water 'H' are ascribed by a free energetic change of -0.050 Kcal mol^{-1} per unit increase in pK and -0.195 kcal mol^{-1} per unit increase in pK, respectively, while the interaction of a 'H' attached to a oxygen or nitrogen heteroatom with the water 'O' is ascribed by a free energetic change of +0.286 and +0.073 kcal mol^{-1} per unit increase in pK respectively; these are shown in **Figures 1C-F**.

The differences arising between heteroatomic lone pair and associated hydrogen interaction strength dependence on lone pair basicity is in part due to a decreased entropic contribtution for multiple hydrogen bonding, therefore, it might be expected that given sole hydrogen bonding to the attached 'H' the apparent differences in functional dependence may diminish. Future work will contemplate the role of bond structure (eg. ketones vs. ethers) in adjusting the entropy component of the hydrogen bond.

A similar methodological approach was undertaken to examine water-vapor (not shown) and water-octanol, shown as the open squares in **Figures 1A-F**, transfer processes as well.

IV. Discussion

The viability of our method to provide reasonable estimates of the atomic level contributions is demonstrated by its ability to accurately predict ΔG_{tr} of amino acid side chain analogs from water to any of a number of different secondary phases, eg. vapor, alkane, and octanol, as illustrated in **Figures 2A and B**. **Table I** presents the water-hexane and water-octanol ΔG_{tr} data, both calculated and experimental used in the generation of **Figures 2A and B.** Proline was excluded from the results due to the uncertainty of its structural definition.

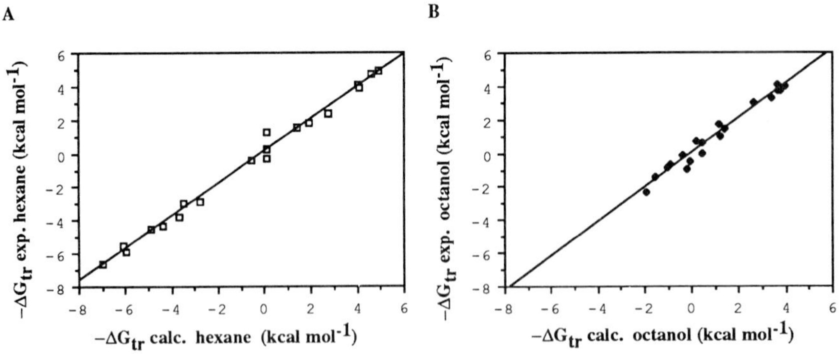

Figure 2: **A**) Calculated -ΔG_{tr} for water-hexane transfer of amino acid sidechain analogs versus their experimentally determined -ΔG_{tr} [6, 41]. **B**). Calculated -ΔG_{tr} for water-n-octanol transfer of amino acid sidechain analogs versus their experimentally -ΔG_{tr} [6, 9, 41]. The slope, intercept and squared regression coefficient from linear regression analysis of the data sets were 0.97, 0.09 and 0.992, and 0.94, 0.025 and 0.970 respectively.

Table I

Amino Acid Analogs	$-\Delta G_{tr}$ calc hex.	$-\Delta G_{tr}$ hex [a,b]	$-\Delta G_{tr}$ calc oct.	$-\Delta G_{tr}$ oct[b]
Ala (methane)	1.90	1.81	1.46	1.38
Arg (n-prguanidinium)	-5.99	-5.89	-2.31	-1.93
Asn (acetamide)	-6.99	-6.61	-1.42	-1.56
Asp (acetic acid)	-4.37	-4.32	-0.94	-0.23
Cys (methanethiol)	0.12	1.27	0.65	0.17
Gln (propionamide)	-6.10	-5.55	-0.68	-0.89
Glu (propionic acid)	-3.48	-2.99	-0.04	0.45
Gly (H)	0.13	0.27	0.58	0.44
His (imidazole)	-4.87	-4.56	-0.51	-0.10
Ile (butane)	4.93	4.94	3.98	3.95
Leu (2-methylpropane)	4.59	4.74	3.69	3.63
Lys (butylamine)	-0.53	-0.41	0.96	1.19
Met (dimethyl sulfide)	1.43	1.49	1.75	1.12
Phe (toluene)	4.05	3.90	4.04	3.61
Ser (methanol)	-3.65	-3.82	-0.84	-1.05
Thr (ethanol)	-2.78	-2.91	-0.11	-0.42
Trp (3-methylindole)	2.73	2.33	3.75	3.78
Tyr (4-methylphenol)	0.09	-0.35	2.99	2.59
Val (propane)	4.05	4.05	3.25	3.37

Calculated and experimental transfer free energies (ΔG_{tr}) for water-hexane and water-octanol transfers for amino acid sidechain analogs, energies are in kcal mol^{-1}. Calculated values were obtained by the procedure described in the text and utilizing the calibration lines presented in **Figure 1 A-F**. [a,b] The experimental transfer free energies were taken from references [6, 9, 41].

Comparing the results of our analysis of the water-hexane and water-octanol transfer process (**Figures 1A-F**) affords an opportunity to clearly address the nature of octanol solvation, from which the following observations can be made:

a) Water saturated octanol ($octanol_{aq}$) achieves approximately 90% of the hydrophobic effect obtained by water saturated hexane ($hexane_{aq}$); studies with dry octanol suggest that the diminished hydrophobic effect in octanol is attributable to the presence of a substantial amount of water (2.3 M) in $octanol_{aq}$ [13, 14].

b) $Octanol_{aq}$ is able to contribute approximately 40% of the electrostatic solvation present in pure water, again largely due to the presence of molar quantities of water. Preliminary reaction field theory calculations suggests that dry octanol should only contribute roughly 10% of the electrostatic contribution of water [29, 30].

c) The lone pairs of either oxygen or nitrogen compounds, seem for the most part to be selectively desolvated within octanol ($\Delta\Delta G_{Nlp}$(oct-hex) approx. -1 kcal mol^{-1}; $\Delta\Delta G_{Olp}$(oct-hex) approx. -0.25 kcal mol^{-1}); in contrast to the heteroatomic attached hydrogens which appear to be fully solvated. The pK dependence of the water-octanol lone pair transfer free energy contribution implies that the functional dependence of the hydrogen bond interaction with lone pair basicity between solute lone pairs and water and those with octanol are different. This difference may arise as the result of an increased steric repulsion between the alkyl chain of octanol and the alkyl group of the solute. Whether these results are influenced by the presence of water has not as yet been considered.

V. Conclusions

From our analysis it is our conclusion that the use of octanol as a suitable secondary phase from which to estimate the energetic contribution of an amino acid residue to the protein stability is not warranted. This is above and beyond the problems implicit in accurately deriving transfer free energies from a transfer process which inherently does not obey the Nernst equation [12, 32]. It is not our intention to simply promote the exchange of partition coefficients obtained with one system, namely octanol, with those of another, eg. hexane, in the estimation of ΔG_S, since it is has been shown that most ΔG_{tr} regardless of their source correlate with ΔG_S to about the same degree of accuracy, but to consider the fundamental forces, hydrophobicity, H-bonding, and electrostatics, that underlie the folding or binding process [1,2-4, 16 and 33].

The energetics of transfer of phenol through a water-vapor interface calculated from the procedures described above are presented as a model for tyrosine in **Figure 3**, which illustrates the orientational and positional dependence of the transfer process [31]. The orientational and positional dependence of ΔG_{tr} has important implications as to the nature of the energetics of sidechain configuration within the three dimensional protein structure, since it is readily apparent from **Figure 3** that there is a large energetic torque favoring the orientation of the phenol such that its hydroxyl group is localized

A

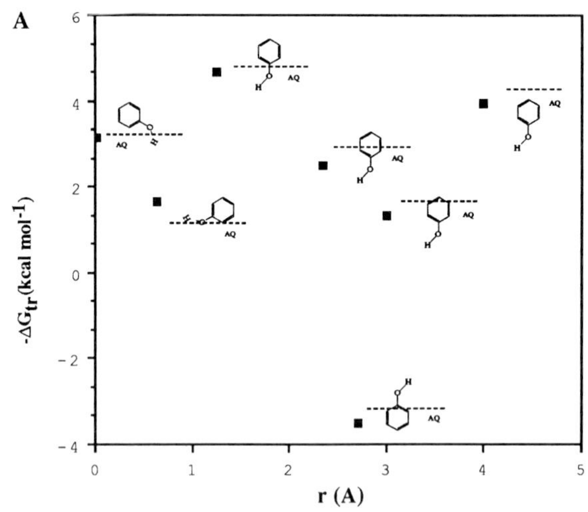

Figure 3: $-\Delta G_{tr}$ calculated for the interfacial movement of phenol from the vapor phase to water as a function of the extent and orientation of its burial, energies are in kcal mol^{-1}. The calculations were carried out using the methodologies presented in the text, although, a corrected water-vapor interfacial tension of 78.8 cal mol^{-1} A^{-2} was used in place of the bulk cavity value[42-45]. The correction was necessary to remove the rotational and translational entropic contributions to the interfacial tension [46].

within the water-vapor interface. Furthermore it is possible utilizing the techniques described previously to make a quantitative distinction between the solvation of the hydroxyl's hydrogen and oxygen lone pairs.

With the refinements described herein to the solvation analysis method presented by the authors previously [16], it is now possible to ascribe quantitative values for the contributions of hydrophobicity, electrostatics, and H-bonding to the overall solvation of a molecule in a multiplicity of solvent environments.

References

1. E. J. Cohn, and J. T. Edsall, Proteins, Amino Acids and Peptides, chap. 9, Reinhold Publ. Corp., New York 1943.
2. W. Kauzmann, (1963) Adv. Protein Chem., **14**, 1.
3. D. Eisenberg and A. D. McLachlan, (1986) Nature, **319**, 199.
4. C. Tanford, (1962) J. Am . Chem. Soc., **84**, 4240.
5. E. J. Cohn, T. L. McMeekin, J. T. Edsall, and J. H. Weare, (1934) J. Am. Chem. Soc., **86**, 2770.
6. C. Hansch and A. Leo, Substituent Constants for Correlation Analysis in Chemistry and Biology, John Wiley & Sons, New York 1979.
7. D. J. Abraham and A. Leo, (1987) Proteins: Structure, Function, and Genetics, 130.
8. S. H. Fleischmann and C. L. Brooks III, (1987) J. Chem. Phys., **87**, 3029.
9. W. J. Dunn III, M. G. Koehler, and S. Grigoras, (1987) J. Med. Chem., **30**, 1121.
10. C. Chothia, (1974) Nature, **248**, 338.
11. J. Janin, (1979) Nature, **277**, 491.
12. A. Ben-Naim and Y. Marcus, (1984) J. Chem. Phys., **81**, 2016.
13. P. Berti, S. Cabani, G. Conti, and V. Mollica, (1986) J. Chem. Soc. Faraday Trans. I, **82**, 2547.
14. M. H. Abraham, (1982) J. Am. Chem. Soc., **104**, 2085.
15. J. Hine and P. K. Mookerjee, (1975) J. Org. Chem., **40**, 292.
16. J. M. Keske, J. M. Bruce, and P. L. Dutton, (1990) Z. Naturforsch., **45c**, 430.
17. M. Matsumura, W. J. Becktel, and B. W. Matthews, (1988) Nature, **334**, 406.
18. J. T. Kellis, K. Nyberg, and A. R. Fersht, (1989) Biochem., **28**, 4914.
19. D. Shortle, W. E. Stites, and A. K. Meeken, (1990) Biochem., **29**, 8033.
20. G. D. Rose, A. R. Geselowitz, G. J. Lesser, R. H. Lee, and M. H. Zehfus, (1985) Science, **229**, 834.
21. K. Sharp, A. Nicholls, R. F. Fine, and B. Honig, (1991) Science, **252**, 106.
22. L. R. DeYoung and K. A. Dill, (1990) J. Phys. Chem., **94**, 801.
23. E. M. Arnett, E. J. Mitchell, and T. S. S. R. Murty, (1974) J. Am. Chem. Soc., **96**, 3875.
24. D. Gurka and R. W. Taft, J. Am. Chem. Soc., **91**, 4794, (1969).
25. R. W. Taft, D. Gurka, L. Joris, P. R. von Scheyler, and J. W. Rakshys, (1969) J. Am. Chem. Soc., **91**, 4801.
26. B. H. Honig, W. L. Hubbell, and R. F. Flewelling, (1986) Ann. Rev. Biophys. Biophys. Chem., **15**, 163.
27. B. Lee and F. M. Richards, (1971) J. Mol. Biol., **55**, 379.
28. L. Onsager, (1971) J. Am. Chem. Soc., **58**, 1486.
29. C. J. F. Bottcher, Theory of Electric Polarization, 2nd ed., vol. I, Elsevier Scientific Publishing Company, Amsterdam 1973.

30. C. P. Smyth, Molecular Interactions, vol. 2, chap. 7, H. Ratajczak and W. J. Orville-Thomas ed., John Wiley & Sons Ltd, New York 1980.
31. A. Pohorille and I. Benjamin, (1991) J. Chem. Phys., **94**, 5599.
32. J. Stecki, (1990) J. Phys. Chem., **94**, 3145.
33. W. P. Jencks, (1975) Adv. Enzymol., **43**, 219.
34. E. M. Arnett, (1963) Prog. Phys. Org. Chem., **1**, 223.
35. CRC Handbook of Biochemistry-Selected Data for Molecular Biology (H. A. Sober, ed.), 2nd ed., pp. J-202-J-221, Chemical Rubber Co., Cleveland, OH. 1970.
36. G. Perdonicn, and G. Scorrano, (1977) J. Am. Chem. Soc. **99**, 6983.
37. A. Levi, G. Modena, and G. Scorrano, (1974) J. Am. Chem. Soc., **96**, 6585.
38. A. Bagno, V. Lucchini, and G. Scorrano, (1991) J. Phys. Chem., **95**, 345.
39. A. Bagno and G. Scorrano, (1988) J. Am. Chem. Soc., **110**, 4577.
40. CRC Handbook of Chemistry and Physics (R. C. Weast, ed.), 66th ed., pp. E-58-E-61, CRC Press Inc., Boca Raton, FL.,1985.
41. A. Radzicka, and R. Wolfenden, (1988) Biochem., **27**, 1664.
42. R. A. Pierotti, (1965) J. Phys. Chem., **69**, 281.
43. J. P. Postma, H. J. Berendsen, and J. R. Haak, (1982) Faraday Symp. Chem. Soc., **17**, 55.
44. C. Y. Lee, J. A. McCammon, P. J. Rossky, (1984) J. Chem. Phys., **80**, 4448.
45. W. D. Harkins, and Y. C. Cheng, (1921) J. Am. Chem. Soc., **43**, 35.
46. D. H. Wertz, (1980) J. Am. Chem. Soc., **102**, 5316.

Acknowledgements

This work has been supported by the National Science Foundation (DMB-8817240) and the Public Health Service (GM 27309).

MOLECULAR CHARACTERIZATION OF AN AGGREGATE FORMED BY A BOVINE GROWTH HORMONE FOLDING INTERMEDIATE

S. Russ Lehrman,[1] Jody L. Tuls, and Henry A. Havel[2]

Control Development, The Upjohn Company,
Kalamazoo, MI 49001

I. INTRODUCTION

Bovine growth hormone (bGH) aggregates following partial denaturation at elevated concentrations (1). Human growth hormone (hGH), on the other hand, does not aggregate under these conditions (2). The difference in the folding behavior of these homologous proteins may account for the observation that hGH does not necessarily form inclusion bodies (3). Folding studies of bGH indicate that the formation of the bGH folding intermediate and its aggregation are separate processes, implying that selective protein modifications may lead to their independent modulation (4). We are interested in determining the structural features of bGH that lead to chemically defined aggregation. Previous studies have suggested that the third α-

[1] Correspondence should be directed to this author at: The Upjohn Company 4861-259-12, 7171 Portage Road, Kalamazoo, MI 49001.
[2] Present address: Eli Lilly and Company, Indianapolis, Indiana 46285-0835

TECHNIQUES IN PROTEIN CHEMISTRY III

helix of bGH, spanning amino acid residues 106 to 128, stabilizes the aggregation of partially denatured protein through the formation of intermolecular contacts (3,5). Since the primary structures of bGH and hGH differ at eight positions, we have suggested that a bGH analogue containing the human sequence within this region (8H-bGH) would have a lower tendency to aggregate. Comparative studies of the folding and aggregation behavior of bGH, 8H-bGH and hGH show that although the wild-type and analogue proteins fold similarly at low concentration, aggregation of the chimeric protein is significantly lower at elevated protein concentration (6). We are interested in determining how the individual amino acid substitutions of 8H-bGH contribute to this altered behavior. In this paper, we report the results of solubility and spectral experiments involving R108N and E126R. The solubility studies suggest that both analogues aggregate to the same extent as wild-type bGH. It therefore appears that neither of these substitutions individually accounts for the destabilization of the partially denatured, aggregated species observed for 8H-bGH.

II. MATERIALS AND METHODS

Protein analogues. The synthesis and purification of bGH and the bGH analogues (8H-bGH, R108N, and E126R) were carried out as previously described (6,7,8). The primary structures of R108N and E126R were verified by tryptic mapping in conjunction with amino acid and N-terminal sequence analysis (9). Pituitary-derived human growth hormone was purchased from Dr. Alan Parlow (UCLA, Harbor Medical Center) and used without further purification.

Peptide analogues. bGH(96-133), 8H-bGH(96-133), E126R(96-133), R108N(96-133) were isolated following mild trypsin digestion of the intact proteins dissolved in 0.1 M ammonium bicarbonate, pH 7.8 (10). The digests were incubated for 1 hour at ambient temperature. Under these conditions, the peptide bonds between residues 95 and 96 and 133 and 134 are selectively cleaved. The reaction was terminated by the addition of guanidinium chloride (GdmCl), and the partial digests were purified by preparative reversed phase HPLC. The structures of these peptides were verified by N-terminal sequence analysis.

Solubility studies. The amount of aggregated bGH that forms under partially denaturing conditions can be determined by use of a solubility assay since this species has significantly lower aqueous solubility than native or unfolded species. This assay has been described previously (4). In brief, soluble aggregates of bGH and related proteins form when these compounds are dissolved at concentrations above 50 µg/mL (2.5 µM) in 3.7 M GdmCl at ambient temperature. The aggregate is precipitated upon dilution of GdmCl to less than 1 M. Samples are incubated for an additional 30 minutes, at which time scattered light is measured at 450 nm. Alternatively, the amount of precipitation can be indirectly measured by filtering the precipitate and determining the amount of soluble protein by using the BCA assay (11).

BGH(96-133) has also been found to inhibit the aggregation of partially denatured bGH (4). Similarly, the inhibition of bGH aggregation by bGH(96-133) analogues parallels the aggregation behavior of the corresponding intact bGH analogues (12). We therefore tested the ability of the bGH(96-133) analogue peptides to inhibit the aggregation of partially denatured bGH. Each peptide (0 to 500 µM) was coincubated with bGH (82 µM) in 3.7 M GdmCl at 4° overnight. These peptides do not precipitate when incubated in the absence of protein.

Near UV CD studies. The near-UV circular dichroism signal intensity of these proteins was monitored at 300 nm by using a Jasco J-500C circular dichroism spectrophotometer (13). The cell pathlength used in these experiments was 2.0 cm.

III. RESULTS

Relative Aggregation of Partially Denatured bGH, 8H-bGH, and hGH. As indicated above, the data from solubility experiments indicate that these proteins differ in their propensities for aggregation following partial denaturation (Figure 1) (6). The solubility assay is a useful indicator of aggregation since it has been established that the aggregated folding intermediate is

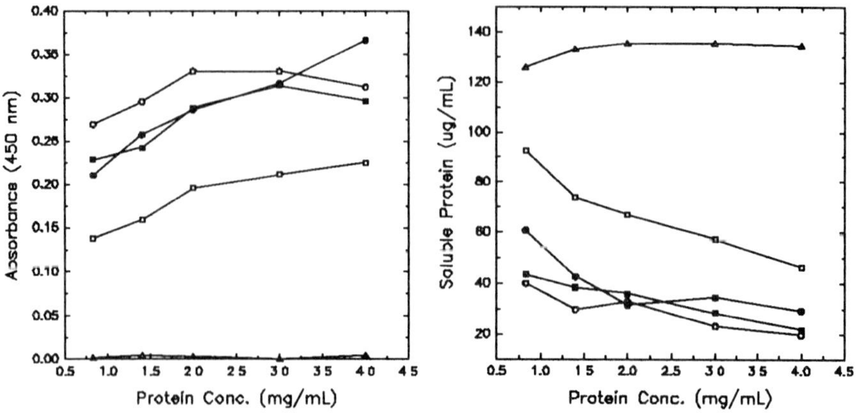

Figure 1. Aggregation of partially denatured bGH, 8H-bGH, hGH, R108N, and E126R using the two step solubility assay. This procedure is briefly described in Materials and Methods. Precipitation was monitored either by light scattering (A) or following filtration and quantitation of the remaining soluble protein (B). bGH (○); 8H-bGH (□); hGH (Δ); R108N (■); E126R (●).

significantly less soluble than native and denatured forms in aqueous buffers (4). Observation of the precipitate via light scatter and the determination of soluble protein following removal of the precipitate indicates that the propensity for aggregation decreases in the following order: bGH > 8H-bGH > hGH. In addition, the partially denatured, aggregated bGH species displays a unique spectral signal in the near-UV CD (300 nm). For bGH, 8H-bGH, and hGH, the intensity of this signal parallels the results of the two-step solubility assay (6). The results of the latter experiments are shown in Figure 2.

Identification of amino acid residues that contribute to formation of the protein aggregate. Since the partially denatured aggregate of bGH: 1) has unique spectral properties that distinguish it from the native and denatured species, and 2) only forms when the partially denatured protein retains about 50% of its α-helical content, it appears to be a chemically well-defined species stabilized by specific intermolecular interactions. Is the lower propensity for aggregation observed for partially denatured 8H-

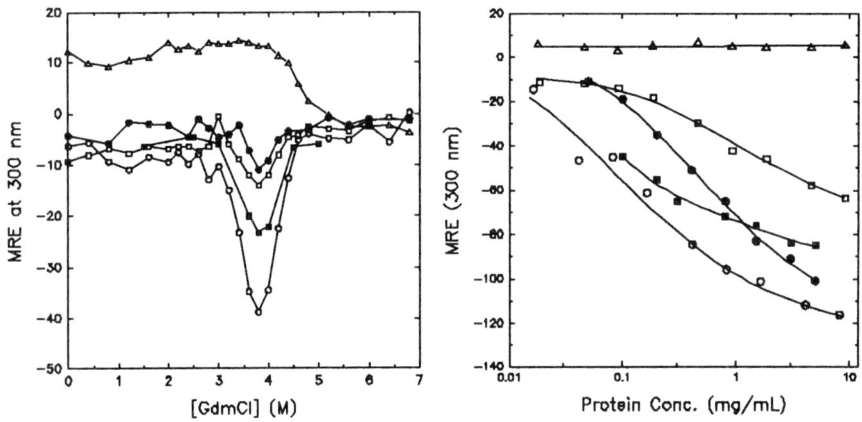

Figure 2. Monitoring of bGH aggregation using near-UV CD signal intensity at 300 nm as a function of GdmCl (A) and protein (B) concentrations. Solution conditions are described in Materials and Methods. Legend: see Figure 1.

bGH due to the loss of stabilizing interactions or indirectly due to the alteration of conformation? Since the fluorescence, far-UV, and second derivative UV absorbance spectra of the partially denatured 8H-bGH and bGH species are not significantly different, it appears that the relative aggregation observed for these proteins is likely due to the direct loss of stabilizing interactions. Of the eight amino acid substitutions in 8H-bGH, three are conservative (E111D, A122T, and T131S) and are not likely to significantly affect aggregation. A fourth substitution, K112L, has been shown to increase the stability of the aggregated species (14). This stabilization must be overcome since net destabilization of the partially folded aggregate is observed for 8H-bGH. The remaining substitutions that could be singly or collectively responsible for net destabilization of the aggregated species are R108N, L121Q, R125G, and E126R.

For this study, we have studied the aggregation of partially denatured R108N and E126R. Both aggregated species are about as soluble as partially denatured, aggregated bGH in the two-step solubility assay (Figure 1). In addition, fragments of R108N and E126E that span residues 96 to 133 inhibit the aggregation of partially denatured bGH to the same extent as the corresponding bGH fragment (Figure 3). Since the inhibition of bGH aggregation by bGH(96-133) analogues closely parallels the

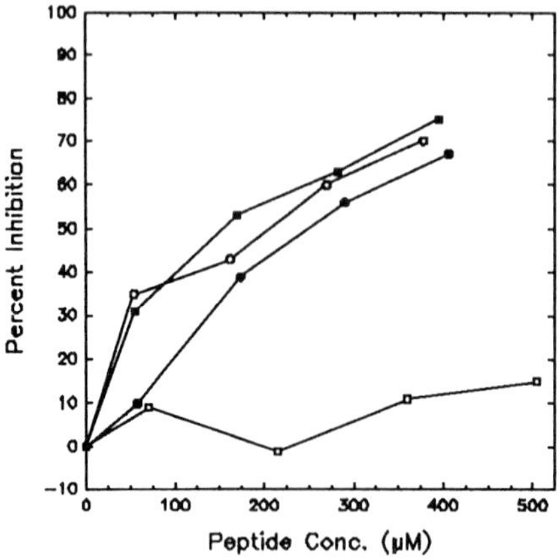

Figure 3. Inhibition of the aggregation of partially denatured bGH using 96-133 fragments of bGH, 8H-bGH, hGH, R108N, and E126R. Legend: see Figure 1.

propensity for aggregation by the corresponding intact protein analogues (12), this finding confirms that these substitutions do not notably affect aggregation.

In contrast to the results of these solubility studies, we find that the signal intensity of R108N and E126R at 300 nm in the near-UV CD is lower than that observed for the wild-type protein when these proteins are dissolved at high concentration in the presence of 3.7 M GdmCl (Figure 2). For example, the spectral intensity of partially denatured E126R and R108N at this wavelength is decreased 80 and 40%, respectively, relative to that formed by partially denatured bGH. Since changes in the near-UV CD is an indirect indicator of aggregation, resulting from alterations in the environment of the Trp^{86} sidechain, these results may be due to structural changes that are unrelated to protein aggregation.

IV. DISCUSSION

These experiments were undertaken in order to identify structural elements of bGH that help stabilize the aggregation of partially denatured protein. The data presented from solubility experiments show that R108N and E126R, two of the amino acid substitutions included within 8H-bGH, do not appear to appreciably destabilize the formation of the partially denatured, aggregated species. These substitutions alter the charge and bulk of the respective sidechains, and therefore should significantly alter aggregation were these sites directly involved in this process. Due to the considerations mentioned above, this finding suggests that positions 121 and/or 125 of the bGH primary structure are more important in stabilizing the aggregated protein species.

As suggested above, the observed differences in the solubility and spectral behavior of R108N and E126R may be the result of environmental changes near the tryptophan sidechain that are not related to aggregation. However other explanations cannot be discounted at this time. For example, the solubility and spectral behavior of the partially denatured proteins may differ because they represent different portions of the aggregation process. The spectral assay monitors the formation of soluble aggregates, that is, early intermediates on the pathway to precipitation. On the other hand, solubility encompasses these and subsequent steps that ultimately result in precipitation. Further experiments are planned to distinguish between these possibilities.

Since the substitutions included in partially denatured 8H-bGH does not entirely eliminate aggregation, other regions of bGH must help stabilize this species. Possibly, α-helix 3 of bGH forms intermolecular, native-like interactions with α-helices 1 and 2. This concept is supported by our ability to observe inhibition of bGH aggregation in the presence of excess bGH(12-34), a fragment that includes a significant portion of α-helix 1 (Lehrman and Tuls, unpublished data).

These experiments suggest that bGH residues 108 and 126 do not, by themselves, appreciably contribute to the formation of the aggregated protein species. We plan to continue these studies with the primary goal of determining how defined intermolecular interactions contribute to the aggregation of partially denatured bGH.

ACKNOWLEDGEMENTS

We gratefully acknowledge the efforts of Drs. C-S. C. Tomich, J. Mott, R. Kirschner, M. Sears, and Dr. J. Hoogerheide for their efforts in synthesizing and characterizing the bGH analogues, and Marilyn Lund for expert technical assistance. We also thank Sharon Andrews for secretarial assistance.

REFERENCES

1. Brems, D. N., Plaisted, S. M., Havel, H. A., Kauffman, E. W., Stodola, J. D., Eaton, L. C., & White, R. D. (1985) *Biochemistry* 24, 7662-7668.
2. Brems, D. N., Brown, P. L., & Becker, G. W. (1990) *J. Biol. Chem.* 265, 5504-5511.
3. Goeddel, D.V., Kleid, D.G., Bolivar, F., Heyneker, H.L., Yansura, D.G., Ross, M.J., Miozzari, G., Crea, R., Hirose, T., Kraszewski, A., Itakura, K., and Riggs, A.D. (1979) *Proc. Natl. Acad. Sci. U.S.A.* 76, 106-110.
4. Brems, D. N. (1988) *Biochemistry* 27, 4541-4546.
5. Brems, D. N., Plaisted, S. M., Kauffman, E. W., Lund, M. E., & Lehrman, S. R. (1987) *Biochemistry* 26, 7774-7778.
6. Lehrman, S.R., Tuls,J.L., Havel, H.A., Haskell, R.J., Putnam, S.D., and Tomich, C-S.C. (1991) *Biochemistry* 30, 5777-5784.
7. Tomich, C-S. C., Olson, E. R., Olsen, M. K., Kaytes, P. S., Rockenbach, S. A., and Hatzenbuhler, N. T. (1989) *Nucleic Acids Res.* 17, 3179-3197.
8. Evans, T. W. & Knuth, M. W. *World Patent* WO 8700204, January 15, 1987.
9. Hartman, P. A., Stodola, J. D, Harbour, G. C., & Hoogerheide, J. G. (1986) *J. Chromatogr.* 360, 385-395.
10. Graf, L., & Li, C.H. (1974) *Biochemistry* 13, 5408-5415
11. Smith, P. K., Krohn, R. I., Hermanson, G. T., Mallia, A. K., Gartner, F. H., Provenzano, M. D., Fujimoto, E. K., Goeke, N. M., Olson, B. J., & Klenk, D. C. (1985) *Anal. Biochem.* 150, 76-85.
12. Lehrman, S.R., Tuls, J.L., & Lund, M. E. (1990) in *Peptides: Proceedings of the Eleventh Peptide Symposium* (Rivier, J.E., & Marshall, G.R., Eds.) pp. 571-574, ESCOM Publishers, Leiden, Netherlands.

13. Havel, H. A., Kauffman, E. W., Plaisted, S. M., & Brems, D. N. (1986) *Biochemistry* 25, 6533-6538.
14. Brems, D.N., Plaisted, S.M., Havel, H.A., and Tomich, C-S.C. (1988) *Proc. Natl. Acad. Sci. USA* 85, 3367-3371.

USE OF SEQUENCE HYDROPHOBIC MOMENT TO ANALYZE MEMBRANE INTERACTING DOMAINS OF BOTULINUM, TETANUS AND OTHER TOXINS

Bal Ram Singh and Xu-hai Be
Department of Chemistry, University of Massachusetts-Dartmouth,
North Dartmouth, MA 02747

I. Introduction

Botulinum and tetanus neurotoxins, closely related high molecular weight (150,000) proteins, are produced by *Clostridium botulinum* and *C. tetani*, respectively. The neurotoxins are synthesized as single chain polypeptides but are nicked subsequently by endogenous proteases resulting in dichain forms which consist of a light chain (50 kDa) and a heavy chain (100 kDa) linked through a disulfide bond (Fig. 1). The dichain structures of botulinum and tetanus neurotoxins resemble those of other toxins, such as diphtheria, cholera, pseudomonas exotoxin A, ricin, etc.

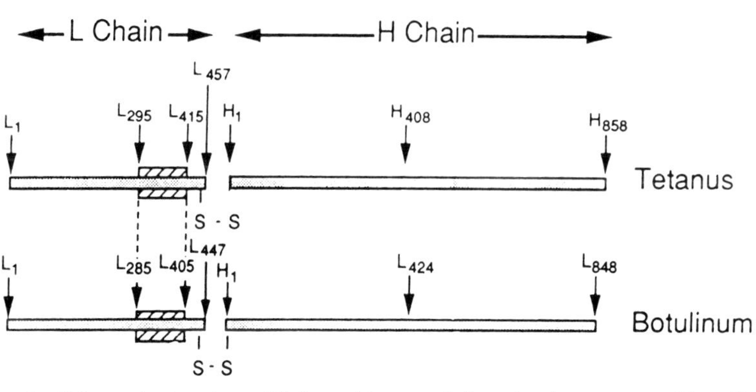

Fig. 1. Schematic drawing of light and heavy chain subunit structure of botulinum and tetanus neurotoxins. The subscripted numbers refer to the residue numbers on the light (L) and heavy (H) chains. Peptide segments on L chain surrounded with slanted bar box represent the "catalytic" domains as described in Singh (1990). Heavy chains are also demarcated for N-terminal and C-terminal halves.

TECHNIQUES IN PROTEIN CHEMISTRY III

The common mode of action of botulinum and tetanus neurotoxins is proposed (Simpson, 1981, 1986; 1989) to involve binding of the heavy chain to presynaptic membranes through an as yet unidentified receptor, and translocation of the whole or a part (possibly light chain) of the neurotoxin in the nerve cell, where the neurotoxin interferes with an as yet unknown biochemical step to block the neurotransmitter release.

According to the working hypothesis based on some experimental evidence and analogies with other dichain toxins listed above, the C-terminal half of the heavy chain of the neurotoxin binds with presynaptic membranes. Upon binding, the neurotoxin is internalized through endocytosis resulting in neurotoxin containing endosomes. The inside pH of the endosome is lowered to ~ 4.0 presumably upon interaction with primary lysosomes. Low pH induces conformational changes in the neurotoxin that leads to its incorporation in the lipid bilayer to form a membrane channel. The N-terminal half of the neurotoxin is involved in the membrane channel formation (Boquet and Duflot, 1982; Blaustein et al., 1987).

Membrane channel formation by a protein involves delicate and specific interactions between membrane lipids and the non-polar domains of the protein. While transmembrane proteins have adequate hydrophobic segments for this purpose, water soluble proteins generally lack a continuous stretch of hydrophobic segments capable of forming channels. However, there are water soluble toxins which form membrane channels as part of their mode of action. A unique feature of these proteins is to form the channels at low pH. There are other examples of proteins which get incorporated in the lipid bilayer at low pH. These proteins include colicin E and influenza hemagglutin HA_2.

A major question is how these proteins get integrated in the lipid bilayer? Do these proteins have hydrophobic segments that are 'hidden' in the aqueous solution? Does the lower pH induce enough conformational change to expose the hydrophobic segments for interaction with non-polar lipid bilayer? Do the sequences of these proteins contain amphiphilic structure that make them compatible for interaction with the lipid bilayer?

Answers to the above questions are not clear for even those proteins such as Influenza hemagglutinin HA_2 and *Pseudomonas* exotoxin A that have been analyzed for their crystal structure (Wilson et al., 1981; Allured et al., 1986). An analysis of the molecular basis of membrane channel formation by toxins is important not only for the understanding of their mode of toxic action but also for the basic knowledge of protein-protein and protein-lipid interactions.

Various approaches used to investigate the formation of membrane channels have included determination of pH-induced conformational changes (Blewitt et al., 1985, DeWolf et al., 1987; Singh et al., 1990), low pH-induced change in hydrophobicity (Idziorek et al., 1990) and low pH-induced change in the interaction with membranes (Papini et al., 1987a; 1986b; Jiang et al., 1989; O'Keefe and Collier, 1989). However, none of the studies have identified a given amino acid sequence that may interact with the membrane lipid bilayer to form channels.

We have used hydrophobic moment calculations to identify the domains of several toxins, including botulinum and tetanus neurotoxins having potential to form membrane channels. Although hydrophobic moment calculation reveal only possibility of membrane interaction sites, such analyses have provided accurate prediction to membrane interaction sites of several known proteins (Eisenberg et al., 1982; 1987).

II. Hydrophobicity and Hydrophobic Moment

Hydrophobicity, measured as a free energy change for the transfer of an amino acid from an apolar solvent such as n-octanol to water has been used, in general, to describe protein segments containing non-polar amino acid residues compatible for interaction with lipid bilayer (Kyte and Doolittle, 1982). Hydrophobicity is basically a measure of the degree of affinity between water and the side chain of an amino acid. Amino acids with non-polar side chains have higher hydrophobicities compared to amino acids with polar or charged side chains. To estimate the hydrophobicity of a protein segment, hydrophobicity of individual amino acids are averaged, assuming that the transfer of a chemical group from an apolar solvent is independent of the larger molecule to which the group is attached.

Prediction of membrane spanning segments of proteins works well with hydrophobicity calculations (Kyte and Doolittle, 1982) only for proteins having continuous stretches of peptide segments having significant hydrophobicity. Several protein segments although known to interact with membrane lipid bilayer have lower average hydrophobicities. It has been noticed that such protein segments have α-helical foldings which allow an asymmetric distribution of hydrophobic and hydrophilic amino acid residues resulting into an amphiphilic structure compatible for interaction with non-polar lipid bilayer. A quantitative estimation of the degree of amphiphilicity or the asymmetric distribution of the hydrophobicity is carried out by calculating the hydrophobic moment (Eisenberg et al., 1982; 1984; 1987). Hydrophobic moment is similar to the electric dipole moment. Electric dipole moment estimates asymmetry of electric charge whereas hydrophobic moment estimates asymmetric distribution of hydrophobicity.

When atomic coordinates of a protein segment are known, the hydrophobic moment (called the structural hydrophobic moment) can be calculated as follows (Eisenberg et al., 1987):

$$\mu = \sum_{n} H_n S_n \qquad (1)$$

where, H_n is the hydrophobicity of the nth residue of the protein segment, and S_n is the unit vector pointing from the α-carbon atom to the center of the residue's side chain. Thus, hydrophobic moment is a measure of the sum of the directions of amino acid side chains, each weighted by its hydrophobicity. Hydrophobic moment can be estimated for a protein segment even if the atomic coordinates of amino acid residues in its sequence are not known. Such hydrophobic moment (called the sequence hydrophobic moment) can be estimated as follows (Eisenberg et al., 1982; 1984) assuming an α-helical or ß-sheet folding of the segment.

$$\mu = \{[\sum_n H_n \mathrm{Sin}(\phi_n)^2] + [\sum_n H_n \mathrm{Cos}(\phi_n)^2]\}^{1/2} \quad (2)$$

where, Hn is the hydrophobicity of the nth residue, ø is the angle (in radians) at which successive side chains emerge from the central axis of the structure, ø for α-helical structures is 100° whereas for ß-sheet structures, it is 160°. Both in structural and sequence hydrophobic moments, an amino acid residue with high hydrophobicity contributes strongly to the moment, whereas a hydrophilic residue having negative hydrophobicity contributes in the opposite direction. However, when hydrophobic residues protrude on one side of the structure and hydrophilic residues on the opposite side, this results into a large hydrophobic moment because both hydrophobic and hydrophilic residues contribute to the sum with the same sign. Thus, a highly amphiphilic peptide segment will have relatively large hydrophobic moment.

III. Methods

In order to calculate the hydrophobic moments of various polypeptides, we have used the normalized consensus hydrophobicity scale (Eisenberg et al. 1987). Calculations were carried out on IBM PC/XT computer using a program provided by William Parker of University of Nebraska based on the equation (2). Hydrophobicity and hydrophobic moments were calculated using an 11-residue moving window for the estimation of the mean values (Eisenberg et al., 1987).

We have assumed α-helical folding of peptide sequence because such a structure provides the largest hydrophobic moment suitable for membrane interaction (Eisenberg et al., 1984). In fact, hydrophobic moment has been used to identify α-helical segments of several proteins (Parker and Song, 1990).

IV. Results and Discussion

As mentioned earlier in the introduction, hydrophobicity calculations alone have limitations in identifying the potential membrane interacting segments unless such segments contain an overwhelming number of hydrophobic amino acids. For example, hydrophobicity calculations of tetanus neurotoxin sequence revealed only one peptide segment from Asn-660 to Ala-691 on the heavy chain having potential to span the membrane lipid bilayer (Eisel et al., 1986). Another segment was observed on the light chain of the tetanus neurotoxin, but here we are focussing only on the heavy chain which is known to form membrane channels (Duflot and Boquet, 1982; Hoch et al., 1985). An examination of the hydrophobicity vs. amino acid residue number plot of tetanus heavy chain (Fig. 2) reveals that the only segment capable of membrane spanning is H-210 (equivalent to residue number 685 based on the sequence of whole neurotoxin) to H-233 (H stands for heavy chain). Considering the level of toxic activity which is almost completely dependent on the membrane channel formation for the translocation of a large (50 kDa) toxic domain, this segment alone may not explain the degree of toxin integration in the nerve membranes. One could argue

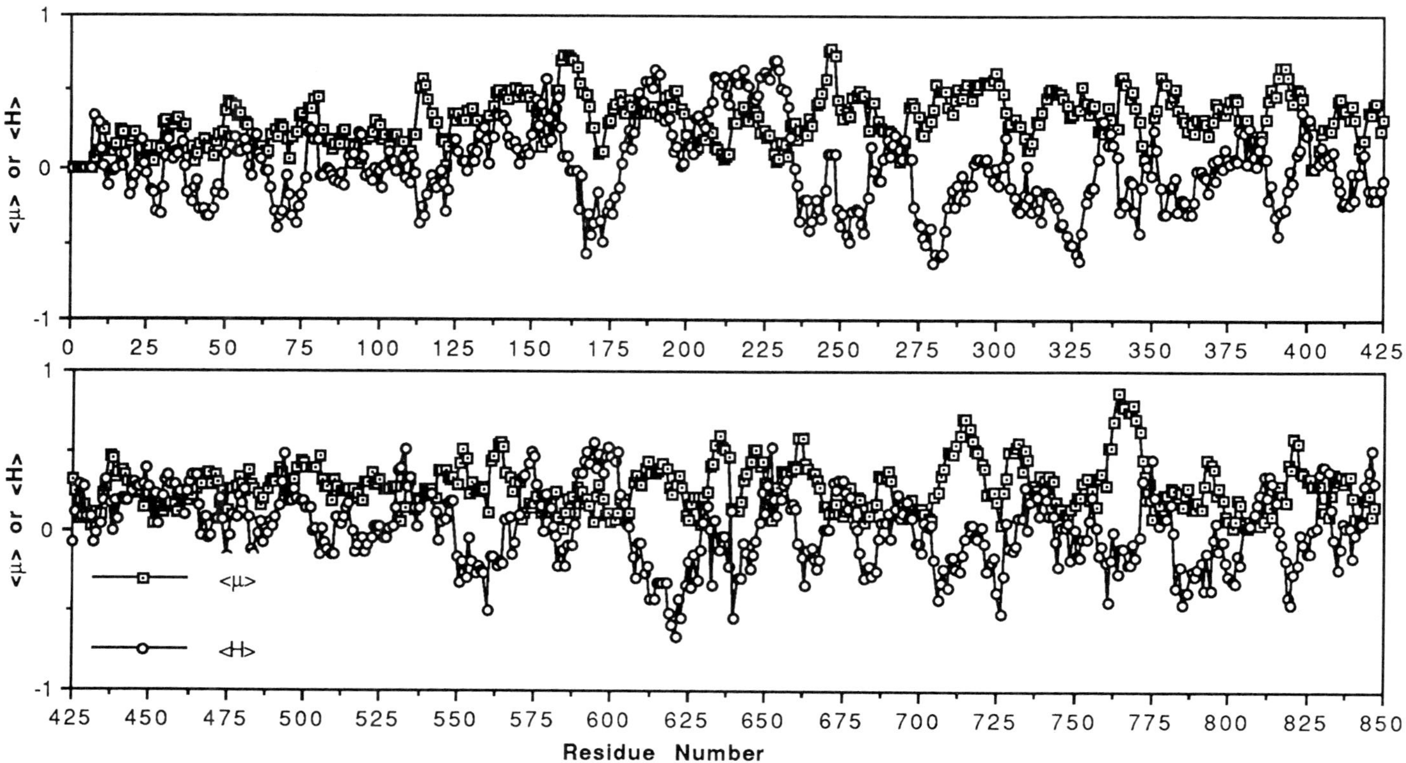

Fig. 2. Residue number versus hydrophobic moment (μ) and hydrophobicity (H) plot for the tetanus heavy chain. Each point represent an average of 11-residue window. The data shows that several peptide segments without any significant hydrophobicity have large hydrophobic moments suggesting amphiphilic character compatible for membrane interaction.

the role of receptors or other factors that may be involved in assisting the integration of the heavy chain. However, it has been demonstrated that the toxin is capable of forming membrane channel in artificial membranes (Hoch et al., 1985; Blaustein et al., 1987). Thus, the toxin is likely to have more structure motifs compatible for interaction with lipid bilayer. Hydrophobic moment calculations provide a way to further analyze the sequence data for identifying the segments with potential to interact with membranes. Hydrophobic moments plotted for the amino acid sequence of the tetanus heavy chain are shown in Fig. 2 It is obvious from Fig. 2 that several peptide segments of heavy chain contain significantly high hydrophobic moments even though mean hydrophobicity of these segments is quite low. Hydrophobic moment being a measure of amphiphilicity suggests that such peptide segments have potential to interact with non-polar lipid bilayer.

Eisenberg et al. (1987) have devised an empirical method to identify peptide segments which could be transmembrane or membrane surface seeking by plotting hydrophobic moment vs. the hydrophobicity. One such plot for heavy chain of tetanus neurotoxin is shown in Fig. 3. Seven different segments have been identified to have potential for interaction with the membrane bilayer. Five of these segments are present in the N-terminal half of the heavy chain which is known to form membrane channels and two segments are located in the C-terminal half of the heavy chain which is known to initially bind with the nerve membrane through gangliosides (Montecucco, 1986). Two notable features of this analysis are as follows: (a) the two transmembrane segments (H-209-232 and H-187-191) are present in the N-terminal half of the heavy chain, (b) the two surface seeking segments of the C-terminal half of the heavy chain (H-709-724 and H-759-776) are closely spaced and located towards the end of the polypeptide sequence.

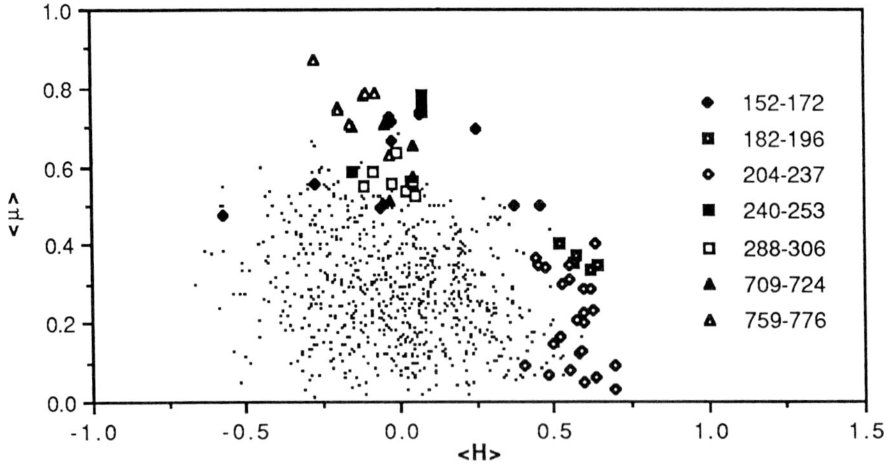

Fig. 3. Hydrophobicity (H) versus hydrophobic moment (μ) plot of tetanus heavy chain. Data is derived from Fig. 2. The seven peptide segments which have hydrophobicity and hydrophobic moment in the range that classifies them as membrane surface seeking or transmembrane domains.

Table I. Membrane and surface seeking peptide segments of various proteins identified by hydrophobic moment calculation

Protein	Residue number	Character	Other evidence
Mellitin	whole chain	surface, membrane	X-ray
Influenza B-chain(HA2)	1-24	surface, membrane	X-ray
	44-74	surface	
	181-213	membrane	X-ray
Diphtheria	25-42	surface	
	118-139	surface	
	271-285	membrane	
	300-323	membrane	
	328-372	surface, membrane	
	423-440	membrane	
Pseudomonas exotoxin A	48-60	surface	
	95-108	membrane	
	135-158	surface	
	164-178	membrane	
	195-209	surface	
	278-296	surface	X-ray
	300-319	surface	X-ray
	433-456	membrane	
	459-475	surface	
	468-481	membrane	
	519-537	surface	
Tetanus heavy chain	152-172	surface	
	182-196	surface, membrane	
	204-237	membrane	
	240-253	surface	
	288-306	surface	
	709-724	surface	
	759-776	surface	
Botulinum heavy-chain	80-92	surface	
	149-164, 178-198,	membrane	
	202-239	membrane	
	351-363, 412-428	surface	
	439-450, 495-510	surface	
	547-558, 555-569	surface	
	618-631	surface	
	704-716	surface	
	803-816	surface	
	820-831	surface	

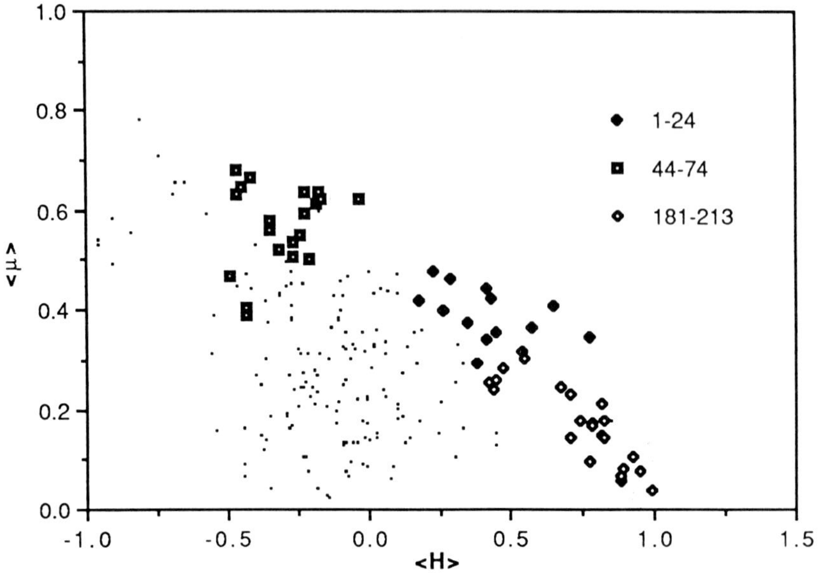

Fig. 4. Hydrophobicity (H) versus hydrophobic moment (μ) plot of influenza virus hemagglutinin HA2 chain. Two peptide segments, 1-24 and 181-213 clearly have surface seeking and transmembrane characteristics, a prediction consistent with X-ray analysis (Wilson et al., 1981).

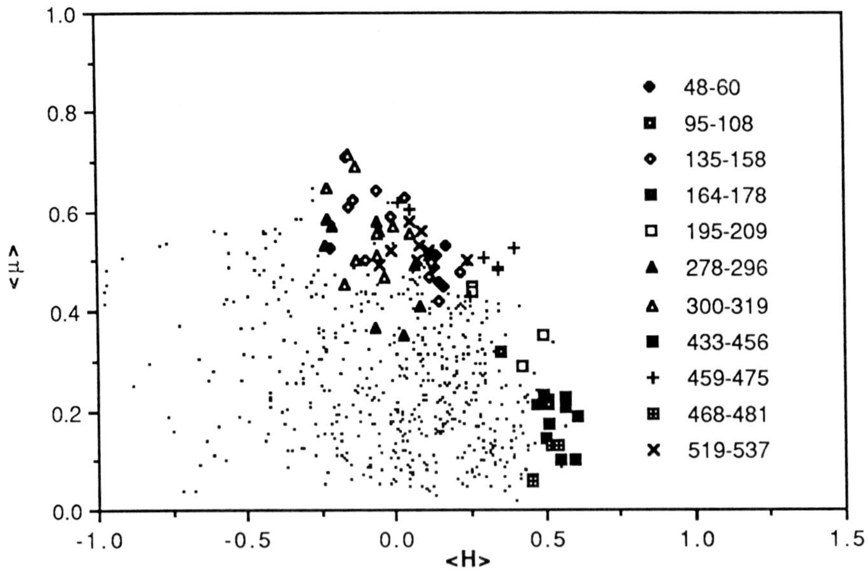

Fig. 5. Hydrophobicity (H) versus hydrophobic moment (μ) plot of *Psedomonas* exotoxin A. The analysis shows peptide segments compatible for membrane interaction not only in the domain II which is known to be involved in the channel formation but also in Ia and III.

Analysis of the hydrophobicity vs. hydrophbic moment plot of botulinum heavy chain based on the sequence published recently (Binz et al., 1990; Thompson et al., 1990) reveals several membrane and surface seeking peptide segments (Table 1). It may be noted that botulinum heavy chain has far more number of segments compatible for membrane interaction than tetanus heavy chain. Four of the peptide segments identified on the respective heavy chains of botulinum and tetanus are located at similar positions in their sequences, and share homologous amino acid residues. This observation is interesting in that botulinum and tetanus neurotoxins have several structural and functional similarities (Binz et al., 1990; Thompson et al., 1990; Simpson, 1986; 1989; Singh, 1990).

In order to provide validity to our conclusions about the prediction of botulinum and tetanus peptide segments, we have used two other proteins, *Pseudomonas* exotoxin A and influenza virus hemagglutinin of known X-ray structure, for hydrophobic moment calculations. Both these proteins have a common feature with botulinum and tetanus neurotoxins of low pH-induced insertion in the lipid bilayer as part of their mode of action. Hydrophobicity and hydrophobic moment calculations were carried out for the polypeptide chains known to participate in the membrane interaction. Fig. 4 shows such plots for influenza virus hemagglutinin HA_2. Two segments identified to be surface seeking or membrane spanning were located as residues 1-24 and residues 181-213, respectively (Table I). This prediction matched exactly with the X-ray structure analysis (Wilson et al., 1981).

Hydrophobicity vs. hydrophobic moment plot of *Pseudomonas* exotoxin A (Fig. 5) revealed 10 peptides segment with characteristics compatible for the interaction with membranes (Table I). Notable observations for the exotoxin A has been the absence of any amphiphilic segment in the domain Ib (Fig. 6) which appears to act as a linker peptide between domain II and III according to the X-ray crystal structure (Allured et al., 1986). Other important points revealed by hydrophobic moment analysis are the presence of 5 amphiphilic transmembrane segments in the domain Ia and two amphiphilic segments in the domain II. These two domains are known to be involved in the interaction of the exotoxin A with the cellular membranes (Pastan and FitzGerald, 1989). One of the segments (residue 278-296), identified as amphiphilic in nature, lies next to two critical amino acids Arg-276 and Arg-279 for the toxic action of the exotoxin A (Jinno et al., 1989).

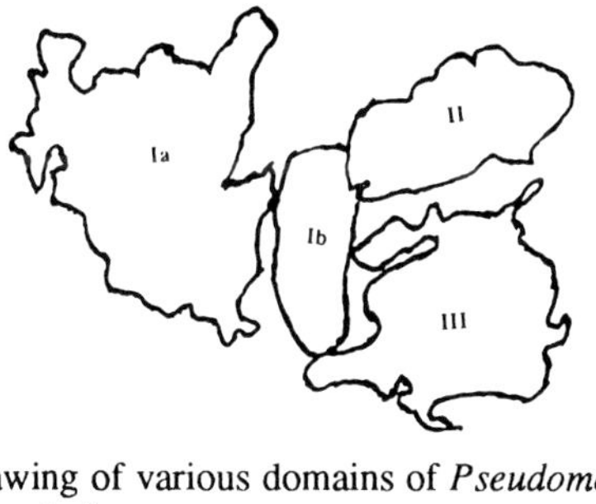

Fig. 6. Cartoon drawing of various domains of *Pseudomonas* exotoxin A as derived from X-ray analysis.

We have also noticed an interesting combination of 3 peptide segments with amphiphilic/transmembrane nature in domain III which contains the enzymatic activity. Two of these segments (residues 433-456, 459-481) have transmembrane characteristics and lie next to the domain II in their three dimensional structure (Allured et al., 1986). The corresponding segments of domain II (residues 278-296 and 300-319) also have amphiphilic nature. Therefore, the identified peptide segments in domain II and III by virtue of being in the same topographical region of the protein molecule may play a role in the membrane channel formation and the translocation of the toxin.

V. Concluding Remarks

A group of bacterial toxins including botulinum and tetanus neurotoxins, although water soluble, interact with membrane bilayers for their translocation across the cell. Primary amino acid sequences of such toxins have not provided any obvious clues for the involvement of specific peptide segments for interaction with membrane. Even three dimensional X-ray structural analysis has failed to definitely identify such segments (for example, Pseudomonas exotoxin A; Allured et al., 1986). Hydrophobicity calculation alone has been inadequate to explain the membrane channel formation activity of these toxins.

A combination of hydrophobicity and hydrophobic moment calculations have proven effective in identifying potential peptide segments of toxins such as *Pseudomonas* exotoxin A, botulinum, tetanus and influenza virus hemagglutinin HA_2. Because the conclusions drawn based on this approach are compatible with the known X-ray structures of at least two examples presented, the method appears to have great promise for analyzing other toxins and proteins with similar mode of action.

Acknowledgements

This work supported by funds from University of Massachusetts-Dartmouth.

References

Allured, V. S., Collier., Carroll, S. F., and McKay, D. B. (1986) *Proc. Natl. Acad. Sci. USA* **83,** 1320-1324.

Binz, T., Kurazono, H., Wille, M., Frevert, J., Werners, K., and Niemann, H. (1990) *J. Biol. Chem.* **265,** 9153-9158.

Boquet, P., and Duflot, E. (1982) *Proc. Natl. Acad. Sci. USA* **79,** 7614-7618.

Blaustein, R. O., Germann, W. J., Finkelstein, A., and DasGupta, B. R. (1987) *FEBS Lett.* **226,** 115-120.

Blewitt G., Chung, L. A., and London, E. (1985) *Biochemistry* **24,** 5454-5456.

DeWolf, M. J. S., Van Dessel, G. A. F., Lagrou, A. R., Hilderson, H. J. J., and Dierick, W. S. H. (1987) *Biochemistry* **26,** 3799-3806.

Eisel, U., Jarausch, W., Goretzki, K., Henschen, A., Angels, J. Weller, U., Hude, M., Habermann, E., and Niemann, H. (1986) *EMBO J.* **5,** 2495-502.

Eisenberg, D., Weiss, R. M., and Terwilliger, T. C. (1982) *Nature* **299,** 371-374.

Eisenberg, D., Weiss, R. M., and Terwilliger, T. C. (1984) *Proc. Natl. Acad. Sci. USA* **81**, 140-144.

Eisenberg, D., Wilcox, W., and Eshita, S. (1987) *In* "Protein Structure and Function" (J. J. L'ltalien, ed.), Plenum Press, New York and London. p. 425-436

Hoch, D. H., Romero-Mira, M., Ehlrich, B. E., Finkelstein, A., DasGupta, B. R., and Simpson, L. L. (1985) *Proc. Natl. Acad. Sci., USA* **82,** 1692-1696.

Idziorck, T., FitzGerald, D., and Pastan, I. (1990) *Infect. Immun.* **58,** 1415-1420.

Jiang, G., Solow, R., and Hu, V. W. (1989) *J. Biol. Chem.* **264,** 13424-13429.

Jinno, Y., Ogata, M., Chaudhary, V. K., and Willingham, M. C. (1989) *J. Biol. Chem.* **264,** 15953-15959.

Kyte, J., and Doolottle, R. F. (1982) *J. Mol. Biol.* **157,** 105-132.

Montecucco, C. (1986) *Trends in Biochem. Sci.* **11,** 314-317.

O'Keefe, and Collier, R. J. (1989) *Proc. Natl. Acad. Sci. USA* **86,** 343-346.

Papini, E., Colonna, R., Cusinato, F., Montecucco, C., Tomasi, M., and Rappuoli, R. (1987a) *Eur. J Biochem.* **169,** 629-635.

Papini, E., Schiavo, G., Tomasi, M., Colombatti, M., Rappuoli, R., and Montecucco, C. (1987b) *Eur. J. Biochem.* **169,** 637-644

Pastan, I., and FitzGerald, D. (1989) *J. Biol. Chem.* **264,** 15157-15160.

Parker, W., and song, P.S. (1990) *J. Biol. Chem.* **265,** 17568-17575

Simpson, L. L. (1981) *Pharmacol. Rev.* **33,** 155-188.

Simpson, L. L. (1986) *Ann. Rev. Pharmacol. Toxicol.* **26,** 427-453.

Simpson, L. L. (ed.) (1989) *Botulinum Neurotoxin and Tetanus Toxin.* Academic Press, San Diego.

Singh, B. R., Fuller, M. P., and Schiavo, G. (1990) *Biophysical Chem.* **36,** 155-166.

Singh, B. R. (1990) *Toxicon* **28,** 992-996.

Thompson, D. E., Brem, J. K., Oultram, J. D., Swingfielk, T. J., Shone, C. C., Atkinson, T., Melling, J., and Minton, N. P. (1990) *Eur. J. Biochem.* **189,** 73-81.

Wilson, I. A., Skehel, J. J., and Wiley, D. C. (1981) *Nature* **298,** 366-373.

Fourier Transform Infrared Spectroscopic Analysis of Proteins in Terms of Detectability, Conformation and Surface Adsorption Density

Bal Ram Singh, Fen-Ni Fu
Department of Chemistry, University of Massachusetts-Dartmouth, North Dartmouth, MA 02747

Michael P. Fuller
Nicolet Instrument Corporation, Madison, WI 53711

I. Introduction

Folding of the primary amino acid sequence into a native protein or a polypeptide is of great interest to protein chemists for the understanding of their functions. Secondary structural content has been one of the most commonly investigated protein folding parameters. Changes in protein folds at the tertiary level are also derived by monitoring intrinsic probes such as aromatic amino acids and attached chromophores (e.g., cytochromes), or by labeling the proteins with extrinsic probes. Protein folding can ideally be derived from x-ray diffraction studies of suitable protein crystals. However, for most proteins, it is very difficult to grow suitable crystals. In addition, even if protein crystals can be grown under one condition, it is not practical to prepare crystals in several relevant conditions to understand structure-function relationship. In many cases, various spectroscopic techniques are used to derive polypeptide folding of proteins.

Spectroscopic methods commonly used to analyze the polypeptide folding of proteins include fluorescence, circular dichroism, Fourier transform infrared (FT-IR), Raman, and nuclear magnetic resonance (NMR) spectroscopies. Although intrinsic fluorescence is quite commonly used to detect conformational

---

ATR : attenuated total reflectance.

CD : circular dichroism.

DTGS : deuterated triglycine sulfate.

FSD : Fourier self deconvolution.

FT-IR : Fourier transform infrared.

IgG : (human) immunoglbin G.

IRE : internal reflection element.

NaPB : sodium phosphate buffer.

TECHNIQUES IN PROTEIN CHEMISTRY III

changes in proteins (Lakowicz, 1983), it is imperative that the protein sequence contain a high enough concentration of the chromophores capable of emitting absorbed light to make fluorescence a useful technique. Circular dichroism is commonly used to determine the secondary structure of polypeptides (Johnson, 1988). Other spectroscopic techniques mentioned above require a high concentration of proteins for their structural analysis, which could become a limiting factor for many proteins. This is especially true for proteins with less solubility, or proteins which are less abundant and difficult to purify. This includes many proteins which are prepared through genetic engineering.

FT-IR spectroscopy requires a high concentration of proteins when employing the transmission technique, due to the relatively weak absorptivities of the amide functional groups used for structural analysis. Use of infrared spectroscopy for protein structural analysis was first noted by Elliott and Ambrose (1950). Subsequent experimental and theoretical studies of several researchers (Miyazawa et al., 1956; Krimm, 1962; Krimm and Abe, 1972; Susi, 1972; Byler and Susi, 1986; Krimm and Bandeker, 1986; Kaiden et al., 1987; Surevicz and Mantsch, 1988, Dong et al., 1990) have led to identification of three major spectral regions of interest for protein structure.These regions reflect the vibrational transitions of the amide bond, and are listed below:

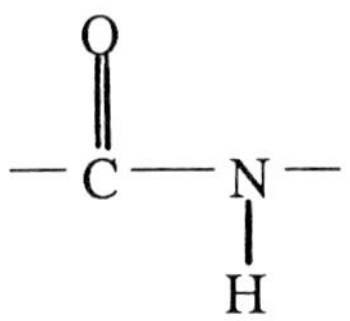

Amide bond

Amide I (1600-1700 cm^{-1}): C=O stretch weekly coupled with C-N stretch and N-H bending.

Amide II (1450-1550 cm^{-1}): C-N stretch strongly coupled with N-H bending .

Amide III (1240-1320 cm^{-1}): N-H in plane bending coupled with C-N stretching; also C-H/N-H deformation vibrations.

FT-IR spectroscopy of polypeptides is being more frequently used to determine their secondary structures in recent years. This is partly due to the availability of computer programs for Fourier transformation of the interferograms into absorption or transmission spectra, and for other data processing. Despite these advances, FT-IR analysis is limited to only those proteins that are available at high concentrations, especially when used in the transmission mode.

In our studies, we have used attenuated total reflectance (ATR) technique for sample handling to test the limits of the sensitivity of FT-IR spectroscopy for protein structural analysis. After successful detection of nanomolar concentrations of protein for secondary structural analysis (a concentration of protein that could not be detected by either circular dichroism or fluorescence spectroscopy), we have utilized this approach to determine adsorption density of proteins on a zinc selenide surface from aqueous solutions on a real time basis.

II. Materials and Methods

Pure chicken egg white lysozyme and human immunoglobin G (IgG) were purchased from Sigma Chemical Co. and were used without further purification. The protein solutions were prepared in 10 mM sodium phosphate buffer (NaPB), pH 6.0, as 2.0 mg/ml stock. Stock solutions were then diluted to the desired concentration using the same buffer.

FT-IR spectra were recorded on a Nicolet Model 8210 spectrometer, which is equipped with a horizontal cell compartment. A 60° horizontal zinc selenide crystal ATR (attenuated total reflectance) accessory was used for the sample compartment. All of the spectra were recorded using a room temperature DTGS detector. The spectrometer was operated at 4 cm^{-1} resolution with 256 scans typically co-added to yield each spectrum.

The instrument was purged with dry air for at least four hours to remove any water vapor that could interefere with the protein band at 1650 cm^{-1}. In order to obtain the protein spectra, an open beam background spectrum through the clean zinc selenide crystal was first collected. This was followed by measuring the spectrum of the aqueous buffer solution (1.5 ml of 10 mM NaPB, pH 6.0) which was layered onto the surface of the crystal. The buffer spectrum is used to eliminate the interfering water absorptions from the protein/buffer solution spectra. The protein solutions were measured at 0, 5, 10, 15, 20, 25 and 30 min incubation periods. Where effects of solvent (methanol) were to be studied, the bulk protein solution was first replaced with the aqueous buffer solution and the spectrum of the adsorbed protein measured. The buffer was then replaced with methanol and the spectrum of the adsorbed protein exposed to methanol was obtained. The adsorbed protein was exposed to methanol for 5 min before the spectra were recorded. Finally, to observe the reversibility of methanol-induced changes, methanol was replaced with the aqueous buffer and incubated for 5-10 min before measuring the spectrum. When a change of solvent was involved, the adsorbed protein was rinsed 5-6 times with the solvent prior to its final addition. To eliminate the absorptions due to methanol a reference spectrum of pure methanol was obtained and subtracted for the adsorbed protein/methanol spectra.

In every case, a ratio between the single beam spectrum of the buffer, methanol, or a protein solution and the background single beam spectrum was calculated to correct for the background. Corrected spectra were then converted into absorbance spectra. To obtain the protein spectra, the solvent (buffer or methanol) spectra were subtracted in an iterative manner by varying the subtraction factor until the broad water absorption around 800 cm^{-1} was totally removed. In the case of methanol subtraction, we used a straight base line, between 1700 and 1800 cm^{-1}, to fix the subtraction factor.

After eliminating the interfering absorption due to water or methanol, protein spectra showed an absorption of only a few thousandths of an absorbance unit in intensity. All the spectra were smoothed using a 9-point Savitsky-Golay algorithm to minimize the noise level before finally processing the data for structural analysis. The spectra were then multiplied by 100 to yield absorbances of a more convenient magnitude. The spectral analysis was carried out using the Nicolet FOCAS™ software package. First the spectral band positions were identified using Fourier self deconvolution (FSD). The bands identified by FSD were also confirmed by derivatizing the spectra to the second order. In the second step, peak widths, positions, intensities and percent Lorentzian/Gaussian shape were varied to obtain resolved spectral bands. The accuracy of the resolved bands was checked by generating a composite spectrum of the resolved bands,then comparing it with the original recorded spectrum by calculating the residual of the two spectra. Lowest possible residual indicated the accuracy of the curve fitting. Relative areas of the resolved bands were used to derive the secondary structural contents.

Adsorption density, Γ (mol/cm^2) of protein on a ZnSe crystal surface was calculated based on an equation developed by Sperline et al. (1987) for adsorption of detergent molecules on surface:

$$\Gamma = [d_p(A/N - C_b \varepsilon d_e)]/2000\ \varepsilon d_e \qquad (1)$$

where d_p is the depth of penetration (cm), d_e is the effective depth (cm), A is the IR absorbance intensity of the protein, N is the number of internal reflections in the ZnSe crystal, ε is the molar absorptivity, and C_b is the bulk equilibrium concentration (M). The d_p and d_e can be calculated(Harrick,1979)from refractive indices, n_{IRE} (2.4) and n_{H_2O} (1.0), and the incident angle, θ (60°) of the IR beam. The number of reflections can be calculated from geometric considerations according to the following:

$$N = (\ell/t)\cot\theta \qquad (2)$$

where ℓ and t are the length and thickness of the internal reflection element (IRE). The ZnSe crystal used in our system had an active reflection number of 7.

Infrared molar absorptivities of proteins were determined by recording transmission spectra of high concentration (20-30 mg/ml) of proteins using a 15 micron BaF_2 sealed cell.

Circular dichroic spectra of proteins in the far UV region (250-190 nm) were recorded on a JASCO J-20A at room temperature as previously described (Singh et al., 1988; Singh and Betley, 1989).

Fluorescence measurements were carried out on a SLM 8000 'smart' spectrofluorometer at room temperature (23-25 °C) as described by Singh and

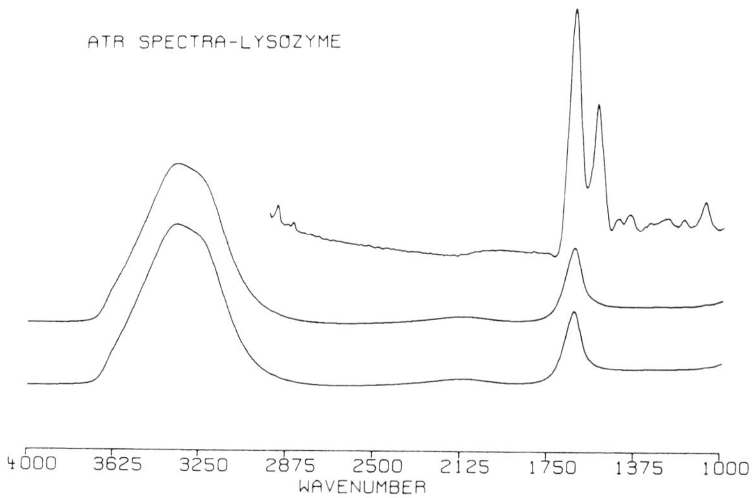

Figure 1. Spectrum of dilute solution of lysozyme in buffer (middle), buffer spectrum (bottom) and the difference spectrum obtained by subtracting the spectrum of the buffer from that of the protein/buffer solution.

Betley (1989), except that the gain control and the sensitivity were set to a maximum when the lowest concentration of protein was used, and excitation wavelength was fixed at 280 nm.

III. Results and Discussion

This study focused on three major points of FT-IR spectroscopy of proteins: detectability of low concentration protein solutions, structural analysis, and the adsorption density estimation. The detectability was checked for two proteins, lysozyme and IgG. Lysozyme and IgG were used in the concentration range of 2 mg/ml to 0.0002 mg/ml. To analyze the spectral features related to the protein structure the strong interfering absorptions due to the high concentration of water in the buffer must be eliminated. The ATR spectra of a dilute solution of lysozyme in buffer, the buffer spectrum and the resulting difference spectrum obtained by subtracting the buffer spectrum from that of the buffer/protein solution are shown in Figure 1. For all the concentrations tested, it was observed that the IR absorbance intensity increased with time. This was especially notable with low concentrations of proteins. Figure 2 shows increase in the Amide I band intensity of IgG as a function of time. The data indicate that, initially, absorbance increases with time, but it levels off at about 20-25 min reaching an equilibrium state.

If secondary structural information is to be obtained from the infrared spectra of adsorbed proteins, care must be taken to assure that the conformation of the protein is not affected during the adsorption process. The spectrum of IgG was

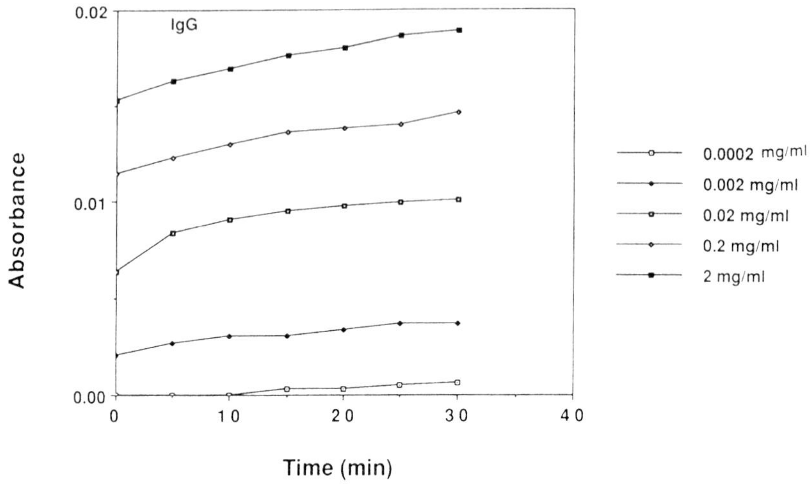

Figure 2. The absorption intensity of the amide I band at 1653 cm^{-1} versus time for the adsorption for a series of solutions containing IgG at 2.0, 0.2, 0.02 and 0.002 and 0.0002 mg/ml (top->bottom)

measured in transmission mode using a 15 micron BaF_2 sealed cell and the horizontal ATR approach on the Nicolet 8210 spectrometer. In the case of the ATR measurements the protein/buffer solution was placed in the trough for 15 minutes and then solution replaced with fresh buffer. At this point the spectrum of the adsorbed protein was measured. The protein spectra corrected for the absorptions due to the buffer are shown in Figure 3. This set of experiments was carried out to determine if adsorption of proteins to the zinc selenide crystals induce any structural change. Adsorption of proteins to certain surfaces is known to induce structural changes (Pitt et al., 1987; Jakobsen and Wasacz, 1990). The IR spectra of lysozyme are quite similar when recorded in transmission and ATR modes, suggesting that structural analysis results obtained using the ATR approach are comparable to those obtained using transmission techniques. Because of the low sensitivity of the transmission technique the proteinconcentration was fifty times that used for the ATR measurement. It may be mentioned here that even if the conformation of a protein changes upon adsorption to the crystal surface, the utility of the FT-IR analysis can still be appreciated for the relative conformational changes, an important parameters to understand the biochemical properties of proteins.

For comparison of the detectability of protein solutions by FT-IR to other commonly used techniques, we attempted to record fluorescence and circular dichroism (CD) spectra of the low concentration of lysozyme and IgG. A CD analysis of both lysozyme and IgG revealed that the protein signal below the 0.2 mg/ml protein concentration is not detectable in the far UV region. The proteins at 0.2 mg/ml concentration showed fairly good signal to noise ratios (Fig. 4). Fluorescence measurements of 0.02 mg/ml immunoglobulin G solution revealed

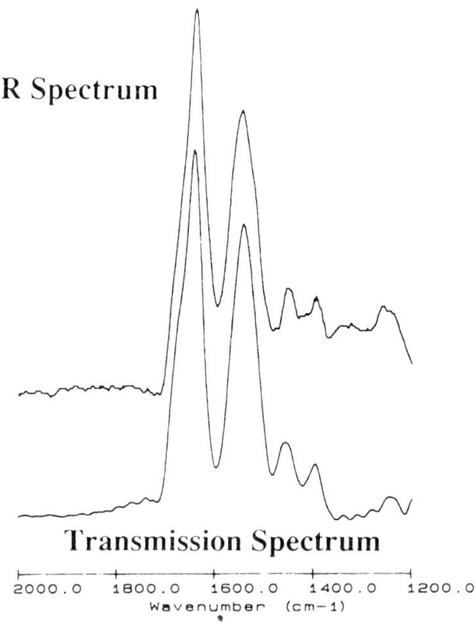

Figure 3. The infrared spectrum of a solution of IgG measured by ATR (top) and using a 15 micron BaF_2 transmission cell (bottom). In both cases the spectrum of the buffer has been spectrally subtracted from the sample spectra.

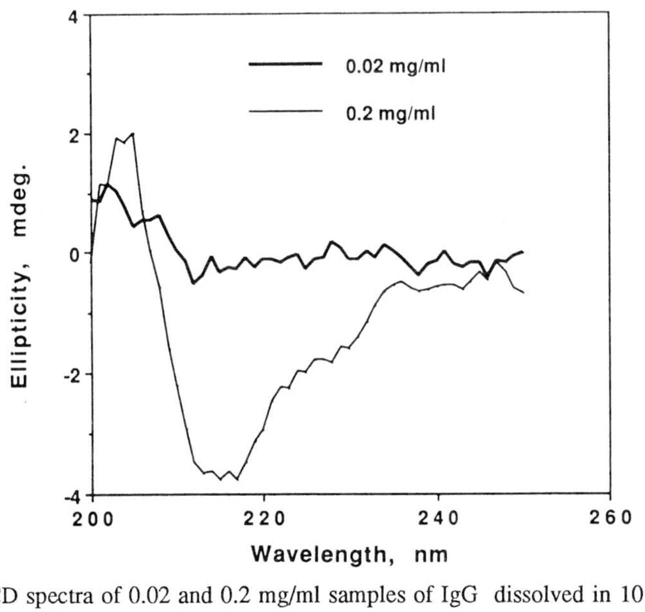

Figure 4. CD spectra of 0.02 and 0.2 mg/ml samples of IgG dissolved in 10 mM sodium phosphate buffer, pH 6.0. Spectra were recorded at room temperature (23-25 °C) using a 1 mm path length quartz cuvette. The figure derived from Singh and Fuller (1991).

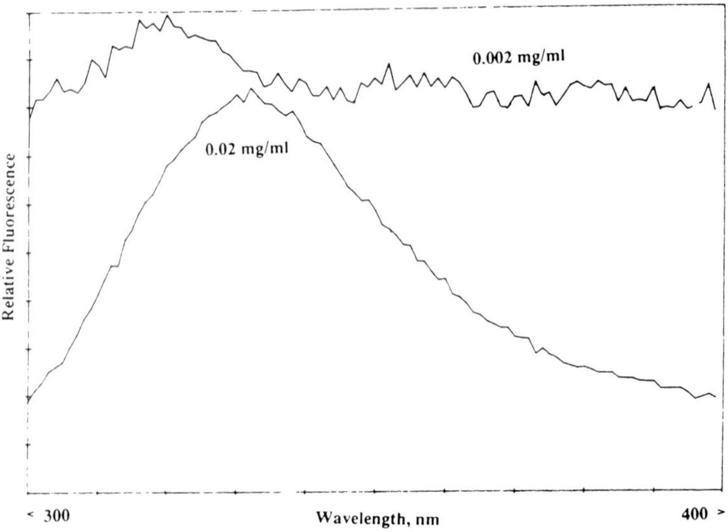

Figure 5. Intrinsic fluorescence spectra of 0.02 mg/ml and 0.002 mg/ml IgG dissolved in 10 mM sodium phosphate buffer, pH 6.0. Excitation wavelength was set at 280 nm. The spectral resolution at excitation and emission monochromators were fixed at 4 and 16 nm, respectively. All the spectral recordings were carried out at room temperature (23-25 ^{o}C). The figure derived from Singh and Fuller (1991).

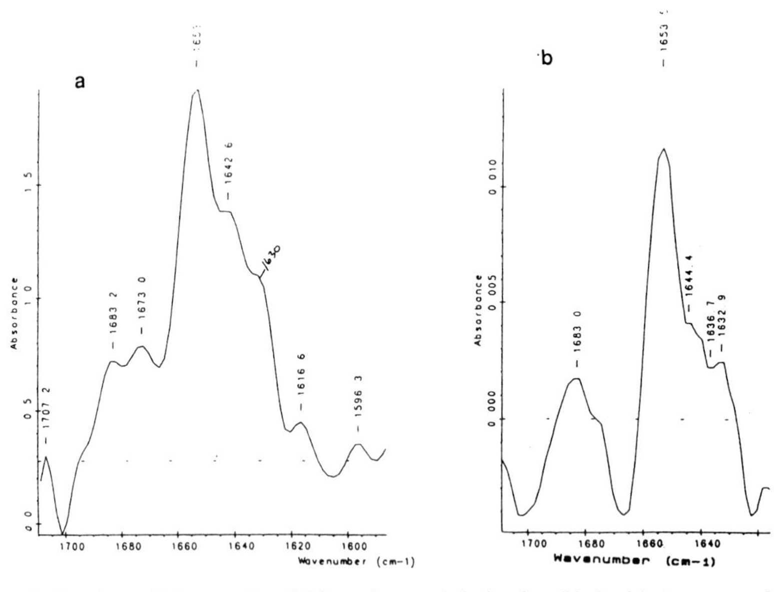

Figure 6. Fourier self deconvoluted (a) and second derivative (b) Amide I spectra of tetanus neurotoxin.

good signals (Fig. 5a). However, at 0.002 mg/ml concentration the signals were not detectable above the noise level (Fig. 5b), even with full gain and at highest sensitivity. Similar observations were made for lysozyme (data not shown).

For the estimation of secondary structure contents, the band positions of a protein spectrum are identified by Fourier self deconvolution and a second derivatization. A typical example of such an analysis is shown in Fig. 6 for tetanus neurotoxin, a protein which has also been examined by circular dichroism for its secondary structure (Singh et al., 1990a). A curve fitting of the identified bands of IR spectrum typically results in several bands as shown in Fig. 7. Amide I band is most prominent of the protein IR bands that are affected by the secondary structural folding. Many researchers have therefore used this band to derive secondary structural contents. One problem of this approach is the main band at 1650-1654 cm^{-1} being the contribution of amide bonds in α-helical conformation as well as in the unordered forms. A band at 1654 cm^{-1} could be assigned to α-helix, and at 1650 cm^{-1} to unordered structures (Susi and Byler, 1986; Dong et al., 1990). However, most commercial available equipments can not reliably distinguish such a small difference. While it is relatively easy to assign this band to α-helix or unordered structure when using standard proteins of known x-ray structure, for unknown proteins, it is not clear what conformation should be derived from this band. To address this dilemma, we also used the Amide III frequency region to derive secondary structure contents based on the assignments made by Kaiden et al. (1987). We assigned the curve-fitted band at 1652±2 cm^{-1} to a combination of α-helix and the unordered structures, and resolved the content of the two

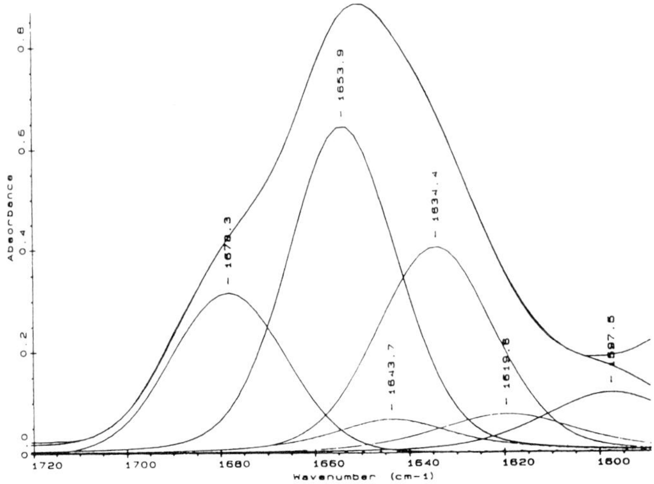

Figure 7. The curve-fitted infrared spectrum of tetanus neurotoxin. The top curves are original recorded curve and the synthetic curve derived from the composite of resolved bands.

Table I. FT-IR spectral band positions and fractional areas of Amide I and Amide III bands of tetanus neurotoxin. Bands are also assigned for secondary structural conformation, and the estimated secondary structural contents are compared with the CD derived data*

Secondary structure	band positions, cm^{-1} (% band area) Amide I	Amide III	Estimations (%) IR	CD
α-helix	1653(50) 1642(1)	1317(8.7); 1302(4.2) 1293 (0.1); 1288(9.4)	22	20
β-pleated sheets	1678(20) 1633(29)	1247(18.3); 1241 (0.3) 1232(32.2)	49	51
β-turns	1642(1)		0	0
unordered	1653(50) 1642(1)	1272(10.4); 1259(16.4)	29	30

*Data derived from Singh et al. (1990a).

Table II. FT-IR-ATR detected Amide I bands of lysozyme and IgG.

Condition*	Lysozyme Amide I bands (nm)	Lysozyme Band area (%)	IgG Amide I bands (nm)	IgG Band area (%)
Exposed to buffer	1678	18.2	1688	1.7
	1653	36.1	1661	24.9
	1642	0.6	1633	69.4
	1629	45.1		

*Lysozyme and IgG solutions used for adsorption contained 0.002 mg protein/ml of 10 mM sodium phosphate buffer, pH 6.0. The Data derived from Singh and Fuller (1991).

structures from the relative areas of their respective bands in the Amide III region. Result of such an approach is shown in Table I.

The resolved band area from curve fitting results of the spectra of lysozyme and IgG at 0.002 mg/ml are shown in Table II. The peak intensity at about 1630 cm^{-1} was less than 0.002 absorbance units. At that concentration, the signal-to-noise-ratio was still adequate to obtain useful secondary structural information of the adsorbed protein (Singh and Fuller, 1991). Exposure of the adsorbed protein to methanol causes a shift of the band at 1653 to 1657 cm^{-1}, and an increase of 50% in the absorbance intensity (Fig. 8). This effect is reversed by replacing the methanol with a portion of the original buffer solution (Fig. 8). Methanol has been used in the past to perturb the conformation of proteins, and is known to increase α-helical contents (Wasacz et al., 1987; Singh et al., 1990b). Similar results were obtained for the set of IgG[5] samples. The results presented here therefore suggest that a concentration of a 13 nM protein can be used not only to analyze its secondary structures, but also conformational changes can be monitored as a function of the microenvironment of the protein. We also attempted to record IR spectra for 7 nM IgG, and were able to detect the protein bands after leaving the protein solution on the ATR crystal for about one hour (spectra not shown, see Fig. 2). It may also be mentioned that only 1.5 ml of 0.001 mg/ml IgG was required for this measurement. Thus, with the FT-IR[1] technique in combination with ATR[7] accessory, we needed less than 15 picomole protein sample for the structural analysis.

A possible reason for the higher detectability of protein by the IR-ATR spectroscopic approach, compared to CD, is due to a higher extinction coefficient for IR absorption than for the circularly polarized light in the far UV region which is used for secondary structural analysis. In addition, use of the Fourier approach permits recording of a large number of scans at a relatively high energy throughout which allows a high signal-to-noise-ratio. In case of fluorescence measurements, the signal can be observed only from a limited number of chromophores (viz. aromatic amino acids) in proteins. Out of the three types of aromatic amino acids, Phe has very low quantum yield for fluorescence (Teale and Weber, 1957). For IR spectroscopy, every amide bond acts as a chromophore and, thus, the technique can be used for structural analysis of proteins that have virtually no Tyr or Trp, or these chromophores are not efficiently fluorescent.

A higher detectability of the low concentration protein solution by FT-IR/ATR technique permitted us to determine adsorption density of lysozyme on ZnSe crystal. This method is unique in the sense that we observed a direct measure of the protein adsorption from the aqueous solution on a real time basis. An adsorption isotherm for lysozyme is shown in Fig.9. Our future plan includes estimation of adsorption density of proteins on different types of surfaces relevant to biological processes.

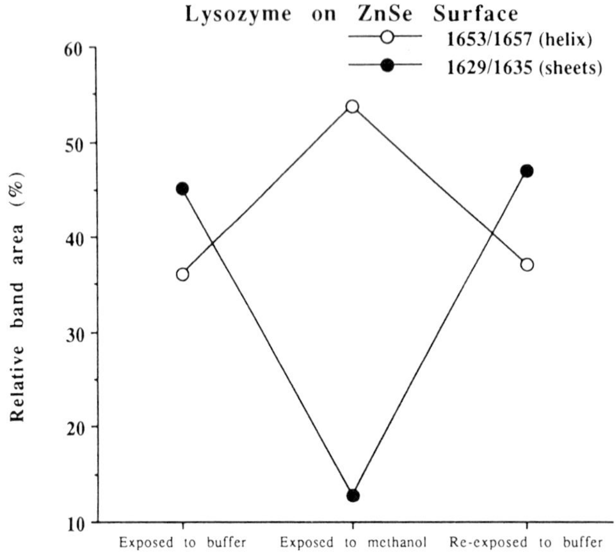

Figure 8. Effect of methanol on the resolved bands of lysozyme. IR spectra were recorded after the lysozyme solution (0.002 mg/ml) was first incubated for 30 min for adsorption.

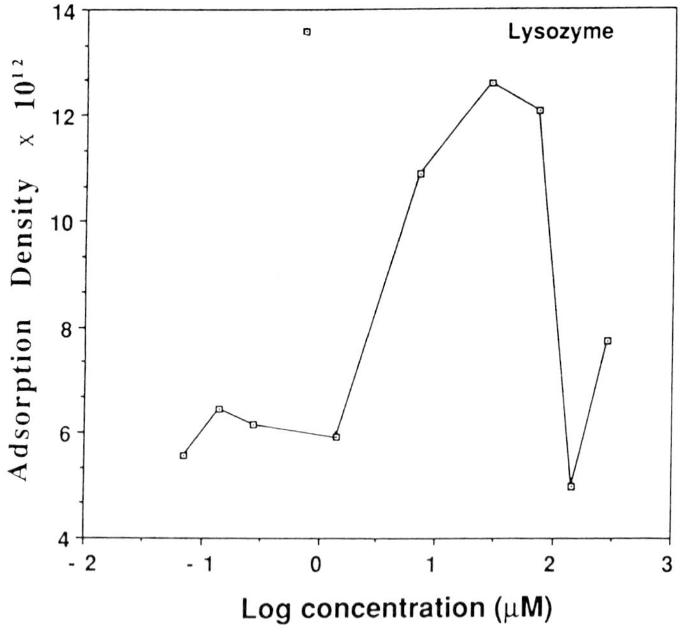

Figure 9. Adsorption isotherm of lysozyme on a ZnSe crystal surface as calculated from the FT-ATR spectroscopic data.

IV. Concluding Remarks

Use of FT-IR-ATR for the conformational analysis of low concentration proteins can be extremely important to biochemists and molecular biologists who have a minute amount of proteins purified, or who have produced small amounts of proteins by genetic engineering. Furthermore, for proteins or protein fragments that are not readily soluble in aqueous solutions, the FT-IR[1]-ATR approach may provide a useful means to analyze their structure. The technique should also provide a way to analyze the structure of proteins at higher concentrations, which is not possible with CD or fluorescence. A comparison of protein structure at high concentration such as in precipitate on crystals, and high concentration aqueous solution versus low concentration aqueous solution may provide information about the relationship between structure and biological activity. For example, botulinum neurotoxin at nanomolar concentrations shows different pharmacological activity on the nerve muscle preparation of mouse phrenic nerve hemidiaphragms than at micromolar concentrations (Maisey et al., 1988). We are working on this project to further elaborate the utility of the high sensitivity of FT-IR-ATR technique for the structure-function analysis of proteins.

The estimation of adsorption density on surfaces should allow to estimate protein binding to lipid layers or other surfaces such as those being tested for artificial organs.

In conclusion, FT-IR-ATR spectroscopy of polypeptides has a high potential for their conformational analysis at concentration levels that cannot be accurately analyzed by commonly used techniques such as circular dichroism or intrinsic fluorescence.

References

Byler, D. M., and Susi, H. (1986) *Biopolymers* **25**, 469.

Dong, A., Huang, P., and Caughy, W. S. (1990) *Biochemistry* **29**, 3303 .

Elliot, A., and Ambrose, E. J. (1950) *Nature* **165**, 921

Harrick,N.J.(1979) "Internal Reflection Spectroscopy" Harrick Scientific Corp., New York

Jakobsen, R. J., and Wasacz, F. M. (1990) *Appl. Spectroc.* **44**, 1478 .

Johnson, Jr., W. C. (1988) *Ann. Rev. Biophys. Biophys. Chem.* **17**, 145.

Kaiden, K., Matsui, T., and Tanaka, S. (1987) *Appl. Spectrosc.* **41**, 180.

Krimm, S. (1962) J. Mol. Biol. **4**, 528.

Krimm, S., and Abe, Y. (1972) *Proc. Natl. Acad. Sci. USA* **69**, 2788.

Krimm, S., and Bandeker, J. (1986) *Adv. Protein Chem.* **38**, 181.

Lakowicz, J. R. (1983). "Principles of Fluorescence Spectroscopy ", Plenum Press, New York.

Maisey, E. A., Wadsworth, J. D. F., Poulain, B., Shone, C. C., Melling, J., Gibbs, P., Tauc, L., and Dolly, J. O. (1988) Eur. *J. Biochem.* **177**, 683 .

Miyazawa, T., Shimanouchi, T., and Mizushima, S. I. (1956) *J. Chem.*

Phy. **24** ,408.

Pitt, W. G., Spielberg, S. H. and Cooper, S. L. (1987) In "Protein at Interfaces: Physicochemical and Biochemical Studies", ACS Symposium Series 343,(J. L. Brash and T. A. Horbett, Eds.), American Chemical Society, Washington, DC. pp. 324-338.

Singh, B. R., and Betley, M. J. (1989) *J. Biol. Chem.* **264**, 4404 .

Singh, B. R., and Fuller, M. P. (1991) *Applied Spectrosc.* **46**, in press

Singh, B. R., Kokan-Moore, N. P., and Bergdoll, M. S. (1988) *Biochemistry* **27**, 8730.

Singh, B. R., Schiavo, G.,and Fuller, M. P. (1990a) *Biophysical Chem.* **36**, 155

Singh, B. R., Wasacz, F. M., Strand, S., and Jakobsen, R. J. (1990b) *J. Protein Chem.* **9**, 705.

Sperline, R. P., Muralidharan, S., and Freiser, H. (1987) Langmuir **3**, 198.

Surevicz, V. K., and Mantsch, H. H. (1988) *Biochim. Biophys. Acta* **952**, 115.

Susi, H., and Byler, D. M. (1986) *Meth. Enzymol.* **26**, 445.

Teale, F. W., and Weber, G. (1957) *Biochem. J.* **65**, 476 .

Wasacz, F. M., Olinger, J. M., and Jakobsen, R. J. (1987) *Biochemistry* **26**,1464

RECOMBINANT HEMOGLOBINS FOR THE ELUCIDATION OF A NEW MECHANISM OF OXYGEN AFFINITY MODULATION BY CL^- IONS[1]

Clara Fronticelli

Department of Biochemistry, University of Maryland School of Medicine, 660 W. Redwood Street, Baltimore, Maryland 21201

I. INTRODUCTION

There is a group of hemoglobins like the one of ruminants and feloidea, who have developed a mechanism of oxygen affinity modulation by the Cl^- ions of the solvent as an alternative to the 2,3-diphosphoglycerate (DPG)[2] modulation found in human Hb. Hybrids of human and bovine Hb have been obtained: they indicated that the Cl^- regulation is produced by the β-chains of bovine Hb. Hydropathy plots of bovine and cat β-chains indicated the presence of two regions of different hydrophobicity with respect to β-chains human Hb, one at the N-terminal end, the other comprising part of the E-helix and the EF corner. Comparison of the amino acids sequence of these regions showed that constant amino acids substitutions are present in bovine and cat β-chains. A recombinant human Hb with the same β-N-terminal residues of bovine Hb failed to show an increased sensitivity to Cl^-. Porcine Hb which has in the β-E-helix the same substitutions of bovine and cat Hbs, also failed to show an increased sensitivity to Cl^-. A model is proposed in which this new mechanism of oxygen affinity modulation is produced by the concerted interaction of residues in the A and E helices which define a new oxygen linked Cl^- binding site.

[1]This research was supported in part by NIH grants HL13164 and HL33629.

[2]Abbreviations: DPG, diphosphoglycerate; Hb, hemoglobin; HbA, human hemoglobin; P_{50}, oxygen partial pressure at 50% hemoglobin saturation.

II. RESULTS AND DISCUSSION

Structure-function relationships in proteins can be explored at the molecular level by replacing, via site directed mutagenesis, specific amino acids with others of different characteristics.

This approach is being used in Hb to investigate the molecular mechanism of ligand binding to the heme. A series of mutant myoglobins and Hbs have been constructed in which His-E7 and/or Val-E11, in the distal site of the heme pocket, were replaced by a variety of amino acids of different size and polarity (1-6). These studies indicated that the same amino acid replacement affected ligand binding to myoglobin, α-chains and β-chains differently, suggesting the presence of a different molecular mechanism for the modulation of oxygen binding to the heme in these proteins.

A. Heterotropic ligands: Cl^- modulation of oxygen affinity

Another aspect of the structure-function relationship in Hb is the role of heterotropic ligands on the modulation of oxygen affinity. These effectors interact preferentially with the deoxy-form of Hb, increasing its stability and as a result decreasing the oxygen affinity of the system. This mechanism allows an efficient release of oxygen by Hb at the level of the tissues. Examples of heterotropic ligands are protons, Cl^- ions and 2,3-DPG.

There are some Hbs such as those of ruminants and feloidea whose oxygen affinity is not sensitive to 2,3-DPG (7,8). Notably, *in vivo* these species do not produce 2,3-DPG inside the erythrocytes: this suggests the presence in their Hbs of a different molecular mechanism for the modulation of oxygen affinity. This observation prompted us to investigate the effect of Cl^- ions on bovine and cat Hbs, which belong to the ruminant and to the feloidea family respectively. In Fig. 1 the open symbols show the titration curves of the oxygen affinity of human, bovine and cat Hbs in 50 mM Hepes buffer at pH 7.4 at increasing concentration of Cl^-. The close symbols show the oxygen affinity of the same Hbs in the presence of 1 mM 2,3-DPG. The data show that the oxygen affinity of human Hb is more sensitive to 2,3-DPG than to Cl^-. In contrast, bovine and cat Hbs are more sensitive to Cl^- than to 2,3-DPG and, in the presence of Cl^- at physiological concentration (100 mM), the oxygen affinity of bovine and cat Hbs is insensitive to 2,3-DPG. The strong correlation between P_{50} and Cl^- concentration, especially evident near the physiological range of 100 mM Cl^-, suggests that these Hbs have developed a mechanism of oxygen binding modulation extremely sensitive to the concentration of Cl^- in the medium (9-12).

B. Hybrids of human and bovine hemoglobin

In order to investigate whether this new mechanism of

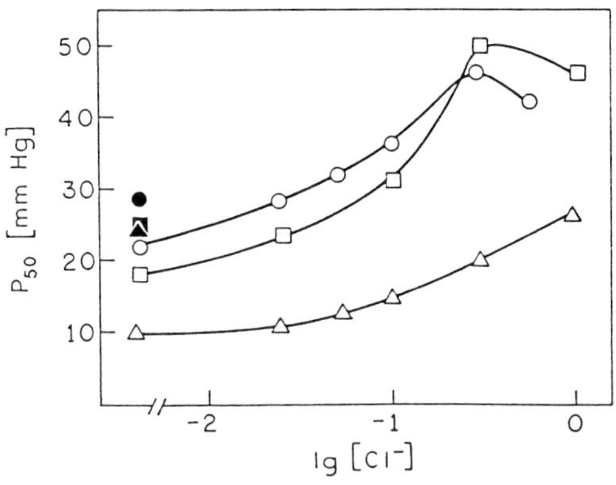

Fig. 1. Oxygen affinity, at 37°C, of human (△), bovine (□), cat (○) hemoglobins in 50 mM Hepes buffer at pH 7.4 and in the presence of increasing concentrations of Cl^-. Oxygen affinity of human (▲), bovine (■), cat (●) Hbs in 50 mM Hepes and 1 mM 2,3-DPG. The pH of the solutions was 7.4. These measurements have been done using the Hemox-Analyzer (TCS Medical Products), using a Hb concentration of 2 mg/ml.

oxygen affinity regulation is a function of the entire Hb tetramer or is confined to either the α or β chains, we have constructed two human-bovine Hb hybrids, one $\alpha^A a^A - \beta^{Bv}\beta^{Bv}$ has the α chains of human Hb and the β-chains of bovine Hb, the other $\alpha^{Bv}\alpha^{Bv} - \beta^A\beta^A$, has the α-chains of bovine Hb and the β-chains of human Hb. The rationale was that if the entire molecule participates in the mechanism of Cl^- modulation the two hybrids would, in the presence of Cl^-, have a similar oxygen affinity, otherwise only the hybrid containing the chains sensitive to Cl^- would have a much reduced oxygen affinity. Fig. 2 shows the oxygen affinity of the two hybrid Hbs at increasing concentration of Cl^-. The data indicate that in the absence of Cl^- the two hybrids had a very similar oxygen affinity, while in the presence of Cl^-, the oxygen affinity of the hybrid containing the β-chains of bovine Hb had a much lower oxygen affinity than the other. This experiment indicates that, in analogy with the allosteric modulation of 2,3-DPG in human Hb, the Cl^- modulation is produced by the β-chains of bovine Hb.

C. Hydropathic characteristics of the hemoglobin molecule

In a search for structural differences between human, bovine and cat β-chains we examined their hydrophobicity plots. They showed that the β-chains of bovine and cat Hbs have two regions with different hydrophobicity than the β-chains of human Hb. One region is more hydrophobic and comprises the N-terminal end, the other region is more hydrophilic and comprises portions of the E helix and the EF corner. The hydrophobicity values derived by Kyte and Doolittle (13) were used for these plots (Fig. 3).

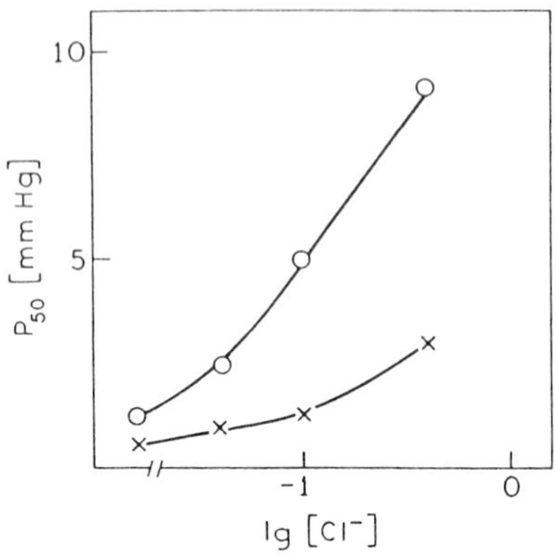

Fig. 2. Oxygen affinity, at 25°C, of human-bovine hybrids $\alpha^A\alpha^A$-$\beta^{BV}\beta^{BV}$ (O) and $\alpha^{BV}\alpha^{BV}$-$\beta^A\beta^A$ (X) in 50 mM Hepes buffer at pH 7.37 and in the presence of increasing concentration of Cl^-. These measurements have been done using the thin layer dilution method (16) using a Hb concentration of 20 mg/ml.

In order to correlate the different hydrophobicity with differences in the amino acids sequence, we compared the sequences of the NA1-A7 and E8-EF5 residues of the β-chains of several primate Hbs, which are 2,3-DPG sensitive and of ruminants and cat Hbs which are Cl^- sensitive. The results in Table I show the presence of constant amino acid substitutions between the primate and ruminant β-chains. In cat β-chains other hydrophobic substitutions are present in position NA1 and NA2 while in the E helix the substitutions are as for bovine Hb, except for the presence of a leucine at position E19 as in human β-chains.

To see whether either the hydrophobic characteristic of the N-terminal end or the hydrophilic characteristics of the E-helix could alone introduce in Hb the mechanism of oxygen affinity modulation by Cl^- ions, we have investigated the functionality of a recombinant human Hb with the N-terminal end of bovine Hb and of porcine Hb which has, in the E-helix of the β-chains, the amino acid substitutions present in bovine and cat β-chains.

D. Recombinant hemoglobin

An expression system for human β-globin has been developed by Nagai and Thogersen in which the β-globin is expressed in E. coli as a fusion protein CII-FX-globin (14). A recognition sequence of blood coagulation Factor Xa, is located precisely at the junction between the two polypeptide chains. Cleavage of the fusion protein with Factor Xa releases the β-globin. We have a plasmid of the Nagai-Thogersen type in which the NS1 gene from influenza replaces the CII gene in the Nagai-Thogersen plasmid. The construction of this plasmid and the reconstitution and characterization of

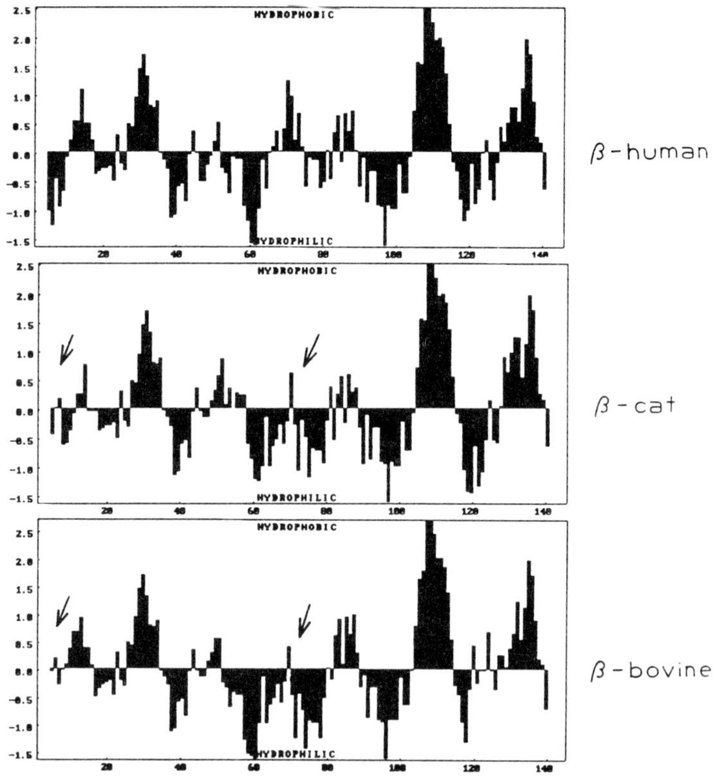

Fig. 3. Hydropathy plots of human, bovine and cat β-chains. These plots were obtained using the hydrophobicity values derived by Kyte and Doolittle (13). In bovine and cat β-chains, the arrows point to the regions which have a modified hydrophobicity with respect to human β-chains.

recombinant HbA is described elsewhere (15). The recombinant HbA has conformational and functional characteristics very similar to native Hb.

We have reconstituted a mutant, Hb-βNA1,2Met, in which the NA1 valine of human β-chains was replaced by a methionine and the histidine deleted. In Table II are listed the functional parameters of this mutant Hb and of natural HbA Hb, recombinant HbA, and bovine Hb. The data indicate that in the presence of 100 mM Cl^- the oxygen affinity of natural and recombinant human HbA was very similar; in Hb-βNA1,2Met the oxygen affinity was decreased, but not to the value present in bovine Hb. They also show that the oxygen affinity of Hb-βNA1,2Met was sensitive to 2,3-DPG and in its presence had the same value as natural HbA.

It can be concluded that although the increased hydrophobicity of the N-terminal end of the β-chains reduces the oxygen affinity of hemoglobin, it does not produce the high sensitivity to Cl^- found in bovine Hb. It will be

TABLE I

Amino acid differences in the regions NA1-A7 and E8-EF5 in primate β-chains and ruminant, cat and porcine ß-chains

	PRIMATES	RUMINANTS	CAT	PORCINE
NA1	Val	Met	NH3-Gly	Val
NA2	His	deleted	Phe	His
A2	Pro	Ala	Ala	Ala
E13	Gly	Asp	Asn	Asn
E14	Ala	Ser	Ser	Ser
E19	Leu	Met	Leu	Leu
E20	Ala	Lys	Lys	Lys
EF1	His	His	Asn	His
EF2	Leu	Leu	Ile	Leu

interesting to see whether substitution of A2 proline with themore flexible alanine will further modify the oxygen affinity of Hb-βNA1,2Met.

E. Porcine Hb

As shown in Table I the β-chains of this Hb have an amino terminal end as human Hb, however in the E-helix they have an aspartic acid at position E13, a serine at position E14 and a lysine at position E20 as in bovine and cat β-chains. The functional parameters shown in Table II indicate that porcine Hb has an oxygen affinity regulation very similar to human Hb.

F. Possible mechanism of oxygen affinity modulation by Cl^-

Both the substitutions in the βNA1-A7 and in the βE8-EF5 sequences are probably necessary for introducing in the hemoglobin a higher sensitivity to Cl^-. Many published considerations support the presence of an extra oxygen linked Cl^- binding site in bovine Hb (9-12). This site is probably part of the molecular mechanism of oxygen binding modulation in ruminant and feloidea Hbs.

The model illustrated in Fig. 4 may explain how the structural and conformational characteristics of the βNA1-A7 and βE8-EF5 regions combine to form an oxygen linked Cl^- binding site. In deoxy Hb, Perutz and Imai (8) have proposed that when the amino terminal end of the β-chains becomes hydrophobic it moves toward the interior of the protein, producing a distortion of the A-helix. In bovine Hb the distortion of the A-helix would result in the formation of a cluster of three positively charged amino acid side chains Lys-A5β, Lys-E20β and His-EF1β which could define a Cl^- binding site. The Cl^- is oxygen linked because only in deoxy

TABLE II

Oxygen binding parameters of natural HbA, recombinant HbA, recombinant Hb-βNA1,2Met, Hb bovine and Hb porcine. Conditions: 0.015 M Hepes, 100 mM NaCl at pH 7.35, 25°C, 25 mg/ml. Measurements using the thin layer dilution method (16).

Protein	P_{50} (torr)	n (Hill)	2,3-DPG (1 mM)
HbA	4.90	2.9	-
	8.26	2.9	+
HbA recomb.	4.53	2.5	-
Hb-βNA1,2Met	5.52	2.4	-
	8.30	2.4	+
Hb bovine	12.4	2.7	-
	12.4	2.7	+
Hb porcine	5.02	2.7	-
	8.40	2.7	+

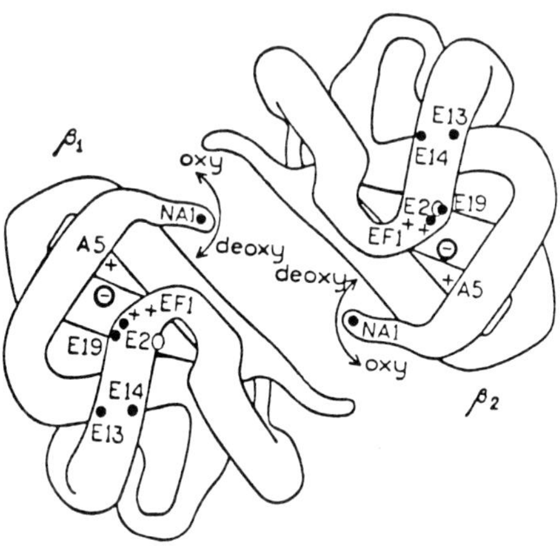

Fig. 4. β-chains of human hemoglobin viewed down the two-fold symmetry axis. The full circles represent the position of the constant substitutions of amino acids with different hydrophobicity in primate and ruminant β-chains. The positive sign identifies the residues involved in the Cl^- binding site. The chloride ions are represented by the open circles containing the negative sign.

hemoglobin does the distortion of the A-helix bring the Lys-A5β toward Lys-E20β and His-EF1β (it may be noted that residue His-EF1β may be redundant in bovine Hb, because in cat Hb there is an asparagine). The increased hydrophobicity of the E-helix may help the formation of the cluster by introducing a distortion of the helix which helps to correctly juxtapose the three residues of the cluster.

The only approach to the elucidation of the mechanism of oxygen affinity modulation by Cl^- is through genetic engineering. If our hypothesis is correct the construction of mutant human Hbs with different combinations of these amino acid substitutions will produce a system representing various stages in the formation of oxygen-linked Cl^- binding sites in human Hb.

ACKNOWLEDGMENTS

The site directed mutagenesis of HB-βNA1,2Met was done by Dr. William S. Brinigar at Temple University. I thank Dr. Enrico Bucci for the critical reading of the manuscript.

REFERENCES

1. Nagai, K., Perutz, M.F. and Poyart, C., (1985) Proc. Natl. Acad. Sci. USA, 82, 7552-7557.
2. Rohlfs, R.J., Mathews, A.J., Carver, T.E., Olson, J.S., Springer, B.A., Egeberg, K. D., Sligar, S.G. (1990), J. Biol. Chem. 256, 11788-11795.
3. Olson, J.S., Mathews, A.J., Rohlfs, R.J., Springer, B.A., Egeberg, K.D., Sligar, S.G., Tame, J., Renaud, J., Nagai, K. (1988) Nature 329, 265-266.
4. Mathews, A.J., Rohlfs, R.J., Olson,J.S., Tame, J., Renaud, J., Nagai, K. (1989), J. Biol. Chem. 264, 16573-16583.
5. Egeberg, K.D., Springer, B.A., Sligar, S.G., Carver, T.E., Rohlfs, R.J., Olson, J.S. (1990), J. Biol. Chem. 265:11788-11795.
6. Nagai, K., Luisi, B., Shih, D., Miyazaki, G., Imai, K., Poyart, C., De Young, A., Kwiatkowsky, L., Noble, R.W., Lin, S.-H., Yu, N.-T., (1987), Nature, 329, 858-860.
7. Bunn, H.F. (1971), Science 172, 1049.20.
8. Perutz, M.F., and Imai, K., (1980), J. Mol. Biol. 136, 183-191.
9. Fronticelli. C., Bucci, E., and Orth, C.(1984) J. Biol. Chem.259, 10841-10844
10. Fronticelli, C., Bucci, E., Razynska, A. (1988), J. Mol. Biol. 202, 343-348.
11. Fronticelli, C.,(1990) Biophys. Chem. 37, 141-146.
12. Razynska, A., Fronticelli, C., Di Cera E., Gryczynski, Z. and Bucci, E. (1990) Biophys. Chem. 38, 111-115.
13. Kyte, J. and Doolittle, R.F. (1982), J.Mol.Biol. 157, 105-132.
14. Nagai, K. and Thogersen, C., (1984) Nature 309, 810.
15. Fronticelli, C. O'Donnel K.S., Brinigar W.S. Accepted for publication in J. Protein Chemistry
16. Dolman, D. and Gill, S.J., (1978) Anal. Biochem. 87, 127-132

AN IMPROVED STRATEGY FOR THE DETERMINATION OF THE ROLE IN CATALYSIS OF AMINO ACID RESIDUES IN THYMIDYLATE SYNTHASE[1]

James R. Appleman
Department of Biochemical and Clinical Pharmacology
St. Jude Children's Research Hospital, Memphis, TN 38101

J. Ernest Villafranca
Agouron Pharmaceuticals, Inc.
La Jolla, CA 92037

I. Introduction

Enzymes are unparalleled as catalysts with respect to both rate acceleration and specificity. Thus one of the major questions of interest to enzymologists remains: How do enzymes work? More precisely, what is the physical arrangement (three-dimensional structure) of a particular enzyme of interest and how does this arrangement, including positioning of specific functional groups, catalyze the particular chemical reaction that is the enzyme's raison d'être? Herein we will discuss the strategy that we are employing to address these two questions for thymidylate synthase from *Escherichia coli* (TS). We believe that this strategy is substantially better than less rigorous methods similarly based upon site-directed mutagenesis and analysis of mutant enzymes.

TS catalyzes the conversion of dUMP and [6R]-5,10-CH_2-H_4folate (CH_2H_4folate) to dTMP and H_2folate by reductive methylation, in which CH_2H_4folate serves as both methyl donor and reducing agent. This is the only known intracellular reaction in which any of the forms of H_4folate are oxidized to H_2folate. TS plays a key role in pyrimidine biosynthesis, providing the only *de novo* source of dTMP which is required for DNA synthesis. Thus, maintaining adequate levels of dTMP is particularly important in

[1]This work was supported in part by United States Public Health Service Research Grant R29 GM 45367 (to JRA) and by American Lebanese Syrian Associated Charities (JRA)

TECHNIQUES IN PROTEIN CHEMISTRY III

rapidly growing cells where DNA synthesis is rapid. The enzyme is therefore an important target of chemotherapeutic agents so that TS is of interest both as a model for enzymatic catalysis and for its clinical relevance.

In our effort to understand how TS works, we are the beneficiaries of a tremendous wealth of information derived from the work of a host of others for thymidylate synthase from *E. coli* and other sources. The individual steps involved in catalysis by TS may be divided into two groups; i) those that compose the so-called kinetic mechanism which consists of steps involved in association and dissociation of substrates and products, and ii) the series of chemical reactions in which enzyme-bound substrates are converted to bound products. Based upon initial velocity and product inhibition studies (reviewed in Santi and Danenberg, 1984), the kinetic mechanism is ordered as shown in Scheme 1. Once the substrates are bound to enzyme, at least five distinct chemical reactions are postulated to occur as substrates are converted to products (Scheme 2). Upon binding of dUMP to TS, an equilibrium (Reaction 2 in Scheme 2) is established between a complex in which dUMP is noncovalently bound and one in which this substrate is covalently attached to enzyme via Cys^{146} (B). Binding of CH_2H_4folate shifts the equilibrium in favor of the covalent complex (B) (Moore *et al.*, 1986 a,b). It is believed that the binding site for the folate is constructed so that it facilitates formation of the N5-iminium ion $CH_2^+{=}H_4$folate (A) from bound CH_2H_4folate (Matthews et al., 1990b). The ion (A) is subsequently attacked by the dUMP enolate ion to form a covalent ternary complex (C). This complex has been isolated by trichloracetic acid precipitation of protein within a few seconds of initiation of catalysis (Moore et al., 1986b). Stereochemical considerations indicate that this intermediate must undergo a conformational isomerization before eliminating H_4folate and a proton, generating an exocyclic methylene on the enzyme-bound nucleotide (D) (Slieker and Benkovic, 1984; Matthews *et al.*,1990b). The methylene is reduced by bound H_4folate and the 5,6 double bond is regenerated by elimination to form dTMP, H_2folate, and enzyme (Reaction 5) (Slieker and Benkovic, 1984). The means by which TS carries out each of these steps remains controversial. For example, some steps are believed to involve conformational changes by the enzyme. Analysis is complicated because of the structure of TS. The enzyme consists of a dimer of identical subunits. Each bears an active site that in some way communicates with the other, presumably through movement of protein backbone and sidechains, so that the sites do not act independently.

In order to elucidate the relationship between the structure of TS and its ability to catalyze each of the discrete steps involved in the overall reaction we are employing the following strategy:

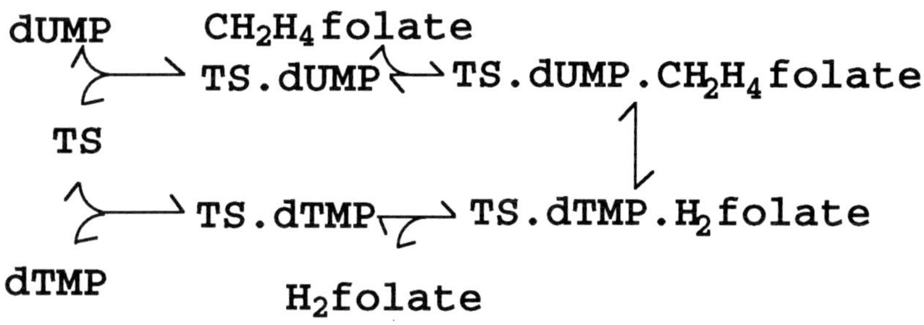

Scheme 1. Order of substrate binding and product release during the catalytic cycle of TS.

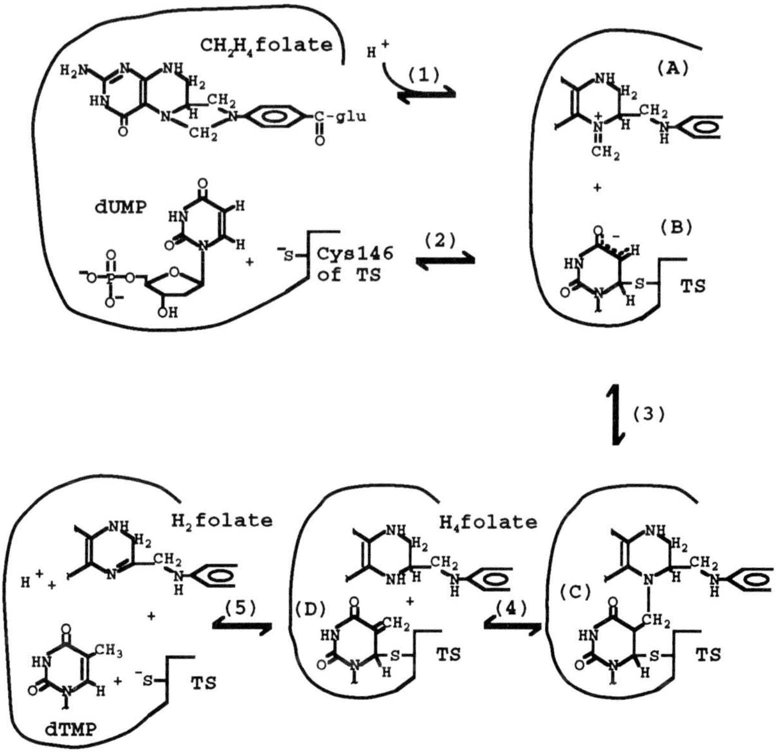

Scheme 2. Chemical reactions involved in conversion of enzyme-bound substrates to bound products.

Step 1) The three-dimensional structure of wild-type (wt) TS is defined by x-ray crystallography.

Step 2) A detailed scheme for all definable steps in the catalytic cycle with wt TS is constructed.

Step 3) Based upon three-dimensional structural data for TS and comparison of known primary sequences of thymidylate synthases, amino acid residues postulated to be important in the catalytic function of the enzyme are selected.

Step 4) Altered enzymes with single amino acid substitutions at these sites are prepared by oligonucleotide-directed mutagenesis.

Step 5) The structures of mutant enzymes are determined by x-ray crystallography to ensure integrity.

Step 6) Studies are carried out in order to define what steps in the catalytic cycle (if any) have been altered by the amino acid substitution. Typically these include investigation of the kinetics of catalysis preceding steady state so that individual reaction steps can be observed. Rate and equilibrium constants governing altered steps are determined.

Step 7) The role of the original amino acid residue is suggested from the change in behavior of the enzyme produced by the amino acid substitution.

The defining principle of this strategy is that all mutants with interesting properties must be scrutinized for structural perturbations and that the kinetics of individual steps in the catalytic cycle rather than just overall reaction rates must be examined. This commitment to detailed kinetic studies in combination with structural characterization sets this strategy apart from others that also are based upon analysis of mutant enzymes.

II. Procedures

Experimental conditions. All results were obtained at 20° C in a buffered solution containing 100 mM Tris, 50 mM 2-mercaptoethanol, at pH 7.4.

Enzyme assays. TS was assayed by monitoring the increase in absorbance at 340 nm accompanying conversion of CH_2H_4folate to H_2folate ($\Delta\epsilon_{rxn}$ = 6.4 $mM^{-1}cm^{-1}$). For wild type TS, k_{cat} is 1.6 s^{-1} and K_m values are 3.0 and 13 μM for dUMP and CH_2H_4folate respectively.

Equilibrium dissociation constants (K_ds). Dissociation constants were determined by titrating TS or TS.dUMP with dUMP or MTX+Glu_4, respectively and monitoring the change in enzyme fluorescence that accompanies complex formation. Corrections for ligand fluorescence, when present, and inner filter effects were calculated from parallel titrations in which enzyme was absent or replaced, respectively, with a quantity of tryptophan with a fluorescence intensity equal to that of the enzyme. In order to determine the value of K_d, the dependence of fluorescence on ligand and protein concentration was fitted as described by Appleman *et al.* (1988).

Stopped-flow spectroscopy. Measurements during the pre-steady-state phase of catalysis were made with either a Hi-Tech Scientific PQ-SF53 or Applied Photophysics SF.17MV stopped-flow spectrophometer interfaced to microcomputers. After collection, data were transferred to a Macintosh II computer for further analysis.

III. Results

The following example illustrates the application of our strategy to the study of TS.

Step 1) One of us was involved in the elucidation of the three-dimensional structure of TS in a complex with 5-fluoro-2'-deoxyuridylate and CH_2H_4folate. This complex is believed to be analogous to (C) in Scheme 2, albeit in a somewhat altered conformation (Matthews *et al.*, 1990b).

Step 2) A scheme for all kinetically definable steps in catalysis with wt enzyme has been derived. Steps include but are not confined to all those depicted in Schemes 1 and 2. For example, some steps involving change in protein conformation but no chemical reaction have been detected. Rate and equilibrium constants governing many of the steps in the complete scheme for wt enzyme have been measured utilizing techniques appropriate for rapid reactions (Appleman and Villafranca, unpublished results). Examples of some of the experimental measurements are described in *Step 6* below.

Step 3) One residue of particular interest is Asp^{169}, which is conserved in all sequences of thymidylate synthase described to date. High-resolution structures from x-ray crystallography (Matthews et al., 1990a,b; Montfort, 1990; Finer-Moore, 1990) indicate that this residue makes extensive contacts with both substrates. These include a side-chain hydrogen bond to N3 of the pteridine ring of CH_2H_4folate and a main chain hydrogen bond to O2 of dUMP. Furthermore, Asp^{169} makes hydrophobic contacts with the ribose moiety of dUMP (via main and sidechain) and with the pteridine ring of CH_2H_4folate (via sidechain). It was postulated that Asp^{169} served as a general acid during catalysis, indirectly donating a proton to N5 of the pteridine ring of CH_2H_4folate in the ternary covalent complex and thus promoting elimination (Matthews *et al.*, 1990b). Replacement of Asp^{169} with a residue that is incapable of proton donation should block or greatly slow the reaction at this step (Reaction 4 in Scheme 2).

Step 4) A mutant enzyme (D169N) was constructed in which Asp^{169} was replaced by Asn in order to test this hypothesis. Such a substitution should maintain most of the interactions with substrates that are important in binding but the catalytic rate should be profoundly altered. As anticipated, the mutant enzyme is completely inactive.

Step 5) Since a number of trivial explanations including the inability to correctly fold or bind substrates could account for such a "dead" enzyme, the following characterization was performed to ensure the integrity of the mutant. The three-dimensional structure was determined by x-ray crystallography when dUMP and 10-propargyl-5,8-dideazafolate (an inhibitor that is structurally analogous to CH_2H_4folate) were both bound to enzyme. The position of protein backbone and sidechains and orientation of bound ligands were essentially identical to that reported previously for wt enzyme (Matthews *et al.*, 1990a).

Step 6) Dissociation constants were determined from the dependence of the quenching of protein fluorescence upon ligand concentration as depicted in Figure 1 for dUMP. The affinity of wt and D169N TS for both this substrate and 4-amino-4-deoxy-10-methylPteGlu$_5$ (MTX+Glu$_4$), another inhibitor that is an analog of CH_2H_4folate, were measured and found to be similar (Table 1). The increase in dissociation constant for each ligand caused by the mutation corresponds to a decrease in the free energy of binding of only about 1 kcal/mol. Furthermore, the decrease in protein fluorescence observed upon ligand binding was approximately the same for wt and mutant. These results provide confidence that the loss of activity in the mutant enzyme is not caused by gross structural perturbations or by inability to bind substrates, but is instead solely attributable to a complete block of one of the steps in the catalytic cycle.

The following measurements were therefore carried out in order to determine more precisely which step is blocked. When preformed wt TS.dUMP complex is mixed with CH_2H_4folate to initiate catalysis in a stopped-flow spectrophotometer, a burst of absorbance increase at 340 nm (A_{340}) is observed (Figure 2). This is followed by a linear increase in A_{340} with time until substrate depletion and/or product accumulation leads to a decrease in the rate (not shown). The linear time course in Figure 2 corresponds

Table I. Comparison of binding of ligands to wt and D169N TS

Enzyme Binding Species	Ligand	K_d (μM)		D169N/wt
		wt	$D^{169}N$	
TS	dUMP	3.9	28.	7.2
TS.dUMP	MTX+Glu$_4$	1.4	7.9	5.6

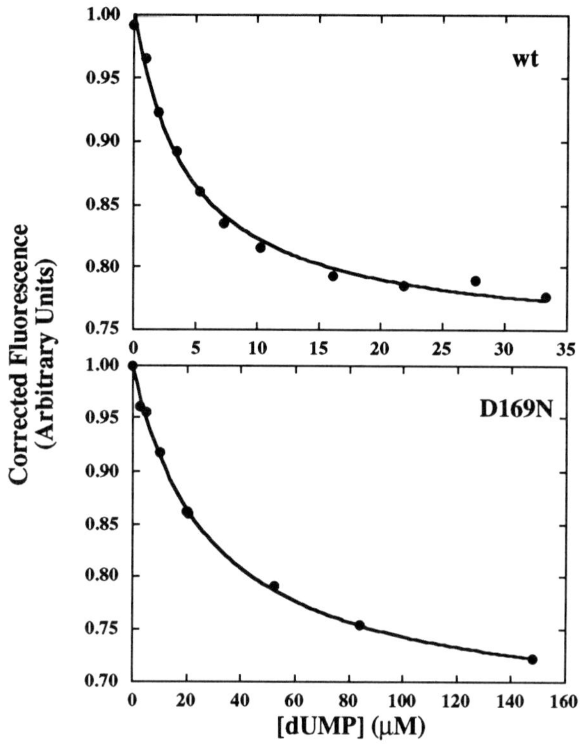

Figure 1. Dependence of fluorescence upon dUMP concentration as this substrate binds to TS for wt and D169N.

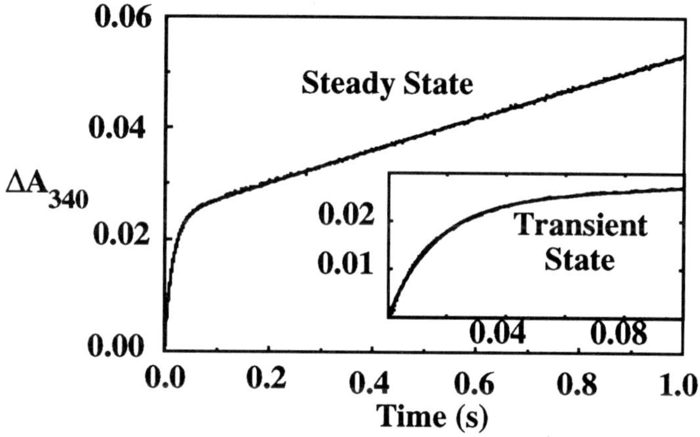

Figure 2. Time course of absorbance change at 340 nm when wt TS preincubated with dUMP is mixed with CH_2H_4folate to initiate catalysis. $\Delta A_{340} = A_t - A°$ as defined in the text. Both experimental data points and a best-fit curve are shown. Concentrations in the final reaction mixture are: TS, 3.11 μM dimeric enzyme; dUMP, 250 μM; CH_2H_4folate, 75 μM.

to the steady state, where the concentration of the various enzyme species that are present is not changing and the rate of appearance of products is the only measurable experimental parameter. In contrast, the burst in A_{340} preceding steady state (i.e. during the transient state) reflects absorbance changes that accompany some of the individual steps that occur between binding of substrates and release of products. At high substrate concentrations, the rate of the burst is independent of ligand concentration and the time course of increase in A_{340} is biphasic: $A_t = A^\circ + V_{ss}t + \Delta A_1(1-e^{-k_1 t}) + \Delta A_2(1-e^{-k_2 t})$ where A_t is A_{340} at time t, A° is the absorbance at t=0, V_{ss} is the linear rate of increase in A_{340} during steady-state, ΔA_1 and ΔA_2 are amplitudes of two phases that precede steady-state, and k_1 and k_2 are observed first-order rate constants governing each phase. From ten independent experiments, the value of k_1 was found to be $50 \pm 10\ s^{-1}$. The amplitude of the slower phase (ΔA_2) is small (~10% of ΔA_1) and the value of k_2 is about $8\ s^{-1}$. Based upon a number of lines of evidence to be discussed elsewhere, we believe that the fast phase corresponds to a reorientation of the C-terminus of the enzyme that seals the substrates into the active site after binding whereas the slow phase corresponds to formation of the covalent ternary complex followed by its rapid breakdown (Reactions 3 and 4 in Scheme 2). When D169N is substituted for wt in such a stopped-flow experiment, no increase in absorbance is observed.

Step 7) If catalysis were blocked prior to breakdown of the ternary covalent complex (Reaction 4), then a burst in A_{340} but no subsequent linear rate should have occurred. The absence of such a burst implies that, contrary to our expectations, Asp^{169} plays a critical role in at least one of the earliest chemical reaction in the catalytic cycle rather than a relatively late step. We are currently testing the hypothesis that Asp^{169} indirectly donates a proton to N10 of CH_2H_4folate as part of ring opening in Reaction 1 of Scheme 2. Modeling suggests that ring opening may be a prerequisite for the closure of the active site by movement of the C-terminus.

IV. Discussion

These results demonstrate how our strategy works in practice for determining the catalytic function of TS residues. A rational selection was made for both the residue of interest and the substitution to be examined, and in this case we picked a key residue. Unfortunately, the detailed kinetic and structural analyses of mutant enzymes that we believe are essential to understanding function are sufficiently time-consuming that only a limited number of residues and substitutions can be examined. Thus selection becomes critical, with the resulting disadvantage that we must be either wise or lucky. Three-dimensional structural information is fundamental. In its absence, we would primarily investigate residues that are conserved over a wide number of species. At least for TS, however, this would be a poor strategy. Michaels *et al.* (1990) have reported that four sites in TS (Gln-33, Arg-35, Asp-81, and Arg-127) that are either absolutely or very highly conserved can accommodate many other amino acids when assayed for ability to support bacterial growth. Climie *et al.* (1990) have reported similar results with TS from *Lactobacillus casei*. It is thus currently unclear why some residues are conserved. Michaels *et al.* (1990) and Climie *et al.* (1990) both employed strategies in which mutants with many substitutions at myriad different residues were examined, and they were guided by both residue conservation and structural information. The primary measure of activity in their studies was the ability of the altered TS to support bacterial growth. Such a broad gauge is useful in indicating the range of substitutions that can be accommodated at a particular site. However, little information concerning why a residue is important in catalysis is obtained. For example, the role of a residue that is essential in such an assay may be entirely related to maintenance of the folded three-dimensional structural. Even if the residue is integral to one of the catalytic steps, the nature of this role is not readily apparent. As demonstrated by our experience with D169N, only by defining and examining individual steps in the catalytic cycle with these mutant enzymes can the step(s) in which a residue participates be definitively determined.

A further advantage to our strategy of examining the involvement of residues in individual reaction steps is that the step does not have to limit the steady-state catalytic rate for us to quantitatively evaluate the effect of amino acid replacement. This is not true of many published studies describing investigation of mutant enzymes. The kinetic constants commonly determined by biochemists, k_{cat} and K_m, may provide only limited and perhaps misleading information. For example, Dev *et al.* (1989) examined a series of mutants at His^{147}. This residue had been postulated to act as a general base and facilitate deprotonation of Cys^{146}, which then carries out a nucleophilic attack upon dUMP (Reaction 1, Scheme 2). However, most of the mutant enzymes that these authors examined decreased k_{cat} by only about ten-fold so that this role was rejected. Such a conclusion may be premature. We have previously demonstrated that the slow step in catalysis is the reduction of the exocyclic methylene (Reaction 4, Scheme 2) (Appleman et al., 1990). The decrease in k_{cat} observed by Dev *et al.* with the mutant enzymes might be an underestimation of the effect of the amino acid substitution on the reaction step of primary interest (activation of the cysteine), perhaps by many orders of magnitude.

There are limitations to the information that can be obtained by mutagenesis. For example, a residue may be involved in the chemistry of two reaction steps. If mutagenesis at this site completely blocks an early reaction step as with D169N, then the role of the original residue in subsequent steps will not be revealed. For this reason we cannot reject the original role proposed for Asp^{169} but can only conclusively state that this is not its only role. Further, the study of mutant enzymes does not in itself provide the answer to the role of a particular amino acid side-chain in enzyme function. Instead, only the extent to which amino acid side-chains can substitute for one another is determined. Nonetheless, it should be possible from such comparisons to correctly describe the functions of individual residues in catalysis based upon our present knowledge of protein chemistry if care is taken to ensure that the structure of the mutant enzyme is like that of wt and if individual catalytic steps are examined.

References

Appleman, J.R., Howell, E.E., Kraut, J., Kühl, M., and Blakley, R.L. (1988) *J. Biol. Chem.* **263,** 9187.

Appleman, J.R., Villafranca, J.E., Matthews, D.A., and Blakley, R.L. (1990) *FASEB J.* **4**, A2301.

Climie, S., Ruiz-Perez, L., Gonzalez-Pacanowska, D., Prapunwattana, P., Cho, S.-W., Stroud, R., and Santi, D.V. (1990) *J. Biol. Chem.* **265**, 18776.

Dev, I.K., Yates, B.B., Atashi, J., and Dallas, W.S. (1989) *J. Biol. Chem.* **264**, 19132.

Finer-Moore, J.S., Montfort, W.R., and Stroud, R.M. (1990) *Biochemistry* **29**, 6977.

Matthews, D.A., Appelt, K., Oatley, S.J., and Xuong, Ng.H. (1990a) *J. Mol. Biol.* **214**, 923.

Matthews, D.A., Villafranca, J.E., Janson, C.A., Smith, W.W., Welsh, K., andFreer, S. (1990b) *J. Mol. Biol.* **214**, 937.

Michaels, M.L., Choll, W.K., Matthews, D.A., and Miller, J.H. (1990) *Proc. Natl. Acad. Sci. USA* **87**, 3957.

Montfort, W.R., Perry, K.M., Fauman, E.B., Finer-Moore, J.S., Maly, G.F., Hardy, L., Maley, F., and Stroud, R.M. (1990) *Biochemistry* **29**, 6964.

Moore, M.A., Ahmed, F., and Dunlap, R.B. (1986a) *J. Biol. Chem.* **261**, 12745.

Moore, M.A., Ahmed, F., and Dunlap, R.B. (1986b) *Biochemistry* **25**, 3311.

Santi, D.V. and Danenberg, P.V. (1984) *In* "Folates and Pteridines Vol. 1" (R.L. Blakley and S.J. Benkovic, eds), p. 345, Wiley-Interscience, New York.

Sliecker, L.J. and Benkovic, S.J. (1984) *J. Am. Chem. Soc.* **106**, 1833.

PROTEIN CONFORMATIONAL CHANGES IN THE μs TIME REGION INVESTIGATED WITH A LASER PULSE PHOTOLYSIS TECHNIQUE[1]

Andrew P. Billington, Norio Matsubara[2], Watt W. Webb[3] and George P. Hess[4]

Section of Biochemistry, Molecular and Cell Biology, Division of Biological Sciences, 217 Biotechnology Building, Cornell University, Ithaca, New York 14853-2703

I. INTRODUCTION

A superfamily of membrane-bound proteins regulates the transmission of signals at the junction between billions of cells in the mammalian nervous system (Stroud et al., 1990; Betz, 1990). The basic function of these receptor proteins is to form transient (~ 1 ms) ion-specific channels through the membrane of nerve or muscle cells (Kandel & Schwartz, 1985) upon binding a chemical signal (neurotransmitter) specific for each of the proteins (for instance, glycine, glutamate, aspartate, γ-aminobutyric acid (GABA), acetylcholine). By far the best characterized of these proteins is the nicotinic acetylcholine receptor from the electric organ of electric fish, *Torpedo* species and *Electrophorus electricus*. Reviews of the biochemical (Ochoa et al., 1989; Changeux, 1989), electrophysiological (Steinbach, 1989), structural (Stroud et al., 1990; Unwin, 1989), and kinetic (Hess et al., 1987) aspects, and the molecular biology (Claudio, 1989), of this protein have appeared.

---

[1]We gratefully acknowledge financial support by grants (GM04942 and RR04804) awarded to G.P.H. by the National Institutes of Health .

[2]Present address: Himeji Institute of Technology, Faculty of Science, Kanaji, Kamigori-cho, Ako-gun, Hyogo-ken, 678-12 Japan.

[3]School of Applied and Engineering Physics, Cornell University.

[4]Address correspondence to this author at 217 Biotechnology Building, Cornell University, Ithaca NY 14853 U.S.A.

The protein consists of five (α_2, β, γ, δ) homologous subunits (Reynolds & Karlin, 1978; Lindstrom et al., 1979). The receptors in the membrane of the *Torpedo* electric organ form ordered two-dimensional arrays (Stroud, 1983) and crystallize with dimeric symmetry (Brisson & Unwin, 1984). The application of a combination of techniques, including electron microscopy and x-ray diffraction, has led to a proposed structure at a resolution of between 17 and 12.5 Å (Toyoshima & Unwin, 1988; Mitra et al., 1989). The receptor protein protrudes about 60 Å above and 20 Å below the planes of the cell membrane, and forms an opening 20 to 25 Å wide on either side of the membrane. Between these openings, and passing through the hydrophobic core of the bilayer, is a much narrower portion about 30 Å long and less than 10 Å wide, presumably the closed transmembrane channel (Toyoshima & Unwin, 1988). The open transmembrane channel is highly selective for cations and allows the passage of about 10^4 ions ms^{-1}.

Measuring the rate of the conformational change of the protein form involved in the formation of open channels has remained a challenging task for many laboratories during the last decade (Jackson, 1988; reviewed by Madsen & Edeson, 1988). The information is of interest because the channel-opening rate provides a guideline for interpretation of structural models of the channel-opening process and sets limits to the rates with which signals between cells can be initiated and to the speed with which the nervous system can function. In all previous attempts to determine the rate at which the receptor-channel opens, measurements were made after the protein-mediated reaction, which involves ligand-binding steps, inactivation of the receptor by its own ligand, as well as the opening of the transmembrane channel, had reached a *quasi-* equilibrium (Jackson, 1988; reviewed by Madsen & Edeson, 1988). This approach has resulted in reported values for the rate constant for opening the transmembrane channel that differ from each other by a factor of as much as 25-fold (Jackson, 1988; Madsen & Edeson, 1988).

The objective of the chemical kinetic technique described here is to equilibrate the receptor with neurotransmitter rapidly enough so that the rate of formation of the open receptor-channel can be measured. To achieve this aim, not only for the acetylcholine receptor but also for other neurotransmitter receptors, we have synthesized biologically inactive precursors of several amino-group-containing compounds: the neurotransmitters glycine (Wilcox et al., 1990; Billington et al., 1991), γ-aminobutyric acid, glutamic acid, serotonin (Ramesh, Carpenter, & Hess, unpublished), and an analog of acetylcholine, carbamoylcholine (Walker et al., 1986; Milburn et al., 1989). The approach we used is based on investigations in which photolabile precursors of biologically inactive phosphates were synthesized (Kaplan et al., 1978; reviewed by McCray & Trentham, 1989). The mechanism for photolysis of the caged carbamoylcholine, *N*-(-α-carboxyl-2-nitrophenyl) carbamoylcholine, that we use in kinetic measurements has been described (Milburn et al., 1988):

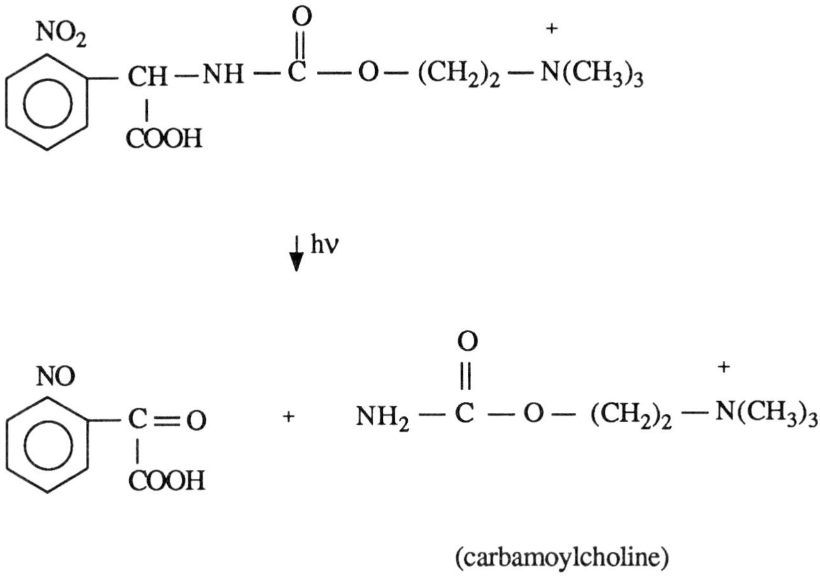

A cell containing the receptor protein is first equilibrated with the inactive neurotransmitter precursor, and a laser pulse of ultraviolet light is then used to liberate the neurotransmitter. Under physiological conditions, and with a laser pulse of 15 ns at 308 nm, the compound shown above photolyzes with a half-time less than 100 μs and a quantum yield of 0.8 (Milburn et al., 1989; Matsubara et al., 1991). The released neurotransmitter binds to its specific receptor protein, leading to the formation of transmembrane channels the concentration of which is measured by recording the resulting whole-cell current (Hamill et al., 1981; Marty & Neher, 1983). This approach has enabled us to investigate in the μs time region the conversion of the closed acetylcholine-receptor channel to the open form.

In this article we describe two experimental techniques. One is used to determine the rate of photolysis and the quantum yield of the photolabile precursors. These measurements are important because the properties of the photolytic reaction (rate, quantum yield) depend on pH (Walker et al., 1986; Milburn et al., 1989), the buffer concentration (Matsubara et al., 1991), the substituents on the α-carbon of the photolabile protecting group (Walker et al., 1986; Milburn et al., 1989), and the leaving group (Billington et al., 1991). The second technique is used to measure the rate at which the closed-channel form is converted to the open-channel form of the protein.

II. EXPERIMENTAL PROCEDURES

A. Measurement of the Photolysis Rate and Quantum Yield

The rate of photolysis and the quantum yield of photolabile compounds in which the *o*-nitrobenzyl group is utilized, can be measured by studying a uv photolysis-induced transient chromophore with an absorbance maximum between 400 and 440 nm, the *aci*-nitro intermediate (McCray & Trentham, 1989). The components of the apparatus to perform these measurements are given in Figure 1 and consist of

1. An excimer laser (Lumonics TE-861 M) used to provide 15-ns pulses of 308-nm light with a pulse energy of 50 mJ. A Gentec ED200 Joulemeter is used to measure the pulse energy.
2. A monitor beam, which is produced by a tungsten light source (Newport 780) and then passed through a quartz photolysis cuvette (Spectrocell Inc., 160 μL) at right angles to the laser beam, giving a path length for both beams of 2 mm. The monitor beam projects on to the detection apparatus, a photomultiplier-monochromator combination (EMI 9635QB, McPherson 275) arranged so as to view at 90° the laser beam immediately after it enters the cuvette, which is where the maximum transient is observed. A current-to-voltage convertor (EMI Gencom A1) is used to treat the photomultiplier output, and is in turn connected to a storage oscilloscope (Tektronix 549) where the results are displayed and recorded. Signal capture is optimized by using a variable time delay trigger (World Precision Instruments A310).

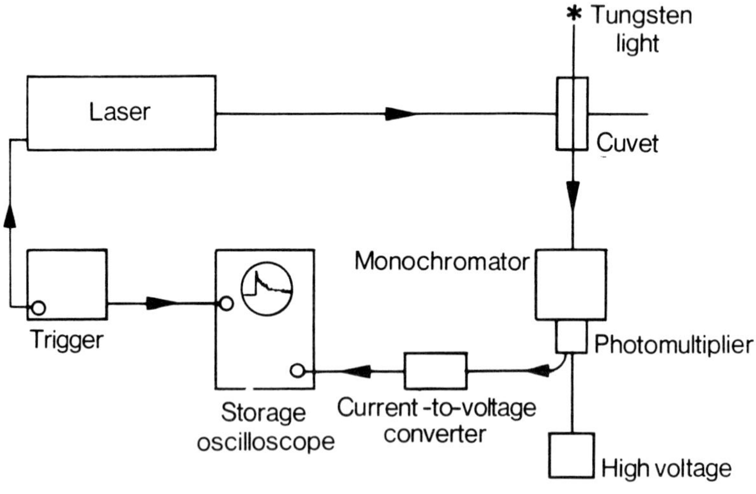

Figure 1. Laser flash photolysis apparatus (for details see the text).

B. Photolysis Procedure

Initially, a linearity test is performed on the detection apparatus, by attenuating the monitor beam with a known optical filter. The high voltage value of the photomultiplier is held constant, so that the detection apparatus is constrained to operate at constant gain. A sample of buffer solution (0.01 to 0.2 M, aqueous) is used as a control for artefacts. The selection of the concentration of photolabile compounds depends upon the signal level but is typically between 0.1 and 10 mM. The buffered photolabile compound is subjected to photolysis by a single laser pulse and the data recorded on the storage oscilloscope. Irradiation depletion of the amount of photolabile compound in the cuvette can be used in the following way to measure the product yield. The *aci* -nitro absorbance signal is considered to be a measure of the concentration of the product released (McCray et al., 1980; Walker et al., 1986, 1988); thus by measuring the absorption transient as a function of the number of laser pulses, the product yield can be established. The amount of laser light absorbed per pulse can be found by measuring with a Joulemeter the difference between the energy transmitted by the sample and that transmitted by buffer alone. Since we can measure the change in *aci*-nitro absorbance and the total energy absorbed, we can use the equation in the legend to Figure 4 to calculate the product yield and, therefore, the quantum yield.

C. Photolysis in situ with Cell Receptors

The components of the apparatus used to perform these measurements are depicted in Figure 2, and are comprised of

1. An inverted microscope (Zeiss IM35).
2. An apparatus for measuring the whole-cell current that has been described in detail elsewhere (Hamill et al., 1981; Marty & Neher, 1983).
3. A flash lamp pumped dye laser system, frequency-doubling crystal, and beam splitter which have been previously described (Milburn et al., 1989; Billington et al., 1991).
4. A cell-flow method with 10-20 ms resolution (Udgaonkar & Hess, 1987), including the flow device developed by Krishtal and Pidoplichko (1980), have been described in detail (reviewed by Hess et al., 1987).

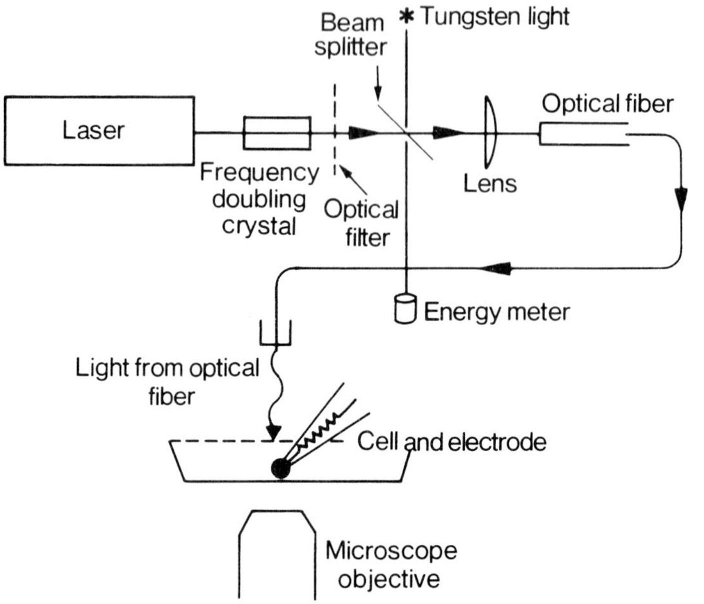

Figure 2. Apparatus for *in situ* photolysis (for details see the text).

D. *Experimental Procedures for in situ Photolysis*

By way of example, we will describe an experiment performed with a mammalian muscle cell line (BC3H1) (Schubert *et al.*, 1974) that contains the nicotinic acetylcholine receptor. These measurements were made with an earlier (Milburn et al., 1989) version of the apparatus than is described above but the results are the same as those obtained with the apparatus presently used (Matsubara et al., 1991). The cell to be used was located on the culture dish with the microscope, and the optical fiber moved into its position approximately; since the tungsten light source is projected through the fiber, it can be used to arrange the fiber position so that the diameter of the irradiated area is between 300 and 400 μm. A microelectrode containing intracellular buffer (145 mM KCl, 10 mM NaCl, 2 mM $MgCl_2$, 1 mM EGTA, and 25 mM HEPES buffer, pH 7.4; Udgaonkar & Hess, 1987) was used to achieve the whole-cell recording configuration and the cell was detached from the dish containing extracellular buffer (140 mM NaCl, 5 mM KCl, 1.8 mM $CaCl_2$, 1.7 mM $MgCl_2$, and 25 mM HEPES buffer, pH 7.4). A flow of carbamoylcholine at known concentration was directed at the cell from the flow device, and the resultant current recorded. The solution bathing the cell was then exchanged for extracellular buffer containing photolabile carbamoylcholine (0.1 to 1.0 mM), a uv pulse delivered, and the data recorded. The flow device was then used to remove solution containing photolyzed material from the culture dish and the same cell was then used for subsequent experiments. From the maximal current responses induced by the carbamoylcholine and from photolysis one can calculate the concentration of photo-generated carbamoylcholine since the relationship between the maximal current amplitude and carbamoylcholine concentration is known (Udgaonkar & Hess, 1987; Matsubara et al., 1991). Furthermore, the current amplitude both before and after the delivery of the uv pulse was determined by the cell-flow method and used to evaluate any possible effect of uv radiation upon receptor function.

II. RESULTS AND DISCUSSION

A representative photolysis trace of caged carbamoylcholine in Figure 3 shows the transient change in absorbance at 435 nm characteristic of the *aci*-nitro intermediate. The disappearance of this species is considered to reflect product formation (McCray et al., 1980; Walker et al.,1986, 1988) and in this case the half time for the appearance of liberated carbamoylcholine is, therefore, 50 μs. An experiment designed to calculate the quantum yield of this compound is illustrated in Figure 4. The peak *aci*-nitro signal declines exponentially (see Figure legend) as successive laser pulses are introduced; from the slope of the semi-logarithmic plot, one can calculate the product quantum yield (see the equation in Figure 4 legend).

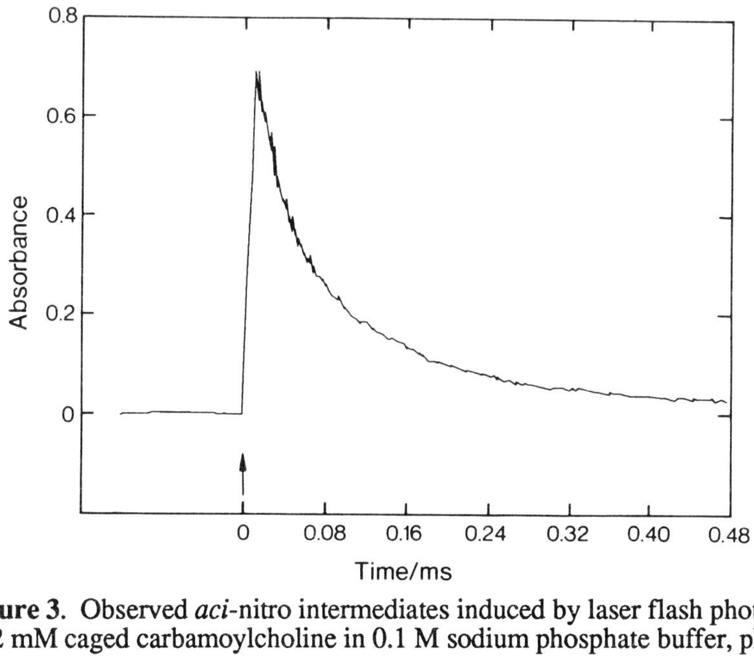

Figure 3. Observed *aci*-nitro intermediates induced by laser flash photolysis of 2 mM caged carbamoylcholine in 0.1 M sodium phosphate buffer, pH 7.5, 23 °C. The observation wavelength was 435 nm.

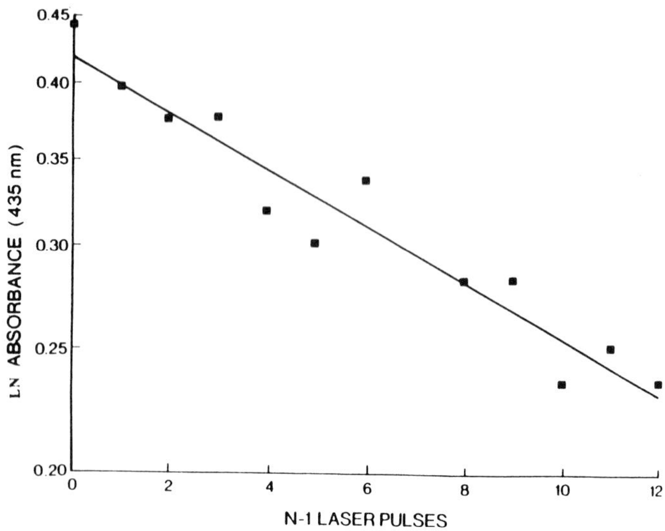

Figure 4. Semi-logarithmic plot of the peak *aci*-nitro absorbance, A, at 435 nm for caged carbamoylcholine as a function of the number of laser pulses N. The predicted variation of the absorbance is related to the absorbance observed after the first laser pulse, A_1, by $\ln A_N = \ln A_1 - \gamma\phi\,(N - 1)$; γ is the ratio of absorbed photons per laser pulse to the total number of molecules present and ϕ is the quantum yield.

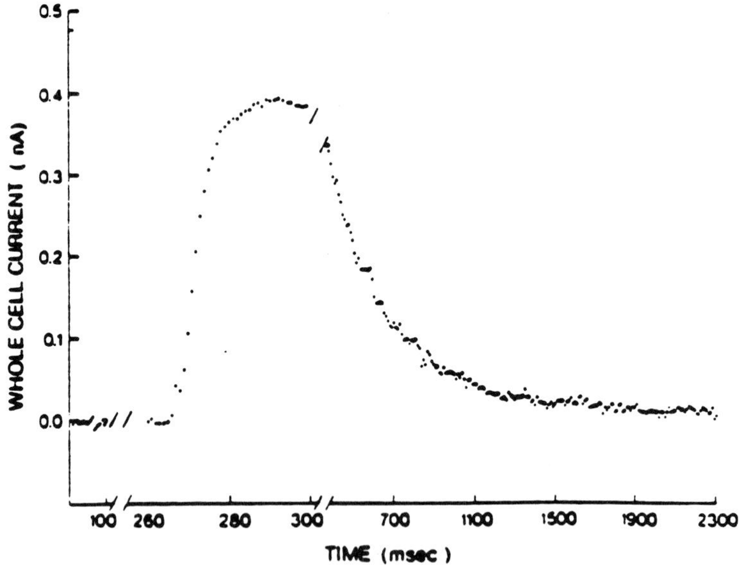

Figure 5. Laser flash experiment using caged carbamoylcholine (0.1 mM) *in situ* with a BC3H1 cell, pH 7.4, 23 °C. The laser pulse (600 ns) was delivered at t = 265 ms .

An experiment utilizing the laser pulse photolysis technique *in situ* with BC3H1 cells is shown in Figure 5. The whole-cell current induced by photolysis and consequent liberation of carbamoylcholine is due to the formation of open transmembrane channels and hence the transition of the receptor protein from the closed- to the open-channel form. The phases of the reaction are well resolved. The rising phase of the current occurs in the millisecond time region, and from the effect of carbamoylcholine concentration on the observed rate of this phase, one can deduce the rate constants for the closed- to open-channel transition and the reverse process. From the effect of carbamoylcholine concentration on the observed maximal current amplitude, one can evaluate the equilibrium constants that determine the concentration of open channels (Udgaonkar & Hess, 1986). From the falling phase of the current, shown on a different time scale, one can obtain the rate coefficients for the inactivation (desensitization) of the receptor protein induced by carbamoylcholine (Udgaonkar & Hess, 1986).

The integrated function of the nervous system depends on the rapid formation of transmembrane channels initiated by the binding of specific chemical signals (neurotransmitters) to receptor proteins. Using the apparatus and methods given here one can investigate the rate of formation of open transmembrane channels, and hence of the transformation of protein conformations that give rise to them. and therefore obtain information about the rates by which signals can be transmitted between cells of the nervous system and information can be received, stored and transmitted. The techniques described can be used not only with the acetylcholine receptor in cultured cells (Matsubara et al., 1991) but also with other receptors in cells from specific areas of the nervous system. A series of photolabile precursors of neurotransmitters have been synthesized for investigation of the channel-opening process of the glycine (Wilcox et al., 1990; Billington et al., 1991), γ-aminobutyric acid, glutamate, *N*-methyl-D-aspartate, and serotonin receptors (Ramesh, Carpenter, & Hess, unpublished). Furthermore, these photolabile precursors of neurotransmitters that we have synthesized can also be used in the identification of receptors associated with specific neurons in intact organisms. For instance, nerve cells associated with specific responses of the nematode *Caenorhabditis elegans* are known, but the receptors associated with these cells are not (Horvitz et al., 1982). Perfusing *C.elegans* with the photolyzable precursors has been tried so far only with carbamoylcholine and no deleterious effects were observed (Hess, unpublished). Photolysis of neurotransmitter precursors on the cell surface of a specific cell in *C.elegans* and observation of the response (for instance muscle contraction) can be used to identify the receptors on a specific cell. The spatial resolution required in such experiments can now also be obtained. Two photon excitation of the photolabile neurotransmitter precursors by the focused laser beam of a laser scanning microscope allows three dimensionally resolved neurotransmitter release with about 1 μm resolution at the focus of the exciting laser beam (Denk et al., 1990).

REFERENCES

Betz, H. (1990) *Neuron* **5,** 383-392.

Billington, A. P., Walstrom, K. M., Ramesh, D., Guzikowski, A. P., Carpenter, B. K., & Hess, G. P. (1991) (Unpublished results).

Brisson, A. & Unwin, N. (1984) *Nature,* **315,** 474-477.

Changeux, J.-P. (1989) *Fidia Research Foundation Neuroscience Award Lectures* **4,** 21-168.

Claudio, T. (1989) In *Frontiers in Molecular Biology* D. M. Glover & Hammes, B. D. (eds) IRL Press, pp 63-142.

Denk, W., Strickler, J. H., & Webb, W. W. (1990) *Science* **248,** 73-76.

Hamill, O. P., Marty, A., Neher, E., Sakmann, E., & Sigworth, F. J. (1981) *Pfluegers Arch.* **391,** 85-100.

Hess. G. P., Udgaonkar, J. B., & Olbricht, W. L. (1987) *Ann. Rev. Biophys. Biophys. Chem.* **16,** 507-534.

Horvitz, H. R., Chalfie, M., Trent, C., Sulston, J., & Evans, P. D. (1982) *Science* **216,** 1012-1014.

Jackson, M. B. (1988) *J. Physiol.* **397,** 555-583.

Kandel, E. & Schwartz, J. H. (1985) *Principles of Neuronal Science*, Second Edition (Elsevier, New York).

Kaplan, J. H., Forbush, B., & Hoffman, J. F. (1978). *Biochemistry* **17,** 1929-1935.

Krishtal, O. A. & Pidoplichko, V. I. (1980) *Neurosciences* **5,** 2325-2327.

Lindstrom, J., Merlie, J., & Yogeeswarani, G. (1979) *Biochemistry* **18,** 4465-4470.

Madsen, B. W. & Edeson, R. O. (1988) *Trends in Pharm. Sci.* **9,** 315-316.

Marty, A. & Neher, E. (1983) In *Single-Channel Recording*, eds. Sakmann, B. & Neher, E. (Plenum, New York), pp. 107-122.

Matsubara, N., Billington, A. P., & Hess, G. P. (1991). (Unpublished results).

McCray, J. A. & Trentham, D. R. (1989) *Ann. Rev. Biophys. Biophys. Chem.* **18,** 239-270.

McCray, J. A., Herbette, L., Kihara, T., & Trentham, D. R. (1980) *Proc. Natl. Acad. Sci. USA* **77,** 7237-7241.

Milburn, T., Matsubara, N., Billington, A. P., Udgaonkar, J. B., Walker, J. W., Carpenter, B. K., Webb, W. W., Marque, J., Denk, W., McCray, J. A., & Hess, G. P. (1989) *Biochemistry* **28,** 49-55.

Mitra, A.K., McCarthy, M. P., & Stroud, R. M. (1989) *J. Cell. Biol.* **109,** 755-774.

Ochoa, E. L. M., Chattopadhyay, A., & McNamee, M. G. (1989) *Cellular and Molecular Neurobiology* **9,** 141-177.

Ramesh, D. Carpenter, B. K. & Hess, G. P. (Unpublished results).

Reynolds, J. & Karlin, A. (1978) *Biochemistry* **17,** 2035-2038.

Schubert, D., Harris, A. J., Devine, E. E., & Heinemann, S. (1974) *J. Cell Biol.* **61,** 398-413.

Steinbach,J. H. (1989) *Ann.Rev.Physiol.* **51,** 353-365.

Stroud, R. M. (1983) *Neurosc. Commun.* **1,** 124-138.

Stroud, R. M., McCarthy, M. P., & Shuster, M. (1990) *Biochemistry* **29,** 11009-11023.

Toyoshima, C. & Unwin, N. (1988) *Nature* **336,** 247-250.

Udgaonkar, J. B. & Hess, G. P. (1986) *J.Membrane Biol.* **93**, 93-109.
Udgaonkar, J. B. & Hess, G. P. (1987) *Proc. Natl. Acad. Sci. USA* **84**, 8758-8762.
Unwin, N. (1989) *Neuron,* **3**, 665-676.
Walker, J. W., McCray, J. A., & Hess, G. P. (1986) *Biochemistry* **25**, 1799-1805.
Walker, J. W., Reid, G. P., McCray, J. A., & Trentham, D. R. (1988) *J. Am. Chem. Soc.* **110**, 7170-7177.
Wilcox, M., Viola, R. W., Johnson, K. W., Billington, A. P., Carpenter, J. A., McCray, A. P., Guzikowski, A. P., & Hess, G. P. (1990). *J. Org. Chem.* **55**, 1585-1589.

Macromolecular Dynamics Observed by Intramolecular Energy Transfer Using Frequency-Domain Fluorometry

J.R. Lakowicz
J. Kuśba
I. Gryczynski
H. Szmacinski
W. Wiczk
and M.L. Johnson*

Center for Fluorescence Spectroscopy, Department of Biological Chemistry, School of Medicine, University of Maryland, 660 West Redwood Street, Baltimore, MD 21201

* University of Virginia, Department of Pharmacology, Charlottesville, Virginia 22908

ABSTRACT

We used time-dependent fluorescence energy transfer to recover the site-to-site distance distributions and diffusion coefficients for a flexible donor-acceptor pair and for the proteins melittin (random coil and α-helix states) and troponin I (native and denatured). Energy transfer occurred from single tryptophan residues to dansyl acceptors. The differential equation describing the donor population in the presence of site-to-site diffusion (Biopolymers 17:11-31, 1978) was solved numerically. These data were used to obtain the predicted frequency-response for least squares analysis. Global analysis of progressively quenched samples was used to enhance the resolution of recovered parameters. The validity of the recovered distance distributions and diffusion coefficients was verified by simulations and by studies of the flexible methylene-chain linked donor-acceptor pair in solutions of varied viscosity.

I. INTRODUCTION

In recent years fluorescence resonance energy transfer (FRET) has been used to determine the distributions of donor (D) to acceptor (A) distances (1-6). In the present paper we extend our studies of distance distributions to allow determination of D-to-A diffusion coefficients of proteins and model compounds. Such data can reveal the extent of structural fluctuation of proteins in solution, as seen by the time-dependent changes in distance between the donor and acceptor. In order to effect energy transfer, these fluctuations must occur on the timescale of the lifetime of the donor, which in our case is 2-5 ns. This work is an extension of the promising work of Haas and co-workers (7,8) who were the first to report the effects of end-to-end diffusion on energy transfer in labeled peptides.

TECHNIQUES IN PROTEIN CHEMISTRY III

The effect of D-to-A diffusion is to increase the extent of energy transfer and to alter the form of the time-dependent donor decays. Simulations are used to reveal the form of the time and distance-dependence of the excited state donor populations and to predict the frequency-domain data. These data were used to construct χ^2_R surfaces for the recovered parameters, which reveal the resolution of these parameters. The ability to recover distance distributions and diffusion coefficients near 10^{-6}-10^{-7}cm^2/s was verified by studies of a methylene-chain linked D-A pair in propylene glycol at 20°C.

II. THEORY

We used a numerical solution of the diffusion equation for the distance normalized excited state fluorophore (donor) population $\bar{N}^*(r,t)$ (7,9),

$$\frac{\partial \bar{N}^*(r,t)}{\partial t} = -\frac{1}{\tau_d}\left[1+\left(\frac{R_o}{r}\right)^6\right]\bar{N}^*(r,t)+\frac{1}{N_o(r)}\frac{\partial}{\partial r}\left[N_o(r)D\frac{\partial \bar{N}^*(r,t)}{\partial r}\right] \quad (1)$$

τ_D is the decay time of the fluorophore in the absence of acceptor, R_o is the critical radius for long-range D-A energy transfer, D is the apparent diffusion coefficient, and $N_0(r)$ denotes the initial donor-to-acceptor distance distribution (at t=0) of the excited molecules.
We assume the t = 0 distribution is described by a Gaussian

$$N_o(r) = \frac{1}{Z}\exp\left[-\frac{1}{2}\left(\frac{r-R_{av}}{\sigma}\right)^2\right] \quad (2)$$

where Z is the normalization factor, the mean distance is given by R_{av}, and half-width (fwhm) is given by hw = 2.355σ.
The donor fluorescence decay. I(t) can be calculated by integration of the product $N_0(r)$ N^* (r,t) over the entire r interval

$$I(t) = \int_{r_{min}}^{r_{max}} N_0(r)\,\bar{N}^*(r,t)\,dr \quad (3)$$

The fitting procedure has been described elsewhere (10).

III. MATERIALS AND METHODS

Synthesis of the donor control (TMA) and donor-acceptor pair (TU2D) are described elsewhere (11,12). Synthetic melittin and dansyl melittin were the generous gifts from Professor F.G. Prendergast (Mayo Foundation). The troponin I (TnI) energy transfer data (from Trp 158 to Cys 133, which was labeled with N-(iodoacetyl)-N′-(1-sulfo-5-naphthyl) ethylene diamine (IE)) were obtained previously (13).

Frequency-domain measurements were performed using the instrument described elsewhere in detail (14,15).

IV. RESULTS & DISCUSSION

A. SIMULATIONS

The simulations were performed using assumed values of τ_D = 5ns, R_o = 25Å, R_{av} = 20Å, hw = 15Å, which are comparable to tryptophan-dansyl D-A system. Random noise in the phase and modulation data was δ_p = 0.2° and δ_m = 0.005 Diffusion coefficient was varied from $10^{-8}cm^2/s$ to $10^{-5}cm^2/s$ to cover the range expected for proteins and model compounds. Figure 1 shows frequency responses for different diffusion coefficients.

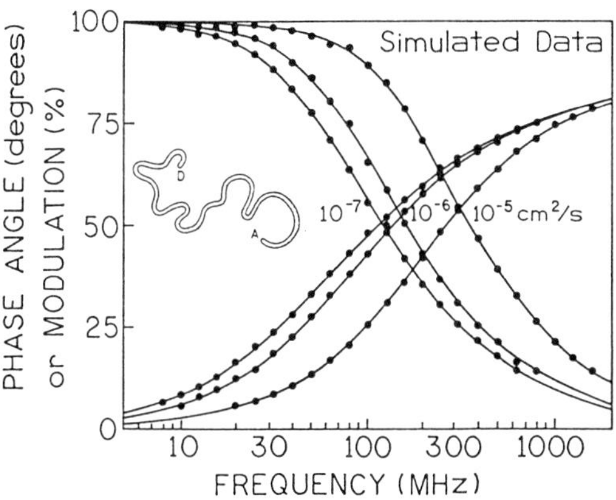

Figure 1. Simulated frequency-domain data for a flexible D-A pair with varying diffusion coefficients. The parameters used were: τ_D = 5 ns, R_o = 25Å, R_{av} = 20Å and hw = 15Å.

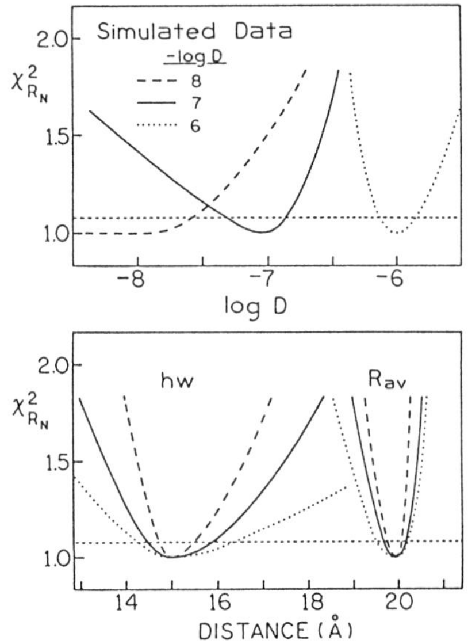

Figure 2. Diffusion coefficient and distance distribution χ^2_R surfaces for simulated data (see Figure 1).

The ability to resolve both, distribution parameters and diffusion was judged by analyzing the χ^2_R surfaces for each parameter (the two other parameters were floating). From the analysis (Figure 2) we have the following information:

- For slow diffusion ($D \leq 10^{-8}cm^2/s$), the resolution of the diffusion coefficient is poor.
- Resolution of the distance distribution parameters is better when diffusion is smaller.
- For the assumed D-A parameters and uncertainties in phase and modulation it is possible to resolve all three parameters using the donor decay only.

B. CAN EXTERNAL QUENCHING INCREASE THE RESOLUTION?

We considered 10-fold quenching, resulting in $\tau_D = 0.5$ ns and $R_o = 17$Å. We simulated quenched and non-quenched files (assuming a rather small diffusion coefficient of 10^{-7}cm^2/s) and then analyzed them separately and simultaneously.

Phase and modulation data for non-quenched and quenched samples are presented in Figure 3. The solid lines represent the best global fit to the data.

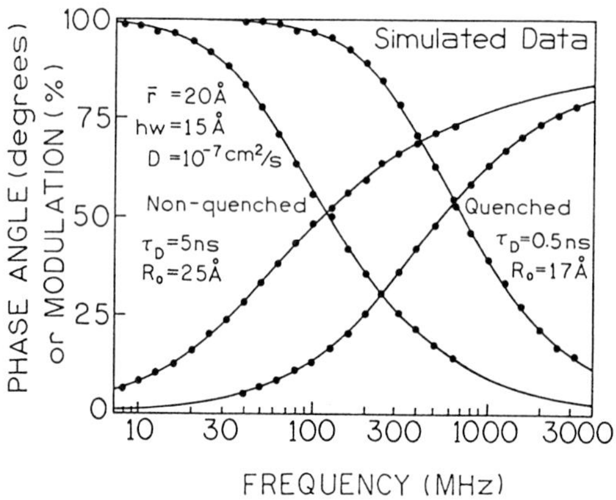

Figure 3. Phase and modulation data for non-quenched and quenched samples. The solid lines represent the best global fit to the simulated data.

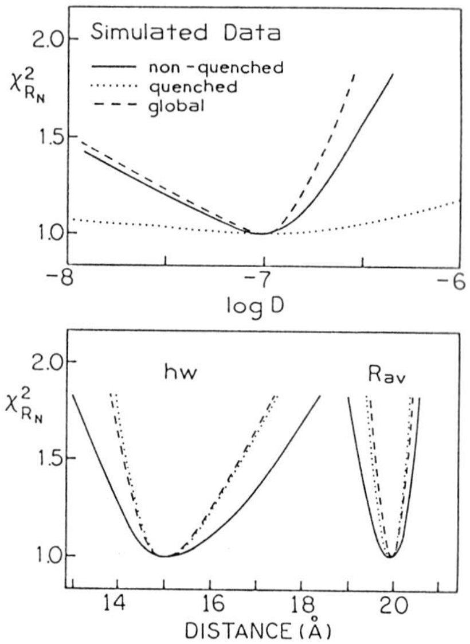

Figure 4. Diffusion coefficient and distance distribution χ_R^2 surfaces for global analysis of quenched and unquenched samples.

The χ_R^2 surfaces for all three recovered parameters are presented in Figure 4. For the quenched sample (••••) the resolution in diffusion is poor and in distribution parameters much better than for non-quenched sample (——). The global fit (– – –) provides increase of resolution in diffusion and gives resolution in distribution as good as quenched sample.

The external quenching and global analysis can increase the resolution in all three parameters, even if the diffusion coefficient is small.

C. TU2D AS A MODEL D-A SYSTEM

The global analysis of phase and modulation data for TU2D in propylene glycol, 20°C, is shown in Figure 5. The best fit to the data (solid lines) reveals a diffusion coefficient of $1.27 \times 10^{-7} cm^2/s$ and distribution parameters are close to those obtained in highly viscous solution at -5°C. (Figure 5, insert).

The χ_R^2 surfaces for the global analysis are presented in Figure 6.

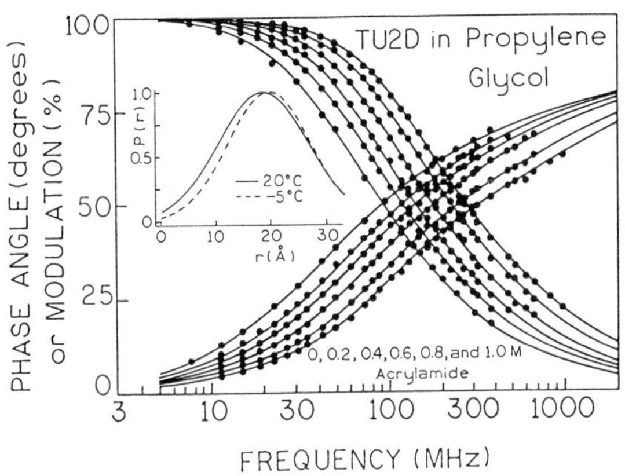

Figure 5. Phase and modulation data for progressively quenched D-A (TU2D) system. The dots are the experimental data, the solid lines the best fit with $D = 1.27 \times 10^{-7}$ cm^2/sec, R_{av} = 18.4Å and hw = 18.8Å.

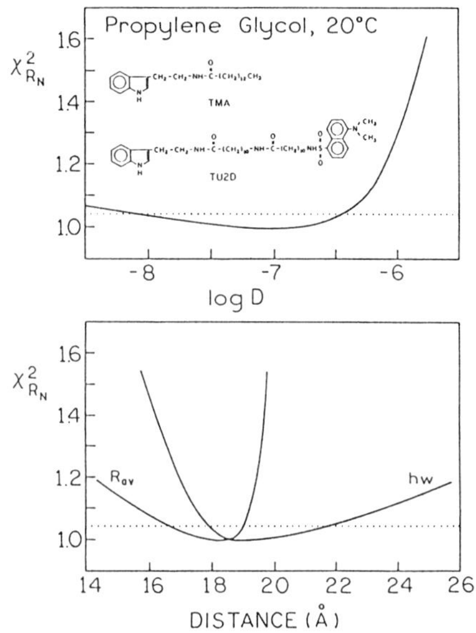

Figure 6. Distance distribution and diffusion coefficient χ_R^2 surfaces for TU2D in propylene glycol at 20°C.

D. END-TO-END DIFFUSION AND DISTANCE DISTRIBUTION IN MELITTIN.

It is known that the structures of melittin in water and methanol are different. In water melittin is almost fully random-coil and in methanol forms an α-helix (Figure 7).

We measured end-to-end diffusion and distance distribution from tryptophan (residue 19) to the N-terminal α-amino group, which was labeled with a dansyl residue. Figure 8 shows frequency responses of the donor (Trp) emission for random-coil peptide in the absence and presence of acrylamide.

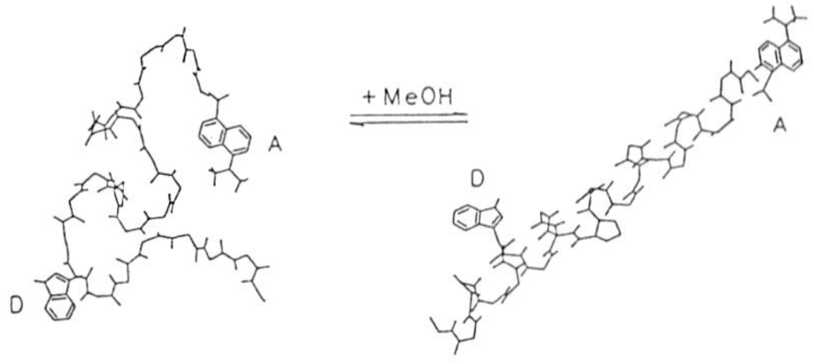

Figure 7. Random-coil and α-helical melittin.

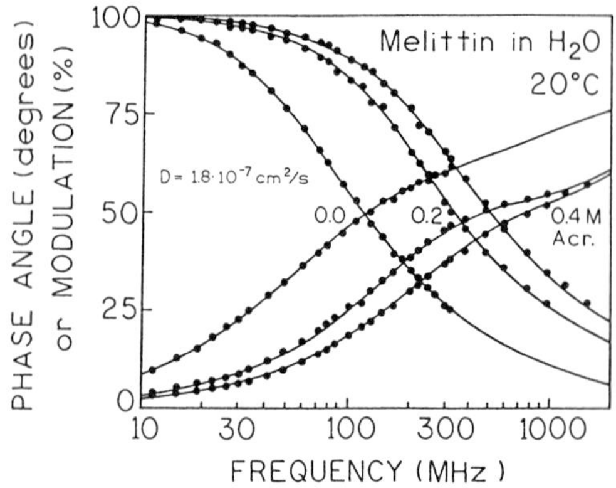

Figure 8. Phase and modulation data for melittin in H_2O.

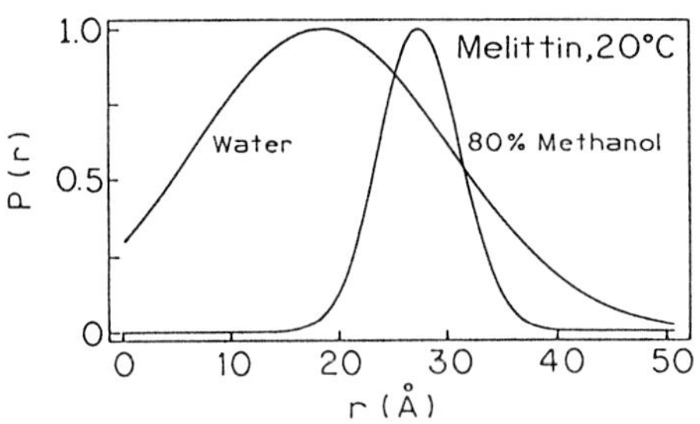

CONDITIONS	R_{av}(Å)	hw(Å)	log D[a]	χ^2_R
Water	18.2	27.2	-6.7	3.2
80% Methanol	27.2	8.2	-5.2	3.8

[a] D in cm^2/s

Figure 9. Distance distribution for random coil and α-helical dansyl melittin.

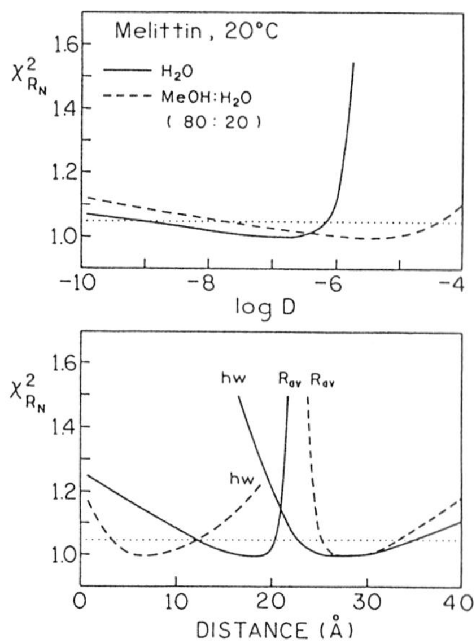

Figure 10. Diffusion coefficient and distance distribution χ_R^2 surfaces for random coil (——) and α-helical (- - -) dansyl melittin.

From the best global fits to the data we recovered diffusion coefficients and distribution parameters for the random-coil and α-helix melittin (Figure 9).

The χ_R^2 surfaces, describing uncertainties of the recovered parameters, are shown in Figure 10. Surprisingly, the diffusion coefficient obtained for α-helix conformation is significantly larger than for random-coil, but it should be noted that in α-helix diffusion occurs only in limited range of about 8Å.

E. END-TO-END DIFFUSION AND DISTANCE DISTRIBUTION IN TROPONIN I.

Figure 11 shows frequency responses of the donor (Trp) emission for native and denatured TnI in the absence and presence of acrylamide. From the best global fits to the data (solid lines) we recovered diffusion coefficients and distribution parameters for the native and denatured protein. The distribution has been changed dramatically in the presence of 5M GuHCl, from relatively narrow in native to very wide in denatured state. The diffusion coefficient of denatured TnI is almost an order of magnitude higher than if obtained for native protein.

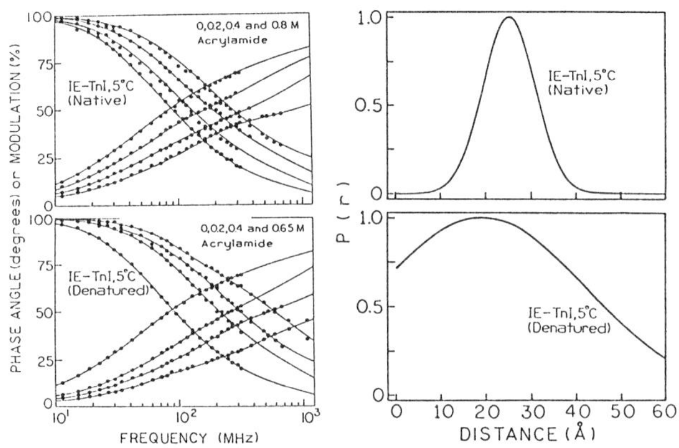

CONDITIONS	R_{av} (Å)	hw (Å)	log D	χ_R^2
BUFFER	24.8	13.2	-7.13	1.4
5M GuHCl	18.7	53.6	-6.30	2.2

Figure 11. Phase and modulation data for native and denatured TnI (left) and recovered distance distribution (right).

V. CONCLUSIONS

- Simulations indicate that for typical tryptophan-acceptor proteins and for model compounds the diffusion coefficient and distance distribution can be resolved from the donor decay only, if diffusion is in the range of 10^{-7}-10^{-5}cm^2/s.

- External quenching of the donor emission can be used to increase resolution in the recovered parameters, as was illustrated by simulations and studies of model compound TU2D.
- The apparent donor-to-acceptor diffusion coefficient is larger for α-helical melittin, than for random coil melittin. This rapid diffusion may reflect local mobility of the indole and dansyl motion.
- The apparent donor-to-acceptor diffusion coefficient is one order of magnitude higher in denatured than in native state of TnI.

VI. REFERENCES

1. Haas, E., Wilchek, M., Katchalski-Katzir, E., and Steinberg, I.Z., 1975, Proc. Acad. Sci. 72:1807-1811.
2. Grinvald, A., Haas, E., and Steinberg, I.Z., 1972, Proc. Acad. Sci. 69:2273-2277.
3. Gryczynski, I., Wiczk, W., Johnson, M.L., and Lakowicz, J.R., 1988, Chem. Phys. Lett. 145:439-446.
4. Lakowicz, J.R., Gryczynski, I., Cheung, H.C., Wang, C., and Johnson, M.L., 1988, Biopolymers. 27:821-830.
5. Valeur, B., Mugnier, J., Pouget, J., Bourson, J. and Santi, F., 1989, J. Phys. Chem. 93:6073-6079.
6. Albaugh, S., and Steiner, R.F., 1989, J. Phys. Chem. 93:8013-8016.
7. Haas, E., Katchalski-Katzir, E., and Steinberg, I.Z., 1978, Biopolymers 17:11-31.
8. Katchalski-Katzir, E., and Steinberg, I.Z., 1981, Ann. N.Y. Acad. Sci. 366:44-61.
9. Beechem, J.M. and Haas, E., 1989, Biophys. J. 49:1225-1236.
10. Lakowicz, J.R., Laczko, G., Cherek, H., Gratton, E. and Limkeman, M., 1984, Biophys. J. 46:463-477.
11. Lakowicz, J.R., Johnson, M.L., Wiczk, W., Bhat, A. and Steiner, R.F. 1987, Chem. Phys. Lett. 138:587-593.
12. Gryczynski, I., Wiczk, W., Johnson, M.L., Cheung, H.C. Wang, C. and Lakowicz, J.R., 1988, Biophysical Journal 54:577-586.
13. Lakowicz, J.R., Gryczynski, I., Cheung, H.C., Wang, C-K., Johnson, M.L., and Joshi, N., 1988, Biochem. 27:9149-9160.
14. Lakowicz, J.R., Laczko, G. and Gryczynski, I., 1986, A 2-GHz frequency-domain fluorometer. Reviews of Scientific Instrumentation. 57:2499-2506.
15. Laczko, G. Lakowicz, J.R., Gryczynski, I., Gryczynski, Z., and Malak, H., 1990, Reviews of Scientific Instrumentation. 61:2331-2337.

ACKNOWLEDGEMENTS

This work was supported by Grants GM 35159 from the NIH and DM8 8804939 from the NSF, with support for instrumentation from NSF Grant DIR 8710401. The authors thank Professor Herb Cheung for use of the TnI data, which was obtained under NIH Grant AR-31239 and Professor Frank Prendergast for the labelled melittin.

STOPPED-FLOW CIRCULAR DICHROISM AND ^{19}F NMR AS PROBES FOR THE FOLDING OF RAT INTESTINAL FATTY-ACID BINDING PROTEIN (IFABP)[1]

Ira J. Ropson†, Jeffrey I. Gordon‡, David P. Cistola† and Carl Frieden†

†Department of Biochemistry and Molecular Biophysics
‡Department of Molecular Biology and Pharmacology
Washington University School of Medicine
St. Louis, MO 63110

I. INTRODUCTION

The mechanism of protein folding continues to be one of the major unsolved questions in biochemistry. Although considerable progress has been made in the study of the contributions of disulphide bond rearrangements (Creighton, 1988) and proline *cis-trans* isomerization (Brandts et al. 1975) to slow events in protein folding, the more rapid processes of hydrophobic collapse and secondary structure formation remain poorly understood. Recent studies of the kinetics of protein folding, primarily by stopped-flow hydrogen exchange methods, have shown the existence of specific intermediates containing secondary structures similar to those found in the native protein for cytochrome C (Roder et al. 1988), ribonuclease A (Udgaonkar & Baldwin, 1988), barnase (Bycroft et al. 1990) and lysozyme (Miranker, et al. 1991). However, for the great majority of proteins studied thus far, the kinetically identified intermediates have not been associated with specific secondary or tertiary structures. Here, we describe the use of stopped-flow circular dichroism and ^{19}F NMR as probes of secondary and tertiary structure formation, respectively, during the folding of rat intestinal fatty-acid binding protein (IFABP).

IFABP is a small (131 residues, 15 kDa) predominantly β-structure protein (Fig. 1) that contains two tryptophans but neither cysteine nor proline, thereby avoiding the folding complications associated with those residues. The equilibrium and kinetic folding properties of this protein have been examined, primarily by monitoring fluorescence and circular dichroism changes during the transition (Ropson et al., 1990). Although no significant concentration of intermediates were detected at equilibrium, at least one transient intermediate was shown to be present during both unfolding and

---

[1]Supported by NIH research grants DK13332 to C.F. and DK30292 to J.I.G.

TECHNIQUES IN PROTEIN CHEMISTRY III

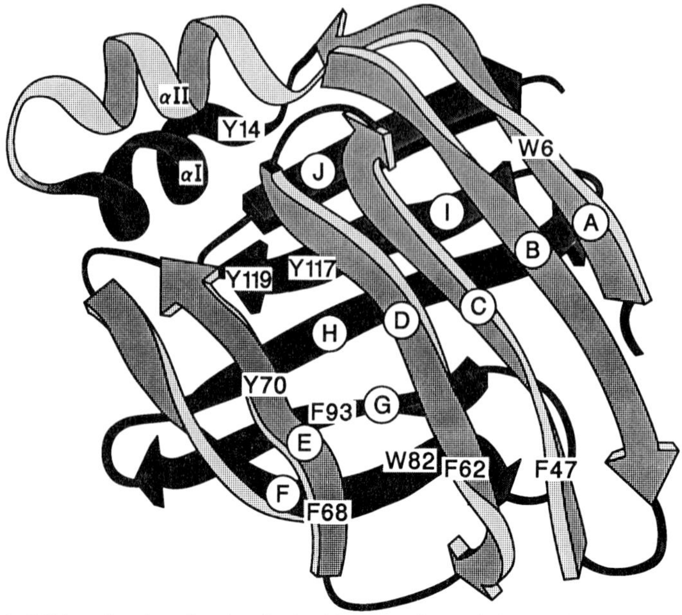

Figure 1. Ribbon drawing showing the backbone tracing and the location of several aromatic residues of IFABP.

refolding of the protein. Unfortunately, the data could not determine which region(s) of the protein were responsible for the observed fluorescence changes or whether any significant secondary structure was associated with the intermediate(s).

Circular dichroism (CD) in the far UV has been used to measure secondary structure content in proteins since the spectra can be deconvoluted to obtain information on α-helical, β-sheet and turn content. Since the CD for an asymmetric molecule is only about 10^{-4} of the absorption (Johnson, 1988), the signal to noise ratio in the far UV is usually quite low. Historically, this data collection problem has been a major barrier to the use of stopped-flow CD to study secondary structure formation during protein folding. Recent advances in both CD and stopped-flow instrumentation have allowed some proteins to be studied by this technique (Kuwajima et al. 1987; 1988; Sugawara et al. 1991; Mo et al. 1991). The simple apparatus described below has been used to measure the disappearance of secondary structure during unfolding and its reappearance during refolding for IFABP.

Proton NMR as a probe of protein folding is problematic since the sheer number of resonances generated by a protein are difficult to resolve and tedious to assign. Site specific incorporation of fluorine through the use of labeled amino acids and the appropriate *E. coli* auxotrophic strain represents one means by which this problem can be avoided. Fluorine is a spin-1/2 nucleus that is present in 100% natural abundance and is similar in size to hydrogen. Fluorine NMR is nearly as sensitive as ^{1}H NMR and the resonances occur over an extremely wide chemical shift range (reviewed by Gerig, 1989). Furthermore, the chemical shift is extremely sensitive to changes in solvent environment, as would be expected during the folding process. Since fluorine is not a naturally occurring atom in biological systems, the observed signals come only from the incorporated amino acids. Some caution must be exercised in the interpretation of ^{19}F NMR because the strength of the carbon-fluorine dipole or some other property of fluorine may perturb structure and/or function, although this seems not to be the case in many systems examined thus far (Gerig, 1989). To date, the folding of proteins has not been examined by this technique. The data collected thus far with 6-fluorotryptophan substituted IFABP (6FTrp-IFABP) are promising and there is some hope that the folding of this protein can be observed in real time by ^{19}F NMR.

II. MATERIALS AND METHODS

IFABP was synthesized in *E. coli*, isolated and delipidated as described elsewhere (Ropson et al, 1990). 6FTrp-IFABP was produced by the induction of protein synthesis in the tryptophan auxotroph *E. coli* strain W3110 trpA33 under conditions identical to those used for the synthesis of fluorotryptophan substituted cellular retinol-binding protein II (Li et al. 1989). The purification and delipidation protocols for 6FTrp-IFABP were identical to those of the unsubstituted protein. Unless stated otherwise, all chemicals and preparation methods were identical to those used previously (Ropson et al. 1990).

A Jasco J600 CD-spectropolarimeter was used for all CD measurements. The system assembled for stopped-flow CD consisted of two parts. A fused silica lens (SPX025, Newport Corporation) with lens positioner (LP-1-XYZ, Newport Corporation) was used to focus the large (12 by 7 mm) collimnated beam produced by this instrument on the stopped-flow cell located in the center of the sample compartment. The focal length of the lens (at 220 nm) was chosen to be approximately half the length of the sample compartment. A RX1000 stopped-flow apparatus was purchased from Applied Photophysics. This stopped flow system has a standard size 1 by 1 cm cuvette attached via a long thermostatted hose to the drive and stop syringes, allowing any spectrometer with a fast sampling time to be used as a stopped-flow instrument. Access to the sample compartment was gained through the front panel, allowing the larger area 2 mm pathlength cell to be used. The CD changes were monitored at 217 nm. The deadtime of the instrument is approximately 20-30 msec. Unfolding experiments involved the mixing of IFABP in 20 mM PO_4, 0.1 mM EDTA at pH 7.2 with solutions containing guanidine-HCl (GdnHCl) in the same buffer. Refolding experiments involved the mixing of IFABP unfolded with 2 M GdnHCl in the same buffer. A wide range of final denaturant concentrations was achieved through the use of different size drive syringes (1 and 2.5 mls) and several concentrations of GdnHCl in buffer. The final protein concentration was 0.2 mg/ml (13 μM) for all experiments and 5 transients were averaged for each folding and unfolding jump. Data were fit to mono- and biexponential decay equations using the NLIN procedure of SAS as described elsewhere (Ropson et al. 1990).

NMR data were collected on a Varian VXR-500 spectrometer operating at 470.3 MHz for ^{19}F using a 5 mm Varian Hydrogen/Fluorine probe. The ^{19}F spin-lattice relaxation times (T_1) were determined by the inversion recovery method. The T_1 values for 6FTrp, 6FTrp containing peptides, 6FTrp-IFABP and 2-fluoro-phenylalanine were less than 0.4 s for all samples. The pulse interval was 2 s (> 5 T_1 values). Ninety degree flip angle pulses (ranging from 7 to 19 μsec) were calibrated for each sample. An external standard, 2-fluoro-phenylalanine (-42.53 ppm from TFA), was used in a coaxial insert with all samples for intensity measurements. All solutions contained 20 mM PO_4, 0.1 mM EDTA and 10% D_2O adjusted to pH 7.2 at 25 ^oC. No correction to pH was made for D_2O content. All spectra were collected at 25 ^oC and processed with 5 Hz of line broadening.

In an attempt to assign the resonances of unfolded 6FTrp-IFABP, two 5 residue peptides were synthesized that were identical in sequence of IFABP around the two tryptophans. The peptides were Gly-Thr-Trp-Thr-Met and Gly-Thr-Trp-Lys-Val, corresponding to the sequences about Trp 82 and Trp 6, respectively. The 6-fluoro-L-tryptophan (6FTrp) incorporated into these peptides was purchased from Fluka. *t*-BOC-6FTrp was prepared and purified using BOC-ON (Aldrich) by the methods contained in technical information bulletin AL-125 supplied by Aldrich. The 6FTrp containing peptides were synthesized by Mark Zuper of the Monsanto Company using standard solid-phase synthesis techniques for *t*-BOC amino acids. Each amino acid was added in 2-fold excess except 6FTrp which was added in stoichiometric amount. Kaiser tests after coupling indicated complete coupling for each step except for 6FTrp, which coupled at 90-95% efficiency. The peptides were acetylated, deprotected with TFA and cleaved from the resin with HF. The crude peptides were purified on a semipreparative Deltapak C-18 column using a linear water/acetonitrile gradient containing 0.05% TFA. Confirmation of the peptide identity was obtained at the Monsanto mass spectrometry facility. The peptides were dissolved and lyophilized several times with 0.1% HCl to remove remaining TFA and other fluorine-containing contaminants. The concentration of the peptides was estimated using the extinction coefficient at 280 nm for unmodified tryptophan.

III. RESULTS AND DISCUSSION

Stopped-flow Circular Dichroism. Figure 2 shows the raw data for two stopped-flow CD unfolding transitions as well as native protein and unfolded protein baseline values. It is clear that for these unfolding experiments the entire loss of secondary structure is observed. The data were best fit by a single exponential decay that accounted for the entire amplitude of the transition. Similarly, a single exponential decay best described the refolding data, but the entire amplitude of the transition was not observed (data not shown). When unfolding was monitored by fluorescence (Ropson et al. 1990), a biexponential decay equation was required to fit both the folding and unfolding transitions. Simulations of the CD unfolding transitions have shown that if as little as 10% of the amplitude change occurred at the slower rate found by fluorescence, the fitting algorithm would have detected it. The rate constants determined by CD and fluorescence are compared in Figure 3.

The close correspondence between one of the observed rate for fluorescence and the rate observed by CD indicates that the same physical process is being followed by both techniques. As discussed previously (Ropson et al. 1990), a single pathway for unfolding containing one intermediate is the simplest explanation for the unfolding transitions. Since the CD rates correspond to the faster rate of the fluorescence change for unfolding, the slower rate of fluorescence change appears to be caused by some tertiary interaction involving one or both of the tryptophans that exists in the absence of secondary structure. A similar argument can be made for the changes observed for folding transitions.

^{19}F NMR. Figure 4 shows the ^{19}F NMR spectra of free 6FTrp, two 5-mer peptides corresponding to the local sequences about the two tryptophan residues in IFABP, native 6FTrp-IFABP, and 6FTrp-IFABP unfolded in 3.5 M GdnHCl. As expected, two resonances were observed for native 6FTrp-IFABP, implying that the two tryptophan residues are in different environments within the protein. Unexpectedly, two separate resonances were observed for unfolded protein, only one of which might correspond to 6FTrp in solution. This implied that the two tryptophans of IFABP are actually in

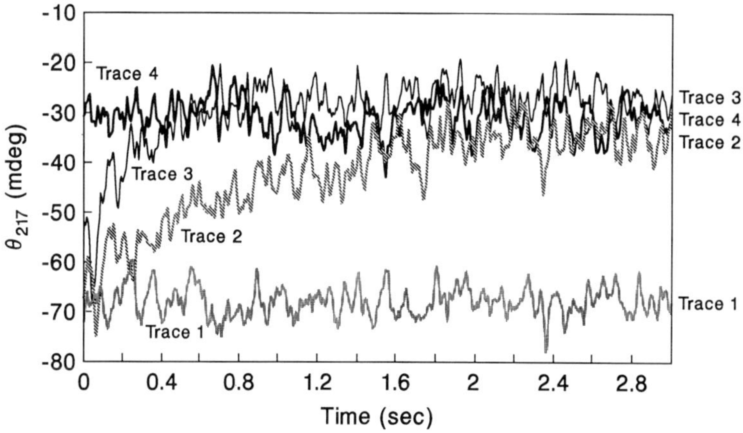

Figure 2. Stopped-flow CD data. Four experiments are shown. Trace 1. Native baseline (IFABP in buffer mixed with buffer). Trace 2. Unfolding transition - final concentration of 2.5 M GdnHCl. Trace 3. Unfolding transition - final concentration of 3.6 M GdnHCl. Trace 4. Unfolded baseline (IFABP in 2.5 M GdnHCl mixed with 2.5 M GdnHCl).

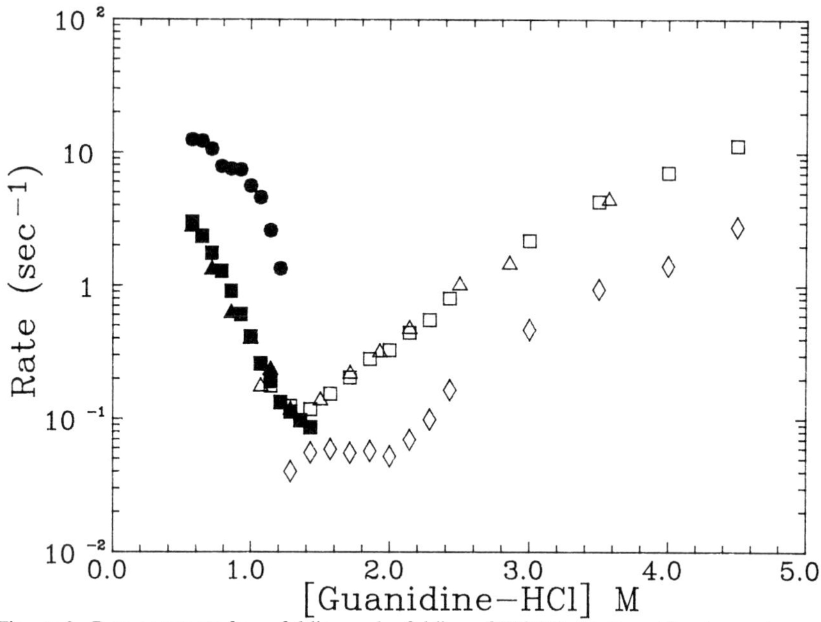

Figure 3. Rate constants for unfolding and refolding of IFABP. (●,■,□,◇) rate constants measured by fluorescence. (▲,△) rate constants measured by CD. Filled and open symbols are for refolding and unfolding transitions, respectively.

different environments in the "unfolded" state. The 6FTrp-containing peptides were synthesized in the hope of assigning the resonances to specific sequences, but the peptides also resulted in two distinguishable resonances, both of which were different from those observed for unfolded protein and free 6FTrp. Thus, we are currently unable to identify which tryptophan is responsible for each resonance, even in the unfolded protein, but site-directed mutants should make this distinction possible.

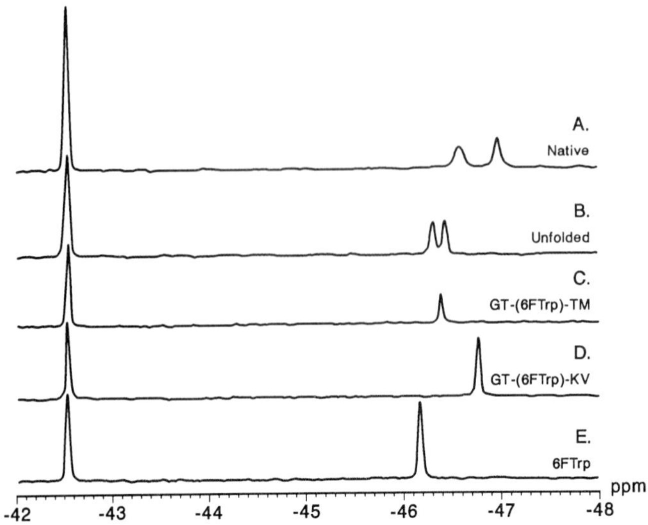

Figure 4. ^{19}F NMR spectra. The resonance at -42.53 ppm (relative to TFA) is the 2-fluorophenylalanine external reference. (A) 0.7 mM 6FTrp-IFABP, 256 transients (native protein spectrum). (B) 0.7 mM 6FTrp-IFABP, 256 transients (unfolded protein spectrum in the presence of 3.5 M GdnHCl). (C) 2 mM GT-(6FTrp)-TM, 64 transients (Peptide corresponding to residues 80-84 of IFABP). (D) 4 mM GT-(6FTrp)-KV, 64 transients (Peptide corresponding to residues 4-8 of IFABP). (E) 5 mM 6-fluorotryptophan, 64 transients.

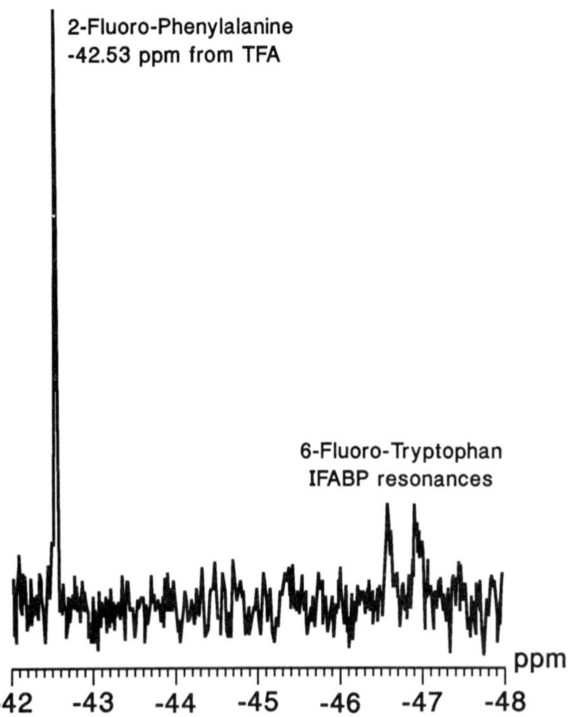

Figure 5. ^{19}F NMR spectra of 1 mM 6FTrp-IFABP in buffer, 1 transient.

The clear differences in the ^{19}F NMR spectra of folded and unfolded 6FTrp-IFABP show the utility of this method for the study of protein folding at equilibrium. Since these spectra report changes in environment only at tryptophan residues, and tryptophan residues are frequently involved in tertiary contacts in the hydrophobic core of proteins, ^{19}F NMR should report on tertiary structure changes at specific locations in the structure.

Figure 5 shows a single transient spectra of 6FTrp-IFABP at 1 mM concentration (15 mg/ml) collected in 0.4 sec. Both the approximate chemical shift and the relative intensity of the two resonances can be determined from this minimal amount of data. The sensitivity and resolution of these spectra appear to be sufficient for stopped-flow NMR studies of protein folding in real time. The ability to monitor changes in structure in real time at specific sites is very appealing.

IV. CONCLUSIONS

The problems associated with determining the mechanism of protein folding have proven to be difficult to overcome. As such, any techniques that can follow changes in secondary structure or at specific sites during folding are useful. Both stopped-flow CD and ^{19}F NMR have proven important to the study of the folding of IFABP, and may be helpful in the study of other proteins as well.

REFERENCES

Brandts, J.F., H.R. Halvorson, and M. Brennan. (1975). *Biochemistry* **14**:4953-4963.

Bycroft, M., A. Matouschek, J.T. Kellis, L. Serrano and A.R. Fersht. (1990). *Nature* **346**:488-490.

Creighton, T.E. (1988). *Biophys. Chem.* **31**:155-162.

Dill, K.A. (1990). *Biochemistry* **29**:7133-7155.

Gerig, J.T. (1989). *Meth. Enzymol.* **177**:3-23.

Johnson, W.C. (1988). *Ann. Rev. Biophys. Biophys. Chem.* **17**:145-166.

Kuwajima, K., H. Yamaya, S. Miwa, S. Sugai and T. Nagamura. (1987). *FEBS Letters* **221**:115-118.

Kuwajima, K., A. Sakuraoka, S.Fueki, M. Yoneyama and S. Sugai. (1988). *Biochemistry* **27**:7419-7428.

Li,E., S. Qian, L. Nader, N.C. Yang, A. d'Avigon, J.C. Sacchettini and J.I. Gordon. (1989). *J. Biol. Chem.* **264**:17041-17048

Miranker, A., S.E. Radford, M. Karplus and C.M. Dobson. (1991). *Nature* **349**:633-636.

Mo, J., M.E. Holtzer and A. Holtzer. (1991). *Proc. Natl. Acad. Sci. USA* **88**:916-920.

Roder, H., G.A. Elöve and S.W. Englander. (1988). *Nature* **335**:700-704.

Ropson, I.J., J.I. Gordon and C. Frieden. (1990). *Biochemistry* **29**:9591-9599.

Sugawara, T., K. Kuwajima and S. Sugai. (1991). *Biochemistry* **30**:2698-2706.

Udgaonkar, J.B. and R.L. Baldwin. (1988). *Nature* **335**:694-699.

SECTION VIII

Mass Spectrometry

Analysis of Protein-Protein Binding with Laser Desorption Mass Spectrometry: Can Multimeric Forms be Studied?

Marc J.-F. Suter, William T. Moore, Terry B. Farmer, John S. Cottrell[1], and Richard M. Caprioli

The Analytical Chemistry Center and the
Department of Biochemistry and Molecular Biology
University of Texas Medical School, Houston, Texas 77030

[1]Finnigan MAT Ltd., Hemel Hempstead HP2 4TG, UK

I. Introduction

Since its introduction in 1987(1), matrix-assisted laser desorption mass spectrometry (LD/MS) has become a potentially important tool in the field of biochemistry because of its capability of providing spectra at molecular weights up to and above 100,000 daltons. Although the laser/time-of-flight instrumentation is relatively simple, the sample preparation and choice of matrix compound is not straightforward for many applications. A wide variety of matrices and laser wavelengths have been investigated with the intention of increasing the sensitivity and applicability of the method to different classes of molecules. The most widely used matrices to date are nicotinic acid (266 nm, Nd-YAG laser) and sinapinic acid (337 nm, N_2-laser; 355 nm Nd-YAG laser) and analogues of these compounds.

The capability of measuring high molecular weights in a facile manner suggests the use of this instrumentation in experiments involving the study protein-protein interactions, subunit composition of protein multimers, verification of hybridization, identification of ligands and detection of chemical modifications. With respect to the measurement of protein multimers, only a few examples of protein dimers, trimers, etc., that have survived the ionization process as intact multimers have been reported (2-4). One important

TECHNIQUES IN PROTEIN CHEMISTRY III

parameter appears to be the choice of matrix and laser wavelength. Another important aspect is the method of sample preparation itself. *Karas et al.* [(2)] demonstrated the influence of the alcohol content of the matrix mixture on the dissociation of multimers into subunits. For example, nicotinic acid is water soluble, providing a less denaturing environment, whereas sinapinic acid is not soluble in water and requires organic solvents.

This report will describe experiments that investigate the potential of LD/MS for studying protein-protein interactions using the sinapinic acid/ 337 nm nitrogen laser combination. Several proteins that are known to form multimers were analyzed, i.e., alcohol dehydrogenases from both yeast and horse liver, avidin, and streptavidin. Several proteins that are known to be monomers were also analyzed, including carbonic anhydrase and myoglobin.

II. Experimental Conditions

A. Mass Spectrometry

The laser desorption time-of-flight mass spectrometer (Finnigan-MAT Lasermat) has a 50 cm flight tube and is operated at an accelerating voltage of 20kV[(5)]. A N_2-laser yields an adjustable power density of approximately 10^6 W/cm^2.

Electrospray spectra were recorded on a Finnigan-MAT TSQ 70 with an Analytica source. The sample was infused directly at a concentration of approximately 10 pmol/μL (in 50% methanol and 5% aqueous acetic acid) and a flow rate of 2.25 μL/min. Q3 was scanned from M/Z 600 to 2000 in 1 second. Deconvolution of the spectra were performed on a DECstation 2100 using Finnigan-MAT software.

B. Sample Preparation

Yeast alcohol dehydrogenase (y-ADH) was dissolved in water at a concentration of 20 pmol/μL and horse liver alcohol dehydrogenase (h-ADH) at a concentration of 10 pmol/μL in 0.1% TFA. Avidin was dissolved in a 1 mmolar aqueous d-biotin solution at 10 pmol/μL. One μL each of the analyte mixture and a saturated sinapinic acid solution in 50% ethanol (avidin, y-ADH) or 70% acetonitrile (h-ADH) was mixed and subsequently dried on the sample holder. The 50% ethanol mixture gave a better signal-to-noise ratio, but essentially the same relative intensities for avidin and y-ADH (monomer:dimer), compared to the preparation with the 70% acetonitrile mixture. Interleukin-2 was prepared in the same manner as y-ADH. Reduction was performed with a hundredfold molar excess of mercaptoethanol and subsequent pyridylethylation with a hundredfold molar excess of reagent relative to the thiol groups present in the mixture. All chemicals were purchased from SIGMA (St. Louis, MO), all solvents were HPLC grade or better, and the water used had a resistance of 18 M Ω.

Gel filtration was performed on a Sephadex G25 column (1.4 x 29.0 cm) at a flow rate of 0.9 ml/min with 1% NH_4HCO_3 at pH 8. The sample concentrations were 10 pmol/μL, 700 μL were loaded. y-ADH was denatured in 3M guanidine hydrochloride (10 min. at room temperature) or 1% NH_4HCO_3 (overnight at room temperature). UV detection was at 280 nm.

III. Results and Discussion

A. Alcohol Dehydrogenase

Bakers yeast ADH exists as two isozymes that differ in 23 amino acids. Both are acetylated at the N-terminus. Each isozyme is reported to consist of 4 identical subunits of average molecular weights 36,748 (y-ADH I) or 36,774 (y-ADH II) as calculated from the amino acid sequence and contain 3-4 tightly bound zinc ions. Horse liver ADH is a dimer of two different subunits (E- and S-chain), both are acetylated at the N-terminus and bind two zinc ions each. Their respective molecular weights are 39,847 (E-chain) and 39,726 (S-chain). Comparison of y-ADH and h-ADH shows a non-homologous primary structure. The tertiary structures are similar with the exception of the C-terminal end of the molecules and a segment in h-ADH that corresponds to an additional loop in the catalytic domain. These structural aspects affect the subunit interaction and in part explain the different quaternary structures of the two proteins [(6)].

The LDMS spectrum of y-ADH is shown in Figure 1. The upper panel of the figure shows a relatively intense signal for the dimer as well as one for the monomer. Treatment of the native y-ADH with 1% NH_4HCO_3 dissociates the complex into monomers, as can be seen in the lower panel of Figure 1. The low intensity dimeric species seen in this panel are generated by the desorption/ionization process itself and are seen in nearly all spectra of proteins. Using classical procedures, verifications of the multimeric forms of the y-ADH samples was done with gel filtration, employing bovine serum albumin and carbonic anhydrase as molecular weight markers. The dimeric, but no tetrameric form, was seen for the native enzyme preparation. Subsequent treatment of y-ADH with denaturing agents (3M guanidine hydrochloride, 1% NH_4HCO_3) dissociates the dimer predominantly into monomeric subunits, indicating that the dimeric form does not contain covalently linked monomers. Electrospray spectra of y-ADH also showed a mixture of dimer and monomer present in the untreated sample (spectrum not shown). The average molecular weight determined for y-ADH by LD/MS is 36,915 (σ_{18}= 37) for the monomer and 73,540 (σ_{18}= 195) for the dimer. It is not known at this time whether these molecular weights include the zinc ions.

The LDMS spectrum of native horse liver ADH is given in Figure 2 and shows that the monomer is the major species desorbed. It is concluded from these results that either the sample preparation procedure or the

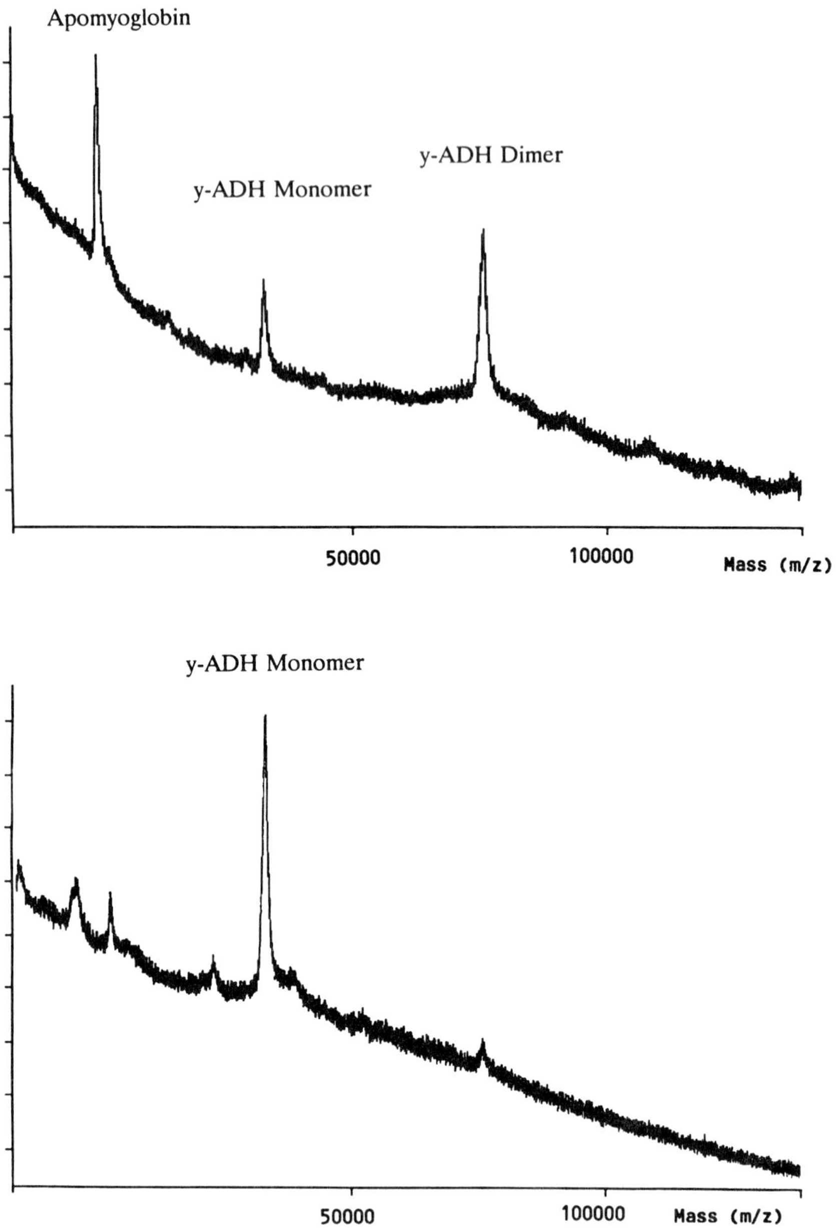

Figure 1. LDMS spectrum of yeast alcohol dehydrogenase. top: native form, bottom: native form after treatment with 1% NH_4HCO_3.

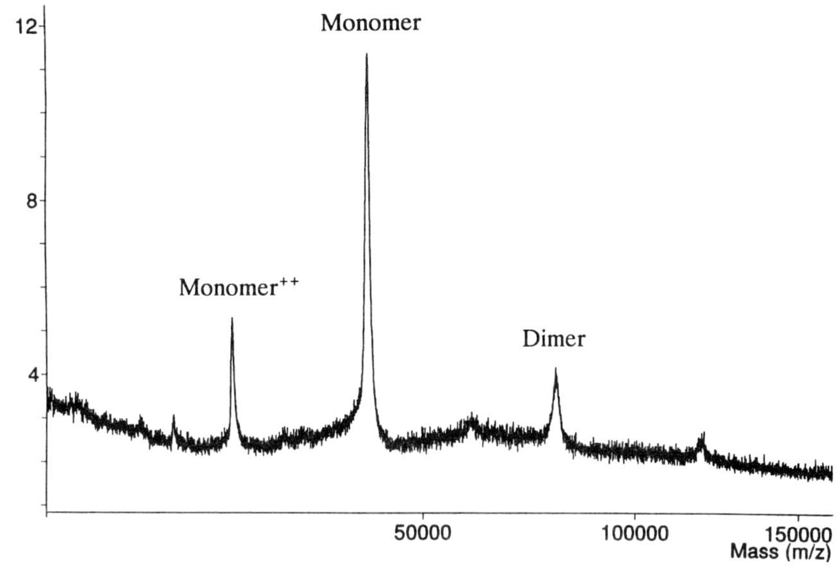

Figure 2. LDMS spectrum of horse liver alcohol dehydrogenase.

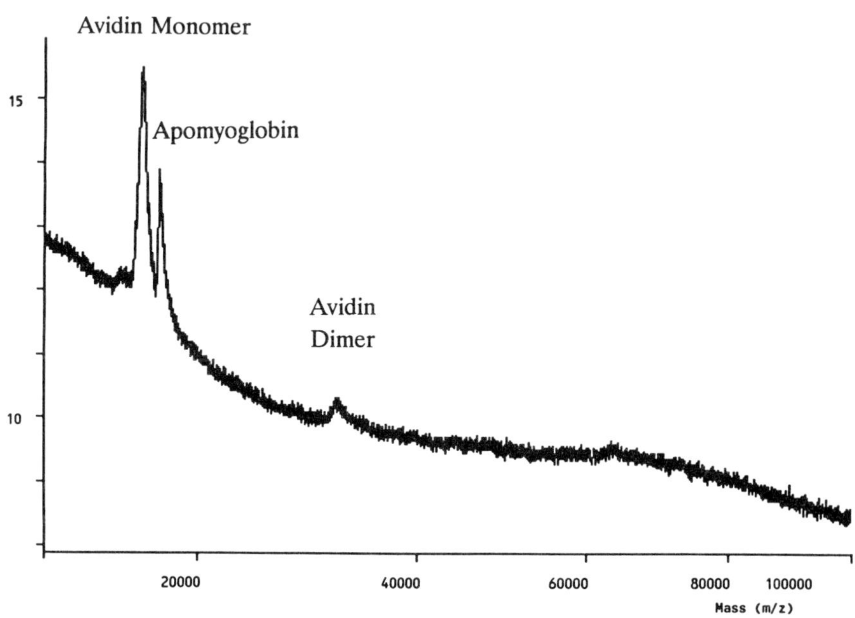

Figure 3. LDMS spectrum of avidin/biotin.

desorption process itself led to the dissociation of the two protein chains. It is also noted that since the linear time-of-flight mass spectrometer has a resolution of typically 150 to 300 (50% valley definition), the S- and E-chain monomers of h-ADH cannot be resolved.

B. Avidin

Avidin is a glycoprotein composed of 4 essentially identical polypeptide chains (half the chains have Thr-34 instead of Ile-34) with a carbohydrate chain attached to Asn-17 (4 glucosamine, 5 mannose residues). The molecular weight of the monomer is 15,786 as calculated from the sequence (with Thr-34). Avidin forms a very strong complex with biotin (MW 244), a fact that has been widely exploited for the design of affinity columns. The binding constant is in the order of 10^{-15} M (25°C) and the biotin binding ability occurs between pH 2 and 13, and at guanidine hydrochloride concentrations of up to 3 M, reflecting the stability of the complex. It is suspected that biotin assists in the cohesion of the subunits (thermal dissociation of avidin: ΔH = 76 kcal/unit, thermal dissociation of avidin/biotin: ΔH = 270 kcal/unit, where one unit binds 1 μg of d-biotin at pH 8.9). Binding of biotin by single subunits was observed by *Green et al.* with a binding constant K = 10^{-7} M [7,8].

The LDMS spectrum of avidin, prepared as described above, is given in Figure 3. No evidence of multimers and biotin binding was observed. The relatively low intensity multimer peaks seen in the spectrum are those produced by the laser desorption process itself. The measured average molecular weight for avidin with no biotin present (15,926, σ_9= 25) is slightly higher than the expected value, probably due to inhomogeneity of the carbohydrate group. No significantly different values were obtained with biotin present (15,939, σ_{10}= 15). Increasing the ratio biotin/avidin, aimed at preventing the dissociation into subunits, did not show any change in relative intensities of the multimers. Also no change was seen when reducing the sinapinic acid concentration. Sperm whale apomyoblobin was used as internal standard (MW_{avg} 17,199). Experiments with streptavidin gave the same results.

C. Interleukin-2

Interleukin-2 (IL-2) plays an important role in immune response[9]. The human IL-2 is a protein of 133 amino acids and has been well characterized by x-ray diffraction[10]. Two of its three cysteines (Cys-58, Cys-105) are internally linked by a disulfide bridge that gives the native form a rigid tertiary structure, essential for bioactivity[11]. The recombinant protein produced by *Escherichia coli* possesses an additional methionine residue at the N-terminus (MW_{avg} 15,549).

Even though native recombinant interleukin-2 is described as monomeric in the literature, it gives an intense signal with LD/MS for the $(M_2H)^+$ dimeric form, as seen in Figure 4. Only a small fraction of this signal

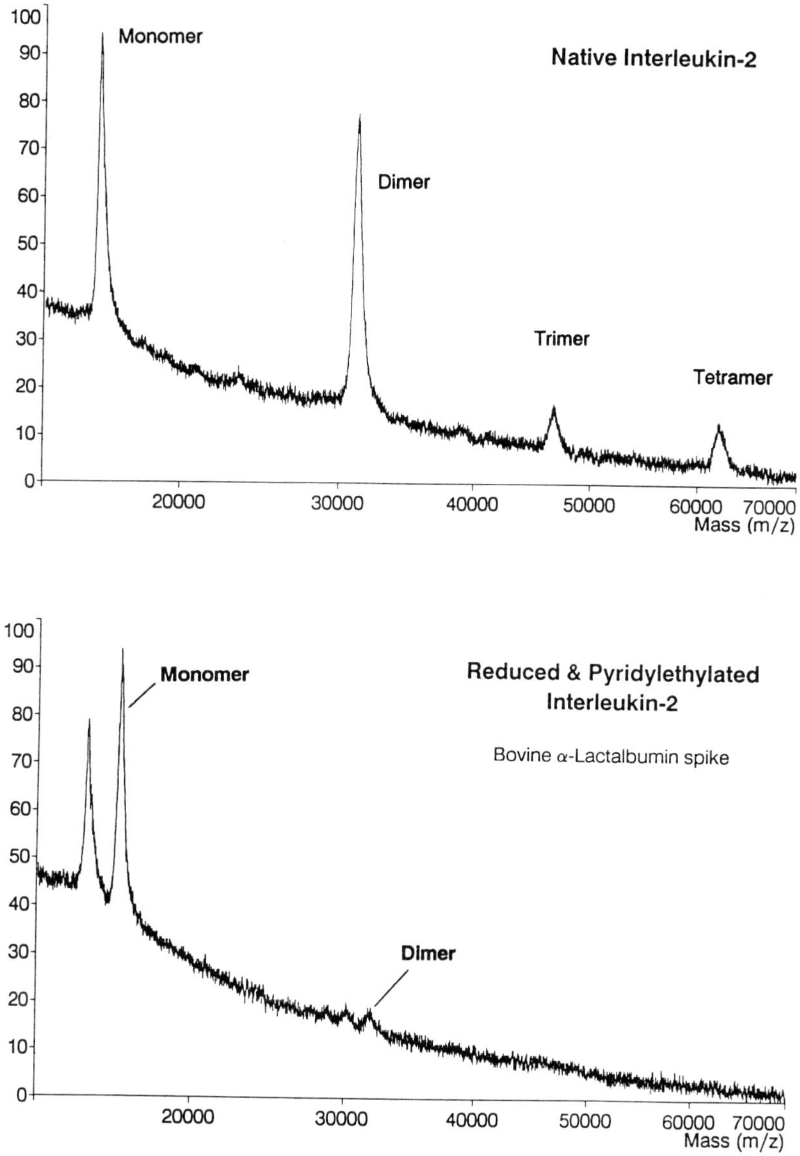

Figure 4. LDMS spectrum of recombinant Interleukin-2. top: native form, bottom: native form after reduction with mercaptoethanol and pyridylethylation of cys residues.

is attributed to cluster formation which can occur during the desorption process, as shown in Figure 2 for horse liver ADH. When the sample was treated with mercaptoethanol followed by pyridylethylation of the cysteinyl residues, the dimeric species was significantly reduced in intensity. In view of the stability of the dimer under a range of sample preparation conditions, it is most probable that the dimer is formed through covalent disulfide linkage as a result of disulfide interchange between two monomers. The mass measurement accuracy of 0.02-0.05% is not sufficient to detect the loss of 2 mass units for formation of a disulfide from two reduced sulfhydryl groups. Nevertheless, the analysis and identification of a dimeric species by LD/MS shows that such an interaction can be easily detected.

IV. Conclusion

Matrix-assisted laser desorption mass spectrometry, using the sinapinic acid/337 nm combination, can be used to determine multimeric forms of proteins under certain conditions. In the case of y-ADH, the hydrophobic forces holding the subunits together were sufficiently strong to prevent dissociation under the conditions used for sample preparation and analysis. This was not the case for avidin/biotin binding, which showed only monomeric avidin on LD/MS analysis. Dimeric interleukin 2 was also found, although the nature of the protein-protein interaction in this case is unknown. It is clear that sample preparation plays a major role in the success of such studies. Current experiments are aimed at elucidating the critical parameters of these and other protein systems to better understand conditions necessary for the observation of multimeric protein complexes.

Acknowledgment

This work has been supported by a grant from the National Institute of Health (grant # GM43783-02).

References

1. Karas, M., Bachmann, D., Bahr, U., and Hillenkamp, F. (1987). *Int.J.Mass Spectrom. Ion Processes* **78**, 53.
2. Karas, M., Bahr, U., Ingendoh, A., Nordhoff, E., Stahl, B., Strupat, K., and Hillenkamp, F. (1990). *Anal. Chim. Acta* **241**, 175.
   Karas, M., Bahr, U., Arndt, I., and Hillenkamp, F. (1989). *Angew. Chem. Int. Ed. Engl.* **28**, 759.
3. Beavis, R.C., and Chait, B.T. (1989). *Rapid Commun. Mass Spectrom.* **3**, 432.
4. Perera, I.K., Candy, J.M., Håkansson, P., Oakley, A.E., Brinkmalm, G., and Sundqvist, B.U.R. (1990). *Rapid Commun. Mass Spectrom.* **4**, 527.

5. Caprioli, R.M., Whaley, B., Mock, K.K., and Cottrell, J.S. (1991). *In* "Techniques in Protein Chemistry II", (J. Villafranca, ed.), p. 496. Academic Press, San Diego, CA.
6. Jörnvall, H., Eklund, H., and Brändén, C.-I (1978). *J.Biol.Chem.* **253**, 8414.
7. Green, N.M. (1963a). *Biochem. J.* **89**, 599, 609.
8. Green, N.M. (1975). *Advan. Protein Chem.* **29**, 85.
9. Morgan, D.A., Ruscetti, F.W., and Gallo, R. (1976). *Science* **193**, 1007.
10. Brandhuber, B.J., Boone, T., Kenney, W.C., and McKay, D.B. (1987). *Science* **238**, 1707.
11. Yamada, T., Fujishima, A., Kawahara, K., Kato, K., and Nishimura, O. (1987). *Arch. Biochem. Biophys.* **257**, 194.

Applications of Liquid Chromatography-Electrospray Mass Spectrometry (LC-ES/MS)

Kristine M. Swiderek, Shiuan Chen, Gottfried J. Feistner, John E. Shively & Terry D. Lee

Division of Immunology, Beckman Research Institute, City of Hope,
Duarte, CA 91010

Abbreviations: HPLC, high performance liquid chromatography; (LC-)ES/MS, (liquid chromatography) electrospray mass spectrometry; TFA, trifluoroacetic acid

INTRODUCTION

Mass spectrometry has been used as an analytical method in chemistry for many years. Over the last decade it has gained increasing attention by the protein chemist. The introduction of instruments capable for determining molecular weights higher than 100,000 (1, 2, 3) as well as the ability to determine amino acid sequences and posttranslational modifications in peptides has opened a new era of mass spectrometry (4, 5). One of the newer techniques especially useful to the protein chemist is the electrospray mass spectrometry (ES/MS) originally described by Dole and coworkers (6). This type of mass spectrometer produces multiple charged ion species without significant sample fragmentation of high molecular weight molecules. Molecular masses can be determined within a deviation of 0.01 % of the actual mass. On line connection of ES/MS with an HPLC system allows samples

TECHNIQUES IN PROTEIN CHEMISTRY III

with critical solubility properties to be introduced into the mass spectrometer more easily (7, 8). It is not necessary to redissolve them in specific solvents which are compatible with the instrument. Complex peptides and protein mixtures do not need to be prepurified into single components simplifying preparation procedures and minimizing sample loss. Also the liquid chromatography step can be used to concentrate dilute samples allowing the analysis of low concentrations of sample.

Here we describe several applications of an electrospray triple quadrupole mass spectrometer interfaced with a micro HPLC separation system. We present data from low and high molecular mass analyses of proteins as well as samples applied at low picomole levels.

EXPERIMENTAL

Mass spectra were recorded in the positive ion mode using a TSQ-700 triple quadrupole instrument (Finnigan-MAT, San Jose, CA) with an electrospray ion source operating at atmospheric pressure. The electrospray needle was operated with a voltage differential of 3-5 kV and a sheath flow of 2 µl/min of a 1:1 mixture of 2-methoxy-ethanol/iso-propanol. The heating coil for the drying gas was held to 240°C. The actual temperature of the drying gas was around 80°C. The nebulization and focusing gas was N_2 set at a pressure of 30 psi. Scans were continously acquired every three seconds between m/z 470 and 2000 in the centroid mode. The conversion dynode was set at -15 kV.

Micro capillary HPLC was carried out using a model 140A Applied Biosystem microgradient syringe pump and a Rheodyne injection valve with a 20 µl sample loop. Fused silica capillary columns with an inside diameter of 160 µm and a length of about 40 cm were packed in our laboratory with Vydac 3 µm C18 RP support (9). Chromatography was performed using linear gradients between 0.1 % TFA (solvent A) and 90 % acetonitrile, 0.07 % TFA (solvent B).

A solvent flow of 2-3 µl/min through the capillary column was achieved by splitting the flow before the point of injection and keeping the pressure constant. Elution of the sample was monitored by on-line UV detection at a wavelength of 200 nm before introduction into the electrospray source.

Cytochrome C peptide digests were prepared by digesting 40 pmol/µl of Cytochrome C with endoprotease Lys-C (Boehringer) for 12 h at ph 9.0 in 100 mM Tris-Cl at 37°C. α-Bungarotoxin was purchased from Sigma. Rat pituitary was prepared from one gland extracted with 200 µl of 1 % TFA as described before (10).

RESULTS AND DISCUSSION

Usually for the structural characterization protein digestion mixtures, individual peptides need to be isolated by at least one chromatographic step but often further purification chromatography is necessary. This is time consuming and results in sample loss due to irreversible interaction with sample collection tubes. On-line connection of HPLC and ES-MS avoids many of these problems. Structural analysis may be done in a single step without further sample manipulation. However, certain parameters need to be evaluated before LS-ES/MS can be used in this manner routinely.

1. Flow rate: ES-MS operates typically at flow rates between 1 - 10 µl/min. Commercially available HPLC columns and gradient systems are not suitable for such low flow rates without modifications as described in 7 and 8. The use of fused silica capillary columns with an inner diameter of 160 µm offers the possibility of decreasing the flow rate to 2 - 3 µl/min and allows additionally to carry out analysis on very low amounts of samples. This is an important consideration for many analyses at the low picomole level. To obtain a reliable low flow rate from the HPLC system we introduce a splitter T to divide the flow. The splitter is connected to a stainless steel tube filled with reversed phase material to create the necessary back pressure for a flow rate of 2 - 3 µl/min on the capillary column. The flow rate from the HPLC is regulated by constant pressure and has to be adjusted depending on the column used. The split occurs before the point of injection so that 100 % of the injected sample is loaded onto the column.

2. Sheath liquid: The gradient normally used in reversed phase HPLC is formed by TFA in H_2O/acetonitrile. TFA as well as high concentrations of water interfere with the process of forming droplets in the electrospray source. To overcome this the eluent from the HPLC is mixed with a sheath liquid of 2-methoxy-ethanol or a 50/50 mixture with isopropanol to increase the concentration of organic solvent and to dilute the TFA concentration.

An example of the analysis of an enzymatic digestion mixture is shown in Figure 1. This figure compares the results obtained from an application of 6 or 40 pmoles of cytochrome C digested with Lys-C. The comparision of the UV absorbance, shown from the application of 40 pmoles, and the base peak profile demonstrates that there is only little loss of separation between the UV detector and the mass spectrometer. However, due to the distance between the UV monitor and the mass spectrometer we observe a time delay between the detection of peptides monitored in the UV- and base peak trace. Although not encountered in this example, often a fraction that appears homogenous by UV detection resolves into different peptides by mass spectrometry. In this case the single scans recorded from the ES/MS under each peak are related to the masses of different peptides. As can be seen from Figure 1, there are differences in signal intensities between UV detection and mass spectrometry. Peptides having low UV absorbance are easily detected by ES/MS. Figure 1 also shows that there is essentially no loss of information by applying only 6 pmoles of the digestion mixture. Experiments are in progress to determine the limits of detection.

The analysis of only 10 pmoles α-bungarotoxin with an expected molecular mass of 7,986 gives us an idea what to expect when samples with higher molecular masses than normally obtained from peptide digests are analyzed. Figure 2A shows the reconstructed ion current recorded from scans taken every three seconds. The toxin is separated on a micro capillary C18 reversed phase column using an increasing gradient of 2 %/min from solvent A to solvent B. An average spectrum is computed over scans 478 - 497 (Figure 2B) showing the ion species of α-bungarotoxin $[M+4H]^{4+}$ to $[M+8H]^{8+}$. The deconvolution of this spectrum (insert in Figure 2B) gives the average mass of 7,984.

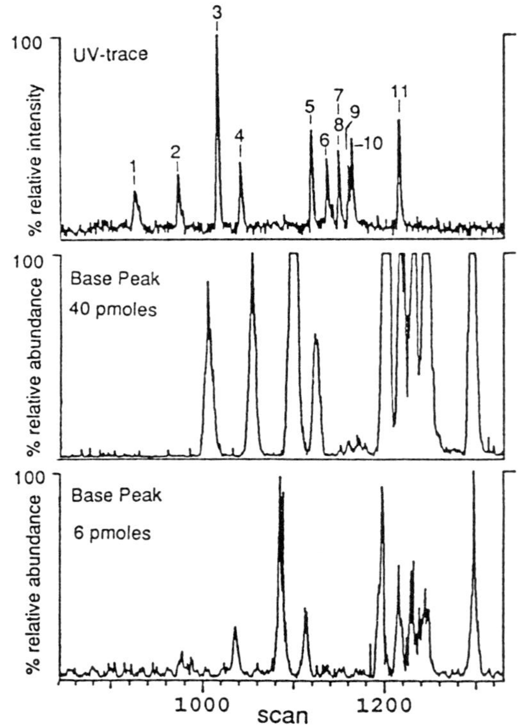

m/z	charge	residue	sequence	MH+ calculated	
678.5	+1	74 - 79	YIPGTK	678.38	(1)
634.5	+1	9 - 13	IFVQK	634.39	(2)
736.4	+2	40 - 53	TGQAPGFTYTDANK	1470.69	(3)
604.5	+1	56 - 60	GITWK	604.35	(4)
649.3	+2	28 - 39	TGPNLHGLFGRK	1296.72	(5)
779.5	+1	80 - 86	MIFAGIK	779.45	(6)
740.2	+2	88 - 99	KTEREDLIAYLK	1478.82	(7)
740.2	+2	89 - 100	TEREDLIAYLKK	1478.82	(8)
676.3	+2	89 - 99	TEREDLIAYLK	1350.73	(9)
818.4	+2	9 - 22	IFVQKCAQCHTVEK	1633.82	(10)
748.5	+2	61 - 72	EETLMEYLENPK	1495.7	(11)

Figure 1

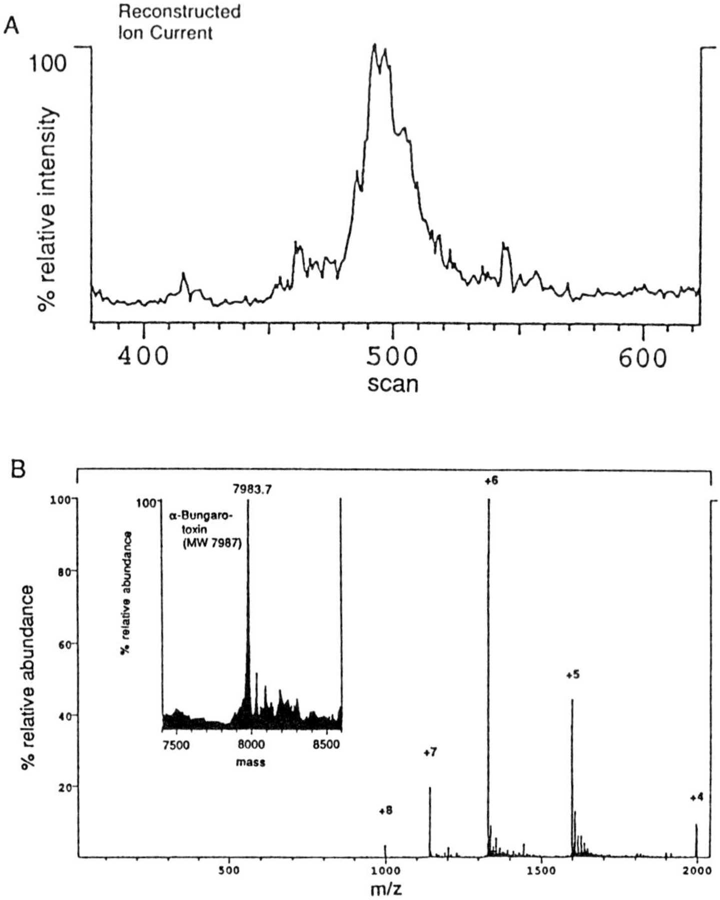
A
Reconstructed
Ion Current
100
% relative intensity
400
500
600
scan
B
100
80
60
40
20
% relative abundance
7983.7
100
α-Bungaro-
toxin
(MW 7987)
% relative abundance
7500
8000
8500
mass
+8
+7
+6
+5
+4
500
1000
1500
2000
m/z

Figure 2

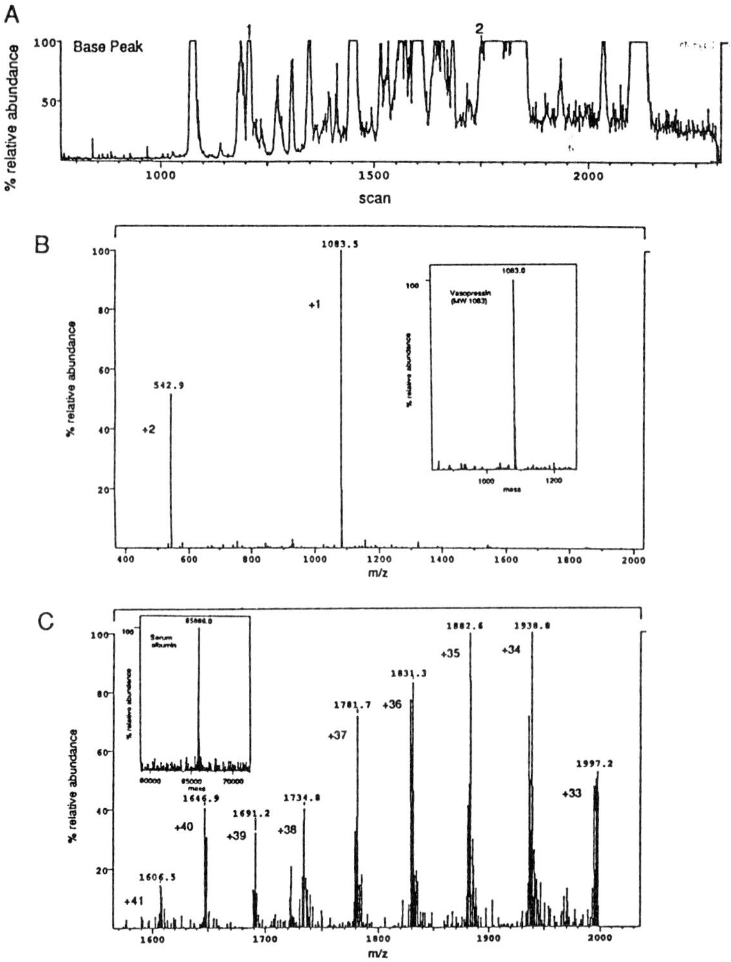
A
Base Peak
% relative abundance
100
50
1
2
1000
1500
2000
scan
B
% relative abundance
1083.5
+1
542.9
+2
400
600
800
1000
1200
1400
1600
1800
2000
m/z
Vasopressin
(MW 1083)
1083.0
C
% relative abundance
Serum
albumin
1882.6
+35
1938.8
+34
1831.3
+36
1781.7
+37
1997.2
+33
1646.9
+40
1691.2
+39
1734.8
+38
1606.5
+41
1600
1700
1800
1900
2000
m/z

Figure 3

As an example of a highly complex mixture we apply a tenth of an extract from a single rat pituitary gland (Figure 3). The extract is injected onto the micro capillary C18 reversed phase HPLC column and eluted with a gradient from 0 % to 60 % solvent B within 60 minutes. Figure 3A shows the base peak profile recorded over the time of separation. To demonstrate the versatility of LC-ES/MS we chose spectra of a low and high molecular mass component separated from the extract. Both spectra are obtained by averaging a number of single scans. Figure 3B shows the average of scan 1203 - 1210 corresponding to the labeled peak **1** in the base peak profile. It shows the singly and doubly charged ion species of vasopressin that deconvolutes to the average mass of 1083 (insert in Figure 3B). Figure 3C is the average spectrum of scan 1747 - 1764 (peak **2**) showing the ion species of $[M+33H]^{33+}$ to $[M+41H]^{41+}$. Deconvolution of this spectrum (see insert) results in the average mass of 65,886 corresponding well to the expected molecular weight of rat serum albumin (11). These examples demonstrate the utility of LC-ES/MS. In a relatively short analysis time (60 - 120 minutes), whole tissue extracts can be charted in the mass range of 500 to more than 60,000 Da.

CONCLUSION

We have presented examples demonstrating that LC-ES/MS is very useful in determining the molecular masses. Low amounts of peptides or proteins in simple and complex mixtures of low and high molecular weight components can be analyzed in a short period of time. We are extending this application by combinating LC-ES/MS/MS to on-line primary structure analysis. In addition, we are working on methodologies for the characterization of posttranslational modifications such as the glycosylation of peptides and proteins.

REFERENCES

1. Mann, M., Meng, C. K. & Fenn, J. B. (1989) Anal. Chem. 61, 1702 - 1708
2. Smith, R. D., Loo, J. A., Edmonds, C. G., Barinaga, C. J. & Udseth, H. R. (1990) Anal. Chem. 62, 882 - 899
3. Spengler, B. & Cotter, R. J. (1990) Anal. Chem. 62, 793 - 796

4. Hunt, D. F., Yates, III, J. R., Shabanowitz, J., Winston, S. & Hauer, C. R. (1986) Proc. Natl. Acad. Sci. 83, 6233 - 6237
5. Covey, T., Sushan, B., Bonner, R., Schroeder, W. & Hucho, F. (1991) Methods in Protein Sequence Analysis Joernvall/Hoeoeg/Gustavson (eds.) Birkhaeuser Verlag Basel, 249 - 256
6. Dole, M., Mack, L. L., Hines, R. L., Mobley, R. C., Ferguson, L. D. & Alice, M. B. (1968) J. Chem. Biol. Phys. 49, 2240 - 2249
7. Hemling, M. E., Roberts, G. D., Johnson, W. & Carr, S. A. (1990) Biomed. Environ. Mass Spectrom. 19, 677 - 691
8. Huang, E. C. & Henion, J. D. (1991) Anal. Chem. 63, 732 - 739
9. Lee, T. D. & Davis, M. T. (1991) manuscript in preparation
10. Feistner, G. J., Barofsky, D. F., Evans, C. J., Faull, K. F. & Roepstorff, P. (1991) Methods and Mechanisms for Producing Ions from Large Molecules; Standing, K. & Ens, W.; Plenum (in press)
11. Sargent, T. D., Yang, M. & Bonner, J. (1981) Proc. Nat. Acad. Sci. USA 78, 243 - 246

ACKNOWLEDGMENTS

This work was supported by the Cancer Center Core Grant CA 33572 and NIH grant GM 37297.

ANALYSIS OF PROTEINS BY MASS SPECTROMETRY

Patrick R. Griffin, Karen Furer-Jonscher, Leroy E. Hood, and John R. Yates, III

Division of Biology, California Institute of Technology, Pasadena, California 91125

Jae Schwartz and Ian Jardine

Finnigan MAT, San Jose, California 95134

I. INTRODUCTION

Complex interactions occur in biological systems between a wide range of biomolecules. Proteins function in a key role in nearly all these processes and the functional diversity they exhibit can be attributed to structural variability. Obtaining primary structural information may allow inferences to be made at the molecular level about structure-function relationships that exist in proteins. The ability to obtain primary sequence information from small quantities of protein aids in locating, isolating, and sequencing the gene which encodes for the protein. Developing an analytical technique for sequencing that could utilize separation techniques such as 2-dimensional gel electrophoresis could revolutionize structural biology. Mass spectrometry would appear to be the logical analytical tool for studying protein structure due to its sensitivity of analysis and high sample throughput.

Shortly after the technique of fast atom bombardment (FAB) was introduced [1], Hunt and coworkers presented methodology for protein sequencing by tandem mass spectrometry [2]. The strategy involved the digestion of the protein sample, either chemically or enzymatically, then fractionation of the resultant peptide mixture by reverse phase high-performance liquid chromatography (HPLC). In the next step, an aliquot of sample was placed on the FAB probe and the molecular masses of all peptides present were recorded. This procedure was repeated for each HPLC fraction. A limitation of this strategy is the additional sample-handling steps required following proteolytic digestion, including fraction collection and concentration of each HPLC fraction prior to mass analysis. Each additional sample-handling step exposes the sample to new surfaces, resulting in loss of material. Clearly a procedure that incorporates direct analysis of HPLC effluent will yield improved sample recovery and increased sample throughput.

Several techniques have been developed, in an effort to minimize sample loss and to improve overall throughput of analysis, that couple liquid chromatography (LC) with mass spectrometry (MS) [3,4]. Coupling microbore HPLC [5,6] or packed capillary HPLC [7-10] with electrospray ionization (ESI)

TECHNIQUES IN PROTEIN CHEMISTRY III

on various types of mass analyzers has emerged as the method of choice for high sensitivity analysis of proteins and peptides. Electrospray ionization has been coupled to several types of mass analyzers [11-15]. However, ESI on the quadrupole ion trap may prove to be the most powerful tool for analysis of protein structure. The application of the quadrupole ion trap to the analysis of biomolecules has yielded extraordinary results [12-14]. Impressive sensitivities have been observed and MS^n experiments [14] and ion-molecule reactions [12] have been performed suggesting that the ion trap will become an important tool for protein analysis.

The following discussion demonstrates the use of ESI for the analysis of proteins. The techniques used involved coupling packed capillary HPLC and open-tubular liquid chromatography (OTLC) with ESI on a triple quadrupole mass spectrometer and on a quadrupole ion trap.

II. EXPERIMENTAL

Protein samples were dissolved in 50 μl of 100 mM ammonium bicarbonate, pH 8.5, and then exposed to trypsin (1-2% w/w (enzyme/substrate)) for 12 h at 37° C. Peptide mixtures were fractionated by packed capillary HPLC with a gradient of 0 to 100 % acetonitrile (0.085% TFA) in 0.1 % TFA using an ABI 140A solvent delivery system. The pump was operated at 100 μl/min and the flow was split prior to the sample injection loop (0.5 μl) at a ratio of 50:1. Capillary C18 columns were purchased from LC Packings (320 μm i.d. x 15 cm). Eluent was monitored at 220 nm using a UV detector equipped with an capillary flow cell (LC Packings, path length = 7 mm). Signal from the UV detector was sent directly to the DECstation 2100 that was used to acquire data on the TSQ-700 and the UV chromatogram was displayed in the CHRO program. Eluent exiting the detector was coupled to the ESI source with 50 μm capillary. For OTLC experiments, 50 μm x 1 m fused silica capillary (Polymicro Technologies) was derivatized with C18 as described by Tock et al [15]. Sample was loaded onto the derivatized capillary using a pressure vessel, then the column was washed with several

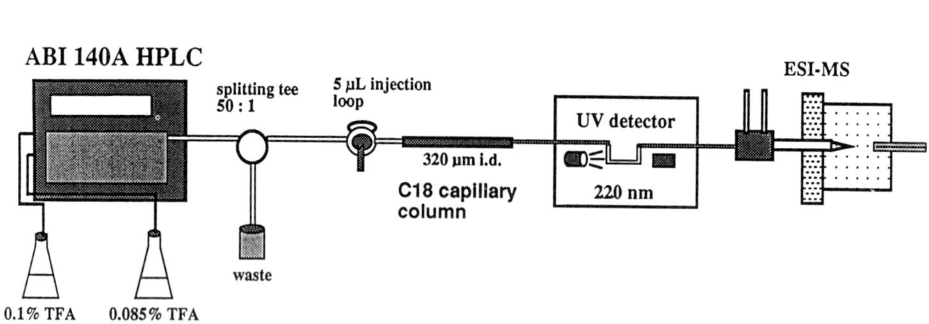

Figure 1. Diagram of the instrumental configuration used for capillary HPLC-MS.

column volumes of 0.1% TFA. Peptides were eluted isocratically into the ESI source using 20% acetonitrile in 0.1 % TFA at flow rates of 100 to 400 nl/min.

Mass spectra were recorded on a Finnigan TSQ-700 triple quadrupole mass spectrometer and on a unique quadrupole ion trap instrument that utilizes components of a Finnigan ITMS and a TSQ-700. Both instruments were equipped with a Finnigan ESI source as described in detail elsewhere [11,16]. The configuration used for capillary HPLC-MS is illustrated in Figure 1. Molecular masses of peptides were recorded and CAD experiments performed on the TSQ-700 as previously described [9]. A mass range of 50 to 1950 m/z was scanned at a rate of 180 (m/z)/μsec on the ion trap with ionization times of 30 msec, and spectra were summed over 6 scans.

III. RESULTS AND DISCUSSION

Excellent sensitivities can be achieved by coupling packed capillary HPLC with ESI [7,9,10]. In our laboratory, proteolytic digests of proteins are analyzed routinely by packed capillary HPLC-MS at the 1 to 5 pmol level, but have been analyzed at sample levels as low as 800 fmol [9]. Typical results from the capillary HPLC system are displayed in Figure 2a. The RIC was obtained from a packed capillary HPLC-MS analysis of a tryptic digest of a DNA polymerase (90 kDa) isolated from a thermophilic bacterium. Inspection of the data showed that the early eluting peaks in the RIC were derived from peptides in the mass range 0.5-5 kDa and the later eluting peaks were derived from large, partially digested fragments, 7-40 kDa. Summing scans 854 to 863 from this analysis afforded the spectrum displayed in Figure 2b. ESI mass spectra are "deconvoluted" using a computer algorithm that is available as part of the TSQ-700 ICIS™ software, resulting in a derived mass spectrum with the predicted molecular mass of the peptide or protein. This algorithm was used to deconvolute the spectrum shown in Figure 2b and the derived spectrum is displayed in Figure 2c.

The charge state of a peptide can be determined directly by slowing the mass scan rate of the ion trap [13], thus resolving the isotopic cluster which is representative of the composition of the peptide. Measuring the difference in molecular mass between each isotope peak will then afford the charge state of that peptide (ie., if mass difference is 1 amu between each peak, the peptide is singly charged and if the mass difference is 0.25 amu, the peptide is quadruply charged). In Figure 3b, the scan rate was attenuated by a factor of 100 (after recording the spectrum shown in Figure 3a) reducing the mass window of the experiment to approximately 3 m/z. The average mass difference between the peaks observed in this electrospray spectrum was 0.25 m/z, indicating a quadruply charged ion. These preliminary results demonstrate the feasibility of using the ion trap for the analysis of biomolecules.

Packed capillary HPLC-MS can also be used to analyze mixtures of native proteins. The RIC of an HPLC-MS analysis of a mixture of three proteins using a C4 column is shown in Figure 4. The protein mixture contained 1.5 pmol of horse heart cytochrome c, 3.5 pmol of γ-bungarotoxin (MW 7.5 kDa) isolated from the snake *Bungarus multicintus*, and 3.5 pmol of uteroferrin. The molecular masses of both cytochrome c and γ-bungarotoxin

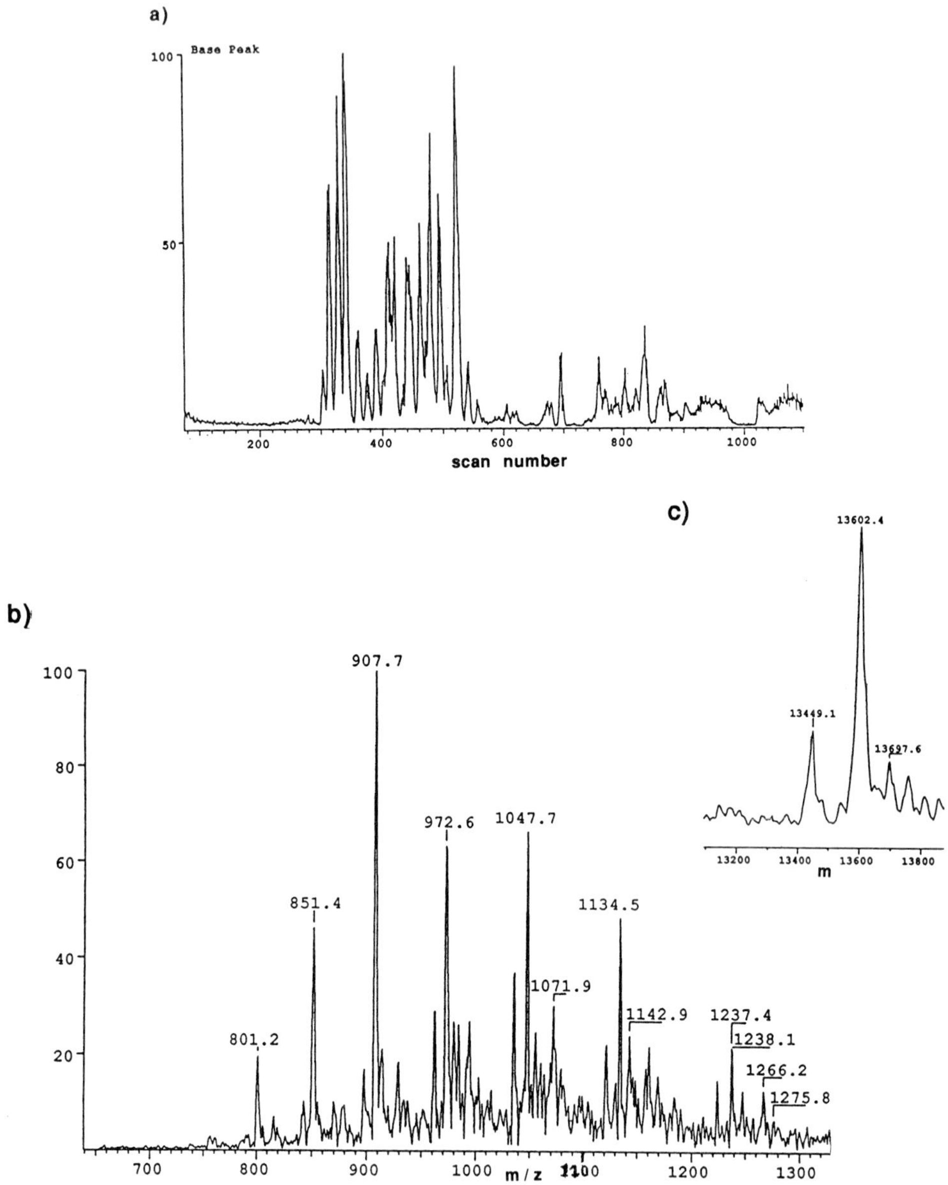

Figure 2. a) RIC from a capillary HPLC-MS analysis of 10 pmol of a DNA polymerase isolated from a thermophilic bacterium. b) ESI spectrum obtained from summing scans 854 to 863 of the analysis described in Figure 2a. c) A derived spectrum for the ESI spectrum displayed in Figure 2b.

were determined to within 1 mass unit in this experiment. Uteroferrin, which is a glycoprotein, afforded an ESI spectrum that was difficult to deconvolute. Using ESI, protein molecular masses are typically determined using ESI by infusing sample (1 to 5 pmol/μl) for several minutes. This procedure consumes tens of picomoles of sample compared with the 1 to 5 pmols of protein required for a capillary HPLC-MS analysis.

Tandem MS or MS/MS has been used to obtain primary sequence information on peptides. The parent ion is selected in the first stage of MS, then in an intermediate region the parent ion is activated with a target gas (CAD), surface, or a photon of light from a laser, and finally the daughter ions are analyzed in the second stage of MS. Alternatively, MS/MS in trapping instruments (ICR or ion trap) allows experiments to be performed tandem in time [12-14]. In this experiment the parent ion of interest is isolated and all other ions are swept from the trap. Following isolation, the parent ion can be activated using various techniques. Although MS/MS offers a rapid and sensitive means for obtaining primary structural information of a protein, the best strategy towards complete characterization of a protein's primary structure may involve obtaining partial amino acid sequence to design synthetic deoxynucleotide probes. These oligonucleotide probes are used to

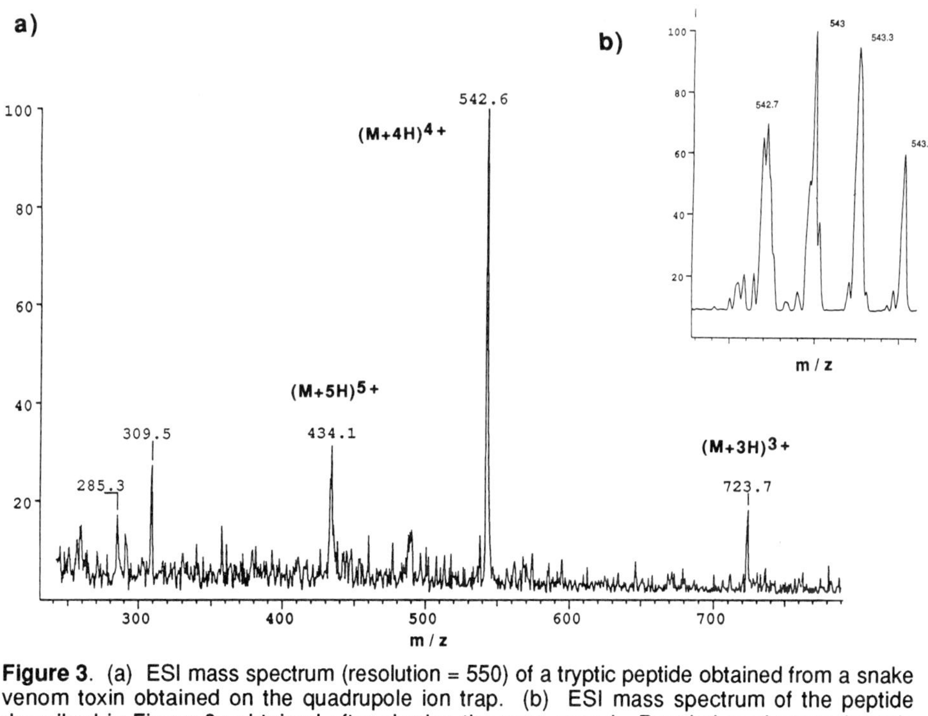

Figure 3. (a) ESI mass spectrum (resolution = 550) of a tryptic peptide obtained from a snake venom toxin obtained on the quadrupole ion trap. (b) ESI mass spectrum of the peptide described in Figure 3a obtained after slowing the scan speed. Resolution of approximately 5650 was obtained allowing separation of the isotope peaks for the quadruply charged ion.

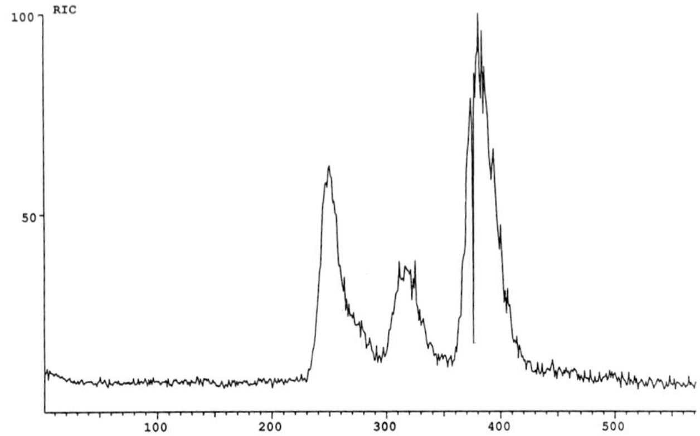

Figure 4. RIC from a capillary HPLC-MS analysis of a mixture of 3 proteins.

isolate the protein's corresponding gene and then the gene can be sequenced [17]. Using capillary HPLC-MS/MS partial amino acid sequences are determined from several different locations in the protein in a single analysis and this allows multiple deoxynucleotide probes to be used for screening cDNA libraries or for performing PCR with increased fidelity [17].

Displayed in Figure 5 is a CAD mass spectrum recorded on $(M+2H)^{2+}$ ions, m/z 696.4, of a 14 residue peptide generated from a tryptic digest of myoglobin. This spectrum was obtained by injecting 5 pmol of the digest into the capillary HPLC-MS system. Subtraction of masses of any two fragment ions that differ by a single amino acid affords a mass that identifies the amino

Table I Predicted fragment ions for the peptide HGVTVLTALGAILK (Figure 5). Ions that are observed in the ESI MS/MS spectrum displayed in Figure 5 are underlined.

No.	Seq	B	Y	No.
1	His	138.1	<u>1393.7</u>	14
2	Gly	<u>195.2</u>	<u>1256.6</u>	13
3	Val	<u>294.3</u>	<u>1199.5</u>	12
4	Thr	<u>395.4</u>	<u>1100.4</u>	11
5	Val	<u>494.6</u>	<u>999.3</u>	10
6	Leu	<u>607.7</u>	<u>900.2</u>	9
7	Thr	<u>708.8</u>	<u>787.0</u>	8
8	Ala	<u>779.9</u>	685.9	7
9	Leu	<u>893.1</u>	<u>614.8</u>	6
10	Gly	<u>950.1</u>	<u>501.6</u>	5
11	Ala	<u>1021.2</u>	<u>444.6</u>	4
12	Ile	<u>1134.4</u>	<u>373.5</u>	3
13	Leu	1247.5	<u>260.4</u>	2
14	Lys	1375.7	<u>147.2</u>	1

acid present at that position. Differentiation of leucine and isoleucine residues present in peptides is currently not possible using multiple, low energy collisions. Predicted masses for fragment ions for type B_n and type Y_n [18] of the tryptic peptide described above are displayed in Table I and fragment ions that are observed in the mass spectrum are underlined. Fragment ions of type Y allowed residues 1-6 and 8-14 to be specified in the present example. Fragment ions of type B allowed residues 2-12 to be assigned. Manual interpretation of CAD mass spectra is extremely time consuming and is currently the limiting step in protein analysis. However, preliminary results utilizing computer aided interpretation of low energy CAD spectra of peptides shows considerable promise for rapid interpretation of MS/MS data [19].

Infusion experiments performed using ESI-MS have demonstrated that comparable ion currents can be obtained for a 5 pmol/μl solution of a standard peptide at flow rates ranging from 5 μl/min to 100 nl/min. Operating at a flow rate of 100 nl/min will consume an order of magnitude less sample than operating at a flow rate of 1 μl/min, thus we have made attempts to utilize low flow rates by coupling open tubular chromatography with electrospray on both the ion trap and on the TSQ-700. Coupling of OTLC with MS has previously been reported by de Wit and coworkers [20]. Displayed in Figure 6a is a RIC obtained from an OTLC-MS analysis of 500 fmol of a synthetic peptide. The peptide was loaded and eluted as described above. Elution of the peptide occurred over several scans allowing both MS (Figure 6b) and MS/MS (Figure 6c) experiments to be performed in a single OTLC-MS experiment.

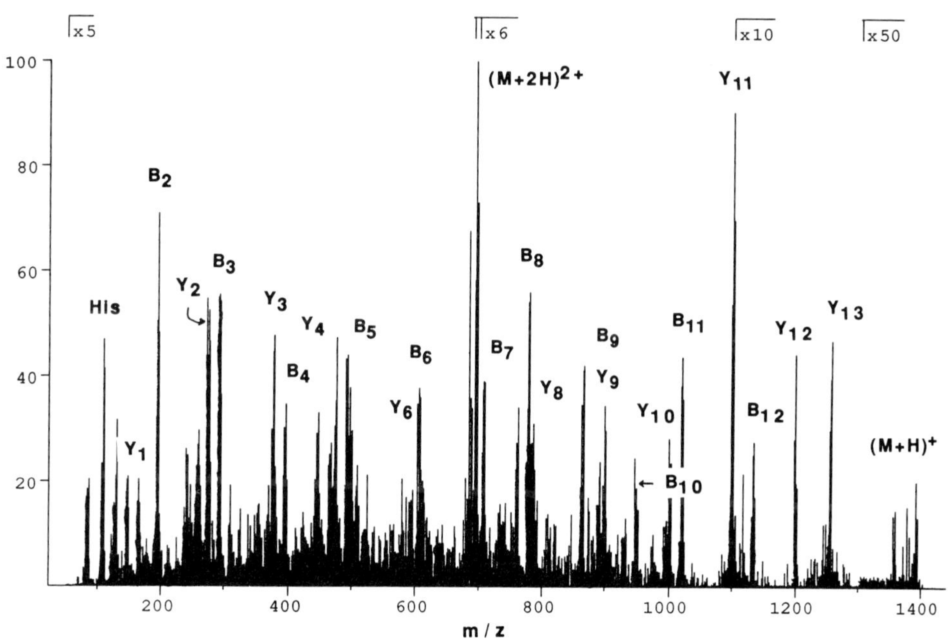

Figure 5. A capillary HPLC-MS/MS spectrum recorded on $(M+2H)^{2+}$ ions of a 14 residue peptide, HGVTVLTALGAILK, generated from a 5 pmol injection of a tryptic digest of myoglobin.

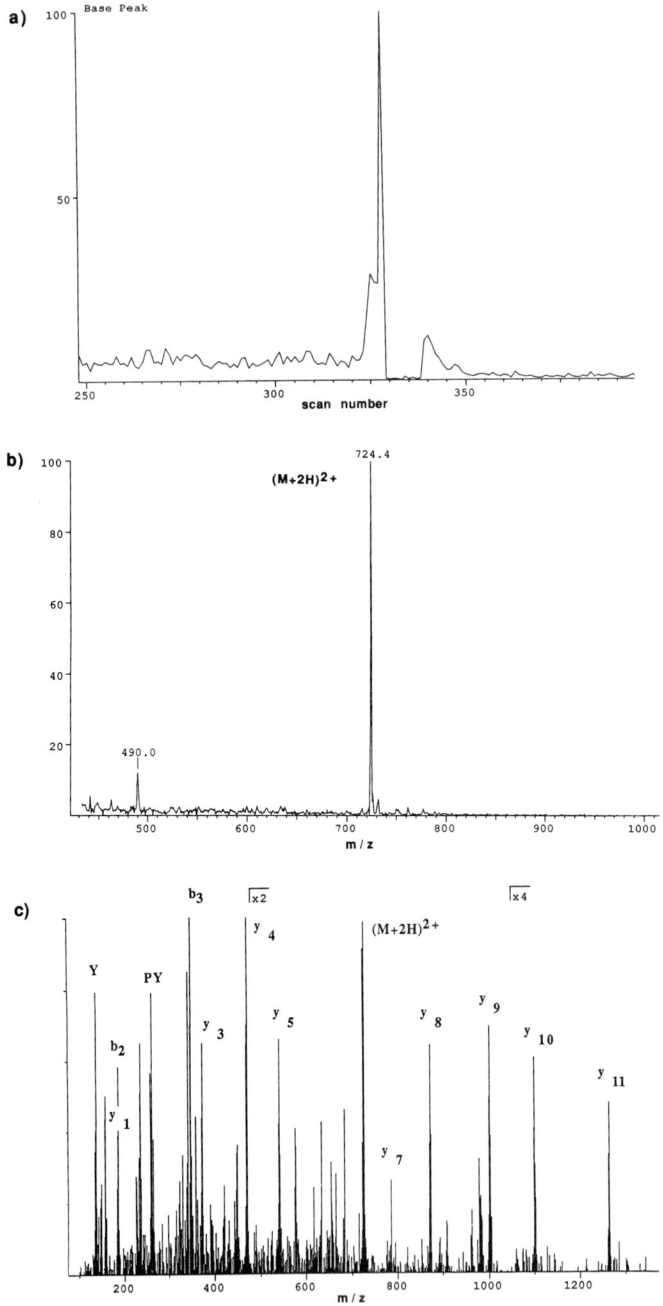

Figure 6. (a) RIC of an OTLC-MS analysis of 500 fmol of a synthetic peptide, SPYVESEDAPPQK. (b) ESI spectrum from the analysis described in Figure 6a. (c) ESI-MS/MS spectrum from the analysis described in Figure 6a.

The combination of packed capillary HPLC with tandem mass spectrometry has the potential to redefine approaches for molecular analysis of proteins. The high throughput and excellent sensitivity of this technique makes possible the analysis of complex biological systems. As much as 10 to 40% of a protein's amino acid sequence may be characterized in the time required to perform several cycles of automated Edman degradation at sample levels below those currently required by commercial sequenators.

ACKNOWLEDGEMENTS

The authors would like to acknowledge Drs. M. Hail, I. Mylchreest for helpful discussions. This work was supported by Finnigan and, in part, by the NSF, Science and Technology Center Cooperative agreement no. 8809710.

REFERENCES

1. Barber, M., Bordoli, R. S., Sedgwick, R. D., and Tyler, (1981) *J. Chem. Soc. Chem. Commum.*, 325-327.
2. D. F. Hunt, W. M. Bone, J. Shabanowitz, J. Rhodes and J. Ballard, (1981) *Anal. Chem.*, 54, 1704-1706.
3. M. L. Vestal, (1990) *Methods in Enzymology*, 193, 107-130.
4. R. M. Caprioli and W. T. Moore, (1990) *Methods in Enzymology*, 193, 214-236.
5. J. A. Loo, H. R. Udseth, and R. D. Smith, (1989) *Anal. Biochem.*, 176, 404-412.
6. S. A. McLuckey, G. J. Van Berkel, G. L. Glish, E. C. Huang, and J. D. Henion, Anal. Chem., 63 (1991) 375-383.
7. D. F. Hunt, J. E. Alexander, A. L. McCormack, P. A. Martino, H. Michel, J. Shabanowitz, N. Sherman, M. A. Moseley, J. W. Jorgenson, K. B. Tomer, in "*Techniques in Protein Chemistry II*," J. J. Villiafranca, Ed., Academic Press, N.Y., 441-454.
8. M. Hail, S. Lewis, I. Jardine, J. Liu, and M. Novotny, (1990) *J. Microcol. Sep.*, 2, 285-292.
9. P. R. Griffin, J. Coffman, L. E. Hood, and J. R. Yates, III (1991) *Int. J. Mass Spectrom. Ion Proc.*, In Press.
10. E. C. Huang and J. D. Henion, (1990) *J. Am. Soc. Mass Spectrom.*, 1, 158-165.
11. J. B. Fenn, M. Mann, C. K. Meng, S. F. Wong, and C. M. Whitehouse, (1989) *Science*, 246, 64-71.
12. G. J. Vanberkel, G. Glish, and S. McLucky, (1990) *Anal. Chem.*, 62, 1284-1295.
13. J. Schwartz, J. E. P. Syka, and I. Jardine, (1991) *J. Am. Soc. Mass Spectrom.*, 2, 198-204.
14. R. G. Cooks and R. E. Kaiser, (1990) *Acc. Chem. Res.*, 23, 213-219
15. P.P.H. Tock, C. Boshoven, H. Poppe, and K. K. Unger, (1989) *J. Chromatography*, 477, 95-106
16. I. Jardine, M. Hail, S. Lewis, J. Zhou, J. Schwartz, and C. M. Whitehouse, (1990) in "*Proceedings of the 38th ASMS Conference on Mass Spectrometry and Allied Topics*," Tucson, 16-17.

17. R. H. Aebersold, J. Leavitt, R. A. Saavedra, L. E. Hood, and S. B. H. Kent, (1987) *Proc. Natl. Acad. Sci., USA*, 84, 6970-6974.
18. P. Roepstorff and J. Fohlman, (1984) *Biomed. Mass Spectrom.*, 11, 601.
19. J. R. Yates, III, P. R. Griffin, L. E. Hood, and J. X. Zhou, (1991) in "*Techniques in Protein Chemistry II*," J. J. Villiafranca, Ed., Academic Press, N.Y., 477-485.
20. J. S. M. de Wit, C. E. Parker, K. B. Tomer, and J. W. Jorgenson, (1987) *Anal. Chem.*, 59, 2400-2404.

Characterization of PEGated Superoxide Dismutase

M. M. Vestling, C. M. Murphy, C. Fenselau
Department of Chemistry and Biochemistry
University of Maryland Baltimore County, Baltimore, MD 21228

J. Dedinas, M. S. Doleman, P. B. Harrsch, R. Kutny, D. L. Ladd, M. A. Olsen
Sterling Research Group, Sterling Drug Inc., Great Valley, PA 19335

I. INTRODUCTION

When polyethylene glycol (PEG) is covalently attached to a protein, the serum stability and circulating half-life of the protein are greatly extended (1). Analytical methods are needed for this class of potential therapeutic agents to support patent registration, clinical testing, and quality control in manufacturing. PEG derivatized superoxide dismutase (SOD) is an antioxidant with potential clinical utility for a variety of conditions from injuries to arthritis (2). This derivatized metal-binding homodimer (2) was chosen for the development of an analytical strategy based on mass spectrometry that would be generally applicable to PEGated proteins.

The procedure (3) for attaching PEG to SOD is outlined in Figure 1. Methoxypolyethylene glycol-succinimidyl succinate, an activated PEG with an average molecular weight of 5000 is mixed at pH 7.8 with bovine SOD. Activated esters such as the succinylsuccinimides react with primary amines (4). Since the amino terminus of SOD is blocked by acetylation, the primary sites of modification by the PEG reagent should be the lysine e-amino groups. The reaction conditions are mild and produce catalytically competent PEG-SOD.

TECHNIQUES IN PROTEIN CHEMISTRY III

Figure 1. Preparation of PEG-SOD.

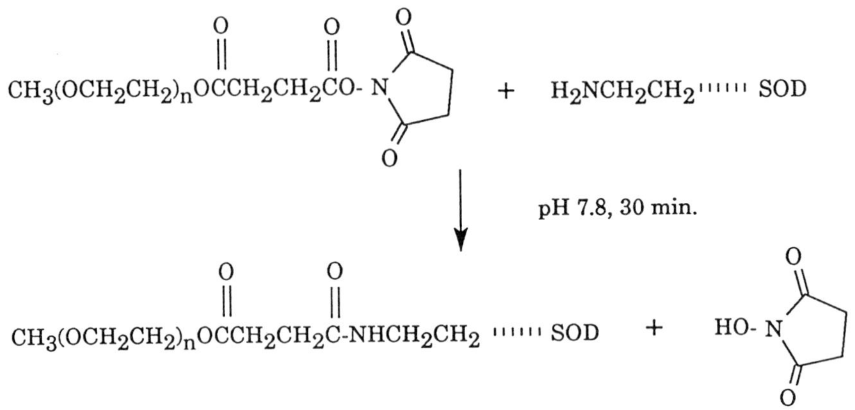

PEG-SOD exhibits heterogeneity deriving from at least two sources. The most obvious source is PEG's wide range of molecular weights. But SOD is also a source of heterogeneity, since each monomer contains ten lysine residues, any or all of which may be modified by PEGation. The primary sequence of the monomer is shown in Figure 2.

Figure 2. The primary sequence of bovine superoxide dismutase (SOD) monomer (Ref. 5).

```
           10        20        30        40
            |         |         |         |
AcATKAVCVLKGDGPVQGTIHFEAKGDTVVVTGSITGLTEGD

           50        60        70        80
            |         |         |         |
  HGFHVHQFGDNTQGCTSAGPHFNPLSKKHGGPKDEERHVG

           90       100       110       120
            |         |         |         |
  DLGNVTADKNGVAIVDIVDPLISLSGEYSIIGRTMVVHEK

          130       140       150
            |         |         |
  PDDLGRGGNEESTKTGNAGSRLACGVIGIAK
```

We have combined several mass spectrometry techniques to characterize PEG-SOD. A determination of the locations at which PEGs are attached to SOD will be described here as well as a strategy for determining the number of PEGs attached to SOD. We have used fast atom bombardment (FAB) mass spectrometry and tandem mass spectrometry of proteolytic digestion products to locate the attachment sites, and electrospray ionization mass spectrometry to estimate the numbers of PEGs attached.

II. EXPERIMENTAL

Chemicals. Iodoacetic acid, and ethylenediaminetetracetic acid (EDTA) were purchased from Aldrich, 2-mercaptoethanol and Tris from Bio-Rad; guanidinium hydrochloride from IBI; ammonium bicarbonate, and acetonitrile from Baker; and trifluoroacetic acid (TFA), dithiothreitol, iodoacetamide, clostripain, α-chymotrypsin, thioglycerol, and 3-nitrobenzyl alcohol from Sigma. Trypsin was purchased from Boehringer-Mannheim and Promega; while *S. aureus* V8 protease was from Boehringer-Mannheim and Pierce. Superoxide dismutase (SOD) was obtained from DDI Pharmaceuticals or Sigma and PEG-SOD from Enzon.

DePEGation of SOD. PEG is released from PEG-SOD by treatment with base. We have used two different procedures to accomplish this. In the first procedure, PEG-SOD (5 mL) (7.5 mg/mL in 50 mM sodium phosphate pH 7.3) was titrated to pH 10 with 5.0 M NaOH and allowed to stand at room temperature for 3 hr. The pH was then adjusted to 7.12 with 1.0 M HCl, and the PEG removed by dialysis against 50 mM sodium phosphate buffer for 72 hr. The second procedure is the reduction and alkylation procedure described below, in which the reduction is carried out at pH 8.5.

Reduction and alkylation. The reduction and alkylation of PEG-SOD and SOD were done in standard ways using 6 M guanidinium hydrochloride, 4 mM EDTA, 0.25 M Tris at pH 8.5 and 2-mercaptoethanol or dithiothreitol at 37° C, followed by addition of iodoacetic acid or iodoacetamide. The excess reagents were removed by dialysis or with a short desalting column (Sephadex G25). The protein was then concentrated or freeze dried.

Protease Digestion. Protease digests with trypsin, clostripain, and α-chymotrypsin were carried out in 100 mM ammonium bicarbonate solutions with enzymes to substrates ratios between 1:40 and 1:100 (wt:wt) at 37° C for several hours. The digestions with *S. aureus* V8 protease were carried out in a similar manner except that 50 mM sodium phosphate at pH 7.8 was used instead of ammonium bicarbonate. The digestions were terminated by freezing, and the products subjected to partial fractionation by high performance liquid chromatography.

High performance liquid chromatography (HPLC). Either C18 (Vydac) regular bore or C8 (Aquapore) narrow bore columns were used. Linear gradients with 0.1% TFA/water and 0.1% TFA/acetonitrile were used, and the effluents were monitored at 214 or 215 nm. The flow rate was either 1.0 mL/min (regular bore) or 200 μL/min (narrow bore).

Mass Spectrometry. Conventional magnetic scans were made with the first two sectors (EB) of a JEOL HX110/HX110 tandem mass spectrometer using either a JEOL FAB gun or a JEOL cesium ion gun and the JEOL DA5000 data system. The FAB gun was operated at 6 kV and the Cs ion gun at 23 kV. MSMS spectra were obtained by utilizing all four sectors (EBEB), mass selecting the precursor ions with MS-1 and linked B/E scanning of product ions with MS-2. Helium at a pressure sufficient to attenuate the precursor ion beam by 80%. The accelerating voltage was 10 kV, and the collision cell was floated at 4 kV. For both scanning modes, the peptides dissolved in 0.1% trifluoroacetic acid were measured from thioglycerol or 3-nitrobenzyl alcohol. Electrospray ionization mass spectra were obtained using a Hewlett Packard 5988A mass spectrometer retrofitted with a Vestec electrospray ionization source and a Phrasor high energy detector. The samples [5 μM in acetic acid/water/acetonitrile solution (3/47/50 v/v/v)] were sprayed at 5 μL/min.

III. RESULTS AND DISCUSSION

Figures 3 and 4 illustrate our strategy for locating the sites at which PEG attaches to SOD. Reduced and carbamoylmethylated SOD and reduced and carbamoylmethylated PEG-SOD clostripain digestions were run in parallel. Figure 3 shows the HPLC (narrow bore reverse phase) profiles for the two digestions. The chromatogram (Figure 3b) of the reduced and alkylated PEG-SOD digest contains every peak seen in the reduced and alkylated SOD digest profile, plus additional peaks, signalling the presence of modified peptides in this sample. As an example, comparison of the FAB mass spectrometric results for the 14.8 min HPLC fractions is shown in Figure 4. The spectrum of the peptide from reduced and alkylated SOD (Figure 4a) is dominated by m/z 3707.5 (monoisotopic (6)), which corresponds to the expected MH^+ ion for a peptide containing the amino acid residues [78-113]. Formation of the peptide is predicted with clostripain, which preferentially catalyzes the hydrolysis of amide bonds whose carbonyls are contributed by arginine. The calculated MH^+ value for [78-113] is 3707.0. This ion is also seen in the spectrum of products from reduced and alkylated PEG-SOD (Figure 4b), but the major ion observed occurs at a m/z value 82 higher. This result is consistent with the dePEGation scheme shown in Figure 5 in which the succinyl linker remains with the protein modifying lysine. Please note that both the modified and unmodified peptide were observed in the protease digest, which indicates that only partial PEG derivatization occurs at lysine-89. Peptides modified by 82 μ were observed in all protease digests of PEG-SOD.

Further support for the retention of the succinimidyl group upon release of PEG is provided in Figure 6. Figure 6 contains the

Figure 3. Reverse phase HPLC profiles of the clostripain digestion of a) reduced and carbamoylmethylated SOD and b) reduced and carbamoylmethylated PEG-SOD.

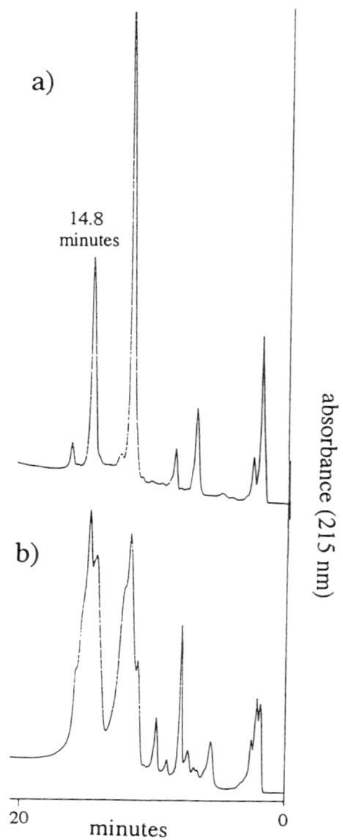

Figure 4. FAB mass spectra for the 14.8 minute fractions of the clostripain digestion of a) reduced and carbamoylmethylated SOD and b) reduced and carbamoylmethylated PEG-SOD.

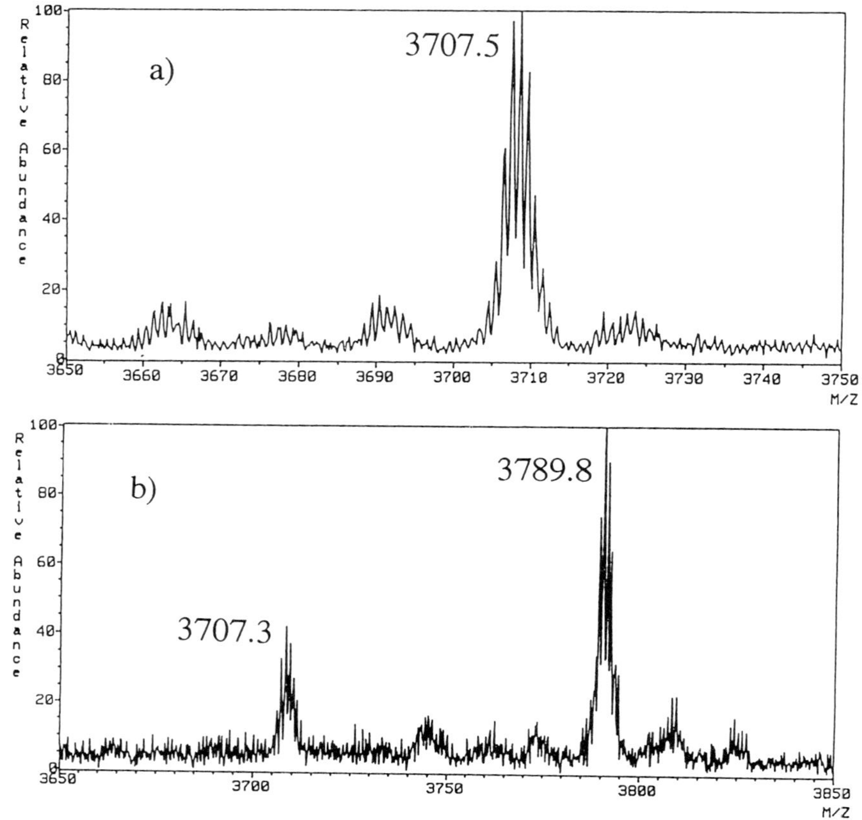

MSMS spectrum of a peptide ion 82 higher than expected for the clostripain peptide with amino acid residues [114-126]. The w series ions (7) as well as the immonium ion at m/z 183 clearly confirm the identity of the peptide and indicate that the mass increase occurs at lysine-120.

Figure 5. The dePEGation reaction of PEG-SOD.

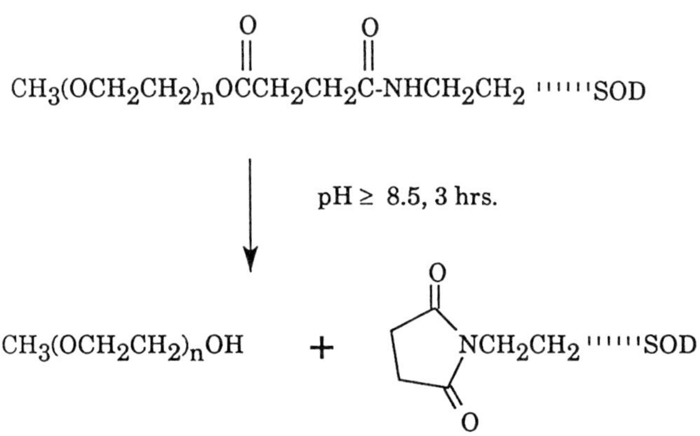

In experiments where PEG-SOD was exposed to strong base or dialyzed extensively, linker markers have been observed with mass increments of 100 instead of 82. This is interpreted as reflecting hydrolysis of the cyclic succinimide shown in Figure 5.

Releasing PEG from PEG-SOD with base treatment leaves succinimidyl markers on the protein. So by examining protease digestion products for the presence or absence of this marker, we can determine where PEG was attached. Table 1 contains a summary of the lysine containing peptides, both derivatized and underivatized that we have observed from several different protease digests.

Ions for peptides containing all ten of SOD's lysines have been identified by FAB mass spectrometry, and ions for peptides containing modified lysines were seen for eight of the ten lysines. Lysines 67 and 68 were not found to be noticeably modified. No evidence was seen that any amino acid residues other than lysines were modified with a succinyl group or a succinyl derivative. Incomplete alkylation of the cysteine residues was consistently noted.

Figure 6. MSMS of m/z 1578.9 showing the location of the derivatized lysine.

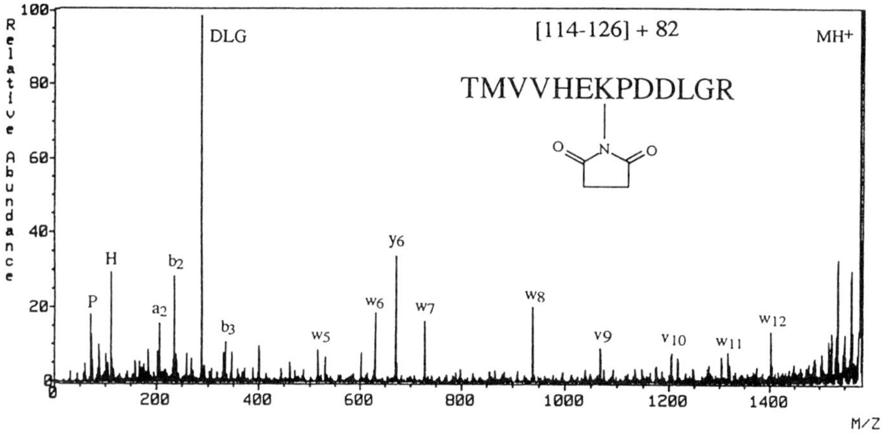

Figure 7. The expansion of the +14, +13, and +12 charge clusters from the electrospray ionization mass spectrum of dePEGated SOD. Five sub-clusters are marked in each cluster.

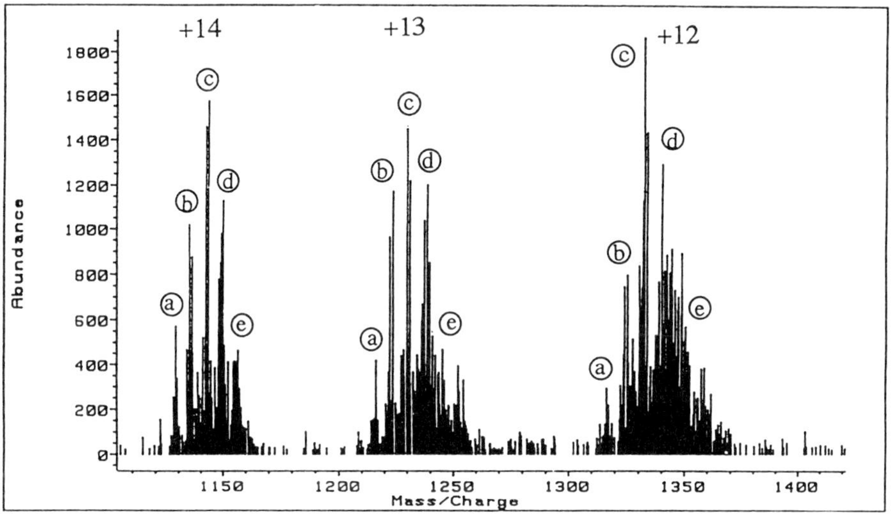

Table 1. The lysine containing peptides observed proteolytic digests of reduced and alkylated PEG-SOD.

position	derivatized		underivatized	
lysine	m/z	peptide	m/z	peptide
3	1056.6	[1-9]	361.5	[1-3]
9	2150.9	[4-23]	632.5	[4-9]
	1337.7	[9-20]	1255.8	[9-20]
23	1731.0	[22-38]	1648.0	[22-38]
	2508.9	[21-44]	2427.0	[21-44]
67	-	-	2593.5	[50-74]
68	-	-	1152.8	[68-77]
73	1106.5	[69-77]	1024.5	[69-77]
89	897.6	[89-96]	815.6	[89-96]
	1568.2	[82-96]	1486.0	[82-96]
	3789.8	[78-113]	3707.3	[78-113]
120	1578.8	[114-126]	1497.0	[114-126]
134	1546.8	[127-141]	1464.8	[127-141]
151	1026.6	[142-151]	944.8	[142-151]

The number of PEGs attached to each SOD was deduced from the distribution of molecular ions recorded in electrospray mass spectra. Electrospray ionization mass spectrometry of SOD confirmed the molecular weight predicted on the basis of the reported sequence; 15,591 calculated; 15,596 daltons ± 0.03% measured. Figure 7 contains part of the electrospray ionization mass spectrum of dePEGated SOD expanded to show the sub-clusters in the +14, +13, and +12 charge clusters. The sub-clusters marked a through e correspond to two through six markers per SOD and indicate heterogeneity in the number of PEGs attached to SOD.

IV. CONCLUSIONS

This study indicates that mass spectrometry can be used to characterize PEG-SOD and provides a general strategy for analyzing other PEG derivatized proteins. Release of PEG by base treatment of PEG-SOD leaves a modified lysine group. Location of the modified lysines can be accomplished by a combination of proteolytic digestion, high performance liquid chromatography, and mass spectrometry. The number of modifications can be seen in the electrospray ionization mass spectrum of the dePEGated protein. Work in progress includes an investigation of the electrospray ionization behavior of intact PEG-SOD and site specific quantitative analysis of partial derivatization. The study reported here indicates that the heterogeneity in PEG-SOD is due to the partial derivatization of the ten lysines in SOD and to the numbers of PEGs attached to each molecule of SOD.

ACKNOWLEGEMENTS

This work was supported by the National Science Foundation and Sterling Drug Inc.

REFERENCES

1. F. F. Davis, A. Abuchowski, T. van Es, N. C. Palczuk, K. Savoca, R. H. Chen, P. Pyatak in *Biomedical Polymers* ed. E. P. Goldberg, A. Nakajima, Academic Press, NY, 1980 pp 441-452.
2. J. V. Bannister, W. H. Bannister, G. Rotilio, *CRC Critical Reviews in Biochemistry* **22**: 111-180 (1987).
3. A. Abuchowski, G. M. Kazo, C. R. Verhoest, Jr., T. van Es, D. Kafkewitz, M. L. Nucci, A. T. Viau, F. F. Davis, *Cancer Biochem. Biophys.* **7**: 174-186 (1984).
4. G. W. Anderson, J. E. Zimmerman, F. M. Callahan, *J. Am. Chem. Soc.* **86**: 1839-1842 (1964).
5. H. M. Steinman, V. R. Naik, J. L. Abernethy, R. L. Hill, *J. Biol. Chem.* **249**: 7326-7338 (1974).
6. J. Yergy, D. Heller, G. Hansen, R. J. Cotter, C. Fenselau, *Anal. Chem.* **55**: 353-357 (1983).
7. K. Biemann, *Biomed. Environ. Mass Spectrom.* **16**: 99-111 (1988).

N-ε-Acetylation Can Occur at Lysine Residues 157, 167, 171, And 180 of Recombinant Bovine Somatotropin

Gary C. Harbour
Robert L. Garlick
Stephen B. Lyle
Frank W. Crow
Russell H. Robins
John G. Hoogerheide[1]

The Upjohn Company
Kalamazoo, Michigan 49001

I. Introduction

Recombinant bovine somatotropin (rbSt) produced by The Upjohn Company contains 191 amino acids and is expressed in recombinant E. coli as an inclusion body protein which needs to be refolded to achieve its proper biological activity. Pituitary-derived bovine somatotropin (pbSt) and rbSt both exist as isoforms which differ in isoelectric point (pI) (1-4), the most abundant being the gene product which has a pI of 8.2. Certain lower pI forms increase with sample manipulation and are found as clusters of closely spaced bands on isoelectric focusing (IEF) at pI values of approximately 7, 6, and 5, in some cases reportedly resulting from deamidation at asparagine residues 13 and 148 and glutamine 140 (5), as well as isoaspartic acid formation and chain cleavage at asparagine residue 99 (3,6).

We have found that a significant fraction of the pI 7 rbSt arises from acetylation of certain lysine residues. This report details the isolation and structural elucidation of the acetylated pI 7 bands in rbSt.

[1]To whom all correspondence should be addressed.

TECHNIQUES IN PROTEIN CHEMISTRY III

II. Materials And Methods

The pI 8.2 and 7 forms of rbSt were isolated by column chromatofocusing either on a Pharmacia FPLC system with a 5/20 Mono-P column or by low pressure chromatofocusing using a 2.5 cm X 36 cm column of Pharmacia PBE-94 chromatofocusing medium. In each case the sample and column were equilibrated with 25 mM ethanolamine-HCl, pH 9.5, after which the pI 8.2 and 7 rbSt fractions were eluted either with a solution containing 1.0 ml Pharmalyte 8-10.5 plus 5.2 ml Polybuffer 96 per 100 ml, titrated to pH 8.0, or with a buffer solution consisting of 1.25 mM each of ethanolamine, Tris, TES (N-tris[Hydroxymethyl]methyl-2-aminoethanesulfonic acid), and ACES (N-[2-Acetamido]-2-aminoethanesulfonic acid), titrated to pH 7.0 with HCl. Following chromatography, Polybuffer and Pharmalyte salts were removed by size exclusion chromatography.

Isoelectric focusing was performed using LKB PAG plates (1804-101) with a pH range of 3.5-9.5.

Tryptic digestion and tryptic peptide mapping on reversed phase HPLC were carried out as described previously (7).

Amino acid sequence analysis was performed on an Applied Biosystems (ABI) 470A sequencer equipped with an on-line ABI 120A PTH analyzer.

RP-HPLC for LC/MS was carried out on a Varian Vista 5560 ternary liquid chromatograph coupled to a Finnigan Model 4600 mass spectrometer. LC conditions were identical to those used in tryptic peptide mapping, with the exception that a flow rate of 1.5 ml/min was used to promote better thermospray conditions. The thermospray jet (or source block) temperature was maintained at 200° C. The vaporizer tip temperature was optimized to 85° C by using flow injections of a model peptide. The instrument was scanned from mass 200-1800 in 2 seconds. Data were acquired and processed with a Finnigan SuperIncos data system.

III. Results And Discussion

Chromatofocusing of rbSt on a low pressure column of Pharmacia PBE-94 separated the protein into a major pI 8.2 component and a minor pI 7 fraction (Figures 1 and 2). Samples were pooled to include five-10 ml fractions each, which were subjected to analytical IEF (Figure 2). Samples from chromatofocusing runs utilizing FPLC and low pressure chromatography were used to characterize the pI 8.2 and 7 rbSt fractions.

On tryptic peptide mapping of the pI 8.2 and 7 rbSt components in the non-reduced form (Figure 3), three peaks appeared in the pI 7 map which did not appear in the pI 8.2 material. One of these peaks (peak B) had been identified previously (3) as peptide T9 containing the isoaspartic acid rearrangement product of the asparagine residue at position 99 of rbSt (Figure 4). This was reconfirmed by amino acid sequence analysis (Table

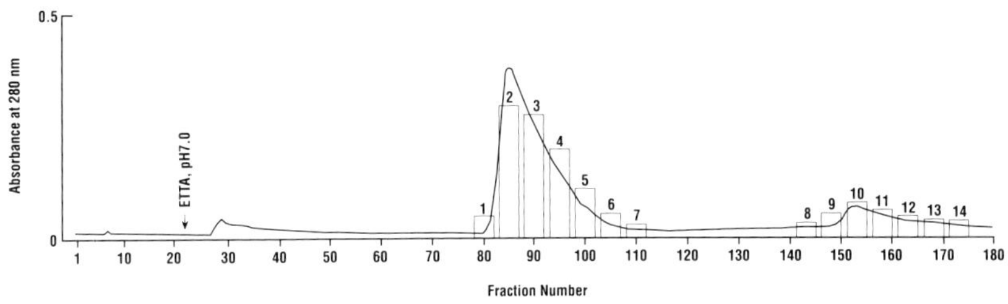

Figure 1. Chromatofocusing of rbSt using Pharmacia PBE-94 medium and the ethanolamine-Tris-TES-ACES buffer.

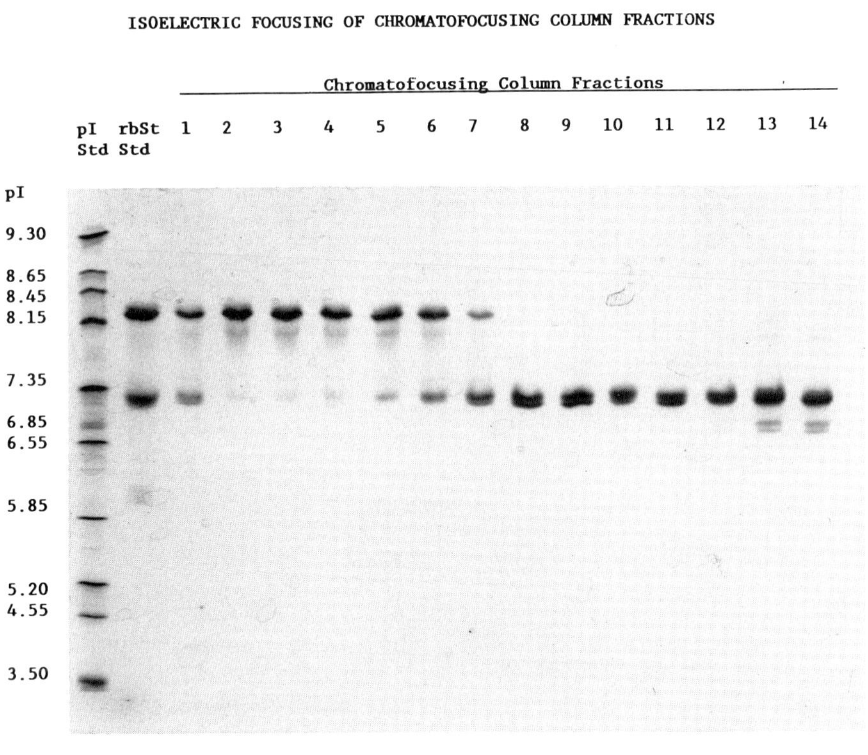

Figure 2. Analytical isoelectric focusing of column fractions from the chromatofocusing run shown in Figure 1.

I). The first three residues in T9 were correctly identified, followed by a loss of recovery due to isoaspartic acid at position 4, a phenomenon which has been reported earlier (3).

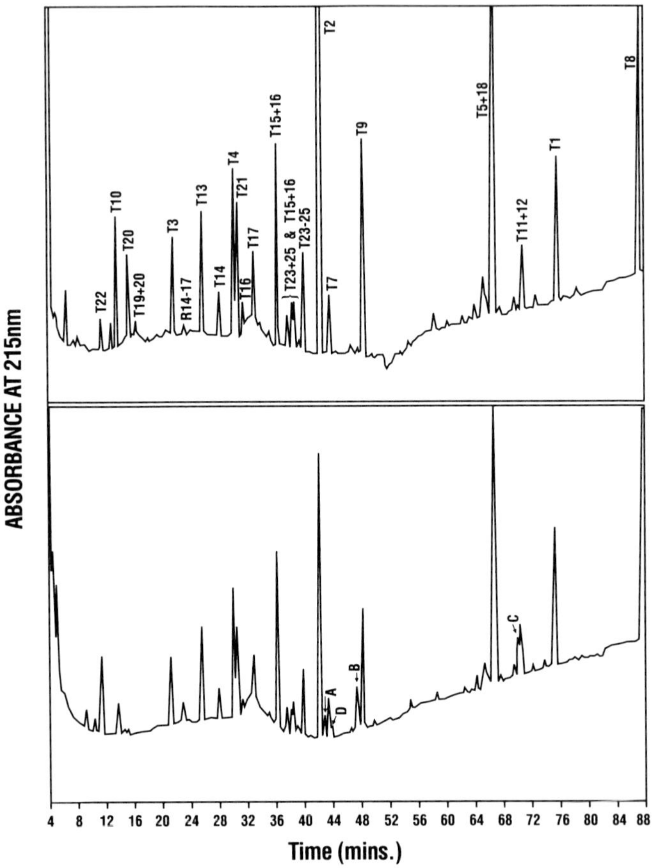

Figure 3. Tryptic peptide maps of pI 8.2 (Top) and pI 7.0 (Bottom) rbSt.

The other two peaks (A and C), which were present in the pI 7.0 but not the pI 8.2 map of rbSt, also were subjected to amino acid sequence analysis (Table I). Sequence analysis of peak A indicated that it was a mixture composed of peptides T22-23 plus T25. Previous unpublished work has shown that T24, a single arginine residue, is partially cleaved after treatment with trypsin; however, cleavage of the lysine-cysteine bond between T22 and T23 was always observed. Results from the current work indicate that the T22-T23 bond in this species has not been cleaved, and the carboxyl-terminal lysine of T22 has been replaced by an unidentified amino acid. Peptide T25 is most likely connected to T23 by a disulfide bond (Figure 4).

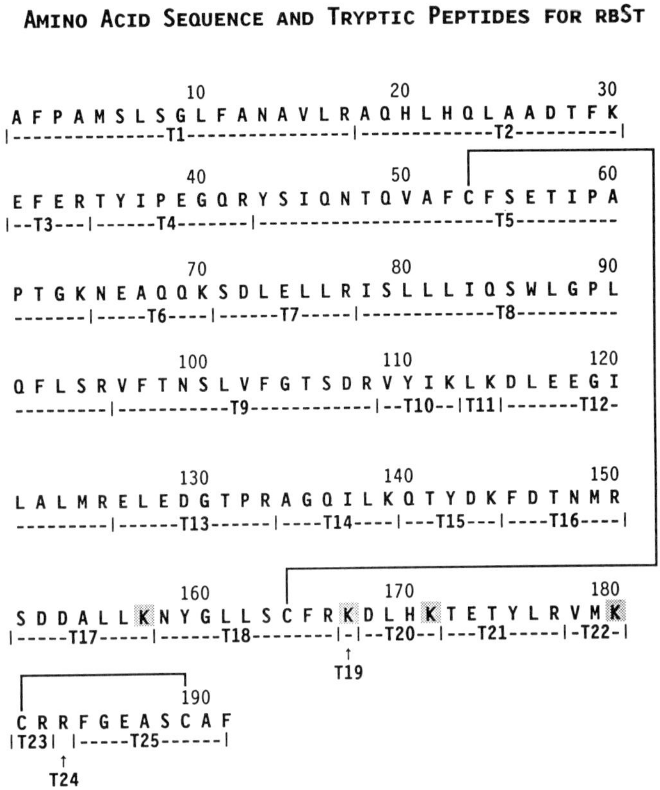

Figure 4. Tryptic peptides from rbSt. The shaded lysine residues are those which were found to be acetylated in the current study.

Sequencing of peak C indicated a mixture containing sequences for peptides T17-18, T5, and T11-12. The same unknown amino acid seen at position 3 in peptide T22-23 was at position 7 in T17-18. On the protein sequencer, an authentic sample of ∊-amino-acetyl lysine was found to have a retention time identical to that of the unknown lysine derivative found in peptides T17-18 and T22-23, confirming the presence of acetyl-lysine. It is thought that peptides T5 and T17-18 are connected by a disulfide bond and T11-12 is an unrelated contaminant. The acetylated lysines in peaks A and C correspond to lysines 180 and 157 in rbSt.

Thermospray liquid chromatography/mass spectrometry (TSP LC/MS) tryptic peptide mapping was used to determine the molecular weights of peptides A and C. A weak protonated molecular ion ($MH^+ = 1506$) and an intense doubly charged molecular ion ($MH_2^{++} = 754$) were detected for peak A (Figure 5), indicating a molecular weight corresponding to T22-T23+25 plus an additional 42 mass units, suggesting the addition of

Table I. Peptide sequencing of Peaks A, B, and C

Cycle	Peak A		Peak B	Peak C		
	T22-23	T25	T9	T17-18	T5	T11-12
1	Val(30)	Phe(42)	Val(64)	Ser(17)	Tyr(20)	Leu(16)
2	Met(27)	Gly(28)	Phe(70)	Asp(13)	Ser(17)	Lys(9)
3	Unk(34)	Glu(27)	Thr(10)	Asp(20)	Ile(18)	Asp(20)
4	Cys	Ala(20)	Isoasp	Ala(10)	Gln(17)	Leu(12)
5	Arg(Tr)	Ser(18)	Ser(-)	Leu(16)	Asn(15)	Glu(7)
6		Cys	Leu(-)	Leu(16)	Thr(3)	Glu(8)
7		Ala(5)	Val(-)	Unk(11)	Gln(13)	Gly(7)
8		Phe(3)	Phe(-)	Asn(11)	Val(6)	Ile(5)
9			Gly(-)	Tyr(7)	Ala(8)	Leu(6)
10			Thr(-)	Gly(9)	Phe(11)	Ala(5)
11			Ser(-)	Leu(12)	Cys	Leu(12)
12			Asp(-)	Leu(10)	Phe(8)	Met(-)
13			Arg(-)	Ser(11)	Ser(11)	Arg(-)
14				Cys	Glu(4)	
15				Phe(5)	Thr(-)	
16				Arg(Tr)	Ile(-)	

The numbers in parentheses represent picomoles of recovered amino acid. Cys was not determined. Isoaspartic acid blocks peptide sequencing. Tr = Trace. Unk = Unknown.

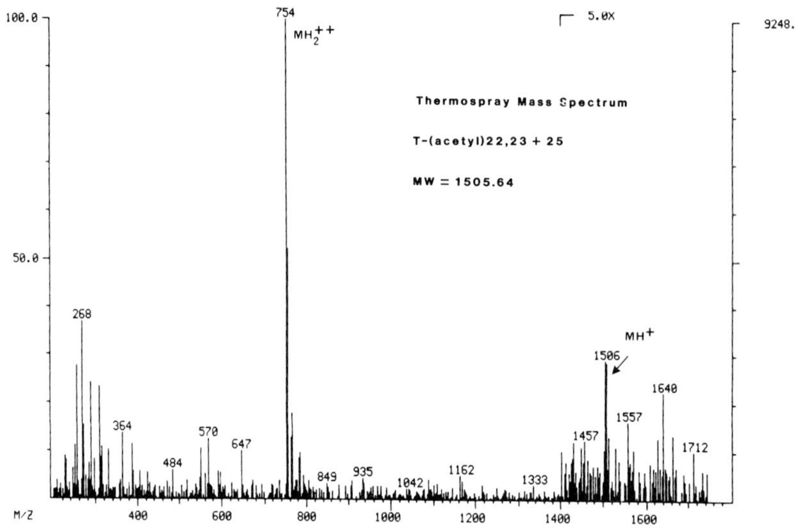

Figure 5. Thermospray mass spectrum of T-(acetyl)22-23+25.

of an acetate group. Analysis of peak C indicates that, buried under the signals for T12 and T11-12 is an ion which corresponds to the triply charged molecular ion (MH_3^{+++} = 1420) for the tryptic peptide T17-

18+T5 plus an additional 42 amu. Peak D did not appear in the TSP LC/MS chromatogram, and has not been identified.

The TSP LC/MS tryptic peptide map of reduced pI 7.0 rbSt was compared to that of the oxidized form (Figure 6). Peaks E, F, G, and H were found to have TSP mass spectra which corresponded to tryptic fragments T19-20, T22-23, T20-21, and T17-18, to each of which were added an additional 42 amu (Table II). These data indicate that potential acetylation has taken place on lysine residues 157, 167, 171, and 180 in

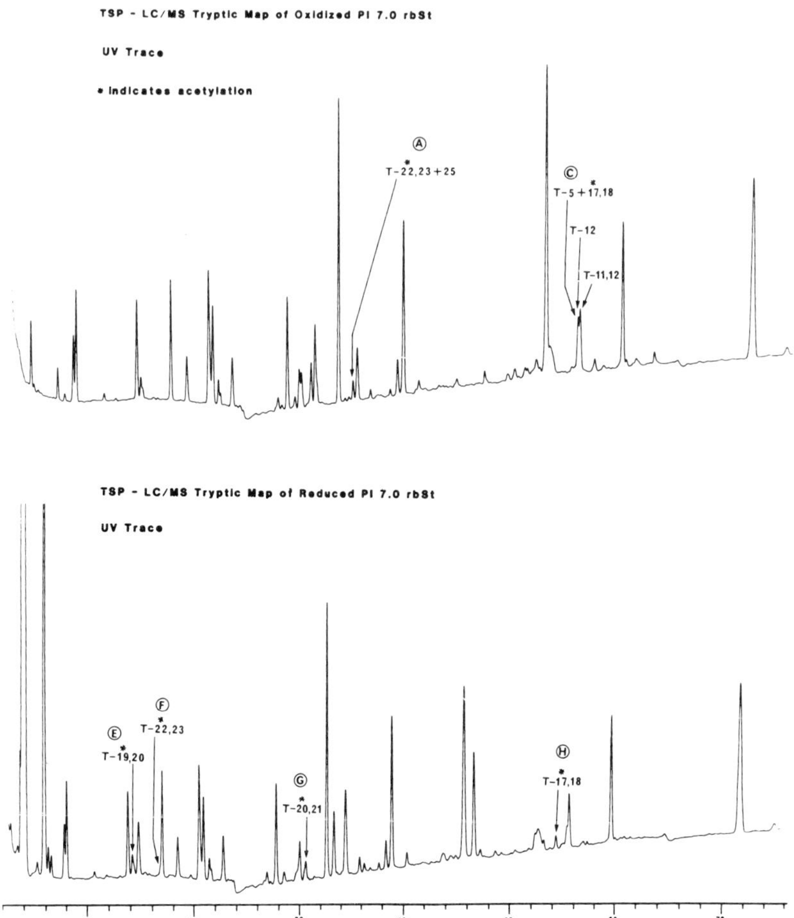

Figure 6. Thermospray LC/MS tryptic maps of oxidized (Top) and reduced (Bottom) pI 7.0 rbSt.

Table II. Acetylated Peptide Assignments from the TSP-LC/MS Tryptic Map of Reduced pI 7.0 rbSt

Peak	Mass of Molecular Ion(s)	Theoretical Mol. wt.[a]	Peptide Assignment
E	682 (MH^+)	681	T-(acetyl)19-20
F	678 (MH^+)	677	T-(acetyl)22-23
G	1318 (MH^+)	1317	T-(acetyl)20-21
	660 (MH_2^{++})		
H	929 (MH_2^{++})	1856	T-(acetyl)17-18

[a] Theoretical molecular weight of the acetylated peptide.

Table III. Amounts of Modified Species Observed

Peptide	Mole Percent
Isoasp T9	28
Acetyl T17-18	11
Acetyl T19-20	22
Acetyl T20-21	21
Acetyl T22-23	13

rbSt. The TSP LC/MS data were checked for the presence of all possible tryptic fragments which might result from acetylation of other lysines elsewhere in the rbSt molecule (lysines 30, 64, 70, 112, 114, 139, and 144). None of these species were found.

Approximate amounts of the acetylated species were estimated from peak heights observed on the peptide maps. Thus, comparing the response of T19-20 (Figure 6, reduced map) to the sum of the responses for all T20-containing peptides (T20; T19-20; T20-21) gives an amount of approximately 22%. Data for the other acetylated peptides are given in Table III. The total contribution of acetylated peptides to the pI 7 mixture is thus estimated to be about 67%. Isoaspartic acid-99 rbSt accounts for an additional 28% of the material.

Acetylation of lysine is a well known post-translational modification mediated by an acetyltransferase enzyme using acetyl-CoA as a substrate. It has been demonstrated with the histones, where it is believed to regulate the binding of these very basic proteins to the negatively charged ribose backbone of the nucleic acids (8). We believe that the acetylation of lysines on rbSt is a post-translational modification carried out by the host E. coli organism. Acetylation of lysines makes them unrecognizable to trypsin and in each case results in the removal of one positive charge in the parent protein, thus lowering the pI.

The data indicate that at least four of the pI 7 species are produced by acetylation of lysines 157, 167, 171, and 180. Another species is formed by the base catalyzed rearrangement of asparagine 99 to isoaspartic acid. These species appear to account for most of the protein in the IEF bands in the pI 7 region. The further possibility of deamidation of asparagines and glutamines has not been ruled out. Acetylation of multiple lysines, or conversion of asparagine 99 to isoaspartic acid and/or deamidation of asparagine or glutamine in molecules already containing one or more acetyl-lysines, would also yield proteins with even lower pI values. These multiply modified species are no doubt responsible for some of the observed pI 6 and 5 bands.

Recent data not reported in this study suggest that other lysine residues in rbSt may become acetylated. The acetylation sites reported in the present study were found in purified rbSt. Preliminary new data suggest that other acetylation sites may exist in rbSt prior to purification which presumably does not co-purify with the bulk of the rbSt.

Acknowledgements

The authors wish to thank Catherine I. Johnson, Carla M. Campbell, and Ward M. Vanderberg for their expert technical assistance. In addition, we thank Richard J. Kirschner and Tim W. Evans for help in providing rbSt and Rita B. Holzman for use of her Fast Protein Liquid Chromatograph (FPLC).

References

1. Wingfield, P.T., Graber, P., Rose, K., Simona, M.G., and G.J. Hughes (1987) *J. Chromat.* **387**, 291.
2. Wingfield, P.T., Graber, P., Buell, G., Rose, K., Simona, M.G., and B.D. Burleigh (1987) *Biochem. J.* **243**, 829.
3. Harbour, G.C., Hoogerheide, J.G., and R.L. Garlick (1987) Int. Patent Application No. PCT/US86/01860, Int. Pub. No. WO87/01708.
4. Wood, D.C., Salsgiver, W.J., Kasser, T.R., Lange, G.W., Rowold, E., Violand, B.N., Johnson, A., Leimberger, R.M., Parr, G.R., Siegel, N.R., Kimack, N.M., Smith, C.E., Zobel,J.F., Ganguli, S.M., Garbow, J.R., Bild, G., and G.W. Krivi (1989) *J. Biol. Chem.* **264**, 14741.
5. Secchi, C., Biondi, P.A., Negri, A., Borroni, R., and S. Ronchi (1986) *Int.J. Pep. Prot. Res.* **28**, 298.
6. Violand, B.N., Schlitter, M.R., Toren, P.C., and N.R. Siegel (1990) *J. Prot. Chem.* **9**, 109.
7. Hartman, P.A., Stodola, J.D., Harbour, G.C., and J.G. Hoogerheide (1986) *J. Chrom.* **360**, 385.
8. Allfrey, V.G., Dipaola, E.A., and R. Sterner (1984) *In* "Methods in Enzymology" Vol. **107**. 224-241.

PLASMA DESORPTION MASS SPECTRAL (PDMS) ANALYSIS OF POLYPEPTIDES CONTAINING SOME POST-TRANSLATIONAL MODIFICATIONS; SCISSION OF COVALENT BONDS

Koji Takio[1] and Koiti Titani[2]

Laboratory for Aging Process Research
Frontier Research Program, RIKEN
Wako, Saitama, Japan 351-01

I. INTRODUCTION

Soft ionization techniques have made it possible to analyze peptides or even proteins with a mass spectrometer without prior derivatization. The principal advantage of the method lies in the analysis of modified or unusual amino acid residues in peptides. Since most, if not all, proteins undergo post-translational modifications, the importance of obtaining information regarding post-translationally modified residues is ever increasing. Although plasma desorption is known for mild ionization especially in combination with nitrocellulose coated sample targets, we observed scissions of specific covalent bonds. One was thioether bonds between the heme and the polypeptide chain of cytochrome c (cyt.c). The other was a sulfate ester bond in sulfated gastrin in the positive mode, where only weak signals of the sodium forms of free peptide were observed. The negative mode, however, gave the molecular ion signal. These results suggest some possi-

---

[1]Present address: Division of Biomolecular Characterization, RIKEN, Wako, Saitama, Japan 351-01.

[2]He is also a professor at Division of Biomedical Polymer Science, Institute for Comprehensive Medical Science, School of Medicine, Fujita Health University, Toyoake, Aichi, Japan 470-11.

TECHNIQUES IN PROTEIN CHEMISTRY III

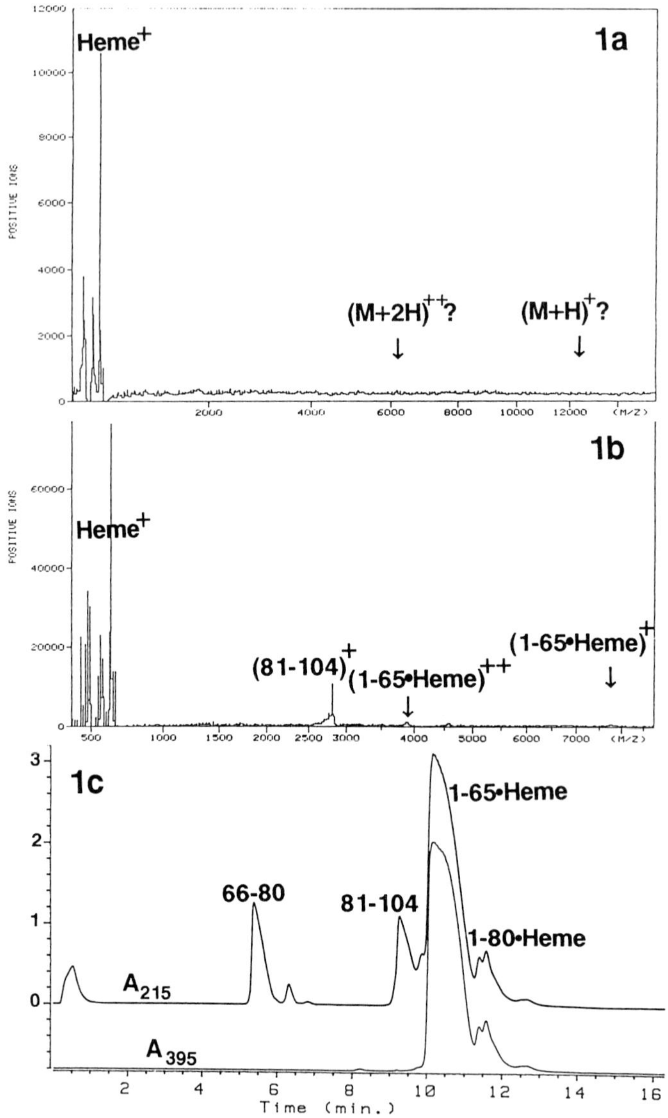

Figure 1. Spectra of horse heart cyt.c (a), its heme-containing BrCN fragment (b), baker's yeast cyt.c (d), and its tryptic heme peptide (e). The signal at M/Z = 2780 in (b)

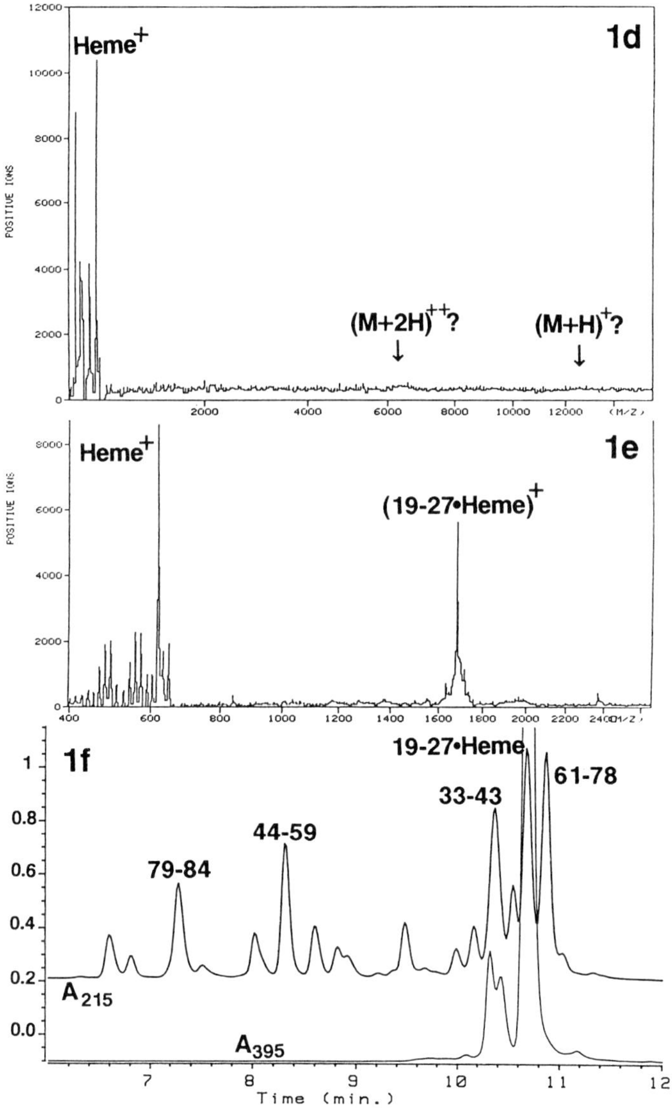

resulted from a trace contaminant. RP-HPLC profiles of the BrCN fragments of horse heart cyt.c (c) and the tryptic peptides of baker's yeast cyt.c (f).

bilities of overlooking modification groups. These specific cleavages may also give some insight into the desorption mechanism.

II. MATERIALS AND METHODS

Gastrin was obtained from Peptide Institute, Inc. (Osaka, Japan). Sulfated gastrin and horse heart cyt.c were purchased from Sigma (St. Louis, Mo.). Baker's yeast (Red Star) cyt.c was a byproduct of cyt.c peroxidase (1) and further purified on a CM-cellulose column. TPCK-trypsin and pyroglutamyl aminopeptidase were products of Cooper Biomedical (Malvern, Pa.) and Suntory (Osaka, Japan) respectively. Other chemicals were of analytical grade.

Gastrin and sulfogastrin (1 nmol) were treated with pyroglutamyl aminopeptidase (1 mU) in 10 mM phosphate buffer pH 7.8 containing 1 mM EDTA and 10 mM 2-mercaptoethanol at 37 ^{o}C for 4 h. Baker's yeast cyt.c (10 nmol) was digested with TPCK-trypsin (1 μg) in 50 mM ammonium bicarbonate at 37 ^{o}C for 4 h. Horse heart cyt.c was cleaved with BrCN in 70 % formic acid (2). Peptides were separated by reversed-phase high-performance liquid chromatography (RP-HPLC) with a Hewlett-Packard model 1090M Liquid Chromatograph on an Aquapore RP-300 column (2.1 x 30 mm, Applied Biosystems) with an aqueous trifluoroacetic acid/acetonitrile gradient system at a flow rate of 0.2 ml/min.

Plasma desorption mass spectral analyses were carried out with a BioIon 20 Biopolymer Mass Analyzer (Applied Biosystems, Foster City, Ca.) on nitrocellulose coated targets (3). Approximately 100 pmols of each protein or peptide was used.

III. RESULTS AND DISCUSSION

Plasma desorption mass spectrometry was found to be a convenient and versatile method for analyzing peptides with modified residues. We have successfully analyzed phosphopeptides, acetylated peptides or pyroglutamyl peptides. Artificially modified peptides including carboxymethyl- or pyridylethyl-cysteine or cysteic acid did not cause any problems.

Therefore, it was a surprise to see only intense signals of free heme in the PD mass spectra of cyt.c (Fig. 1a and 1d) rather than intense signals of multiply charged intact molecule. Cyt.c is a strongly basic protein and contains an ironheme covalently attached to the poly-peptide chain through two thioether bonds which have been thought rather stable. Cyanogen bromide cleavage products (Fig. 1b and c) and tryptic digest (Fig. 1e and f) were analyzed to see whether shorter polypeptide chain helps to maintain integrity of the peptides. With the BrCN fragment (65 residues), signals of the intact heme peptide were barely recognizable, whereas the tryptic heme peptide (9 residues) gave comparable signals of the intact heme peptide and free heme. These results may reflect the efficient energy absorption to heme, conjugated double bonds, and the relative energy required for the desorption of the polypeptide from nitrocellulose and for the cleavage of the two thioether bonds.

Gastrin is a strongly acidic 17-residue peptide containing 6 acidic residues with an amino-terminal pyroglutamic acid. It does not contain any positively ionizable group and is used as a standard peptide for the negative mode calibration. Tyrosine (residue 12) is sulfated in sulfogastrin. In the negative mode, both gastrin (M_γ=2,098) and sulfogastrin (M_γ=2,178) gave singly charged molecular ion signals (Fig. 2a and 2c), although their intensity were not great. In the positive mode, however, both gave identical signals of the mono- and di-sodium forms (Fig. 2b and 2d) of free gastrin indicating scission of the sulfate ester bond in sulfogastrin. It is not likely that the free gastrin signal came from the contaminant of sulfogastrin since the quantity of gastrin in sulfogastrin was minimal if any (Fig. 2h). Considering that the molecules completely lack positively ionizable groups, the effect of a single positively ionizable group was assessed by removing the amino-terminal pyroglutamyl residue. Even with the single positively ionizable group, sulfogastrin failed to give the intact molecular ion signal in the positive mode (Fig. 2f and 2g).

To see whether these two covalent bond scissions are unique to PDMS, cyt.c and sulfogastrin were subjected to fast atom bombardment mass spectrometry (FAB-MS) and to IonSpray-MS. While results obtained with FAB-MS were identical to PDMS, those with IonSpray-MS were different (results not shown). Signals of free heme from cyt.c was very small and intense multiply charged intact molecular ion signals were observed with IonSpray-MS. Sulfogastrin seems to be decomposed in the positive mode even with the IonSpray-MS, although we have not completed interpretation of the signals. Difference in the mode of ionization may be responsible for the results with cyt.c. While PDMS and FAB-MS desorb ions from solid or liquid surface by collision energy, IonSpray-MS forces solute to ionize by evaporation of solvents.

Although mass spectral analysis is rapidly gaining popularity in protein structural analysis, we do not know much

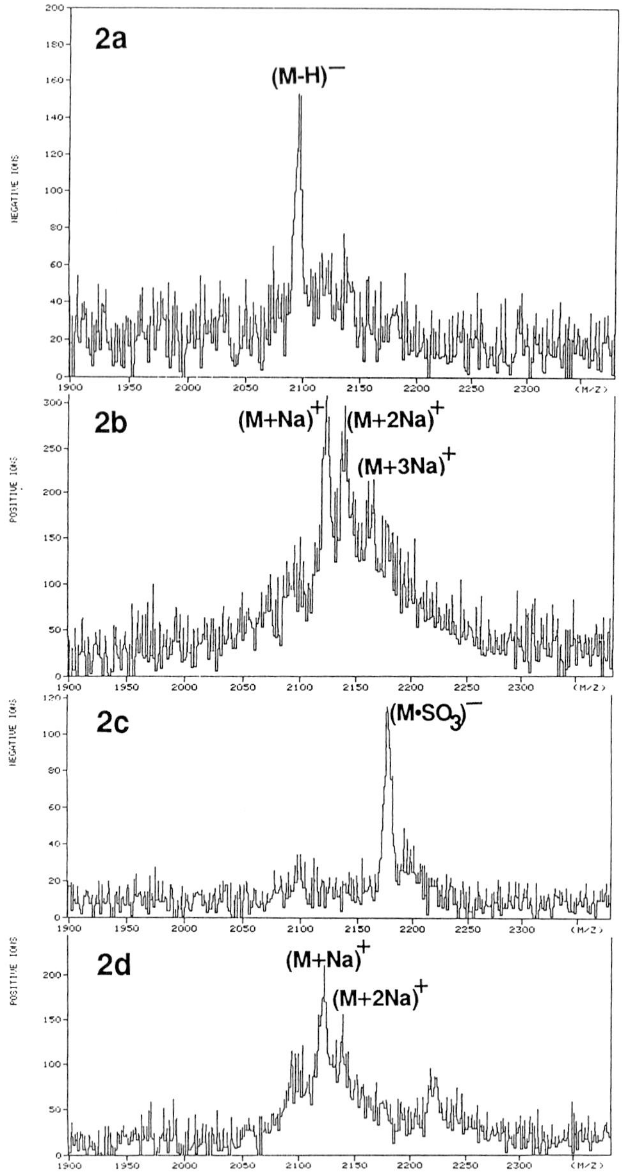

Figure 2. Spectra of gastrin (a and b), sulfogastrin (c and d), gastrin 2-17 (f) and sulfogastrin 2-17 (g) in negative (a and c) and positive (b, d, f and g) modes. Amino

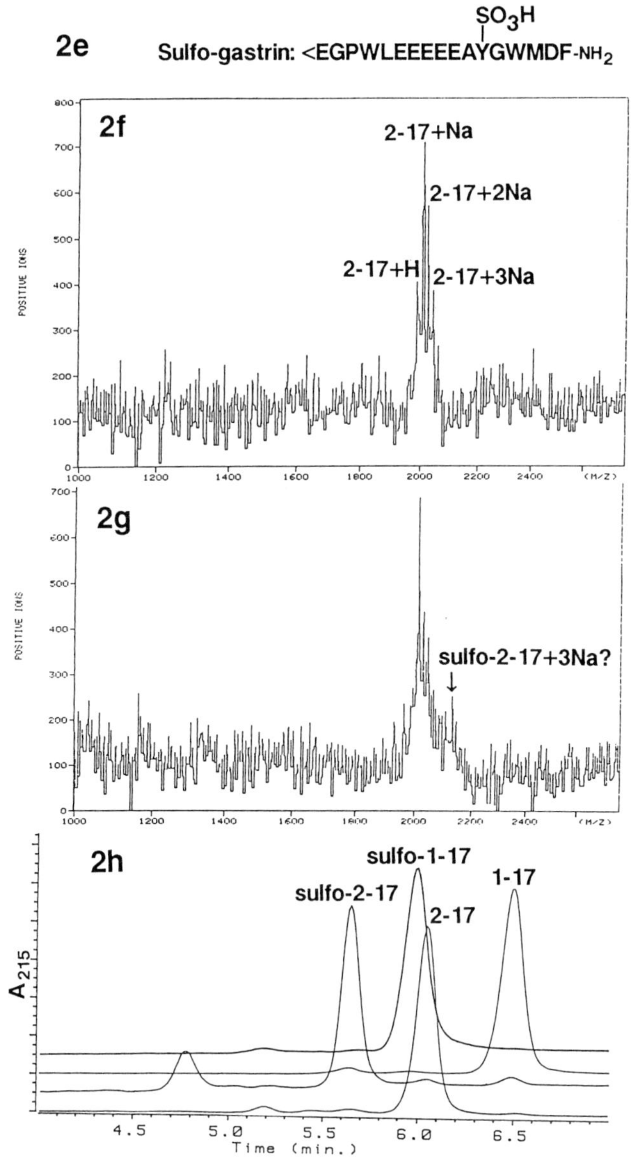

acid sequence of sulfogastrin and RP-HPLC profiles of gastrin and its derivatives are given in (e) and (h), respectively.

about ionization mechanisms (4). Common to those examples of specific covalent bond scissions are existence of conjugated double bonds (heme and benzene) and involvement of sulfur atoms. It may be too early to draw conclusions from only these two examples. Accumulation of information is necessary regarding signal intensity and characteristics of peptides to get a better understanding of the ionization mechanisms. Since some specific covalent bond scissions may be unavoidable, as seen with sulfogastrin in the commonly used positive mode, it is important to get spectra in both positive and negative modes to avoid overlooking unknown modifications.

ACKNOWLEDGEMENTS

Authors are grateful to Applied Biosystems Japan for their generous free loan of BioIon 20 Biopolymer Mass Analyzer for our use. We also thank to Mr. M. Suzuki and to Dr. K. Esumi for mass analyses on PE-Sciex API III (IonSpray) and JEOL JMS-HX110 (FAB) Mass Spectrometers, respectively.

REFERENCES

1. Yonetani, T. & Ray, G.S. (1965) J. Biol. Chem. **240**, 4503-4508.
2. Gross, E. (1967) Methods Enzymol. **11**, 238-255.
3. Jonsson, G.P., Hedin, A.B., Hakansson, P.L., Sundqvist, B.U.R., Save, B.G.S., Nielsen, P.F., Roepstorff, P., Johansson, K.-E., Kamensky, I. & Lindberg, M.S.L. (1986) Anal. Chem. **58**, 1084-1087.
4. Sundqvist, B.U.R. (1990) in "Biological Mass Spectrometry" (Burlingame, A.L. & McCloskey, J.A. Eds.) pp. 1-23, Elsevier Science Publishers, Amsterdam.

Determination of the Cleavage Site of the Amyloid Precursor Protein by Plasma Desorption Mass Spectrometry

Rong Wang, Robert J. Cotter
Middle Atlantic Mass Spectrometry Facility
Department of Pharmacology and Molecular Sciences
The Johns Hopkins University School of Medicine, Baltimore, MD 21205

James F. Meschia, Sangram S. Sisodia
The Neuropathology Laboratory, The Johns Hopkins University School of Medicine
Baltimore, MD 21205

I. Introduction

Alzheimer's disease (AD) is a progressive neurodegenerative disorder which affects a significant population of individuals over 65 years of age (1). A principal neuropathologic hallmark of AD is the deposition of β-amyloid in extracellular senile plaques and in the cerebral microvasculature (2). The principal component of β-amyloid is a 4 kD peptide (β/A4) (3) derived from large amyloid precursor proteins (APP) (4). APP's are membrane bound glycoproteins of between 110 to 130 kD and are encoded by alternatively spliced transcripts derived from a single gene on chromosome 21. β/A4 encompasses 14-15 amino acids of the transmembrane domain and 28 amino acids of the extracellular region of APP. APP mature through the secretory pathway and are translocated to the plasma membrane, where a extracellular region proteolytic cleavage event occurs that results in release of a soluble peptide (SAPP). This cleavage site has been determined to be within the β/A4 peptide using molecular biological (5) and biochemical (6,7) approaches. The biological function of the released soluble peptide (SAPP) and the cellular events responsible for the aberrant generation of β/A4 from APP are not clear. Thus, the study of the primary proteolytic cleavage event for APP will assist our understanding of the mechanism of amyloidgenesis.

TECHNIQUES IN PROTEIN CHEMISTRY III

In our own studies of the proteolytic cleavage of APP, plasmadesorption mass spectrometry (PDMS) has been used to analyze fragments derived from chemical digestion of SAPP. Introduced in 1974 by Macfarlane and co-workers (8), PDMS was used to record a mass spectrum of insulin in 1982 (9), and has since been widely used for protein and peptide research (10). The technique provides an alternative to classical SDS-PAGE for determining peptide and protein molecular weights, with considerably higher mass accuracy. When combined with *in situ* enzymatic and/or chemical digestion, PDMS can also be used to identify the amino acid sequences in proximity to the amino or carboxy terminus of a peptide(11). Thus, recent efforts in our laboratory have been directed toward demonstrating that PDMS is a unique, sensitive and rapid method for studying protein processing. In this paper, we show that PDMS can be used to determine the primary cleavage site of APP and discuss some of the important operational techniques for obtaining different structural information on the constitutive cleavage site of amyloid precursor protein. The influence of protein sample size on the different operations is discussed in detail.

II. Experimental

Preparation of the CNBr digest of soluble peptide from APP - A Chinese hamster ovary (CHO) cell line has been constructed which overexpresses human APP-770. Soluble APP-770 peptide (SAPP-770) was purified from the conditioned medium of these cells using Mono-Q anion exchange FPLC and chemically digested with cyanogen bromide (CNBr). The digested peptides were then separated by reverse phase HPLC. Details of this procedure have been described previously (7).

Plasma desorption mass spectrometry - PDMS spectra of selected CNBr digested peptides and *in situ* enzymatic digests were obtained on a BIO-ION Nordic (Uppsala, Sweden) BIN-10K plasma desorption time-of-flight mass spectrometer equipped with a 10-μCi Californium-252 ionization source. Spectra were collected by a PDP 11/73-based BIO-ION data system. Peptide samples were dissolved in 0.1% TFA in water, and deposited on a nitrocellulose-covered, mylar-backed aluminum foil. Positive ion mass spectra were recorded with an accelerating potential of +18 kV for two to five million primary events. Negative ion mass spectra were recorded with an accelerating potential of -15 kV for five million primary events. The time resolution was 1 nsec/channel. H^+ and Na^+ ions were used for positive ion mass calibration, and H^- and CN^- ions were used for negative ion mass calibration.

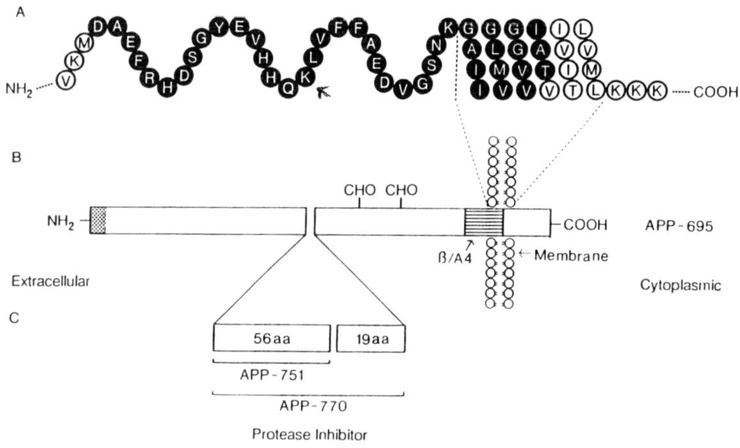

Figure 1. Schematic of the structure of amyloid precursor protein. (a) Partial amino acid sequence of the transmembrane domain and adjacent extracellular domain of APP. The darkened residues indicate the β/A4 peptide region, while the arrow indicates primary cleavage site verified in the present study. (b) Transmembrane orientation and position of β/A4 region of APP. Shaded region depicts the signal peptide. The striped region symbolizes the β/A4 peptide. CHO represents potential N-linked glycosylation sites. (c) Diagram showing the structural relationship of APP-695, APP-751 and APP 770.

In situ Carboxypeptidase Y digestion - Sequence grade Carboxypeptidase Y, *Peptidyl-L-amino acid hydrolase, EC 3.4.16.1* (Boehringer Mannheim Biochemica) was used to digest peptide samples directly on the nitrocellulose foil following the initial molecular weight measurements by PDMS, to confirm the identity of the C-terminal peptide of SAPP-770 from its C-terminal amino acid sequence. Using chemically synthesized β/A$4_{1\text{-}16}$ as a standard, a wide range of enzyme to substrate ratios and incubation times were tested to determine the optimal digestion conditions. 400 ng of Carboxypeptidase Y was used in each case, and the digestions were carried out at room temperature (22 °C).

III. Results and Discussion

Molecular biology studies have previously determined that APP cleavage is a membrane-associated event that occurs at a site located within the β/A4 region (5). The cleavage site within APP-695 and APP-751 has been mapped to the Gln_{15} or Lys_{16} of β/A4 (6). Since Lys_{16} was not detected in either the C-terminus of the soluble portion or the N-terminus of the membrane-associated portion of APP-695, this suggested that the primary cleavage could occur on either side of Lys_{16} (Fig. 1). To determine and clarify the precise APP cleavage site

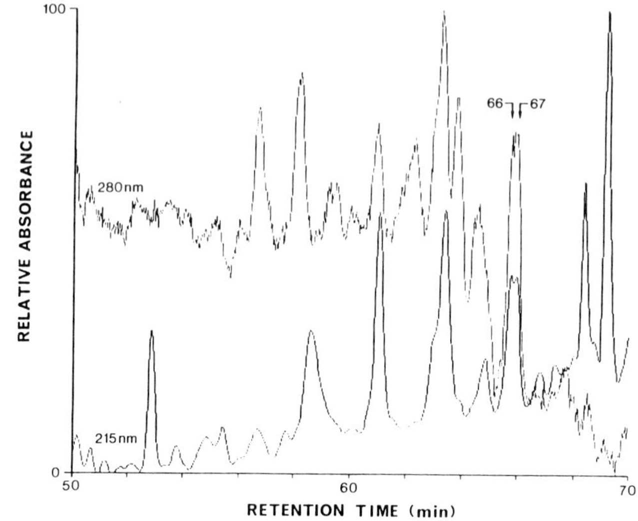

Figure 2. Partial reverse phase HPLC profile of CNBr digests of SAPP-770. The labels indicate fraction numbers which contain the C-terminal peptide derived from SAPP-770.

leading to secretion of the soluble peptide, we chemically digested SAPP-770 using cyanogen bromide (CNBr) and separated the digests by reverse phase HPLC. A partial HPLC profile is shown in Figure 2. From the amino acid sequence deduced from APP-770 cDNA sequence, the C-terminal peptide of SAPP-770 contains aromatic residues and

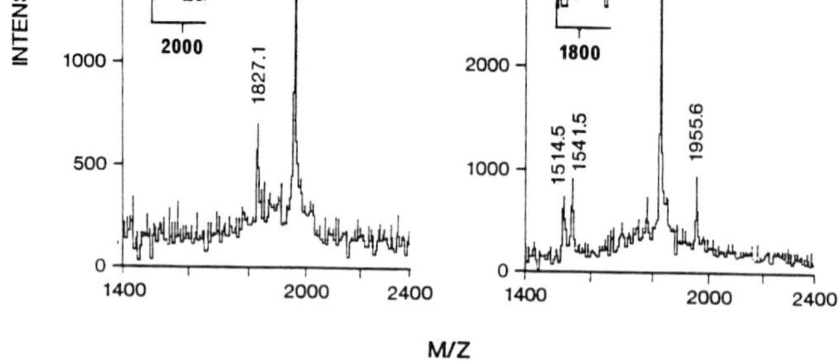

Figure 3. Partial positive ion PDMS spectra of HPLC fractions 66 (A) and 67 (B). The inserts show molecular ion observed in the negative ion PDMS spectra.

allowed us to eliminate those HPLC peaks that did not show absorbance at both 215 nm and 280 nm. Furthermore, using identical HPLC conditions, the retention time for chemically synthesized β/A4$_{1\text{-}16}$ was 65.7 min. Taken together, we focused our attention on the two unresolved fractions at 66 and 67 minutes shown in Figure 2.

Plasma desorption mass spectra were obtained for these two fractions (Figure 3) and for other peaks having absorbances at 215 and 280 nm (data not shown). For fractions 66 and 67, the most intense peaks were observed (respectively) at m/z 1955.6 and 1826.9 in the positive ion spectra, and m/z 1953.7 and 1824.5 in negative ion spectra. The two amu mass difference in the positive and negative ion spectra suggested that these are protonated and deprotonated ions, respectively, and that the molecular weights of the neutral peptides obtained from fractions 66 and 67 are 1954.6 and 1825.7. These agree well with the 16 amino acid sequence: *DAEFRHDSGYEVHHQK* (calculated molecular weight = 1953.9) corresponding to residues 672-687 and the 15 amino acid sequence: *DAEFRHDSGYEVHHQ* (calculated molecular weight = 1825.8) corresponding to residues 672-686 from APP-770. Calculating the concentration of these peptide from their absorbances at 280 and 260nm (12), we estimate that less than 5 picomoles of peptide was required to measure the molecular weight.

Amino acid sequences of the C-terminal peptide from the soluble product from APP-770 were confirmed by Edman sequencing (7), which generally requires about 50 to 100 picomoles of peptide. Furthermore, Edman sequencing revealed that six out of sixteen amino acids in fraction 66 were detectable as the second most abundant and one out of fifteen amino acids in fraction 67 was detectable as the second most abundant. Thus, in addition to high sensitivity, a simple molecular weight measurement by plasma desorption provides high specificity for verification of sequence identity. As suggested by Chait *et al.* (13), PDMS has the additional advantages of low sample consuming, non-destructive condition, and ability to perform enzymatic digestions on

Table I. The Efficiency of *in situ* Carboxypeptidase Y digestion

E/S[a]	Ion Ratio	20min	2 hr	3 hr	14 hr	24 hr
1:1	I_2/I_1[b]	0.00[c]	0.37	0.42	0.91	1.08
	I_3/I_1	0.00	0.00	0.26	0.64	1.09
10:1	I_2/I_1	0.35	0.62	1.32		
	I_3/I_1	0.00	0.47	1.09		

[a]Weight ratio of enzyme to substrate.

[b]I_1, I_2 and I_3 are represent molecular ions at m/z 1956, 1827 and 1699.

[c]The ratio of peak high.

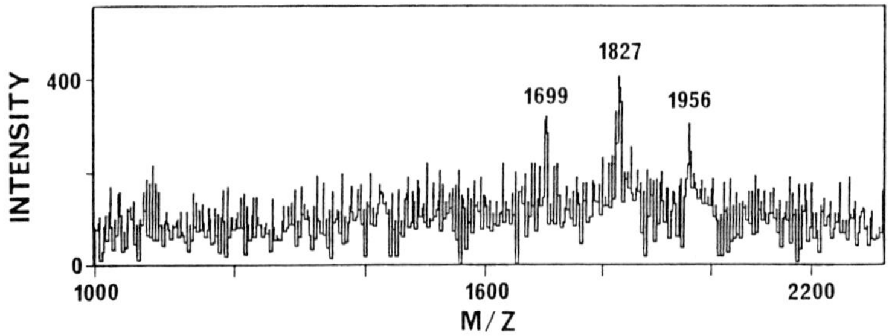

Figure 4. Partial positive ion PDMS spectrum of the Carboxypeptidase Y digestion product of HPLC fraction 66 of SAPP-770. The ratio of enzyme to substrate is 10:1 (w/w). 400 ng of sequence grade Carboxypeptidase Y was used. Incubation was carried out at room temperature for 3 hours.

the nitrocellulose-bound sample after recording a spectrum of the intact molecule. The observed losses of amino acids from the amino or carboxyl termini can provide additional specificity.

In carrying out *in situ* enzymatic digestion, we investigated a numbers of reaction conditions, that included the amount of peptide sample, the ratio of enzyme to substrate, and incubation times. When enzymatic reactions are carried out directly on the nitrocellulose foil

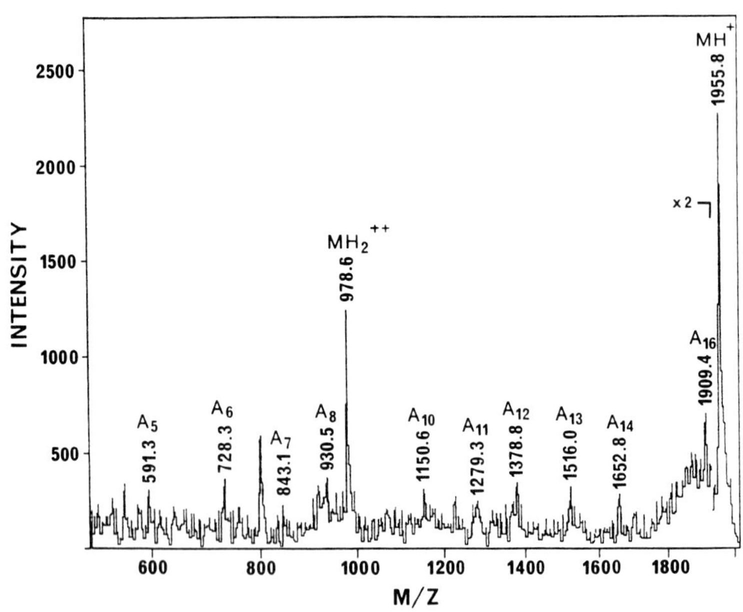

Figure 5. Partial positive ion PDMS spectrum of synthetic β/A4$_{1\text{-}16}$. The sequence is DAEFRHDSGYEVHHQK.

(rather than in solution), we have noted that a higher ratio of enzyme/substrate is required for digestion (11). Using synthetic β/A4 1-16 as a substrate, enzyme/substrate ratios of 1:1, w/w (1:30 in molar ratio) and 10:1, w/w (1:3 in molar ratio) with constant amount of enzyme (400 ng) were tested. The efficiencies of Carboxypeptidase Y digestion are compared in Table 1, using peak intensity ratios of digested to undigested peptides. Although many factors affect absolute peak intensities (14), these data can be used to determine optimal conditions for performing *in situ* enzymatic reactions. While a higher ratio of enzyme to substrate is required of *in situ* digestion, we found a constant amount (400 ng) of Carboxypeptidase Y to be convenient in our experiments. Longer incubation times are required for *in situ* digestion. For example, 200 pmoles of peptide (E/S=1:1, w/w) requires 24 hours to achieve a 1:1 ratio of the 1827/1956 and 1699/1956 peaks, while 20 pmoles of peptide (E/S=10:1, w/w) needs less then 3 hours. In contrast to enzymatic digestion in solution, *in situ* enzymatic digestion has a much lower efficiency that may be due to the immobilization of enzyme and peptide on the nitrocellulose foil leading to a decrease in their effective concentrations, denaturation of the enzyme and concomitant decreases in enzyme activity. At the same time, this procedure enables one to reanalyze the same sample by PDMS, before and after digestion, without sample losses incurred by sample transfer.

Using the procedures outlined above, we were then able to verify the C-terminal amino acid sequence of HPLC fraction 66 of SAPP-770 (Fig. 4). Approximately 15 pmoles of peptide from HPLC fraction 66 was deposited on nitrocellulose foil and incubated with 5 μl of 80 ng/μl Carboxypeptidase Y for 3 hours at room temperature (22 °C). The spectrum shown in Figure 4 confirms the C-terminal amino acid sequence of the ion at m/z 1955.6 as ...QK using the mass differences: 1956-1827 (lysine) and 1827-1699 (glutamine).

We (10b) and others have noted that PDMS spectra can provide peptide sequence information directly from the fragmentation pattern, although the fragment ions are very weak. In contrast with fast atom bombardment (FAB) mass spectrometry, PDMS spectra are generally easier to interpret, since *A* series fragment ions (15) often dominate the spectra. In general, we have observed that larger sample sizes are required for producing sequence fragmentation, and tested that observation using a 100 pmole sample of synthetic $\beta/A4_{1\text{-}16}$ peptide. The spectrum shown in Figure 5 suggests that PDMS may be employed to obtain peptide sequence data directly.

IV. Conclusions

The constitutive cleavage site of a human amyloid precursor protein synthesized in a transfected CHO cell line has been determined using plasma desorption mass spectrometry combined with enzymatic digestion. The results presented above corroborate the earlier molecular biology studies that the primary cleavage site of APP is located within the amyloidogenic β/A4 region. The primary cleavage site for APP-770 in CHO cells occurs at the C-terminus of Lys_{687}. An alternative cleavage site at the C-terminus of Gln_{686} may result from post-enzymatic cleavage removing the basic carboxy terminal Lys_{687}.

The present study illustrates the potential of PDMS as a sensitive method for studying protein processing. Practically, molecular weight measurements require from 1 to 5 pmoles of peptide. Sequence confirmation can be carried out using *in situ* Carboxypeptidase and/or Aminopeptidase digestion with 20 to 50 pmoles sample. For direct amino acid sequencing, sample sizes of about 100 pmoles are necessary.

Acknowledgements

This work was supported by grants (DMB 86-10589) from the National Science Foundation and (NIH AG 05146 and NS 20471) from the U.S. Public Health Service. Mass spectral analyses were carried out at the Middle Atlantic Mass Spectrometry Laboratory, an NSF Shared Instrumentation Facility. The authors thank Ms. B.J. Earl for the peptide synthesis and Edman sequencing.

References

1. Evans, D.A., Funkenstein, H.H., Albert, M.S., Scherr, P.A., Cook, N.R., Chown, M.J., Herbert, L.E., Hennekens, C.H., Taylor, J.O. (1989) *JAMA*, **262**, 2551-1556.
2. Wisniewski, H.M., Narang, H.K. and Terry, R.D. (1976) *J. Neurol. Sci.*, **27**, 173-181; Price, D.L. (1986) *Annu. Rev. Neurosci.*, **9**, 489-512.
3. Glenner, G.G., and Wong, C.W. (1984) *Biochem. Biophys. Res. Commun.*, **120**, 885-890.
4. Goldgaber, D., Lerman, M.I., McBride, O.W., Saffiotti, U., and Gajdusek, D.C. (1987) *Science*, **235**, 877-880.
5. Sisodia, S.S., Koo, E.H., Beyreuther, K., Unterbeck, A. and Price, D.L. (1990) *Science*, **248**, 492-495.
6. Esch, F.S., Keim, P.S., Beattie, E.C., Blacher, R.W., Culwell, A.R., Oltersdorf, T., McClure, D. and Ward, P.J. (1990) *Science*, **248**, 1122-1124.
7. Wang, R., Meschia, J.F., Cotter, R.J., and Sisodia, S.S. (1991) *J. Biol. Chem.*, (in press).
8. Torgerson, D.F., Skowronski, R.P. and Macfarland, R.D. (1974) *Biochem. Biophys. Res. Commun.*, **60**, 616-621.

9. Hakansson, P., Kamensky, I., Sundqvist, B., Fohlman, J., Peterson, P., McNeal, C.J. and Macfarlane, R.D. (1982) *J. Am. Chem. Soc.*, **104**, 2948-
10. (a) Cotter, R.J. (1988) *Anal. Chem.*, **60**, 781A-791A; (b) Cotter, R.J., Chen, L. and Wang, R. (1991) *Mass Spectrometry of Peptide*, Desiderio, D.M., Ed, CRC press, 17-40; (c) Chait, B.T. (1991) *ibid*, 41-64; (d) Roepstorff, P. (1991) *ibid*, 65-86.
11. Woods, A., Cotter, R.J., Yashioka, M., Bullesbach, E. and Schwabe, C. (1991) *Int. J. Mass Spectrom. Ion Phys.* (in press).
12. Warburg, O and Christian, W. (1942) *Biochem. Z.*, **310**, 384-421.
13. (a) Chait, B.T. and Field, F.H. (1986) *Biochem. Biophys. Res. Commun.*, **134**, 420-426; (b) Chait, B.T., Chaudhary, T. and Field, F.H. (1987) *Methods in Protein Sequence Analysis 1986*, Walsh, K.A. Ed., Humana Press, 483-492.
14. (a) Wang, R., Chen, L. and Cotter, R.J. (1990) *Anal. Chem.*, **62**, 1700-1705; (b) Nielsen, P.F., Klarskov, K., Hojrup, P. and Roepstorff, P. (1989) *Biomed. Environ. Mass Spectrom.*, **18**, 131-137.
15. (a) Roepstorff, P. and Fohlman, J. (1984) *Biomed. Mass Spectrom.*, **11**, 601; (b) Biemann, K. (1988) *Biomed. Environ. Mass Spectrom.*, **16**, 99-111.

Large-Scale Protein Mapping Using Infrequent Cleavage Reagents, LD TOF MS, and ES MS

P. C. Andrews
Dept. of Biological Chem.
University of Michigan Medical School
Ann Arbor, MI 48109-0674

M. H. Allen, M. L. Vestal, R. W. Nelson
Vestec Corporation, Houston, TX 77054

I. Introduction

Changes in recombinant DNA methods and improved DNA sequencing techniques in the past decade have resulted in the elucidation of a large number of gene sequences. The determination of the corresponding protein structures has lagged behind. There are several reasons for the slower rate of protein structure analysis, but they are principally related to the chemical complexity of proteins, their generally low levels, and the presence of post-translational modifications. It is important that more rapid and thorough methods be developed for the analysis of proteins if the structural information is to keep pace with the DNA sequence information.

A rapid method for obtaining information on post-translational modifications is to map proteins using mass spectrometric methods in combination with various cleavage methods. This approach has the advantage of also providing a "check-sum" for the DNA sequence. Cleavage techniques which result in many fragments usually require chromatographic separation before mass spectrometry to avoid ion suppression problems. Cleavage methods which result in only a few large fragments have not been very useful until recently because most of the fragments were beyond the mass range of many mass spectrometers.

Recent developments in mass spectrometry such as the electrospray interface and matrix-assisted laser-desorption time-of-flight mass spectrometry have made these larger mass fragments accessible with good accuracy and sensitivity. Electrospray allows the masses of large peptides to be determined on relatively low mass range instruments by producing ions having a large number of charges so the m/z is relatively low (1). Each peptide typically produces several ions, which allow the mass of the peptide to be determined (2). Since each of the peptides in a mixture gives rise to several ions, the spectrum may be too complex to easily interpret for mixtures of moderate complexity. Laser-desorption methods tend to produce primarily singly-charged species, resulting in simpler spectra (3-9). Laser-desorption also

TECHNIQUES IN PROTEIN CHEMISTRY III

appears to be less affected by the ion suppression effects seen for mixtures of peptides using other ionization methods.

These techniques have raised the possibility of cutting proteins into a few large fragments using reagents which cut proteins at relatively rare sites. Several of these reagents could be used separately or in concert to generate overlapping fragments in much the same way that restriction enzymes are used to generate maps of DNA. These maps would allow sites of post-translational modifications to be localized to relatively small regions of the protein, provide useful information as to the nature of the modification, and also provide verification of the DNA sequence.

While this approach to rapid protein mapping is simple in concept, it is not so straightforward to implement. The most useful cleavage techniques in this context are those which cleave at lower abundance amino acids (Met, Trp, Cys, & His). Of these four amino acids, useful chemical reagents are available for the first three. Cleavage of proteins at methionine using cyanogen bromide (CNBr) is the most commonly used protein cleavage method. However, cleavage using most of the reported protocols leads to artifactual chemical modifications which interfere with mapping by mass spectrometry. Many of these modifications are not detectable by other methods of analysis. One of these problems (formylation) has been mentioned previously in this context (5). Use of other chemical cleavage methods can also produce products which are not ideal for mass analysis.

II. Experimental

Horse heart myoglobin, hen egg lysozyme, and bovine insulin were purchased from Sigma Chem. Co. The plasminogen activation inhibitor (PAI-1) was provided by D. Ginsberg and D. Lawrence (Univ. of Mich.). Cyanogen bromide was from Aldrich Chem. Co.and iodosobenzoic acid was from Pierce Chem. Co. The 2-Nitro-5-thiocyanatobenzoic acid was prepared as described by Degani and Patchornik (10).

Laser desorption mass spectra were obtained on a Vestec model VT 2000 laser-desorption time-of-flight mass spectrometer (Vestec, Houston,TX) using sinapinic acid as the matrix. Samples were prepared in matrix as described by Beavis and Chait (3). Both neodymium-YAG (frequency-tripled to 355 nm) and nitrogen lasers were used in this study with similar results. Electrospray mass spectra were obtained using a Vestec electrospray source and a model 201 single quadrupole mass spectrometer with a 2000 amu range. Samples were delivered to the source at 5 microliters/min. in 4 % acetic acid. Other operating conditions for both machines were as recommended by the manufacturer.

Size exclusion HPLC was performed on a 9.4 mm X 25 cm Hydroxyethyl PolyAspartamide column (PolyLC, Columbia, MD) using 50 m$\underline{M}$ formic acid, 25% acetonitrile as the mobile phase. The pore size of the silica was 1,000 Angstroms. The flow rate was 1 ml/min and detection was at 215 nm. Fractions were collected and flow-injected in 4% acetic acid and 50% acetonitrile.

Cyanogen bromide digests were performed at room temperature in the dark for varying periods of time. Concentrations of CNBr used in early experiments were 1 mg/ml. All of the digests in 70% trifluoroacetic acid reported here were carried out at 10 mg/ml in an attempt to decrease reaction times. The cyanogen bromide was stored at -20 degrees until use, it was then allowed to warm to room temperature to avoid condensation of moisture and opened under a chemical hood. To increase the life of the cyanogen bromide, non-metallic spatulas should be used to remove the reagent from the bottle. Other chemical digests (patchorniks reagent and iodosobenzoic acid) were performed as described in the original references (10,11) and then desalted by various methods. Most of the products of these latter reactions are very insoluble and difficult to subject to normal desalting procedures. Data were analyzed using PROCOMP 1.2, a software package which is available from the author (P. C. Andrews).

III. Results and Discussion

One advantage of mass spectrometric methods is that it is possible to detect structural modifications which are not possible to observe using amino acid analysis or Edman degradation. As a result, it has become more evident that some of the reactions that protein chemists have relied on have serious side reactions when used under usual protocols. Cyanogen bromide cleavage of protein backbones at Met residues is usually considered to be one of the most reliable chemical methods applied to protein cleavage. It has been generally recognized that some problems do occur, including incomplete cleavage at Met residues followed to Thr, Ser, or Cys residues (with conversion of Met to homoserine), lack of cleavage at oxidized methionine residues, non-specific acid hydrolysis, and acylation by carboxylic acids, among others. This study examined several chemical cleavage methods, but it quickly became clear that even the simplest of these methods had a number of side reactions. This manuscript describes our observations with CNBr cleavage of proteins in preparation for mass spectral analysis.

A

```
      * 10    *    20         30         40         50     ▼   60         70
GLSDGEWQQV LNVWGKVEAD IAGHGQEVLI RLFTGHPETL EKFDKFKHLK TEAEMKASED LKKHGTVVLT
        80         90        100        110        120        130        140
ALGGILKKKG HHEAELKPLA QSHATKHKIP IKYLEFISDA IIHVLHSKHP GNFGADAQGA MTKALELFRN
       150 153                                                    ▲
DIAAKYKELG FQG
```

B

```
        10         20         30         40     ▼   50         60  ▼      70
VHHPPSYVAH LASDFGVRVF QQVAQASKDR NVVFSPYGVA SVLAMLQLTT GGETQQQIQA AMGFKIDDKG
▼       80   ▼  *  90        100        110▼       120        130        1*0
MAPALRHLYK ELMGPWNKDE ISTTDAIFVQ RDLKLVQGFM PHFFRLFRST VKQVDFSEVE RARFIINDWV
      ▼150        160        170     *  180        190        200 ▼▼     210
KTHTKGMISN LLGKGAVDQL TRLVLVNALY FNGQWKTPFP DSSTHRRLFH KSDGSTVSVP MMAQTNKFNY
       220        230     ▼  240        250        260  *   ▼ 270        280
TEFTTPDGHY YDILELPYHG DTLSMFIAAP YEKEVPLSAL TNILSAQLIS HWKGNMTRLP RLLVLPKFSL
       290     ▼ ▼300        310        320        330        340       ▼350
ETEVDLRKPL ENLGMTDMFR QFQADFTSLS DQEPLHVAQA LQKVKIEVNE SGTVASSSTA VIVSARMAPE
   ▼   360        370   ▼ ▼379
EIIMDRPFLF VVRHNPTGTV LFMGQVMEP
```

FIGURE 1. Primary structures of horse heart myoglobin (A) and plasminogen activator inhibitor (B). The locations of CNBr cleavage sites are indicated by darts. The positions of tryptophan residues are indicated by asterisks.

Several model proteins were used for these studies, including lysozyme, ribonuclease A, beta lactoglobulin, BSA, insulin, myoglobulin, and plasminogen activator inhibitor protein. The results with the latter two proteins will be discussed here since they lack disulfide bonds (Fig. 1) which add further complexity to interpretation of the results. A large number of the disulfide bonds in BSA survive CNBr cleavage using the conditions described here.

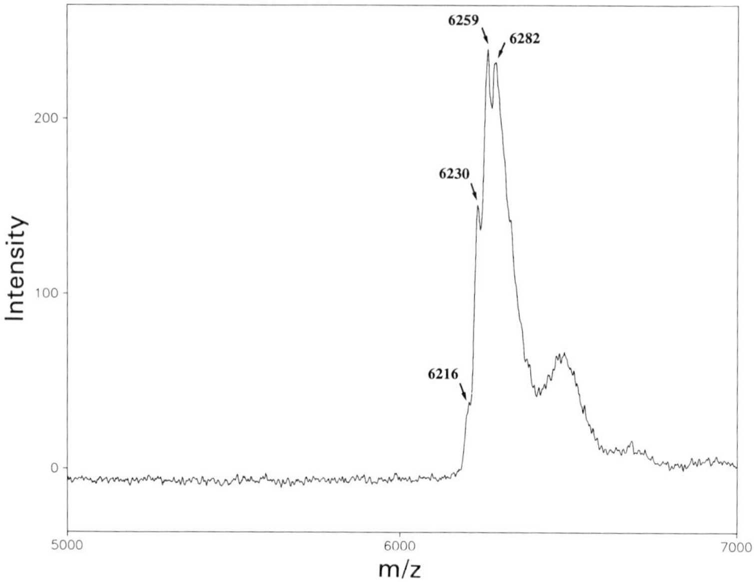

FIGURE 2. A portion of the laser desorption mass spectrum of a CNBr digest of horse heart myoglobin showing the mixture of oxidation products and formylated peptides which are present in the digest after 28 hours at room temperature.

Initial CNBr cleavage of proteins by a commonly used protocol (88% formic, 1 mg/ml CNBr, RT for 12 to 18 hours) resulted in efficient cleavage of the proteins. However, the mass spectra of the fragments produced were difficult to interpret due to formation of multiple adducts (Fig. 2). The heterogeneous nature of the products suggested that multiple processes were at work. The shifts in mass observed for the peptides indicated, not unexpectedly, that for the 1-55 peptide from horse heart myoglobin both oxidation and formylation were involved. Peptides having no oxidizable residues appeared to be formylated only. With these side reactions complicating even a simple cleavage, more complex mapping problems would be inhibited. Substitution of acetic acid for formic acid resulted in efficient cleavage but also led to partial acetylation of the product peptides. The degree of heterogeneity did not appear to be as bad as for formic acid, however. Substitution of 1 N acetic and 0.1 N HCl appeared to result in faster cleavage but also produced partial acid hydrolysis products and did not eliminate the acetylation. Substitution of 70% trifluoroacetic acid for formic acid as the solvent produced the best results, with little or no acid hydrolysis products formed, efficient cleavage, and no apparent trifluoroacetyl adducts. It became apparent from the results for a number model peptides that oxidation of Trp to monohydroxy-tryptophan is a problem for longer incubation times (Fig. 3).

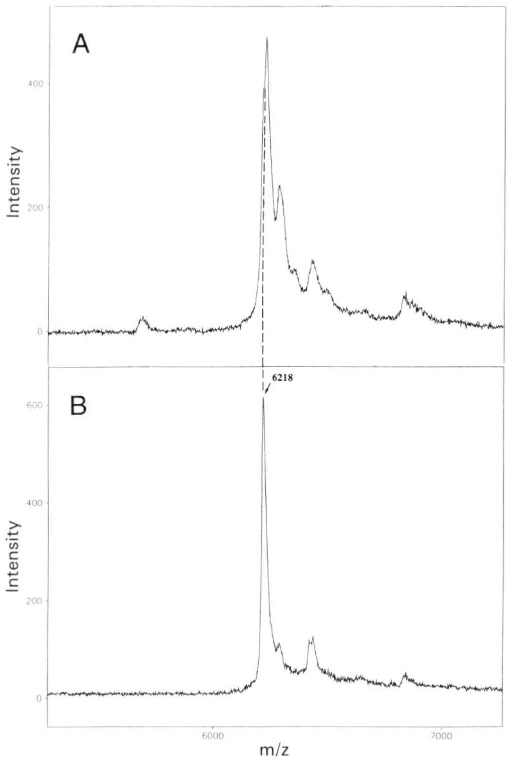

FIGURE 3. The LD mass spectrum of residues 1-55 from horse heart myoglobin after CNBr digestion in 70% trifluoroacetic acid for 18 hours (A) and 2 hours (B).

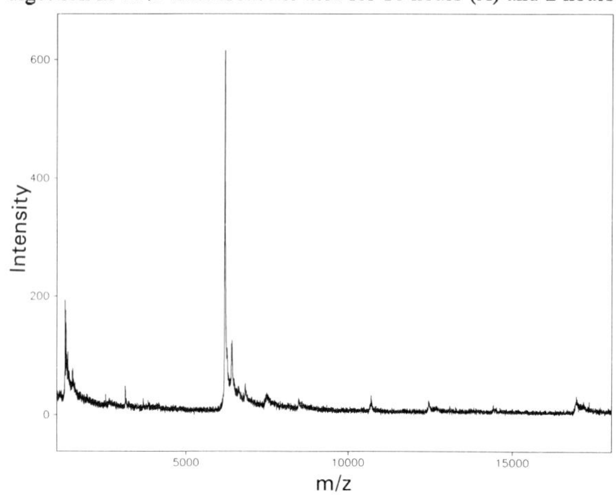

Figure 4. LD mass spectrum of a 2 hour CNBr digest of horse heart myoglobin in 70% trifluoroacetic acid. The major peak represents residues 1-55 (6216 daltons).

Shorter reaction times (< 6 hours) at CNBr concentrations of 10mg/ml appear to be a partial solution. Even short incubation times of 2 hours or less which result in partial cleavages gave laser desorption spectra which were relatively simple (Fig. 4). Beavis and Chait (5) have pointed out that these partial digests are quite useful in constructing protein maps by mass spectrometry. The spectrum shown in Figure 4 for the CNBr cleavage of myoglobin has as its most intense peak the fragment corresponding to residues 1-55. This could reflect a more rapid rate of cleavage at Met-55 or a difference in ionization.

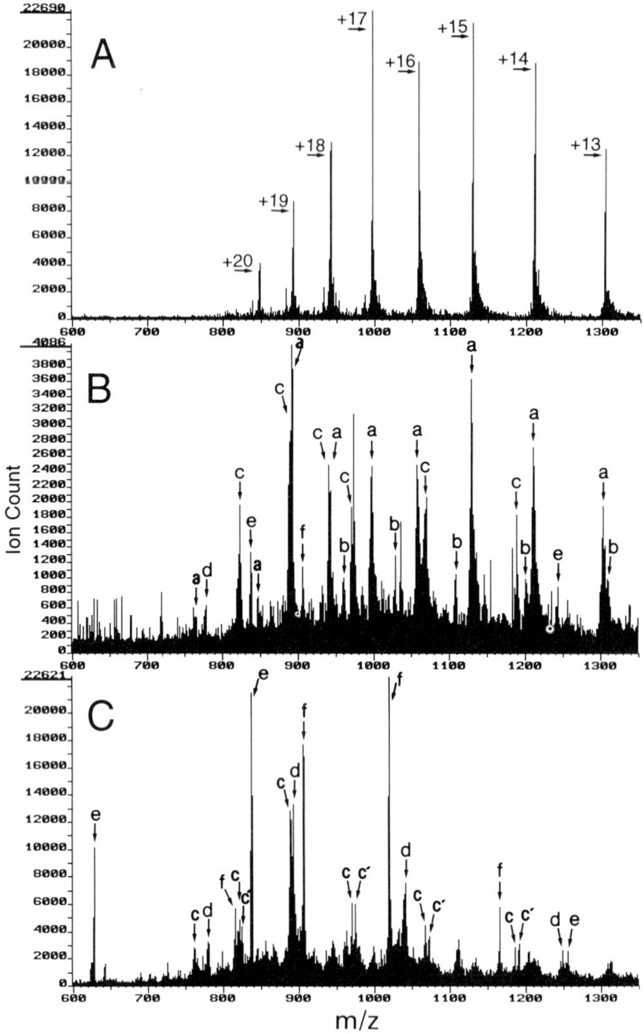

FIGURE 5. Mass spectra of samples introduced by electrospray into a single quadrupole mass spectrometer. A.-intact horse heart myoglobin, B.-the same 2 hour CNBr digest used for the data in FIG. 4 above, C.-18 hour CNBr digest of myoglobin in 70% trifluoroacetic acid. The numbers next to the peaks in spectrum A represent the charge states of the ions. The ions in spectra B and C are labeled as follows: a.) intact myoglobin, b.)residues 1-131 of myoglobin with Met-55 converted to homoserine, c.) residues56-153 with Met 131 converted to homoserine, c'.) residues 56-153 with Met 131 as methionine sulfoxide, d.) residues 1-55 of myoglobin, e.) residues 132-153 of myoglobin,f.) residues 56-131 of myoglobin. Most of the minor, unassigned peaks appear to be a form of intact myoglobin.

Figure 5A shows the spectrum produced for myoglobin ions produced by electrospray. A series of multiply charged ions is produced, the masses of which may be used to determine the mass of the singly charged species. The large number of multiply charged ions produced by this method suggests that more complex mixtures would be difficult to interpret. However, an electrospray spectrum of the same 2 hour CNBr cleavage reaction used to produce the LD MS spectrum in Fig. 4 can be used to identify all the fragments which ionize by this method (Fig. 5B). Points of interest in this spectrum include the presence of intact myoglobin (indicated by the peaks marked by "a"), an ion series corresponding to residues 1-131 of myoglobin with the Met-55 converted to homoserine (indicated by "b"), and ions attributable to partial cleavage at residues 56-153 (indicated by "c") with Met-131 apparently also partially converted to homoserine. All the major ions and many of the minor ones could be assigned in this partial cleavage mixture. A more complete digest after 18 hours under the same conditions is shown in Figure 5C. Of particular interest is the c, c' series of ions which correspond to residues 56 to 153 with Met-131 oxidized to the sulfoxide which is resistant to CNBr cleavage (indicated by "c' "). One final observation is that many of the minor ions which are not marked in Figure 5C correspond to intact myoglobin and may represent a sub-population of myoglobin which has both Met residues oxidized to Met sulfoxide or which are converted to homoserine.

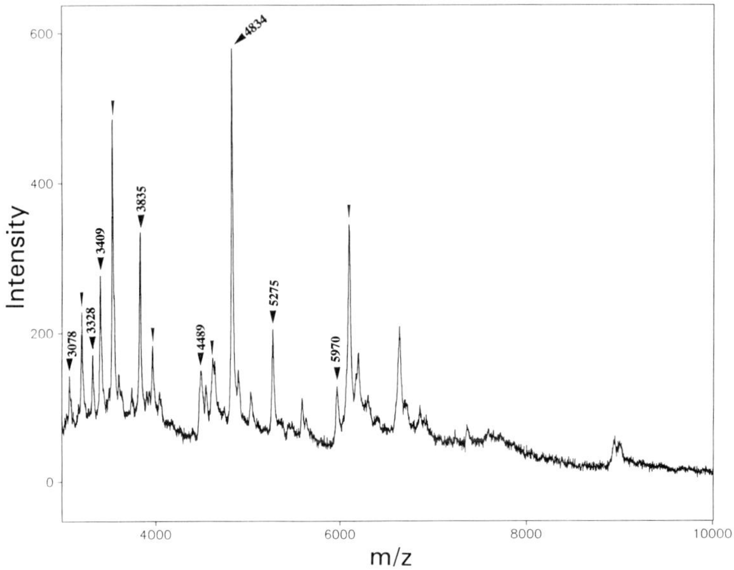

Figure 6. Laser desorption time-of-flight mass spectum of an 18 hour CNBr digest in 70% trifluoroacetic acid of plasminogen activator inhibitor. The assigned masses are indicated by the large darts. Masses representing 131 mass unit adducts are indicated by small darts.

The plasminogen activator inhibitor is a 42 kd protein containing 16 Met residues. A CNBr digest of the protein (shown in Fig. 6) allows assignment of 9 masses (8 are shown in Fig. 6) corresponding to at least 80 % of the protein. Similar results were obtained for BSA (complicated by the presence of disulfide bonds).

Table I. CNBr Fragments of Plasminogen Activator Inhibitor

Residue #	Observed Mass	Calculated Avg. Mass[a]
355-373	2198.5	2198.6
84-110	3078[b]	3077
267-295	3328	3331
236-266	3409[b]	3411
203-235	3835[b]	3835.1
111-147	4489[b]	4492.2
1-45	4834	4835.3
299-347	5275	5274.9
148-201	5970[b]	5968.8

[a] Assuming monohydroxy-tryptophan and homoserine lactone.
[b] Indicated ions also had adducts 131 mass units higher.

The spectrum for this last digest contains another artifact in that many of the ions exhibit a second ion 131 daltons higher. The relative intensities of the pairs of peaks are highly variable. The higher mass peak is likely due to a non-covalent adduct with a lower mass ion present in the sample (unidentified). The intact protein exhibits the same adduct on LD MS.

In summary, both LD MS and ES MS methods proved to be useful for characterizing protein digests consisting of relatively large fragments with good efficiency. Complex protein digests will require partial separation prior to mass analysis using electrospray methods. We have found that size exclusion HPLC in a formic acid/acetonitrile mobile phase to be particularly helpful for simplification of CNBr and other digests prior to mass analysis. Using the conditions described here, it is possible to avoid many of the side reactions associated with the most commonly used protocols for CNBr cleavages. It should be noted that in all cases where the cleavage was allowed to continue for more than 2 hours, a high rate of conversion of Trp to monohydroxy-tryptophan occurred.

IV. References

1. Fenn, J. B., Mann, M., Meng, C. K., Wong, S. F., and Whitehouse, C. M. (1989) Science **246**, 64.
2. Mann, M., Meng, C. K., and Fenn, J. (1989) Anal. Chem. **61**, 1702.
3. Beavis, R. C. and Chait, B. T. (1989) Rapid Comm. in Mass Spectr. **3**, 432.
4. Beavis, R. C. and Chait, B. T. (1989) Rapid Comm. Mass Spectr. **3**, 233.

5. Beavis, R. C. and Chait, B. T. (1990) Proc. Natl. Acad. Sci. **87**, 6873.
6. Karas, M. and Hillenkamp, F. (1988) Anal. Chem. **60**, 2299.
7. Karas, M. Bahr, U., Ingendoh, A., and Hillenkamp, F. (1989) Angew. Chem. **101**,805.
8. Zhao, S., Somayajula, K. V., Sharkey, A. G., Hercules, D. M., Hillenkamp, F. Karas, M. , and Ingendoh, A. (1991) Anal. Chem. **63**, 450.
9. Karas, M. Ingendoh, A., Bahr, U.,and Hillenkamp, F. (1989) Biomed. Mass Spectrom. **18**, 841.
10. Degani, Y. and Patchornik, A. (1971) J. Org. Chem. **36**, 2727.
11. Mahoney, W. C. and Hermodson, M. A. (1981) Biochemistry **18**, 3810.

Acknowledgments

PCA wishes to thank J. E. Hoover for helpful discussions and insightful comments.

Tandem Mass Spectrometric Sequencing of Proteins Isolated From 1- and 2-Dimensional Polyacrylamide Gel Electrophoresis

Steven C. Hall[1], Patrick A. Schindler[1], Frank R. Masiarz[1,3] and Alma L. Burlingame[1,2]

[1]Department of Pharmaceutical Chemistry, Mass Spectrometry Facility and [2]Liver Center
University of California, San Francisco, CA 94143-0446
and
[3]Chiron Corporation, Emeryville, CA 94608

I. INTRODUCTION

In recent years, mass spectrometry has proven to be an essential tool for solving a variety of problems in structural biology [1-6]. These problems range from identifying post-translational modifications, such as N-terminal blocking groups [7, 8], phosphorylation [9] and glycosylation [10-12] sites to sequencing of an entire protein [13, 14]. Successful completion of much of this work has relied heavily upon the use of high performance tandem mass spectrometers equipped with multichannel array detection systems which have the ability to obtain amino acid sequence information on as little as 5-10 picomoles [9, 14-16] of peptide.

One dimensional sodium dodecylsulfate polyacrylamide gel electrophoresis (SDS-PAGE, 1-D) is used routinely to resolve protein mixtures according to the molecular weight of the individual components. Two dimensional (2-D) gel electrophoresis, utilizing isoelectric focusing (IEF) followed by SDS-PAGE, provides the highest resolution for separating a complex mixture of proteins (up to 2000) [17, 18]. Several groups of investigators have developed protocols to obtain amino acid sequence information from proteins purified by these two methods. Successful approaches have included N-terminal sequencing of 1- and 2-D gel purified proteins by electroblotting them onto derivatized glass fiber paper [19-21] or polyvinylidene difluoride (PVDF) membranes [20, 22, 23] and subjecting them to automated Edman degradation. However, since N-terminally blocked proteins can represent at least 40% of proteins isolated from 2-D gels [23], other investigators have resorted to enzymatically digesting their gel resolved proteins and Edman sequencing the resulting peptides [24-26]. Over the past two years, our efforts have focused on development of a generic protocol for obtaining sequence information from proteins isolated from 1- and 2-D polyacrylamide gel electrophoresis using a combination of electroelution, enzymatic digestion, microbore HPLC, liquid secondary ion mass spectrometry (LSIMS) and high performance tandem mass spectrometry with array detection

TECHNIQUES IN PROTEIN CHEMISTRY III

[27, 28]. A major consideration in the development of our protocol was the recovery of as many peptide fragments as possible, representing as much of the primary sequence of the entire protein as possible. Many of the procedures reported in the literature provide only a small amount of sequence information, useful only for DNA probe design. The characterization of proteins which represent a cellular response to external stimuli requires the ability to identify subtle differences in primary structure within the sequence, for example, phosphorylation or glycosylation. Complete digestion and reliable peptide recovery are prerequisites to success. Protocols that do not maximize peptide recovery over the entire sequence do not fulfill this requirement. All aspects of our protocol are compatible with the acquisition of clear fragmentation spectra in a high performance tandem mass spectrometer

II. EXPERIMENTAL

A. *One and Two Dimensional Gel Electrophoresis*

2-D polyacrylamide gel electrophoresis of protein mixtures employed IEF in the first dimension followed by SDS-PAGE [17]. Isoelectric focusing tube gels (125mm x 3mm) were prepared with 4% acrylamide, 2% ampholyte (pH 3.5-7.0), 9.2M urea and 2% NP-40. The anode and cathode solutions were 10mM phosphoric acid and histidine, respectively. All tube gels were prefocused for 15min at 200V, 30min at 300V and 30min at 400V followed by sample loading. Electrophoresis was carried out for 16hrs at 400V followed by 1hr at 800V. Gels were frozen at -70°C in treatment buffer [0.0625M Tris-Cl,(pH 6.8), 2%SDS, 10% glycerol, 100mM dithiothreitol (DTT)] or equilibrated for 20min in treatment buffer and loaded onto 1.5mm thick SDS-polyacrylamide gels. Running gels were prepared with 12.5% acrylamide (16h of polymerization) and the stacking gels with 4.5% acrylamide. The gels were submerged in tank buffer containing 0.025M Tris-HCl (pH 6.8), 0.0192M glycine, 0.1% SDS and 0.1mM sodium thioglycolate as a free radical scavenger and run at 80mA (constant current) for 4h. Protein visualization was accomplished by staining the gels with Coomassie Brilliant Blue R-250 followed by destaining in acetic acid:propanol:water, 10:30:60 (v:v:v).

1-D polyacrylamide gels differed from the second dimension polyacrylamide gels only by their thickness, (0.75mm), their content of acrylamide in running gels (15%) and their running conditions (30mA for 3h). Proteins separated by 1-D gel electrophoresis were dissolved in sample buffer containing dithiothreitol and loaded into the stacking gel wells.

B. *Isolation and Proteolysis of Proteins From 1- and 2-D Gels*

Coomassie Blue stained proteins were excised from the gels with a scapel and stored at -70°C, if not used immediately. Electroelution was performed in an apparatus designed by Hunkapiller [29] or a with a BioRad electroelution apparatus. The electroelution buffer contained 50mM N-ethylmorpholine acetate (pH 8.0), 0.1% SDS and 0.1mM sodium thioglycolate. Gel slices were transfered into the electroelution tubes which contained electroelution buffer supplemented with 0.1% DTT. After lyophilizing the electroeluted solution, a Königsberg acetone precipitation was performed to remove SDS and Coomassie Blue using solvent system A [30].

The purified protein pellets were redissolved in digestion buffer [70mM NH_4HCO_3, (pH 8.5) 0.2mM $CaCl_2$] and digested with TPCK-treated trypsin for 16h at 37°C. Microbore HPLC was employed to separate the peptides. Chromatographic conditions included a 100mm x 1.0mm i.d., C_{18}, 300Å pore

RAT HEART FATTY ACID BINDING PROTEIN TRYPTIC FRAGMENTS

MH^+ = 1036 *ADAFVGTWK LVDSK NFDDYMK 907 SLGVGFATR QVASMTK PTTIIEK

861 NGDTITIK 720** THSTFK NTEISFQLGVEFDEVTADDR KVK 875 SVVTLDGGK

723 LVHVQK 1206 WDGQETTLTR ELSDGK LILTLTHGNVVSTR TYEK EA

Underlined fragments represent peptides for which CID spectra were obtained from 50 picomoles of protein loaded onto a gel.

*acetylated N-terminus

**Identified as the hexyl derivative, MH^+ = 804.

Figure 1. Rat heart FABP sequence cleaved into its putative tryptic peptides. Protonated molecular ion (MH^+) m/z values are represented above the peptides which were sequenced by interpretation of high energy CID spectra obtained by tryptic digestion from 50 pmol of protein loaded onto an SDS-PAGE gel.

size column (Applied Biosystems) and a flow rate of 50µL/min. Fractions were collected in polypropylene microtubes.

C. Mass Spectrometry

The molecular weight of all components of each HPLC fraction was determined by LSIMS on a Kratos MS 50S double-focusing mass spectrometer equipped with a high field magnet and Cs^+ gun operating at a mass resolution of 2000 in the positive ion mode. Approximately one-fifth to one-third of each HPLC fraction was dissolved in a matrix of glycerol/thioglycerol (1:1) with 1% trifluoroaceteic acid (TFA) before spectra were acquired. High energy collision-induced dissociation (CID) mass spectra were obtained for all tryptic components with a Kratos Concept four sector high performance tandem mass spectrometer equipped with a 4% multichannel electro-optical array detection system [15].

III. RESULTS AND DISCUSSION

As a protocol test, a 14kD, 133 amino acid residue rat heart fatty acid binding protein (FABP) was run on SDS-polyacrylamide gels at the 50 picomole level (0.75µg). Figure 1 illustrates the amino acid sequence of the rat heart FABP cleaved into its putative tryptic peptides. Underlined peptides represent those for which CID mass spectra were obtained. In total, amino acid sequences were obtained by tandem mass spectrometry representing 43% of this protein's primary structure. Furthermore, considering losses due to irreversible absorption onto surfaces to which the sample was exposed during the purification procedure, it is likely that CID spectra represented consumption of as little as 20-25 picomoles of any given peptide.

Figure 2 shows the CID mass spectrum of the acetylated N-terminal rat heart FABP tryptic peptide. This spectrum contains the expected, complete *y* and *b* fragment ion series [31] which permits deduction of the sequence from the N-terminal as well as the C-terminal ends. In addition, all of the N-terminal fragment ion masses are shifted 42 mass units higher than anticipated, which identifies the

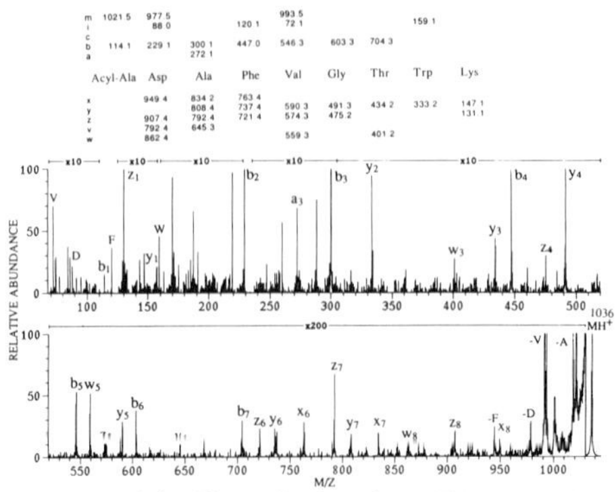

Figure 2. CID mass spectrum of the N-terminal rat heart FABP tryptic peptide Acyl-Ala-Asp-Ala-Phe-Val-Gly-Thr-Trp-Lys (^{12}C MH^+ at m/z 1036) isolated from 50pmol of protein loaded onto an SDS-PAGE gel. The masses of all observed sequence ions are shown in the table.

N-terminal blocking group as an acetyl moiety. Moreover, since this peptide is N-terminally blocked, micro-Edman methods would not have been able to provide the sequence for this peptide.

Having demonstrated the ability to obtain sequence information from a protein isolated from a SDS polyacrylamide gel, we proceeded to use our protocol on proteins which had been separated by two dimensional polyacrylamide gel electrophoresis. We employed a mixture containing four proteins: bovine serum albumin (BSA), rat heart FABP, veal fetuin and chicken ovalbumin, covering the molecular weight range of 14-66kD and pI range of 5.5-6.5. Aliquots of this protein mixture were loaded onto the IEF gel at quantities of 100 or 200 picomoles per protein. Figure 3 shows a typical second dimension polyacrylamide

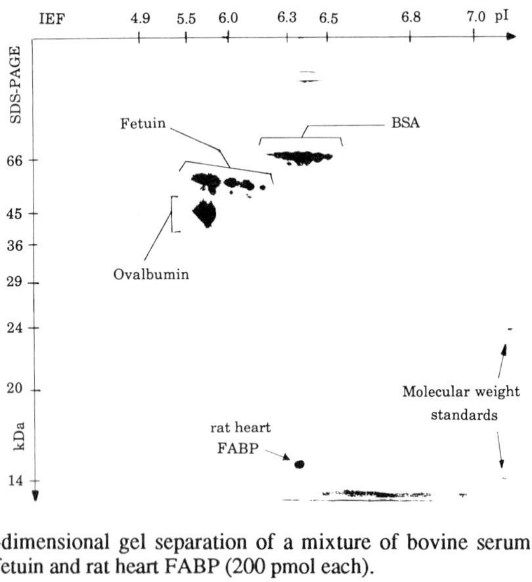

Figure 3. Two-dimensional gel separation of a mixture of bovine serum albumin, chicken ovalbumin, veal fetuin and rat heart FABP (200 pmol each).

gel separation of this mixture. These four proteins are clearly resolved from each other. In addition, the high resolving power of 2-D gel electrophoresis is demonstrated by the ability to resolve charge state heterogeneities present in BSA and fetuin due to their varying degrees of glycosylation. Ovalbumin and rat heart FABP protein spots were excised and subjected to our protocol. Excised areas of the gel that contained BSA and fetuin were stored at -70°C and were not subjected to our protocol. These proteins contain a number of disulfide bridges, which if kept intact, would prevent complete denaturation, thus resulting in incomplete enzymatic digestion. We are currently assessing methods for reduction and S-alkylation that will allow for maximum recovery of low picomole amounts of proteins isolated from gels.

Figure 4 illustrates the microbore reverse phase HPLC chromatogram of the tryptic digest of 2-D gel isolated rat heart FABP. The recovered fractions were analyzed by LSIMS, resulting in molecular weight mapping of 59% of the protein. As indicated in Figure 4, three peptides were sequenced from their CID mass spectra. Figure 5 shows the CID mass spectrum of the tryptic peptide SVVTLDGGK corresponding to amino acid residues 83-91 of 2-D gel isolated rat heart FABP (refer to peak 1 in Figure 4). This spectrum contains complete *y* and *b* fragment ion series along with an extensive number of *a*, *x* and *z* ions, which allows for unambiguous sequencing of the peptide from the N-terminal and C-terminal ends. In addition, the w_6 ion at m/z = 430.2 indicates the presence of leucine rather than its isobaric counterpart isoleucine. Furthermore, during the HPLC separation, this peptide co-eluted with the peptide DGTITIK. By using tandem mass spectrometry, we were able to obtain CID spectra for both of these peptides using the same aliquot of sample and without resorting to further chromatographic purification.

We have also been able to obtain primary structure information from a tryptic digest of approximately 100 picomoles of ovalbumin recovered from the 2-D gel. Molecular weight mapping using LSIMS revealed that seven tryptic peptides gave protonated molecular ions with m/z values that matched those predicted from the known sequence. Figure 6 shows the CID spectrum of one of the ovalbumin tryptic peptides identified as its monohexyl derivative, VTLPR(O-hexyl). Derivatization with hexanol is routinely performed in our laboratory on peptides which otherwise would be too hydrophilic to be sputtered effectively from the glycerol matrix [32]. Again, the presence of complete *x* and *y* fragment

Peaks #	MH+ Da	Matching rat heart FABP peptide sequence
1	875.5	SVVTLDGGK
	862.1	* DGDTITIK
2	1562.7	NGDTITIKTHSTFK
3	1206.5	WDGQETTLTR
4	1545.7	QVASMTKPTTIIEK
5	907.0	SLGVGFATR
6	1523.8	LILTLTHGNVVSTR
7	1036.6	AcADAFVGTWK

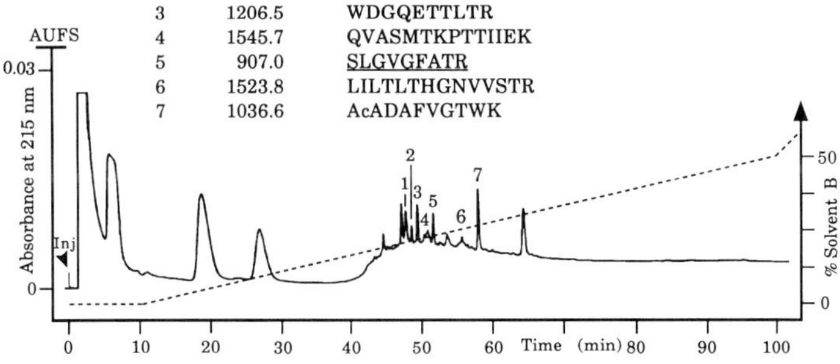

Figure 4. Microbore RP-HPLC chromatogram of a trypsin digest of rat heart FABP isolated from a 2-D gel (200 pmol of FABP loaded onto the IEF gel). The digest was performed in 70 mM NH_4HCO_3 (pH 8.5), 0.2 mM $CaCl_2$ buffer for 16 h at 37°C. In the upper part of the figure, results of the peptide mapping by LSIMS are shown. Underlined sequences have been deduced from CID spectra. *Indicates deaminated peptide NGDTITIK

	Ser	Val	Val	Thr	Leu	Asp	Gly	Gly	Lys
b	88.2	187.2	286.2	387.2	500.2	615.3	672.4	729.2	
a	60.1	159.2	258.1	359.1	472.3			701.1	
x	814.3	715.4	616.3	515.2	402.1		230.1	173.2	
y	788.5	689.4	590.3	489.3	376.2	261.2	204.1	147.2	
z	772.3	673.4	574.4	473.3	360.1	245.2			
v	744.3	645.3	544.3		316.1				
w	757.2	658.4	557.2	430.2	315.1				

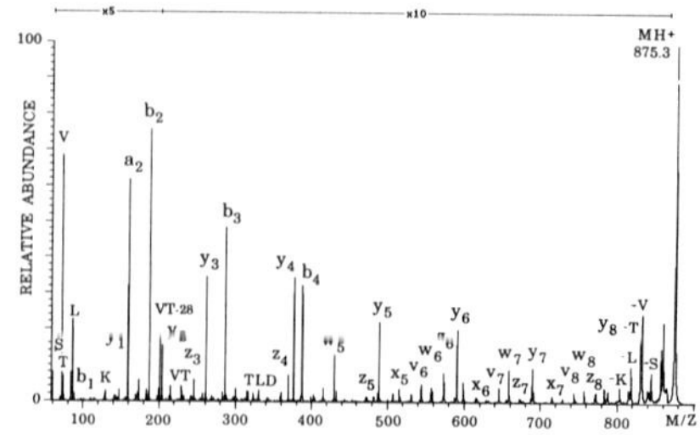

Figure 5. CID mass spectrum of 2-D gel isolated rat heart FABP tryptic peptide (^{12}C MH^+ at m/z 875.3) Ser-Val-Val-Thr-Leu-Asp-Gly-Gly-Lys. The masses of all observed sequence ions are shown in the table; mass differences between adjacent peaks in the same series correspond to the amino acid residue masses.

ion series, as well as sequence redundant *a*, *z*, *v* and *w* ions, allows the entire sequence of this peptide to be deciphered.

Once we had established a functional protocol for using mass spectrometry to sequence proteins isolated from 2-D gels, we set out to apply our procedure to sequencing unknown proteins. Specifically, we are using our protocol to obtain amino acid sequences of cellular proteins whose synthesis is enhanced or suppresed by certain cytokines [33, 34]. Knowing the amino acid sequence of cytokine-induced or -suppressed proteins will allow us to better understand their function in addition to furthering our knowledge of the role cytokines play in regulating cell growth. These data will be published elsewhere.

	Val	Tyr	Leu	Pro	-Arg (O-hexyl)
a	72.1	235.1			
x	658.3	495.4	382.2	285.1	
y	632.4	469.3	356.2	259.1	
z	616.4	453.1		243.1	
v	524.3				
w		410.3	313.2		

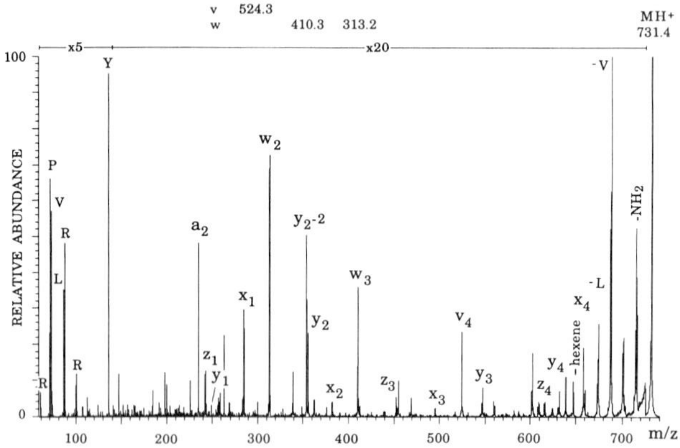

Figure 6. Tandem CID mass spectrum of the 2-D gel isolated ovalbumin peptide (100 pmol digested), Val-Tyr-Leu-Pro-Arg identified as its monohexyl derivative, Val-Tyr-Leu-Pro-Arg (O-hexyl) (^{12}C MH^+ at m/z 731.4).

III. CONCLUSION

This work demonstrates the feasibility of determining primary structures of proteins which have been isolated by one- and two-dimensional polyacrylamide gel electrophoresis using a combination of electroelution, enzymatic digestion, microbore HPLC, LSIMS and tandem mass spectrometric analyses. Furthermore, mass spectrometry offers an attractive and powerful alternative to conventional Edman methods. Advantages include identification of post-translational modifications, sequencing mixtures of peptides and significantly increased speed of sequence analysis.

IV. ACKNOWLEDGEMENTS

We thank Dr. N.M. Bass for providing samples of FABP. This work was supported by NIH National Center for Research Resources Grant RR01614 and NSF Biological Instrumentation Program Grant DIR-8700766 to A. L.B.,and NIH Training Grant NS07219-08 to S.C.H.. Additional support (P.A.S.) was provided by Fondation pour la Recherche Thérapeutique (France).

V. REFERENCES

1. Burlingame, A.L., Maltby, D., Russell, D.H., Holland, P.T. (1988) *Anal. Chem* **60**, 294R-342R.
2. Biemann, K., Martin, S.A. (1987) *Mass Spectrom. Revs.* **6**, 1-76.
3. Schindler, P., Bitsch, F., Klarskov, K., Roepstorff, P., Briand, G.,Wouters-Tyrou, D., Sautière, P., Van Dorsselaer, A. (1991) *Eur. J. Biochem.* **195**, 621-629.
4. Lepage, P., Bitsch, F., Roecklin, D., Keppi, E., Dimarcq, J.-L., Reichhart, J.-M., Hoffann, J., Roitsch, C., Van Dorsselaer, A. (1991) *Eur. J. Biochem.* **196**, 735-742.
5. Arlaud, G., Van Dorsselaer, A., Bell, A., Mancini, M., Aude, C., Gagnon, J. (1987) *FEBS Lett.* **222**, 129-134.
6. Hunt, D.F., Shabanowitz, J., Griffin, P.R., Yates, III, J.R., Martino, P.A., McCormack, A.L. (1990) In *Biological Mass Spectrometry*, (eds.) Burlingame, A.L. and McCloskey, J.A., Elsevier, Amsterdam, pp. 337-362.
7. Gibson, B.W., Yu, Z., Aberth, W., Burlingame, A.L., Bass, N.M. (1988) *J. Biol. Chem.* **263**, 4182-4185.
8. Offner, G.D., Brecher, P.B., Sawlich, W.B., Costello, C.E., Troxler, R.F. (1988) *Biochem. J.* **252**, 191-198.
9. Poulter, L., Ang, S.-G., Gibson, B.W., Williams, D.H., Holmes, C.F.B., Cohen, P. (1988) *Eur. J. Biochem.* **175**, 497-510.
10. Poulter, L., Ernest, J.P., Stroud, R.M., Burlingame, A.L. (1989) *Proc. Natl. Acad. Sci. USA*, **86**, 6645-6649.
11. Carr, S.A., Hemling, M.E., Roberts, G.D. (1988) in *Macromolecular Sequencing and Synthesis*, (ed.) Schlessinger, D.H., A.R. Liss, pp. 83-99.
12. Hefta, S.A., Shively, J.E. (1990) In *Biological Mass Spectrometry*, (eds.) Burlingame, A.L. and McCloskey, J.A., Elsevier, Amsterdam, pp. 259-284.
13. Kaur, S., Medzihradszky, K.F., Yu, Z., Baldwin, M.A., Gillece-Castro, B.L., Walls, F.C., Gibson, B.W., Burlingame, A.L. (1990) In *Biological Mass Spectrometry*, (eds.) Burlingame, A.L. and McCloskey, J.A., Elsevier, Amsterdam, pp. 285-313.

14. Biemann, K. (1990) In *Biological Mass Spectrometry*, (eds.) Burlingame, A.L. and McCloskey, J.A., Elsevier, Amsterdam, pp. 179-196.
15. Walls, F.,C., Baldwin, M.A., Falick, A.M., Gibson, B.W., Kaur, S., Maltby, D.A., Gillece-Castro, B.L., Medzihradszky, K.F., Evans, S., Burlingame, A.L. (1990) In *Biological Mass Spectrometry,* (eds.) Burlingame, A.L. and McCloskey, J.A., Elsevier, Amsterdam, pp. 197-216.
16. Gibson, B.W., Yu, Z., Castro, B., Aberth, W., Walls, F.C., Burlingame, A.L. *In Techniques in Protein Chemistry*, (ed.) Hugli, T., Academic Press, San Diego, CA, pp. 135-151.
17. O'Farrell, P.H. (1975) *J. Biol. Chem.* **250**, 4007-4021.
18. Celis, J.E., Rasmussen, H.H., Leffers, H., Madsen, P.; Honore, B., Gesser, B., Dejgaard, K., Vandekerckhove, J. (1991) *FASEB*, **5**, 2200-2208.
19. Abersold, R.H., Teplow, D.B., Hood, L.E., Kent, S.B.H. (1986) *J. Biol. Chem.*, **261**, 4229-4239.
20. Walsh, M.J., McDougall, J., Wittmann-Liebold, B. (1988) *Biochemistry*, **27**, 6867-6876.
21. Abersold, R.H., Pipes, G.D., Nika, H., Hood, L.E.; Kent, S.B.H (1988) *Biochemistry*, **27**, 6860-6867.
22. Matsudaira, P. (1987) *J. Biol. Chem.*, **21**, 10035-10038.
23. Ward, L.D., Hong, J., Whitehead, R.H., Simpson, R.J. (1990) *Electrophoresis*, **11**, 883-891.
24. Abersold, R.H., Leavitt, J., Saavedra, R., Hood, L.E., Kent, S.B.H. (1987) *Proc. Natl. Acad. Sci. USA.*, **84**, 6970-6974.
25. Kennedy, T.E., Gawinowicz, M.A., Barzilai, A., Kandel, E.R., Sweatt, J.D. (1988) *Proc. Natl. Acad. Sci. USA*, **85**, 7008-7012.
26. Temps, P., Link, A.J., Edmonds, C.G., Barinaga, C.J., Udseth, H.R. (1990) *Electrophoresis*, **11**, 537-553.
27 Hall, S.C., Masiarz, F.R., Walls, F.C., Kaur, S., Gibson, B.W., Evans, S., Burlingame, A.L. In *Proceedings of the 38th Conference on Mass Spectrometry and Allied Topics*, June 3-8, 1990, Tucson, AZ, p. 1505.
28. Hall, S.C., Smith, D.M., Masiarz, F.R., Soo, V.W., Epstein, L.B., Burlingame, A.L. In *Proceedings of the 39th Conference on Mass Spectrometry and Allied Topics*, May 19-24, 1991, Nashville, TN, in press.
29. Hunkapiller M.W., Lujan, E., Ostrander, F., Hood, L.E. (1983) *Methods in Enzymology* **91**, 227-236.
30. Königsberg, W.H., Henderson, L. (1983) *Methods in Enzymology*, **91**, 254-259.
31. Throughout the text the nomenclature first proposed by Roepstrorff, P. and Fohlman, J. (1984) *Biomed. Mass Spectrom.* **11**, 601-603 and modified by Biemann, K. (1988) *Biomed. Environ. Mass Spectrom.*, **16**, 99-111 is used for the peptide fragment ions.
32. Falick, A.M., Maltby, D.A. (1989) *Anal. Biochem.*, **182**, 165-169.
33. Weil, J., Epstein, C.J., Epstein, L. B., Sedmak, J.J., Sabran, J.L., Grossberg, S.E. (1983) *Nature*, **301**, 437-439.
34. Beresini, M.H., Lempert, M.J., Epstein, L.B. (1988) *J. Immuno.*, **140**. 485-493.

Index

B

C

F

G

H

I

K

L

M

N

O

P

T

W

Y